Optoelectronic Devices

Design, Modeling, and Simulation

With a clear application focus, this book explores optoelectronic device design and modeling through physics models and systematic numerical analysis.

By obtaining solutions directly from the physics-based governing equations through numerical techniques, the author shows how to design new devices and how to enhance the performance of existing devices. Semiconductor-based optoelectronic devices such as semiconductor laser diodes, electro-absorption modulators, semiconductor optical amplifiers, superluminescent light-emitting diodes and their integrations are all covered.

Including step-by-step practical design and simulation examples, together with detailed numerical algorithms, this book provides researchers, device designers, and graduate students in optoelectronics with the numerical techniques to solve their own structures.

Xun Li is a Professor in the Department of Electrical and Computer Engineering at McMaster University, Hamilton. Since receiving his Ph.D. from Beijing Jiaotong University in 1988, he has authored and co-authored over 160 technical papers and co-founded Apollo Photonics, Inc., developing one of the company's major software products, "Advanced Laser Diode Simulator". He is a Member of the OSA and SPIE, and a Senior Member of the IEEE.

Optoelectronic Devices

Design, Modeling, and Simulation

XUN LI

Department of Electrical and Computer Engineering
McMaster University
Hamilton, Ontario

CAMBRIDGE
UNIVERSITY PRESS

University Printing House, Cambridge CB2 8BS, United Kingdom

One Liberty Plaza, 20th Floor, New York, NY 10006, USA

477 Williamstown Road, Port Melbourne, VIC 3207, Australia

314-321, 3rd Floor, Plot 3, Splendor Forum, Jasola District Centre, New Delhi - 110025, India

103 Penang Road, #05-06/07, Visioncrest Commercial, Singapore 238467

Cambridge University Press is part of the University of Cambridge.

It furthers the University's mission by disseminating knowledge in the pursuit of education, learning and research at the highest international levels of excellence.

www.cambridge.org
Information on this title: www.cambridge.org/9780521875103

First published 2009

A catalogue record for this publication is available from the British Library

ISBN 978-0-521-87510-3 Hardback

Contents

Preface *page* xi

1 Introduction 1

1.1 The underlying physics in device operation 1
1.2 Modeling and simulation methodologies 1
1.3 Device modeling aspects 3
1.4 Device modeling techniques 3
1.5 Overview 5

2 Optical models 6

2.1 The wave equation in active media 6
2.1.1 Maxwell equations 6
2.1.2 The wave equation 8
2.2 The reduced wave equation in the time domain 9
2.3 The reduced wave equation in the space domain 11
2.4 The reduced wave equation in both time and space domains – the traveling wave model 12
2.4.1 The wave equation in fully confined structures 12
2.4.2 The wave equation in partially confined structures 17
2.4.3 The wave equation in periodically corrugated structures 21
2.5 Broadband optical traveling wave models 31
2.5.1 The direct convolution model 32
2.5.2 The effective Bloch equation model 34
2.5.3 The wavelength slicing model 37
2.6 Separation of spatial and temporal dependences – the standing wave model 40
2.7 Photon rate and phase equations – the behavior model 47
2.8 The spontaneous emission noise treatment 48

3 Material model I: Semiconductor band structures 54

3.1 Single electron in bulk semiconductors 54
3.1.1 The Schrödinger equation and Hamiltonian operator 54
3.1.2 Bloch's theorem and band structure 57

3.1.3 Solution at $\vec{k} = 0$: Kane's model 65
3.1.4 Solution at $\vec{k} \neq 0$: Luttinger–Kohn's model 71
3.1.5 Solution under 4×4 Hamiltonian and axial approximation 76
3.1.6 Hamiltonians for different semiconductors 80
3.2 Single electron in semiconductor quantum well structures 80
3.2.1 The effective mass theory and governing equation 80
3.2.2 Conduction band (without degeneracy) 84
3.2.3 Valence band (with degeneracy) 85
3.2.4 Quantum well band structures 87
3.3 Single electron in strained layer structures 91
3.3.1 A general approach 91
3.3.2 Strained bulk semiconductors 93
3.3.3 Strained layer quantum well structures 95
3.3.4 Semiconductors with the zinc blende structure 96
3.4 Summary of the k–p theory 98

4 Material model II: Optical gain 102

4.1 A comprehensive model with many-body effect 102
4.1.1 Introduction 102
4.1.2 The Heisenberg equation 103
4.1.3 A comprehensive model 104
4.1.4 General governing equations 109
4.2 The free-carrier model as a zeroth order solution 122
4.2.1 The free-carrier model 122
4.2.2 The carrier rate equation 123
4.2.3 The polariton rate equation 126
4.2.4 The susceptibility 127
4.3 The screened Coulomb interaction model as a first order solution 128
4.3.1 The screened Coulomb interaction model 128
4.3.2 The screened Coulomb potential 129
4.3.3 Solution under zero injection and the exciton absorption 133
4.3.4 Solution under arbitrary injection 137
4.4 The many-body correlation model as a second order solution 140
4.4.1 The many-body correlation model 140
4.4.2 A semi-analytical solution 141
4.4.3 The full numerical solution 144

5 Carrier transport and thermal diffusion models 151

5.1 The carrier transport model 151
5.1.1 Poisson and carrier continuity equations 151
5.1.2 The drift and diffusion model for a non-active region 152
5.1.3 The carrier transport model for the active region 154
5.1.4 Simplifications of the carrier transport model 158

5.1.5 The free-carrier transport model 160
5.1.6 Recombination rates 162
5.2 The carrier rate equation model 164
5.3 The thermal diffusion model 165
5.3.1 The classical thermal diffusion model 165
5.3.2 A one-dimensional thermal diffusion model 168

6 Solution techniques for optical equations 172

6.1 The optical mode in the cross-sectional area 172
6.2 Traveling wave equations 173
6.2.1 The finite difference method 173
6.2.2 The split-step method 183
6.2.3 Time domain convolution through the digital filter 188
6.3 Standing wave equations 191

7 Solution techniques for material gain equations 200

7.1 Single electron band structures 200
7.2 Material gain calculations 200
7.2.1 The free-carrier gain model 200
7.2.2 The screened Coulomb interaction gain model 205
7.2.3 The many-body gain model 205
7.3 Parameterization of material properties 211

8 Solution techniques for carrier transport and thermal diffusion equations 214

8.1 The static carrier transport equation 214
8.1.1 Scaling 215
8.1.2 Boundary conditions 216
8.1.3 The initial solution 218
8.1.4 The finite difference discretization 218
8.1.5 Solution of non-linear algebraic equations 228
8.2 The transient carrier transport equation 231
8.3 The carrier rate equation 232
8.4 The thermal diffusion equation 233

9 Numerical analysis of device performance 236

9.1 A general approach 236
9.1.1 The material gain treatment 236
9.1.2 The quasi-three-dimensional treatment 238
9.2 Device performance analysis 240
9.2.1 The steady state analysis 240
9.2.2 The small-signal dynamic analysis 243
9.2.3 The large-signal dynamic analysis 245
9.3 Model calibration and validation 246

10 **Design and modeling examples of semiconductor laser diodes** 251

10.1 Design and modeling of the active region for optical gain 251
10.1.1 The active region material 251
10.1.2 The active region structure 255
10.2 Design and modeling of the cross-sectional structure for optical and carrier confinement 259
10.2.1 General considerations in the layer stack design 259
10.2.2 The ridge waveguide structure 260
10.2.3 The buried heterostructure 265
10.2.4 Comparison between the ridge waveguide structure and buried heterostructure 268
10.3 Design and modeling of the cavity for lasing oscillation 269
10.3.1 The Fabry–Perot laser 269
10.3.2 Distributed feedback lasers in different coupling mechanisms through grating design 271
10.3.3 Lasers with multiple section designs 281

11 **Design and modeling examples of other solitary optoelectronic devices** 288

11.1 The electro-absorption modulator 288
11.1.1 The device structure 288
11.1.2 Simulated material properties and device performance 288
11.1.3 Design for high extinction ratio and low insertion loss 292
11.1.4 Design for polarization independent absorption 297
11.2 The semiconductor optical amplifier 299
11.2.1 The device structure 299
11.2.2 Simulated semiconductor optical amplifier performance 300
11.2.3 Design for performance enhancement 302
11.3 The superluminescent light emitting diode 305
11.3.1 The device structure 305
11.3.2 Simulated superluminescent light emitting diode performance 305
11.3.3 Design for performance enhancement 306

12 **Design and modeling examples of integrated optoelectronic devices** 313

12.1 The integrated semiconductor distributed feedback laser and electro-absorption modulator 313
12.1.1 The device structure 313
12.1.2 The interface 315
12.1.3 Simulated distributed feedback laser performance 315
12.1.4 Simulated electro-absorption modulator performance 317
12.2 The integrated semiconductor distributed feedback laser and monitoring photodetector 321
12.2.1 The device structure 321

12.2.2 Simulated distributed feedback laser performance 325
12.2.3 Crosstalk modeling 326

Appendices 332

A Lowdin's renormalization theory 332
B Integrations in the many-body gain model 334
C Cash–Karp's implementation of the fifth order Runge–Kutta method 347
D The solution of sparse linear equations 348
D.1 The direct method 349
D.2 The iterative method 351

Index 356

Preface

Over the past 30 years, the world has witnessed the rapid development of optoelectronic devices based on III-V compound semiconductors. Past effort has mainly been directed to the theoretical understanding of, and the technology development for, these devices in applications in telecommunication networks and compact disk (CD) data storage. With the growing deployment of such devices in new fields such as illumination, display, fiber sensor, fiber gyro, optical coherent tomography, etc., research on optoelectronic devices, especially on those light emitting components, continues to expand with the pursuit of many experimental explorations on new materials such as group-III nitride alloys and II-VI compounds and novel structures such as quantum wires, dots, and nanostructures.

As the manufacturing technology becomes mature and standardized and few uncertainties are left, design and simulation become the major issue in the performance enhancement of existing devices and in the development of new devices. Recent progress in numerical techniques as well as computing hardware has provided a powerful platform that makes sophisticated computer-aided design, modeling, and simulation possible. So far, the development of optoelectronic devices seems to replicate the history of electronic devices: from discrete to integrated, from technology intensive to design intensive, from trial-and-error experiments to computer-aided simulation and optimization.

The purpose of this book is to bridge the gap between the theoretical framework and the solution to real-world problems, or, more specifically, to bridge the gap between our knowledge acquired on electromagnetic field theory, quantum mechanics, and semiconductor physics and optoelectronic device design and modeling through advanced numerical tools.

Advanced optoelectronic devices are built on compound semiconductor material systems with complicated geometrical structures; they are also operated under varying conditions. For this reason, we can find hardly any easy, intuitive, and analytical solutions to the first-principle-based governing equations that accurately describe the closely coupled physical processes inside such devices. Although solutions are relatively easy to obtain from the equations derived from the phenomenological model, assumptions have to be made in such a model, which often ignores some important effects and fails to achieve quantitative agreement between theoretically predicted and practically measured results.

Therefore, obtaining the solution directly from the physics-based governing equations through numerical techniques seems to be a promising approach to bridge the gap as mentioned above, as not only a qualitative, but also a quantitative matching between the theory and experiment is achievable. This book is intended for readers who want to link their understanding of the device physics through the theoretical framework they have already acquired to the design, modeling and simulation of real-world devices and innovative structures.

This book will focus on semiconductor-based optoelectronic devices such as laser diodes (LDs), electro-absorption modulators (EAMs), semiconductor optical amplifiers (SOAs), and superluminescent light emitting diodes (SLEDs) in various applications. Numerical methods will be used throughout the analysis of these devices.

Derived from physics-based first principles, governing equations will be given for the description of different physical processes, such as light propagation, optical gain generation, carrier transport and thermal diffusion, and their interplays inside the devices. Different numerical techniques will be discussed in detail along with the process of seeking the solution to these governing equations. Discussions on device design optimizations will also be followed, based on the interpretation of the numerical solutions.

The methodology introduced in this book hopefully will help its readers to learn (1) how to extract the governing equations from first principles for the accurate description of their devices; and more importantly, (2) how to obtain the numerical solution to those governing equations once derived. Practical design and simulation examples are also given to support the approaches used in this book.

I am in debt to my colleague and friend, Professor W.-P. Huang, who showed me the prospect of computer-aided design, modeling and simulation in this field 15 years ago, and with whom I had countless stimulating discussions on almost every topic involved in this book, from the material physics to waveguide theory, from the model establishment to result interpretation, and from the modeling methodology to numerical algorithm. I would like to thank Dr. T. Makino (former Nortel), Dr. K. Yokoyama (former NTT), Dr. T. Yamanaka (NTT), Dr. C.-L. Xu (RSoft Inc.), Dr. J. Hong (Oplink Inc.), Dr. A. Shams (former Photonami Inc.), Professor S. Sadeghi (University of Alabama at Huntsville), Professor W. Li (University of Wisconsin at Platteville), Professor Y. Luo (Tsinghua University), Professor Y.-H. Zhang (Arizona State University), Ms. T.-N. Li (InPhenix Inc.), Ms. N. Zhou (AcceLink Co.), Mr. M. Mazed (IP Photonics Inc.), Professor T. Luo (University of Minnesota), Professor C.-Q. Xu (McMaster University), Professor M. Dagenais (University of Maryland at College Park), Dr. J. Piprek (former University of California at Santa Barbara), and many other colleagues and friends in this field, for numerous insightful and inspiring discussions and interactions on various subjects in this book, during and after our research collaborations. I am grateful to Ms. Y.-P. Xi, who helped me with the simulation of SOAs and SLEDs, and Mr. Q.-Y. Xu, who helped me with the simulation of crosstalks in the integrated DFB laser and monitoring photodetector. I am also grateful to Professor S.-H. Chen (Huazhong University of Sci. and Tech.) and her graduate students, who helped me to create most of the schematic diagrams in the first eight chapters and all the three-dimensional device structure drawings in Chapters 10 and 12. I would also like to thank my graduate students and many other graduate students in the Department of Electrical and Computer Engineering at McMaster University who took my course on this subject, for their valuable comments and suggestions. Finally, I appreciate the constant help and great patience of Dr. J. Lancashire and Ms. S. Koch.

1 Introduction

1.1 The underlying physics in device operation

Figure 1.1 shows the major physical processes and their linkages in the operation of optoelectronic devices.

To capture these physical processes, we need the following models and knowledge:

(1) a model that describes wave propagation along the device waveguide (electromagnetic wave theory);
(2) a model that describes the optical properties of the device material platform (semiconductor physics);
(3) a model that describes carrier transport inside the device (quasi-electrostatic theory);
(4) a model that describes thermal diffusion inside the device (thermal diffusion theory).

Therefore, the above four aspects should be included in any model established for simulation of optoelectronic devices.

1.2 Modeling and simulation methodologies

There are two major approaches in device modeling and simulation.

(1) Physics modeling: a direct approach based on the first principle physics-based model.

The required governing equations in the preceding four aspects are all derived from first principles, such as the Maxwell equations (including electromagnetic wave theory for the optical field distribution and quasi-electrostatic theory for the carrier transport), the Schrödinger equation (for the semiconductor band structure), the Heisenberg equation (for the gain and refractive index change), and the thermal diffusion equation (for the temperature distribution).

This model gives the physical description of what exactly happens inside the device and is capable of providing predictions on device performance in every aspect, once the device building material constants, the structural geometrical sizes, and the operating conditions are all given.

This approach is usually adopted by device designers who work on developing devices themselves.

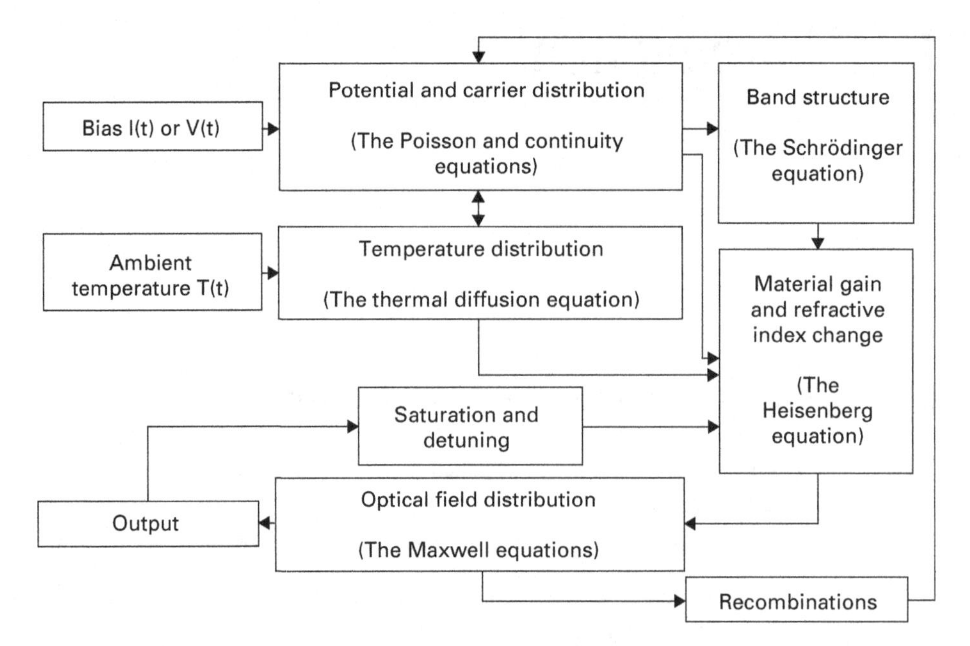

Fig. 1.1. The physical processes and their linkages in the operation of optoelectronic devices. Noted in brackets are the first principle equations that govern these processes.

However, such a modeling technique is usually complex and sophisticated numerical tools have to be invoked in solving the equations involved. Computationally it is usually expensive.

(2) Behavior modeling: an indirect approach based on an equivalent or phenomenological model.

The governing equations in the preceding four aspects are extracted from first principles under various assumptions. Hence they are greatly simplified compared with the equations in the physics-based model. Those frequently used methods in the extraction of the simplified equations include: (1) reducing or even eliminating spatial dimensions; (2) neglecting the dependence that causes only relatively slow or small variation; and (3) ignoring the physical processes that have little direct effect on the aspects of interest. Another method is to replace the original local or discrete variable by a global or integrated variable in the description of the physical process, as the latter usually obeys a certain conservation law, hence a corresponding balance equation can be derived in a simple form.

This model does not give the description of what exactly happens inside the device but is capable of providing the same device terminal performance as the physics-based model. Therefore, if the device is treated as a black box, this model will provide the correct output for any given input.

This approach is usually adopted by circuit and system designers who just use rather than develop devices.

Although this modeling technique is usually simple and computationally inexpensive, it has two major drawbacks that prevent its application in device design and development. The first demerit is that it can give hardly any physical insights. Little information can be obtained on how to make a device work better by improving the design. The second demerit is that it often relies on non-physical input parameters, such as effective constants or phenomenologically introduced coefficients, which are usually difficult to obtain.

In optoelectronic device modeling, we normally take a combination of the preceding two approaches. Depending on different simulation requirements, we usually retain a minimum set of the necessary physics-based equations and replace the rest by simplified ones.

1.3 Device modeling aspects

In device modeling, we normally look at the following aspects.

(1) Device steady state performance.
    No time dependence needs to be considered in this simulation. The device characteristics are usually modeled as functions of the bias.
(2) Device small-signal dynamic performance.
    Based on the small-signal linearization, a direct current (DC) at a fixed bias plus a frequency domain analysis are required in this simulation.
(3) Device large-signal dynamic performance.
    A direct time-domain analysis is required in the simulation.
(4) Noise performance.
    Either a semi-analytical frequency domain analysis or a numerical time-domain analysis is required in this simulation.

1.4 Device modeling techniques

A typical procedure for optoelectronic device modeling and simulation includes:

(1) input geometrical structures;
(2) input material constants;
(3) set up meshes;
(4) initialize solvers (pre-processing);
(5) input operating conditions;
(6) scale variables (physical to numerical);
(7) start looping;
(8) call carrier solver;
(9) call temperature solver;
(10) call material solver;
(11) call optical solver;

(12) go back to step 7 until convergence;
(13) scale variables (numerical to physical);
(14) output assembly (post-processing).

To start this procedure, however, one must have an initial device structure, which relies on one's understanding of the device physics and on one's experience accumulated from analysis and interpretation of the results obtained from device design, modeling and simulation practice.

Other than the initial structure, we still need to collect all the input parameters required by the numerical solvers. These parameters are usually obtained from open literature, experiment, or calibration.

The following are a number of numerical techniques that are often involved in optoelectronic device modeling:

(1) partial differential equation (PDE) solvers (boundary value and mixed boundary and initial value problems);
(2) ordinary differential equation (ODE) solvers (initial and boundary value problems);
(3) algebraic eigenvalue problem solvers;
(4) linear and non-linear system of algebraic equations solvers;
(5) root searching routine;
(6) minimization or maximization routine;
(7) function evaluations, interpolation and extrapolation routines;
(8) numerical quadratures;
(9) fast Fourier transform (FFT) and digital filtering routines;
(10) pseudo-random number generation.

The key issue in device modeling is to establish numerical solvers for PDEs, which usually follows a procedure as shown below.

(1) Scale the variables in given PDEs.
(2) Set up computation window and mesh grids.

(These two steps translate a physical problem into a numerical problem.)

(3) Equation discretization through, e.g., finite difference (FD) scheme.
(4) Boundary processing.

(These two steps translate PDEs into a system of algebraic equations.)

(5) Start Newton's iteration for the system of non-linear algebraic equations.

(This step translates the system of non-linear algebraic equations into a system of linear algebraic equations.)

(6) Find solution to the system of linear algebraic equations.
Direct method (for moderate size or dense coefficient matrix).
Iterative method (for large size sparse coefficient matrix).
Convergence acceleration (for iterative method).

(7) Convergence acceleration for Newton's iteration.

(The numerical solution will be obtained after this step.)

(8) Scale variables and post processing.

(A physical solution will be obtained after this step.)

1.5 Overview

This book is divided into three parts. The first part, comprising Chapters 2, 3, 4, and 5, is on the derivation and explanation of governing equations that model the closely coupled physics processes in optoelectronic devices. The second part, Chapters 6, 7, 8, and 9, is devoted to numerical solution techniques for the governing equations arising from the first part and explains how these techniques are jointly applied in device simulation. Chapters 10, 11, and 12 form the third part, which provides real-world design and simulation examples of optoelectronic devices, such as Fabry–Perot (FP) and distributed feedback (DFB) LDs, EAMs, SOAs, SLEDs, and their monolithic integrations.

2 Optical models

2.1 The wave equation in active media

2.1.1 Maxwell equations

The behavior of the optical wave is generally governed by the Maxwell equations

$$\nabla \times \vec{E}(\vec{r}, t) = -\frac{\partial}{\partial t}\vec{B}(\vec{r}, t), \tag{2.1}$$

$$\nabla \times \vec{H}(\vec{r}, t) = \frac{\partial}{\partial t}\vec{D}(\vec{r}, t) + \vec{J}(\vec{r}, t), \tag{2.2}$$

$$\nabla \cdot \vec{D}(\vec{r}, t) = \rho(\vec{r}, t), \tag{2.3}$$

$$\nabla \cdot \vec{B}(\vec{r}, t) = 0, \tag{2.4}$$

where $\vec{E}$ and $\vec{H}$ indicate the electric and magnetic fields in V/m and A/m, respectively, r and t represent the space coordinate vector and time variable, respectively, $\vec{D}$ the electric flux density in C/m^2, $\vec{B}$ the magnetic flux density in Wb/m^2, $\vec{J}$ the current density in A/m^2, and ρ the charge density in C/m^3.

In semiconductors, the constitutive relation reads

$$\vec{D}(\vec{r}, t) = \int_{-\infty}^{t} \varepsilon(\vec{r}, t-\tau)\vec{E}(\vec{r}, \tau)\mathrm{d}\tau, \tag{2.5}$$

$$\vec{B}(\vec{r}, t) = \mu_0 \vec{H}(\vec{r}, \tau), \tag{2.6}$$

with ε and μ_0 denoting the time domain permittivity of the host medium and permeability in a vacuum in F/m and H/m, respectively.

Noting that

$$\varepsilon(\vec{r}, t) = \varepsilon_0[\delta(t) + \chi(\vec{r}, t)], \tag{2.7}$$

with ε_0 denoting the permittivity in a vacuum in F/m and χ the dimensionless time-domain susceptibility of the host medium, equation (2.5) can also be written as

$$\vec{D}(\vec{r}, t) = \varepsilon_0 \int_{-\infty}^{t} [\delta(t-\tau) + \chi(\vec{r}, t-\tau)]\,\vec{E}(\vec{r}, \tau)\mathrm{d}\tau = \varepsilon_0\vec{E}(\vec{r}, t) + \vec{P}(\vec{r}, t), \tag{2.8}$$

where the induced polarization of the host medium in $\mathrm{C/m^2}$ is defined as

$$\vec{P}(\vec{r},t) \equiv \varepsilon_0 \int_{-\infty}^{t} \chi(\vec{r},t-\tau)\vec{E}(\vec{r},\tau)\mathrm{d}\tau. \tag{2.9}$$

For an electromagnetic field at optical frequencies

$$\rho = 0. \tag{2.10}$$

In a passive area without any radiative recombination process

$$\vec{J} = 0. \tag{2.11}$$

In an active area with a spontaneous emission process

$$\vec{J} = \vec{J}_{\mathrm{sp}}. \tag{2.12}$$

It is worth mentioning that, in an active area, the stimulated emission process will be included in the susceptibility, as it is a purely homogeneous process induced by a given electric field. Therefore, the stimulated emission process is excluded from equation (2.12), as the driven current must be a purely inhomogeneous source.

By using equations (2.5) and (2.6), the electrical and magnetic flux densities $\vec{D}$ and $\vec{B}$ can be eliminated from equations (2.1) and (2.2); hence we obtain

$$\nabla \times \vec{E} = -\mu_0 \frac{\partial}{\partial t}\vec{H}, \tag{2.13}$$

$$\nabla \times \vec{H} = \varepsilon_0 \frac{\partial}{\partial t}\vec{E} + \frac{\partial}{\partial t}\vec{P} + \vec{J}_{\mathrm{sp}}. \tag{2.14}$$

At least in principle, equations (2.13), (2.14) and (2.9) can be solved directly under the given semiconductor material property described by the susceptibility χ over the entire device structure and the spontaneous emission source $\vec{J}_{\mathrm{sp}}$ in the active area. For example, a finite difference time domain (FDTD) approach can be used to discretize equations (2.13) and (2.14) on Yee's unit cells [1]. Each electrical and magnetic field component can therefore be solved through the resulted recursion in the time domain on those cells that fill out the whole device domain. However, although FDTD based numerical solvers have been very successful in dealing with passive structures, they have seldom been employed for solving active structures because of the highly dispersive material property with embedded non-linearity and distributed inhomogeneous driving source. Moreover, every component of the electrical and magnetic fields must be handled as an unknown variable, which exhausts memory capacity and hence makes the computation impossible for devices with sizeable domains. For this reason, a wave equation model with a reduced number of unknown variables is usually more convenient in dealing with active devices.

2.1.2 The wave equation

The duality principle implies that it is not necessary to use both electrical and magnetic fields to describe optical wave propagation: either an electrical or magnetic field will be sufficient. To reduce the number of variables involved, we perform $\nabla\times$ on both sides of equation (2.13) and replace the right hand side (RHS) $\nabla \times \vec{H}$ with equation (2.14) to obtain

$$\nabla(\nabla \cdot \vec{E}) - \nabla^2 \vec{E} = -\mu_0 \left(\varepsilon_0 \frac{\partial^2}{\partial t^2} \vec{E} + \frac{\partial^2}{\partial t^2} \vec{P} + \frac{\partial}{\partial t} \vec{J}_{\rm sp} \right). \tag{2.15}$$

From equations (2.3), (2.8), (2.10) and (2.9), we also have

$$\begin{aligned} \nabla \cdot \vec{E} &= -\frac{1}{\varepsilon_0} \nabla \cdot \vec{P} = -\int_{-\infty}^{t} \nabla \cdot \left[\chi(\vec{r}, t-\tau)\vec{E}(\vec{r}, \tau)\right] \mathrm{d}\tau \\ &= -\int_{-\infty}^{t} \chi(\vec{r}, t-\tau)\left[\nabla \cdot \vec{E}(\vec{r}, \tau)\right] \mathrm{d}\tau - \int_{-\infty}^{t} \left[\nabla\chi(\vec{r}, t-\tau) \cdot \vec{E}(\vec{r}, \tau)\right] \mathrm{d}\tau. \end{aligned} \tag{2.16}$$

If we restrict our model to those structures with

$$\nabla\chi(\vec{r}, t) \cdot \vec{E}(\vec{r}, t) \approx 0, \tag{2.17}$$

we find

$$\nabla \cdot \vec{E} = 0. \tag{2.18}$$

Hence equation (2.15) becomes

$$\nabla^2 \vec{E} = \frac{1}{c^2} \frac{\partial^2}{\partial t^2} \vec{E} + \frac{1}{c^2 \varepsilon_0} \frac{\partial^2}{\partial t^2} \vec{P} + \mu_0 \frac{\partial}{\partial t} \vec{J}_{\rm sp}, \tag{2.19}$$

with

$$c \equiv 1/\sqrt{(\mu_0 \varepsilon_0)}, \tag{2.20}$$

defined as the speed of light in a vacuum in m/s.

Condition (2.17) holds for those structures with weak optical guidance, i.e., χ only changes slightly in the plane perpendicular to the wave propagation direction, as such $\nabla_{\rm T}\chi(\vec{r}, t) \cdot \vec{E}_{\rm T}(\vec{r}, t) \approx 0$. Along the wave propagation direction (assumed to be z), however, $\partial\chi/\partial z$ does not need to be small since $E_z \approx 0$ anyway. Therefore, wave equation (2.19) holds even for devices with non-uniform structures along the wave propagation direction, e.g., distributed feedback (DFB) or distributed Bragg reflector (DBR) lasers, as long as the wave is weakly guided in the cross-section.

Expressions (2.19) and (2.9) form the wave equation model that describes optical wave propagation in a weakly guided device structure. In the wave equation (2.19), the only term on the left hand side (LHS) gives the spatial diffraction of the electrical field, while the first term on the RHS gives the time dispersion of the electrical field. The balance of these two terms gives the inherent property of the optical wave, i.e., the propagation in space. The second term on the RHS denotes the contribution from the wave–media interaction where the convolution reveals that the wave at a certain time t will be affected by the whole past "history" of the media response. This is simply because the media cannot instantaneously respond to the incident wave. The last term on the RHS represents the contribution of the spontaneous emission known as a noise current source. As the only inhomogeneous term, it plays a crucial role as the "seed" in light-emitting devices. Without the inclusion of this inhomogeneous contribution in equation (2.19), a laser will never lase as equation (2.19) would have only zero solution because of its homogeneity.

In comparison with the Maxwell equations in their original form, the wave equation model is physically straightforward and has minimum required unknown variables involved. However, equation (2.19) contains the second order derivatives with respect to time and is in the form of a hyperbolic partial differential equation (PDE). Unlike an elliptical PDE with only static solutions, or a parabolic PDE with solutions exponentially approaching its steady state, a hyperbolic PDE takes harmonic oscillations as its inherent solution and bears no time-invariant steady state. Therefore, stability will always be an issue in looking for its solutions if the initial value is not well posed or not sufficiently smooth.

Knowing that a hyperbolic PDE takes the harmonic wave as its "static" solution, we can therefore write the general solution of the wave equation (2.19) in the form of a "modulated" wave, i.e., a harmonic wave with "slow-varying" envelope. By doing so, we should be able to extract a governing equation for this envelope from equation (2.19). As the envelope changes more slowly, the new equation will take a reduced time derivative and hence become more stable.

2.2 The reduced wave equation in the time domain

We assume that the optical wave is composed of harmonic waves with discrete frequencies and relatively slow-varying envelopes

$$\vec{E}(\vec{r}, t) = \frac{1}{2} \sum_{k} \vec{u}_k(\vec{r}, t) \mathrm{e}^{-\mathrm{j}\omega_k t} + \text{c.c.}, \tag{2.21}$$

with $\vec{u}_k$ and ω_k indicating the kth harmonic wave envelope function in V/m and angular frequency in rad/s, respectively, and where c.c. means complex conjugate. By further assuming the linearity of equation (2.19) (i.e., χ has no explicit dependence on $\vec{E}$), we take only a single frequency ($k = 0$) in the following derivations without losing generality: when multiple frequencies are involved, it is trivial to consider a summation in a linear system because of the superposition principle. For the same reason, we can drop the complex conjugate part by considering

$$\vec{E}(\vec{r}, t) = \vec{u}(\vec{r}, t) \mathrm{e}^{-\mathrm{j}\omega_0 t}, \tag{2.22}$$

only, with ω_0 as the harmonic wave frequency (or reference frequency) and with the subscript of the envelope function omitted. Since equations (2.19) and (2.9) are all real, if we take our real-world optical wave as the real (or imaginary) part of equation (2.22), our real-world result will then be the real (or imaginary) part of the solution obtained from equations (2.19) and (2.9). The reason that we use a complex exponential function to replace the sinusoidal function is that the former is the eigenfunction of any linear and time-invariant system, whereas the latter is not, unless it forms a proper linear combination, i.e., a complex exponential function.

Replacing the optical field in equation (2.9) with equation (2.22) yields

$$\begin{aligned}\vec{P}(\vec{r},t) &= \varepsilon_0 \int_{-\infty}^{t} \chi(\vec{r},t-\tau)\vec{u}(\vec{r},\tau)\mathrm{e}^{-\mathrm{j}\omega_0\tau}\,\mathrm{d}\tau = \varepsilon_0 F^{-1}\left[\widetilde{\chi}(\vec{r},\omega)\widetilde{\vec{u}}(\vec{r},\omega-\omega_0)\right] \\ &\approx \varepsilon_0\widetilde{\chi}(\vec{r},\omega_0)F^{-1}\left[\widetilde{\vec{u}}(\vec{r},\omega-\omega_0)\right] = \varepsilon_0\widetilde{\chi}(\vec{r},\omega_0)\vec{u}(\vec{r},t)\mathrm{e}^{-\mathrm{j}\omega_0 t},\end{aligned} \tag{2.23}$$

with $\widetilde{\vec{u}}$ and $\widetilde{\chi}$ indicating the frequency domain responses of the slow-varying harmonic wave envelope function and susceptibility, respectively, and $F^{-1}[\ldots]$ the inverse Fourier transform. Equation (2.23) holds only when the susceptibility varies much faster than the slow-varying envelope in the time domain; or equivalently, the bandwidth of $\widetilde{\chi}$ is much larger than that of $\widetilde{\vec{u}}$ in the frequency domain. In optoelectronic devices, this is usually true as long as the base-band signal (hence the slow-varying envelope) does not consist of very short pulses. Since the full width half maximum (FWHM) bandwidth of $\widetilde{\chi}$ is usually as broad as 5–10 THz (i.e., around 50–100 nm in a C-band centered at 1550 nm), i.e., χ can respond to any time change slower than sub-picosecond, any base-band signal that varies slower than 10 ps would make equation (2.23) a valid approximation.

We now plug both equations (2.22) and (2.23) into equation (2.19) to obtain

$$\mathrm{j}\frac{2\omega_0}{c^2}\left[1+\widetilde{\chi}(\vec{r},\omega_0)\right]\frac{\partial\vec{u}}{\partial t} = -\nabla^2\vec{u} - \frac{\omega_0^2}{c^2}\left[1+\widetilde{\chi}(\vec{r},\omega_0)\right]\vec{u} + \mu_0\mathrm{e}^{\mathrm{j}\omega_0 t}\frac{\partial}{\partial t}\vec{J}_{\mathrm{sp}}, \tag{2.24}$$

where under the slow-varying envelope assumption ($|\partial^2\vec{u}/\partial t^2| \ll \omega_0^2|\vec{u}|$), $\partial^2\vec{u}/\partial t^2$ is dropped.

Equation (2.24) is the reduced wave equation in the time domain. It governs the slow-varying envelope of an optical field that is assumed in a harmonic wave form with optical frequency ω_0. Compared with equation (2.19), equation (2.24) has the fast-varying harmonic factor ($\mathrm{e}^{-\mathrm{j}\omega_0 t}$) excluded, hence has a reduced time derivative order. Numerically, stable solutions can be obtained through time domain discretization by following the envelope change ($\partial\vec{u}/\partial t$), rather than by following the optical wave change ($\partial\vec{E}/\partial t$) itself. This normally results in a great saving of progressive steps as the former changes much slower than the latter in the time domain.

Actually, equation (2.24) is solved directly only when the device structure does not have any dominant feature in space and hence the wave does not form any time-invariant spatial pattern known as a "mode." In waveguide based optoelectronic and photonic devices, however, the wave is confined at least along one dimension. Therefore, an optical mode can be introduced at least along this dimension and the wave will travel in the reduced spatial dimensions. For this reason, equation (2.24) can be further simplified for waveguide based optoelectronic and photonic devices as shown in Section 2.4.

2.3 The reduced wave equation in the space domain

Because of the symmetry embedded in wave equation (2.19) in respect of time and space, in principle we can also assume that the optical wave is composed of plane waves with discrete propagation constants and relatively slow-varying envelope functions. More specifically, by assuming that the propagation of these plane waves is all in the z direction, we can write

$$\vec{E}(\vec{r}, t) = \frac{1}{2} \sum_k \vec{v}_k(\vec{r}, t) \mathrm{e}^{\mathrm{j}\beta_k z} + \text{c.c.}, \tag{2.25}$$

with $\vec{v}_k$ and β_k indicating the kth plane wave envelope function in V/m and propagation constant in rad/m, respectively. In equation (2.25), the envelope function is only slow-varying in z but can change arbitrarily in the perpendicular xy plane. Again, under the linear assumption of equation (2.19), we need only to consider

$$\vec{E}(\vec{r}, t) = \vec{v}(\vec{r}, t) \mathrm{e}^{\mathrm{j}\beta_0 z}. \tag{2.26}$$

Replacing the optical field in equation (2.9) with equation (2.26) yields

$$\vec{P}(\vec{r}, t) = \varepsilon_0 \int_{-\infty}^{t} \chi(\vec{r}, t - \tau) \vec{v}(\vec{r}, \tau) \mathrm{e}^{\mathrm{j}\beta_0 z} \, \mathrm{d}\tau = \vec{p}(\vec{r}, t) \mathrm{e}^{\mathrm{j}\beta_0 z}, \tag{2.27}$$

with the slow-varying polarization envelope function defined as

$$\vec{p}(\vec{r}, t) \equiv \varepsilon_0 \int_{-\infty}^{t} \chi(\vec{r}, t - \tau) \vec{v}(\vec{r}, \tau) \mathrm{d}\tau. \tag{2.28}$$

We now plug both equations (2.26) and (2.27) into equation (2.19) to obtain

$$\frac{\partial^2 \vec{v}}{\partial x^2} + \frac{\partial^2 \vec{v}}{\partial y^2} + 2\mathrm{j}\beta_0 \frac{\partial \vec{v}}{\partial z} = \frac{1}{c^2} \frac{\partial^2 \vec{v}}{\partial t^2} + \frac{1}{c^2 \varepsilon_0} \frac{\partial^2 \vec{p}}{\partial t^2} + \beta_0^2 \vec{v} + \mu_0 \mathrm{e}^{-\mathrm{j}\beta_0 z} \frac{\partial}{\partial t} \vec{J}_{\text{sp}}, \tag{2.29}$$

where under the slow-varying envelope assumption ($|\partial^2 \vec{v}/\partial z^2| \ll \beta_0^2 |\vec{v}|$), $\partial^2 \vec{v}/\partial z^2$ is dropped.

Equation (2.29) is the reduced wave equation in the space domain. Together with equation (2.28), it governs the slow-varying envelope of an optical field which is assumed in a modulated plane wave form with its propagation constant β_0 in the z direction. Compared with equation (2.19), equation (2.29) has the fast-varying phase factor ($\mathrm{e}^{\mathrm{j}\beta_0 z}$) excluded, hence it has a reduced (from second to first) spatial derivative order in at least one of the three dimensions. However, unlike equation (2.24), equation (2.29) is not well posed [2]. Therefore, no finite difference algorithm in the time domain will lead to a stable solution to equation (2.29). For this reason, we do not use equation (2.29) directly but proceed to reduce the time derivative orders following the method in Section 2.2 to make equation (2.29) a well-posed problem.

2.4 The reduced wave equation in both time and space domains – the traveling wave model

In waveguide based optoelectronic and photonic devices, the optical wave usually propagates along one direction and is fully or partially confined in the cross-sectional plane perpendicular to the propagation direction. Either starting from the reduced wave equation in the time domain (2.24) or starting from the reduced wave equation in the space domain (2.29), we can further simplify the optical governing equation under such a condition.

2.4.1 The wave equation in fully confined structures

If the optical wave propagates only along z and is fully confined in the cross-sectional xy plane, and if the waveguide transverse structure is uniform along z, we can write the optical field in equation (2.19) as

$$\vec{E}(\vec{r}, t) = \frac{1}{2}\vec{s}\phi(x, y)e(z, t)e^{j(\beta_0 z - \omega_0 t)} + \text{c.c.}, \tag{2.30}$$

where

$\vec{s}$ = unit vector along the polarization direction of the optical field $\vec{E}$,
$\phi(x, y)$ = cross-sectional field distribution in 1/m known as the optical mode or the eigenfunction of the optical waveguide,
$e(z, t)$ = longitudinal field envelope function in V taking relatively slow change with time and along the propagation direction,
$e^{j(\beta_0 z - \omega_0 t)}$ = dimensionless propagation factor in the form of $f(z - vt)$ known as a (harmonic) plane wave traveling at a phase velocity defined by ω_0/β_0.

Expression (2.30) clearly shows that the optical field is factorized into a fast-varying plane wave traveling along z, which is also modulated by an envelope function with slow variation in z and t, and a fixed mode profile in the cross-sectional area without any change in the z direction.

A typical example of such a structure, known as the ridge waveguide, is illustrated in Fig. 2.1.

Under the linear assumption made in equation (2.19), we can rewrite equation (2.30) as

$$\vec{E}(\vec{r}, t) = \vec{x}\phi(x, y)e(z, t)e^{j(\beta_0 z - \omega_0 t)}, \tag{2.31}$$

with $\vec{x}$ indicating the unit vector along x. Here we have considered only one polarization direction, taken to be x, without losing generality. Comparing equation (2.31) with equation (2.22), we know that the optical envelope function in equation (2.24) must be in the form

$$\vec{u}(\vec{r}, t) = \vec{x}\phi(x, y)e(z, t)e^{j\beta_0 z}. \tag{2.32}$$

As a consequence of (2.32), again our final result will be the real part of the solution only.

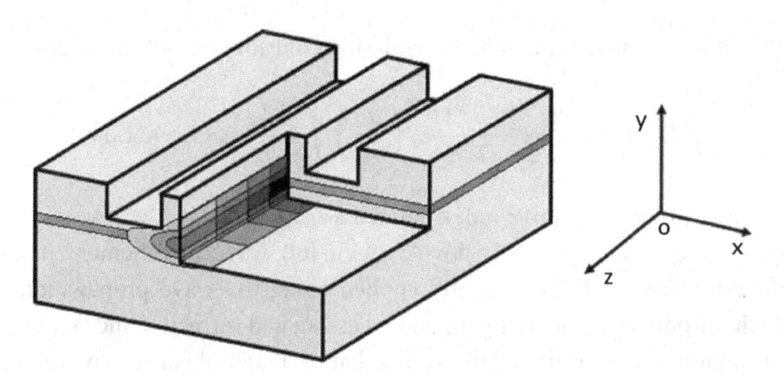

Fig. 2.1. A ridge waveguide structure in which the optical field can be factorized into a modulated plane wave traveling in the z direction and a mode profile in the xy plane without change in z.

Replacing the envelope function in equation (2.24) by expression (2.32), we find:

$$\left\{ \mathrm{j}\frac{2k_0}{c}[1+\tilde{\chi}(\vec{r},\omega_0)]\frac{\partial e(z,t)}{\partial t} + 2\mathrm{j}\beta_0\frac{\partial e(z,t)}{\partial z} \right\}\phi(x,y)$$
$$= -e(z,t)\left\{ \frac{\partial^2}{\partial x^2} + \frac{\partial^2}{\partial y^2} - \beta_0^2 + k_0^2[1+\tilde{\chi}(\vec{r},\omega_0)] \right\}\phi(x,y) + \mu_0 \mathrm{e}^{-\mathrm{j}(\beta_0 z - \omega_0 t)}\frac{\partial J_{\mathrm{sp}x}}{\partial t}, \tag{2.33}$$

where $k_0 \equiv \omega_0/\mathrm{c}$ and $J_{\mathrm{sp}x} \equiv \vec{J}_{\mathrm{sp}} \cdot \vec{x}$. Under the slow-varying envelope assumption ($|\partial^2 e/\partial z^2| \ll |\beta_0|^2|e|$), $\partial^2 e/\partial z^2$ is dropped from equation (2.33). Since the optical field is fully confined in the cross-sectional area with a background material refractive index distribution denoted as $n(x, y, \omega_0)$, we will be able to find the time-invariant optical field distribution in the xy plane, known as the optical mode, by solving the following eigenvalue problem:

$$\left(\frac{\partial^2}{\partial x^2} + \frac{\partial^2}{\partial y^2}\right)\phi(x,y) + k_0^2 n^2(x,y,\omega_0)\phi(x,y) = \beta_0^2\phi(x,y), \tag{2.34}$$

subject to the normalization condition

$$\int_\Sigma \phi^2(x,y)\mathrm{d}x\,\mathrm{d}y = 1, \tag{2.35}$$

with Σ indicating the entire cross-sectional area in m^2 where the optical mode spreads.

Substitute the spatial derivative terms in the xy plane in equation (2.33) by equation (2.34), multiply the optical mode $\phi(x, y)$ on both sides of the equation obtained, and integrate over the entire cross-sectional area to yield

$$\frac{1}{n_{\mathrm{eff}}\mathrm{c}}\left\{\int_\Sigma [1+\tilde{\chi}(\vec{r},\omega_0)]\phi^2(x,y)\mathrm{d}x\,\mathrm{d}y\right\}\frac{\partial e(z,t)}{\partial t} + \frac{\partial e(z,t)}{\partial z}$$
$$= \mathrm{j}\frac{k_0}{2n_{\mathrm{eff}}}\left\{\int_\Sigma [1+\tilde{\chi}(\vec{r},\omega_0) - n^2(x,y,\omega_0)]\phi^2(x,y)\mathrm{d}x\,\mathrm{d}y\right\}e(z,t) + \tilde{s}(z,t), \tag{2.36}$$

where the inhomogeneous spontaneous emission contribution in V/m is given as

$$\widetilde{s}(z,t) \equiv -\frac{\mathrm{j}e^{-\mathrm{j}(\beta_0 z-\omega_0 t)}}{2n_{\mathrm{eff}}\omega_0}\sqrt{\left(\frac{\mu_0}{\varepsilon_0}\right)}\int_\Sigma \frac{\partial J_{\mathrm{sp}x}}{\partial t}\phi(x,y)\mathrm{d}x\,\mathrm{d}y, \tag{2.37}$$

with the dimensionless effective index defined as $n_{\mathrm{eff}} \equiv \beta_0/k_0$.

In optoelectronic and photonic devices with full wave confinement in the cross-sectional area, external bias is usually applied along the wave propagation direction in z, which introduces a material gain and an associated refractive index change inside the active region along this direction. Noting that the material gain per wavelength cycle (i.e., $g/k_0 = gc/\omega_0 = g\lambda_0/2\pi$, with λ_0 denoting the reference wavelength in a vacuum) and the refractive index change (i.e., Δn), both induced by the external bias, appear to be much smaller than the background (i.e., under zero external bias or "cold cavity") material refractive index $n(x, y, \omega_0)$, which is uniform along z, we find

$$\begin{aligned} n^2(\vec{r},\omega_0) &= [n(x,y,\omega_0)+\Delta n(z,\omega_0)]^2 \\ &\approx n^2(x,y,\omega_0)+2n(x,y,\omega_0)\Delta n(z,\omega_0), \end{aligned} \tag{2.38}$$

$$g(\vec{r},\omega_0) = g(z,\omega_0)-\alpha(x,y,z), \tag{2.39}$$

$$\begin{aligned} 1+\widetilde{\chi}(\vec{r},\omega_0) &= \left[n(\vec{r},\omega_0)-\frac{\mathrm{j}}{2k_0}g(\vec{r},\omega_0)\right]^2 \\ &\approx n^2(x,y,\omega_0)+2n(x,y,\omega_0)\Delta n(z,\omega_0)-\frac{\mathrm{j}}{k_0}n(x,y,\omega_0)[g(z,\omega_0)-\alpha(x,y,z)]. \end{aligned} \tag{2.40}$$

In equations (2.38) to (2.40), we have defined the following:

$n(\vec{r},\omega_0)$ = dimensionless material refractive index,
$n(x, y, \omega_0)$ = dimensionless cross-sectional area refractive index under zero bias, also known as the background or "cold cavity" refractive index,
$\Delta n(z, \omega_0)$ = dimensionless bias induced refractive index change, non-zero only inside the active region,
$g(\vec{r},\omega_0)$ = material gain in 1/m,
$g(z, \omega_0)$ = bias induced interband stimulated emission gain (or loss, when it is negative) in 1/m, non-zero only inside the active region,
$\alpha(x, y, z)$ = optical loss in 1/m because of non-interband processes such as free-carrier absorption and scattering.

We have also dropped higher order terms such as $(\Delta n)^2$, $(g/k_0)^2$ and $(\Delta n)(g/k_0)$.

By utilizing equation (2.40) and noting that the interband stimulated emission gain $g(z, \omega_0)$ and the associated refractive index change $\Delta n(z, \omega_0)$ exist only inside the active region, we can derive

$$\int_\Sigma [1+\widetilde{\chi}(\vec{r},\omega_0)]\phi^2(x,y)\mathrm{d}x\,\mathrm{d}y \approx \int_\Sigma n^2(x,y,\omega_0)\phi^2(x,y)\mathrm{d}x\,\mathrm{d}y \equiv \overline{n}^2 \tag{2.41a}$$

and

$$\int_{\Sigma}[1+\tilde{\chi}(\vec{r},\omega_0)-n^2(x,y,\omega_0)]\phi^2(x,y)\mathrm{d}x\,\mathrm{d}y$$
$$\approx\left[2\Delta n(z,\omega_0)-\frac{\mathrm{j}}{k_0}g(z,\omega_0)\right]\int_{\Sigma_{\mathrm{ar}}}n(x,y,\omega_0)\phi^2(x,y)\mathrm{d}x\,\mathrm{d}y$$
$$+\frac{\mathrm{j}}{k_0}\int_{\Sigma}n(x,y,\omega_0)\alpha(x,y,z)\phi^2(x,y)\mathrm{d}x\,\mathrm{d}y$$
$$\approx\bar{n}\Gamma\left[2\Delta n(z,\omega_0)-\frac{\mathrm{j}}{k_0}g(z,\omega_0)\right]+\frac{\mathrm{j}}{k_0}\overline{n\alpha}(z),\tag{2.41b}$$

with Σ_{ar} defined as the cross-sectional area of the active region. Also in deriving equation (2.41b), we have utilized the following approximations:

$$\int_{\Sigma_{\mathrm{ar}}}n(x,y,\omega_0)\phi^2(x,y)\mathrm{d}x\,\mathrm{d}y\approx\bar{n}\int_{\Sigma_{\mathrm{ar}}}\phi^2(x,y)\mathrm{d}x\,\mathrm{d}y=\bar{n}\Gamma\tag{2.42}$$

and

$$\int_{\Sigma}n(x,y,\omega_0)\alpha(x,y,z)\phi^2(x,y)\mathrm{d}x\,\mathrm{d}y\approx\bar{n}\int_{\Sigma}\alpha(x,y,z)\phi^2(x,y)\mathrm{d}x\,\mathrm{d}y=\overline{n\alpha}(z),\tag{2.43}$$

with the optical confinement factor and optical modal loss defined as

$$\Gamma\equiv\frac{\int_{\Sigma_{\mathrm{ar}}}\phi^2(x,y)\mathrm{d}x\,\mathrm{d}y}{\int_{\Sigma}\phi^2(x,y)\mathrm{d}x\,\mathrm{d}y}=\int_{\Sigma_{\mathrm{ar}}}\phi^2(x,y)\mathrm{d}x\,\mathrm{d}y,\tag{2.44}$$

$$\bar{\alpha}(z)\equiv\frac{\int_{\Sigma}\alpha(x,y,z)\phi^2(x,y)\mathrm{d}x\,\mathrm{d}y}{\int_{\Sigma}\phi^2(x,y)\mathrm{d}x\,\mathrm{d}y}=\int_{\Sigma}\alpha(x,y,z)\phi^2(x,y)\mathrm{d}x\,\mathrm{d}y.\tag{2.45}$$

Finally, we plug equations (2.41a) and (2.41b) into equation (2.36) to yield

$$\frac{1}{v_{\mathrm{g}}}\frac{\partial e(z,t)}{\partial t}+\frac{\partial e(z,t)}{\partial z}=\left[\mathrm{j}k_0\Gamma\Delta n(z,\omega_0)+\frac{1}{2}\Gamma g(z,\omega_0)-\frac{1}{2}\bar{\alpha}(z)\right]e(z,t)+\tilde{s}(z,t),\tag{2.46}$$

where $v_{\mathrm{g}}\equiv \mathrm{c}/n_{\mathrm{g}}\approx cn_{\mathrm{eff}}/\bar{n}^2$ and $n_{\mathrm{eff}}\approx\bar{n}$ are assumed, and where n_{g} is the group index and v_{g} is the group velocity.

Equation (2.46) governs the envelope function of the optical wave propagating along $+z$. For the optical wave that propagates along the opposite direction ($-z$), we just need to use $-z$ to replace z in equation (2.46) as both the material property and the spontaneous emission contribution have bidirectional symmetry along $\pm z$. Therefore, we obtain

$$\left(\frac{1}{v_g}\frac{\partial}{\partial t}+\frac{\partial}{\partial z}\right)e^{\mathrm{f}}(z,t)=\left[\mathrm{j}k_0\Gamma\Delta n(z,\omega_0)+\frac{1}{2}\Gamma g(z,\omega_0)-\frac{1}{2}\overline{\alpha}(z)\right]e^{\mathrm{f}}(z,t)+\widetilde{s}^{\mathrm{f}}(z,t), \tag{2.47a}$$

$$\left(\frac{1}{v_g}\frac{\partial}{\partial t}-\frac{\partial}{\partial z}\right)e^{\mathrm{b}}(z,t)=\left[\mathrm{j}k_0\Gamma\Delta n(z,\omega_0)+\frac{1}{2}\Gamma g(z,\omega_0)-\frac{1}{2}\overline{\alpha}(z)\right]e^{\mathrm{b}}(z,t)+\widetilde{s}^{\mathrm{b}}(z,t), \tag{2.47b}$$

where we have used the superscripts f and b to indicate the forward and backward propagating wave envelope functions, respectively. In equation (2.47), because the inhomogeneous spontaneous emission contributes to both the forward and backward propagating waves, we have $\widetilde{s}^{\mathrm{b}}(z,t)=\widetilde{s}^{\mathrm{f}}(-z,t)$ with $\widetilde{s}^{\mathrm{f}}$ given by equation (2.37).

The one-dimensional (1D) slow-varying envelope equation (2.47) along the wave propagation direction (i.e., $\pm z$), together with the two-dimensional (2D) eigenvalue equation (2.34) in the cross-sectional area (i.e., the xy plane), form the governing equations for modeling the optical wave that propagates along $\pm z$ and is fully confined by the waveguide in the cross-sectional xy plane. These equations can be solved subject to certain initial and boundary conditions. Since the initial and boundary conditions are related to the operating conditions and structures of specific devices, we will find the effect of these conditions on device performance through examples in Chapter 10.

Once the optical mode $\phi(x,y)$, the forward (along $+z$) and the backward (along $-z$) slow-varying envelopes $e^{\mathrm{f}}(z,t)$ and $e^{\mathrm{b}}(z,t)$ are solved by equations (2.34) and (2.47), respectively, the real-world optical field is obtained by using

$$\vec{E}(\vec{r},t)=\frac{1}{2}\vec{x}\phi(x,y)[e^{\mathrm{f}}(z,t)\mathrm{e}^{\mathrm{j}\beta_0 z}+e^{\mathrm{b}}(z,t)\mathrm{e}^{-\mathrm{j}\beta_0 z}]\mathrm{e}^{-\mathrm{j}\omega_0 t}+\mathrm{c.c.} \tag{2.48}$$

As seen in the derivation process, we find that this model is valid under the following conditions.

(1) Assumptions on the optical wave.
- Wave propagates along the device in a longitudinal direction only.
- Wave is fully confined in the cross-sectional area perpendicular to the propagation direction.
- Wave has discrete optical frequencies with relatively slow-varying envelopes.

(2) Assumptions on the material.
- Material has linear optical property.
- Material takes no time to respond to any variation of optical wave envelope.

Also, in the above derivations, we have assumed that the optical wave has:

- a single operating frequency (i.e., ω_0);
- a single optical mode (i.e., the waveguide supports a single guided mode only);
- a single polarization state (assumed to be along x).

However, equations (2.34) and (2.47) can readily be expanded to model the optical wave with multiple operating frequencies, or multiple modes, or arbitrary polarization states, by utilizing the linear superposition theory and the mode orthogonality. Therefore, the last three constraints are removable.

By comparing equation (2.31) with equation (2.26), we find that the optical envelope function in equation (2.29) can be written as

$$\vec{v}(\vec{r}, t) = \vec{x}\phi(x, y)e(z, t)\mathrm{e}^{-\mathrm{j}\omega_0 t}. \tag{2.49}$$

Inserting equation (2.32) into equation (2.23) yields

$$\vec{P}(\vec{r}, t) = \vec{x}\varepsilon_0\widetilde{\chi}(\vec{r}, \omega_0)\phi(x, y)e(z, t)\mathrm{e}^{\mathrm{j}(\beta_0 z-\omega_0 t)}. \tag{2.50}$$

Further, comparing equation (2.50) with equation (2.27), we also find

$$\vec{p}(\vec{r}, t) = \vec{x}\varepsilon_0\widetilde{\chi}(\vec{r}, \omega_0)\phi(x, y)e(z, t)\mathrm{e}^{-\mathrm{j}\omega_0 t}. \tag{2.51}$$

By replacing the optical field and polarization envelope functions in equation (2.29) by expressions (2.49) and (2.51), respectively, and utilizing equations (2.34) and (2.35), we obtain equation (2.36) again, as it should be. This confirms that equation (2.47) is the reduced wave equation both in time and space domains; it can be obtained from the full wave equation (2.19) by reducing the time derivative and the space derivative in either sequence. The condition under which the time derivative can be reduced requires the optical field to take an amplitude-modulated harmonic wave in the time domain, with its modulation bandwidth (i.e., the base bandwidth) much smaller than the harmonic wave frequency (i.e., the carrier frequency), as required by the time slow-varying envelope assumption. The condition under which the spatial derivative can be reduced in a certain direction requires the optical field to take an amplitude-modulated plane wave in that direction, with its modulation bandwidth (i.e., the maximum spatial frequency) much smaller than the propagation constant, as required by the spatial slow-varying envelope assumption. Since equation (2.47) has been derived under both conditions, it governs the (slow-varying) envelope function of an optical field in the form of a modulated harmonic plane wave in a certain direction (along z in this derivation). A harmonic plane wave in a certain direction describes a plane wave propagating along that direction. Therefore, equation (2.47) governs the (slow-varying) envelope functions of the two traveling plane waves along $\pm z$, respectively.

2.4.2 The wave equation in partially confined structures

In some applications, the waveguide transverse structure is not uniform along the wave propagation direction. A typical example is a horizontally varied structure such as the horn waveguide [3] shown in Fig. 2.2.

In such a structure, the optical wave is confined only in the vertical direction y, rather than in the entire cross-sectional xy plane. Therefore, instead of equation (2.31), we have

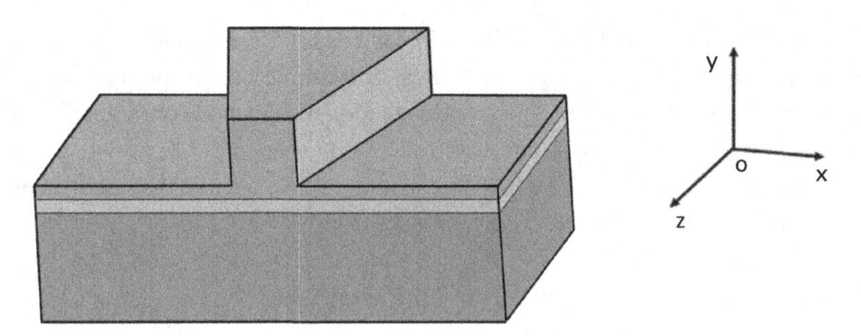

Fig. 2.2. A horn waveguide structure in which the optical wave is confined only in the y direction and is propagating along both z and x directions.

to write the optical field in the form of [4]

$$\vec{E}(\vec{r}, t) = \vec{x}\phi(y)e(x, z, t)e^{j(\beta_0 z - \omega_0 t)}, \tag{2.52}$$

with $\phi(y)$ indicating the optical field distribution (or the 1D optical mode) along y in $1/\mathrm{m}^{1/2}$, and $e(x, z, t)$ the envelope function in $\mathrm{V/m}^{1/2}$. In accordance with equation (2.52), we will use

$$\vec{u}(\vec{r}, t) = \vec{x}\phi(y)e(x, z, t)e^{j\beta_0 z}, \tag{2.53}$$

to replace equation (2.32) as well. Plugging (2.53) into (2.24), multiplying the 1D optical mode $\phi(y)$ on both sides of the equation obtained, and integrating along the vertical direction y yields

$$\begin{aligned}
&\frac{1}{n_{\mathrm{eff}} c}\left\{\int_{\Sigma_y}[1+\widetilde{\chi}(\vec{r}, \omega_0)]\phi^2(y)\mathrm{d}y\right\}\frac{\partial e(x, z, t)}{\partial t} + \frac{\partial e(x, z, t)}{\partial z} \\
&= \frac{\mathrm{j}}{2n_{\mathrm{eff}}k_0}\frac{\partial^2 e(x, z, t)}{\partial x^2} + \frac{\mathrm{j}k_0}{2n_{\mathrm{eff}}}\left\{\int_{\Sigma_y}[1+\widetilde{\chi}(\vec{r}, \omega_0) - n^2(y, \omega_0)]\phi^2(y)\mathrm{d}y\right\} \\
&\quad \times e(x, z, t) + \widetilde{s}(x, z, t),
\end{aligned} \tag{2.54}$$

where the 1D optical mode along y can be found by solving the following eigenvalue problem

$$\frac{\partial^2}{\partial y^2}\phi(y) + k_0^2 n^2(y, \omega_0)\phi(y) = \beta_0^2\phi(y), \tag{2.55}$$

subject to the normalization condition

$$\int_{\Sigma_y}\phi^2(y)\mathrm{d}y = 1. \tag{2.56}$$

In expressions (2.54) to (2.56), Σ_y indicates the entire vertical range along y in m where the 1D optical mode spreads. Note that $n(y, \omega_0)$ denotes the dimensionless background or "cold cavity" material refractive index distribution along y. Again, $\partial^2 e/\partial z^2$

is dropped from equation (2.54) under the slow-varying envelope assumption. Finally, in equation (2.54), the inhomogeneous spontaneous emission contribution in $\mathrm{V/m^{3/2}}$ is given as

$$\widetilde{s}(x, z, t) \equiv -\frac{\mathrm{j}e^{-\mathrm{j}(\beta_0 z - \omega_0 t)}}{2n_{\mathrm{eff}}\omega_0}\sqrt{\frac{\mu_0}{\varepsilon_0}}\int_{\Sigma_y}\frac{\partial J_{\mathrm{spx}}}{\partial t}\phi(y)\mathrm{d}y. \tag{2.57}$$

In such a waveguide structure, the active region usually expands to an entire xz plane within one of the vertically stacked layers in the y direction. In accordance with the active region distribution, the external bias is usually applied on the top xz plane, which introduces a material gain and an associated refractive index change inside the active region. Similarly to equations (2.38) to (2.40), by assuming that the bias induced material gain per wavelength cycle and the associated refractive index change are perturbations in the background material refractive index distribution, which is uniform in the xz plane, we find

$$n^2(\vec{r}, \omega_0) \approx n^2(y, \omega_0) + 2n(y, \omega_0)\Delta n(x, z, \omega_0), \tag{2.58}$$

$$g(\vec{r}, \omega_0) = g(x, z, \omega_0) - \alpha(x, y, z), \tag{2.59}$$

$$\begin{aligned} 1 + \widetilde{\chi}(\vec{r}, \omega_0) &= \left[n(\vec{r}, \omega_0) - \frac{\mathrm{j}}{2k_0}g(\vec{r}, \omega_0)\right]^2 \\ &\approx n^2(y, \omega_0) + 2n(y, \omega_0)\Delta n(x, z, \omega_0) - \frac{\mathrm{j}}{k_0}n(y, \omega_0)\left[g(x, z, \omega_0) - \alpha(x, y, z)\right]. \end{aligned} \tag{2.60}$$

In equations (2.58) to (2.60), we have dropped the higher order terms $(\Delta n)^2$, $(g/k_0)^2$ and $(\Delta n)(g/k_0)$.

By utilizing equation (2.60) and noting that the interband stimulated emission gain $g(x, z, \omega_0)$ and the associated refractive index change $\Delta n(x, z, \omega_0)$ exist only inside the active region, we can further derive

$$\int_{\Sigma_y}[1 + \widetilde{\chi}(\vec{r}, \omega_0)]\phi^2(y)\mathrm{d}y \approx \int_{\Sigma_y} n^2(y, \omega_0)\phi^2(y)\mathrm{d}y \equiv \overline{n}^2 \tag{2.61a}$$

and

$$\begin{aligned} &\int_{\Sigma_y}\left[1 + \widetilde{\chi}(\vec{r}, \omega_0) - n^2(y, \omega_0)\right]\phi^2(y)\mathrm{d}y \\ &\approx \left[2\Delta n(x, z, \omega_0) - \frac{\mathrm{j}}{k_0}g(x, z, \omega_0)\right]\int_{\Sigma_{\mathrm{ary}}} n(y, \omega_0)\phi^2(y)\mathrm{d}y \\ &\quad + \frac{\mathrm{j}}{k_0}\int_{\Sigma_y} n(y, \omega_0)\alpha(x, y, z)\phi^2(y)\mathrm{d}y \\ &\approx \overline{n}\Gamma\left[2\Delta n(x, z, \omega_0) - \frac{\mathrm{j}}{k_0}g(x, z, \omega_0)\right] + \frac{\mathrm{j}}{k_0}\overline{n\alpha}(x, z), \end{aligned} \tag{2.61b}$$

with Σ_{ary} defined as the active region vertical thickness along y. In deriving (2.61b), we have utilized the following approximations:

$$\int_{\Sigma_{ary}} n(y,\omega_0)\phi^2(y)\mathrm{d}y \approx \overline{n}\int_{\Sigma_{ary}} \phi^2(y)\mathrm{d}y = \overline{n}\Gamma, \tag{2.62}$$

$$\int_{\Sigma_y} n(y,\omega_0)\alpha(x,y,z)\phi^2(y)\mathrm{d}y \approx \overline{n}\int_{\Sigma_y} \alpha(x,y,z)\phi^2(y)\mathrm{d}y = \overline{n}\,\overline{\alpha}(x,z), \tag{2.63}$$

with the optical confinement factor and optical modal loss defined as

$$\Gamma \equiv \frac{\int_{\Sigma_{ary}}\phi^2(y)\mathrm{d}y}{\int_{\Sigma_y}\phi^2(y)\mathrm{d}y} = \int_{\Sigma_{ary}}\phi^2(y)\mathrm{d}y, \tag{2.64}$$

$$\overline{\alpha}(x,z) \equiv \frac{\int_{\Sigma_y}\alpha(x,y,z)\phi^2(y)\mathrm{d}y}{\int_{\Sigma_y}\phi^2(y)\mathrm{d}y} = \int_{\Sigma_y}\alpha(x,y,z)\phi^2(y)\mathrm{d}y. \tag{2.65}$$

Finally, we plug equations (2.61a) and (2.61b) into equation (2.54) to yield

$$\begin{aligned}\frac{1}{v_g}\frac{\partial e(x,z,t)}{\partial t} + \frac{\partial e(x,z,t)}{\partial z} &= \frac{\mathrm{j}}{2n_{\mathrm{eff}}k_0}\frac{\partial^2 e(x,z,t)}{\partial x^2} \\ &+ \left[\mathrm{j}k_0\Gamma\Delta n(x,z,\omega_0) + \frac{1}{2}\Gamma g(x,z,\omega_0) - \frac{1}{2}\overline{\alpha}(x,z)\right]e(x,z,t) + \tilde{s}(x,z,t),\end{aligned} \tag{2.66}$$

where again $v_g \equiv c/n_g \approx cn_{\mathrm{eff}}/\overline{n}^2$ and $n_{\mathrm{eff}} \approx \overline{n}$ are assumed.

Equation (2.66) governs the envelope function of the optical wave propagating along $+z$. For the optical wave that propagates along the opposite direction $(-z)$, we just need to use $-z$ to replace z in equation (2.66) because of the material bidirectional symmetry along $\pm z$. Therefore, we obtain

$$\begin{aligned}\left(\frac{1}{v_g}\frac{\partial}{\partial t} + \frac{\partial}{\partial z}\right)e^{\mathrm{f}}(x,z,t) &= \frac{\mathrm{j}}{2n_{\mathrm{eff}}k_0}\frac{\partial^2}{\partial x^2}e^{\mathrm{f}}(x,z,t) \\ &+ \left[\mathrm{j}k_0\Gamma\Delta n(x,z,\omega_0) + \frac{1}{2}\Gamma g(x,z,\omega_0) - \frac{1}{2}\overline{\alpha}(x,z)\right]e^{\mathrm{f}}(x,z,t) + \tilde{s}^{\mathrm{f}}(x,z,t),\end{aligned} \tag{2.67a}$$

$$\begin{aligned}\left(\frac{1}{v_g}\frac{\partial}{\partial t} - \frac{\partial}{\partial z}\right)e^{\mathrm{b}}(x,z,t) &= \frac{\mathrm{j}}{2n_{\mathrm{eff}}k_0}\frac{\partial^2}{\partial x^2}e^{\mathrm{b}}(x,z,t) \\ &+ \left[\mathrm{j}k_0\Gamma\Delta n(x,z,\omega_0) + \frac{1}{2}\Gamma g(x,z,\omega_0) - \frac{1}{2}\overline{\alpha}(x,z)\right]e^{\mathrm{b}}(x,z,t) + \tilde{s}^{\mathrm{b}}(x,z,t),\end{aligned} \tag{2.67b}$$

where again the superscripts f and b distinguish the forward and backward propagating wave envelope functions, and the inhomogeneous spontaneous emission contributions to the forward and backward propagating waves, with $\tilde{s}^{\mathrm{f}}$ given by equation (2.57) and $\tilde{s}^{\mathrm{b}}(x,z,t) = \tilde{s}^{\mathrm{f}}(x,-z,t)$, respectively.

Therefore, the 2D slow-varying envelope equation (2.67) in the xz plane, together with the 1D eigenvalue equation (2.55) in y, form the governing equations for modeling the partially confined (along the vertical direction y) optical wave propagation along $\pm z$. Because of the lack of lateral confinement (in the x direction), the 2D envelope function diffracts laterally as indicated by an extra second-order derivative term in respect of x in the governing equation (2.67). Therefore, as the solution of equation (2.67), the envelope function changes only slowly with z and t, which indicates the wave propagation along $\pm z$; its change in x, however, will be determined by the boundary conditions imposed in the x direction, which is normally related to the device lateral structure, and its rate of change may not slow. Again, equations (2.67) and (2.55) can be solved, once the initial and boundary conditions are specified for a given device.

Finally, the real-world optical field is obtained using

$$\vec{E}(\vec{r}, t) = \frac{1}{2}\vec{x}\phi(y)[e^{\mathrm{f}}(x, z, t)e^{\mathrm{j}\beta_0 z} + e^{\mathrm{b}}(x, z, t)e^{-\mathrm{j}\beta_0 z}]e^{-\mathrm{j}\omega_0 t} + \mathrm{c.c.} \tag{2.68}$$

2.4.3 The wave equation in periodically corrugated structures

In DFB or DBR lasers and other grating based devices, periodically corrugated structures must be employed to provide distributed reflections along with the waveguide. A typical example of such a periodically corrugated waveguide structure is shown in Fig. 2.3.

Unlike the previous structures, in which the forward and backward propagating waves have no interaction until they reach the waveguide ends, such a waveguide allows the forward and backward waves to couple to each other as they propagate through the periodically perturbed structure along the waveguide. Moreover, a periodic structure with period Λ can be expanded as a summation of many harmonic grating orders with their wave numbers ranking as $m2\pi/\Lambda$, where $m = 0, \pm 1, \pm 2, \ldots, \pm\infty$. Assuming that the grating harmonic component in the Mth ($M \geq 1$) order couples the forward and backward propagation waves along the waveguide direction ($\pm z$), the mth ($m > M$) order components will fast decay and hence are negligible. The mth ($m < M$) order components will, however, couple the forward and backward propagating waves to the radiation waves which leave the waveguide at a certain angle to the propagation direction along $\pm z$ [5–7].

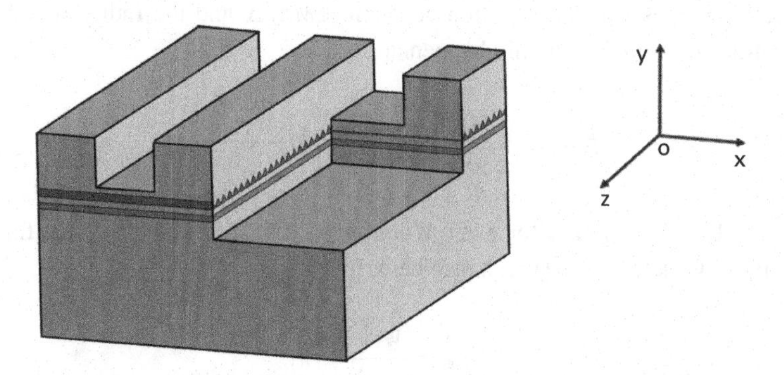

Fig. 2.3. A periodically corrugated waveguide structure in which the propagating waves along $\pm z$ are distributively coupled because of the reflections of the grating.

Since the Mth harmonic order of the grating wave number couples the forward and backward propagation constants, we must have

$$2\delta = \beta_0 - \left(M\frac{2\pi}{\Lambda} - \beta_0\right) = 2\left(\beta_0 - \frac{M\pi}{\Lambda}\right) \ll 2\beta_0, \tag{2.69}$$

where 2δ is the difference of the propagation constants between the original forward and the backward coupled to forward (through the grating backward scattering) propagating waves. Note that -2δ, on the other hand, is the difference in the propagation constants between the original backward and the forward coupled to backward (again through the grating backward scattering) propagating waves. A necessary condition for the forward (and the backward) propagating wave to be sustainable inside the waveguide is apparently that the two forward (and the two backward) propagating wave components have the same propagation constants, known as phase matching. For passive waveguides, we immediately find that the phase matching condition arises at $\delta = 0$. In active waveguides, however, the component with the fastest growing amplitude (because of the gain) does not necessarily correspond to $\delta = 0$. Therefore, the active waveguide may allow sustainable forward and backward propagating waves to have their propagation constants detuned from $\pm M\pi/\Lambda$ (i.e., the Bragg condition), or $\delta \neq 0$. On the scale of β_0 (or π/Λ), the detuning (δ) must be very small, i.e., $\delta/\beta_0 \ll 1$, as otherwise the amplitude loss because of the phase mismatch cannot be compensated for by the amplitude growth from the gain. Therefore, waves with large detuned propagation constants away from the Bragg condition cannot exist. Equation (2.69) addresses such a quasi-phase matching condition in active waveguides.

In accordance with the phase matching condition, we can take the propagation constants of the forward and backward propagating waves as $\pm M\pi/\Lambda$ instead of the previous $\pm\beta_0$ to facilitate deriving the coupled wave equations shown below. However, it is worth mentioning that, by taking $\pm\beta_0$ as the propagation constants of the forward and backward propagating waves in decomposing the total optical field (2.71), we can also get a consistent result.

Also, from the phase matching condition, coupling between the forward and backward propagating waves with propagation constant $\pm M\pi/\Lambda$ and the radiation waves with propagation constant β_r can only happen at

$$\beta_r = \pm\left(\frac{M\pi}{\Lambda} - m\frac{2\pi}{\Lambda}\right) = \pm\frac{M-2m}{\Lambda}\pi, \tag{2.70a}$$

with $m = 1, 2, 3, \ldots, M/2$ for even M and $m = 1, 2, 3, \ldots, (M-1)/2$ for odd M, respectively. Equation (2.70a) can also be written as

$$\beta_r = \frac{M-2m}{\Lambda}\pi, \tag{2.70b}$$

with $m = 1, 2, 3, \ldots, M-1$.

Therefore, in periodically corrugated waveguide structures, we have to retain both forward and backward propagating waves, as well as all the radiation waves that satisfy the phase matching conditions, in decomposing the optical field for deriving the reduced wave equation in both time and space domains. For this reason, we plug

$$\vec{u}(\vec{r},t) = \vec{x}\left\{\phi(x,y)\left[e^{\mathrm{f}}(z,t)\mathrm{e}^{\mathrm{j}\frac{M\pi}{\Lambda}z} + e^{\mathrm{b}}(z,t)\mathrm{e}^{-\mathrm{j}\frac{M\pi}{\Lambda}z}\right] + \sum_{m=1}^{M-1}\hat{e}_m(x,y)\mathrm{e}^{\mathrm{j}\frac{M-2m}{\Lambda}\pi z}\right\}, \tag{2.71}$$

into (2.24) to obtain

$$\begin{aligned}
&\frac{1+\widetilde{\chi}}{n_{\mathrm{eff}}c}\phi\left(\mathrm{e}^{\mathrm{j}\frac{M\pi}{\Lambda}z}\frac{\partial e^{\mathrm{f}}}{\partial t} + \mathrm{e}^{-\mathrm{j}\frac{M\pi}{\Lambda}z}\frac{\partial e^{\mathrm{b}}}{\partial t}\right) + \phi\left(\mathrm{e}^{\mathrm{j}\frac{M\pi}{\Lambda}z}\frac{\partial e^{\mathrm{f}}}{\partial z} - \mathrm{e}^{-\mathrm{j}\frac{M\pi}{\Lambda}z}\frac{\partial e^{\mathrm{b}}}{\partial z}\right) = \\
&\mathrm{j}\frac{1}{2n_{\mathrm{eff}}k_0}\sum_{m=1}^{M-1}\mathrm{e}^{\mathrm{j}\frac{M-2m}{\Lambda}\pi z}\left[\frac{\partial^2}{\partial x^2} + \frac{\partial^2}{\partial y^2} + k_0^2(1+\widetilde{\chi}) - \left(\frac{M-2m}{\Lambda}\right)^2\pi^2\right]\hat{e}_m \\
&+\mathrm{j}\frac{k_0}{2n_{\mathrm{eff}}}\left(1+\widetilde{\chi} - n^2 + \frac{2n_{\mathrm{eff}}\delta}{k_0}\right)\phi\left(e^{\mathrm{f}}\mathrm{e}^{\mathrm{j}\frac{M\pi}{\Lambda}z} + e^{\mathrm{b}}\mathrm{e}^{-\mathrm{j}\frac{M\pi}{\Lambda}z}\right) - \mathrm{j}\frac{\mathrm{e}^{\mathrm{j}\omega_0 t}}{2n_{\mathrm{eff}}\omega_0}\sqrt{\left(\frac{\mu_0}{\varepsilon_0}\right)}\frac{\partial J_{\mathrm{spx}}}{\partial t},
\end{aligned} \tag{2.72}$$

where equations (2.34) and (2.69) have been used while $\partial^2 e^{\mathrm{f}}/\partial z^2$ and $\partial^2 e^{\mathrm{b}}/\partial z^2$ have been dropped under the slow-varying envelope assumption. Strictly speaking, the radiation wave amplitudes $\hat{e}_m$ should have (z, t) dependence as well, since the forward and backward propagating waves are actually the sources of these radiation waves and the former certainly depends on (z, t). However, as will be seen in equation (2.82), the dependence of $\hat{e}_m$ on (z, t) is implicit (i.e., through e^{f} and e^{b} only), therefore, we can ignore the partial derivatives of $\hat{e}_m$ to (z, t) as the changes of $\hat{e}_m$ on (z, t) are adiabatic. For this reason, we only record $\hat{e}_m$ as explicit functions of (x, y).

Usually the grating itself, i.e., the corrugated part, can be viewed as a perturbation in an optical waveguide with full confinement in the cross-sectional area, which has been discussed in Section 2.4.1 and is known as the unperturbed reference waveguide. Following the change in the grating, the material properties, i.e., the refractive index and gain, all change periodically along the wave propagation direction z. Therefore, we can expand the periodically changed material properties into Fourier series by writing

$$\begin{aligned}
n^2(\vec{r},\omega_0) &= \left[n(x,y,\omega_0) + \sum_{m=-\infty,m\neq 0}^{+\infty}\Delta n_m(x,y)\mathrm{e}^{\mathrm{j}m\frac{2\pi}{\Lambda}z} + \Delta n(z,\omega_0)\right. \\
&\qquad \left. + \sum_{m=-\infty,m\neq 0}^{+\infty}\delta n_m \mathrm{e}^{\mathrm{j}m\frac{2\pi}{\Lambda}z}\right]^2 \\
&\approx n^2(x,y,\omega_0) + 2n(x,y,\omega_0)\Delta n(z,\omega_0) \\
&\quad + 2n(x,y,\omega_0)\sum_{m=-\infty,m\neq 0}^{+\infty}\left[\Delta n_m(x,y) + \delta n_m\right]\mathrm{e}^{\mathrm{j}m\frac{2\pi}{\Lambda}z},
\end{aligned} \tag{2.73}$$

and

$$g(\vec{r},\omega_0) = g(z,\omega_0) + \sum_{m=-\infty,m\neq 0}^{+\infty} \Delta g_m e^{jm\frac{2\pi}{\Lambda}z} - \left[\alpha(x,y,z) + \sum_{m=-\infty,m\neq 0}^{+\infty} \Delta\alpha_m(x,y) e^{jm\frac{2\pi}{\Lambda}z}\right]$$

$$= g(z,\omega_0) - \alpha(x,y,z) + \sum_{m=-\infty,m\neq 0}^{+\infty} [\Delta g_m - \Delta\alpha_m(x,y)]\, e^{jm\frac{2\pi}{\Lambda}z}, \tag{2.74}$$

with Δn_m, δn_m, Δg_m, and $\Delta\alpha_m$ denoting the mth Fourier expansion coefficient of the material refractive index, the (bias induced) index change, the (bias induced) stimulated emission gain, and the optical loss, respectively. In these expansions, the Fourier coefficients are obtained through

$$\Delta n_m(x,y) = \frac{1}{\Lambda}\int_0^\Lambda n_p(x,y,z) e^{-jm\frac{2\pi}{\Lambda}z}\, dz, \tag{2.75a}$$

$$\delta n_m = \frac{1}{\Lambda}\int_0^\Lambda \Delta n_p(z) e^{-jm\frac{2\pi}{\Lambda}z}\, dz, \tag{2.75b}$$

$$\Delta g_m = \frac{1}{\Lambda}\int_0^\Lambda g_p(z) e^{-jm\frac{2\pi}{\Lambda}z}\, dz, \tag{2.75c}$$

$$\Delta\alpha_m(x,y) = \frac{1}{\Lambda}\int_0^\Lambda \alpha_p(x,y,z) e^{-jm\frac{2\pi}{\Lambda}z}\, dz, \tag{2.75d}$$

for $m = 0, \pm 1, \pm 2, \ldots$, with n_p, Δn_p, g_p, and α_p denoting the periodically corrugated part of the material refractive index, the (bias induced) index change, the (bias induced) stimulated emission gain and the optical loss, respectively. Also in equations (2.73) and (2.74), the DC components (i.e., the 0th order coefficients Δn_0, δn_0, Δg_0, and $\Delta\alpha_0$) are merged with their corresponding terms in the reference waveguide where the grating does not exist (i.e., n, Δn, g, and α, respectively). Although it is always possible to select the unperturbed reference waveguide in such a way that the average n_p in one period is equal to zero, hence $\Delta n_0 = 0$, it is generally not possible to have all the DC components disappear for a given corrugated structure, no matter how we select our reference. Therefore, we should not forget to include the DC contribution of the corrugated part to the material properties given in equations (2.73) and (2.74).

From equations (2.73) and (2.74), we can derive

$$1 + \tilde{\chi}(\vec{r},\omega_0) = \left[n(\vec{r},\omega_0) - \frac{j}{2k_0} g(\vec{r},\omega_0)\right]^2$$

$$\approx 1 + \tilde{\chi}_0(\vec{r},\omega_0) + 2n(x,y,\omega_0) \sum_{m=-\infty,m\neq 0}^{+\infty} A_m(x,y) e^{jm\frac{2\pi}{\Lambda}z}, \tag{2.76a}$$

with

$$A_m(x,y) \equiv \Delta n_m(x,y) + \delta n_m - \frac{j}{2k_0}[\Delta g_m - \Delta\alpha_m(x,y)], \tag{2.76b}$$

and

$$1 + \widetilde{\chi}_0(\vec{r}, \omega_0) \equiv n^2(x, y, \omega_0) + 2n(x, y, \omega_0)\Delta n(z, \omega_0) - \frac{\mathrm{j}}{k_0} n(x, y, \omega_0)[g(z, \omega_0) - \alpha(x, y, z)], \tag{2.76c}$$

given as the total (i.e., vacuum plus host medium) susceptibility of the unperturbed reference waveguide.

Plugging equation (2.76a) into equation (2.72), and collecting all the forward propagating terms with factor $\mathrm{e}^{\mathrm{j}\frac{M\pi}{\Lambda}z}$ yields

$$\frac{1+\widetilde{\chi}_0}{n_{\text{eff}}c}\phi\frac{\partial e^{\mathrm{f}}}{\partial t} + \phi\frac{\partial e^{\mathrm{f}}}{\partial z} = \frac{\mathrm{j}k_0}{2n_{\text{eff}}}\left(1+\widetilde{\chi}_0 - n^2 + \frac{2n_{\text{eff}}\delta}{k_0}\right)\phi e^{\mathrm{f}} + \frac{\mathrm{j}k_0 n}{n_{\text{eff}}} A_M \phi e^{\mathrm{b}} + \frac{\mathrm{j}k_0}{2n_{\text{eff}}}\sum_{m=1}^{M-1} A_m \hat{e}_m - \mathrm{j}\frac{\mathrm{e}^{-\mathrm{j}(\frac{M\pi}{\Lambda}z - \omega_0 t)}}{2n_{\text{eff}}\omega_0}\sqrt{\left(\frac{\mu_0}{\varepsilon_0}\right)}\frac{\partial J_{\text{sp}x}}{\partial t}. \tag{2.77a}$$

Collecting all the backward propagating terms with factor $\mathrm{e}^{-\mathrm{j}\frac{M\pi}{\Lambda}z}$ yields

$$\frac{1+\widetilde{\chi}_0}{n_{\text{eff}}c}\phi\frac{\partial e^{\mathrm{b}}}{\partial t} - \phi\frac{\partial e^{\mathrm{b}}}{\partial z} = \frac{\mathrm{j}k_0}{2n_{\text{eff}}}\left(1+\widetilde{\chi}_0 - n^2 + \frac{2n_{\text{eff}}\delta}{k_0}\right)\phi e^{\mathrm{b}} + \frac{\mathrm{j}k_0 n}{n_{\text{eff}}} A_{-M} \phi e^{\mathrm{f}} + \frac{\mathrm{j}k_0}{2n_{\text{eff}}}\sum_{m=1}^{M-1} A_{m-M} \hat{e}_m - \mathrm{j}\frac{\mathrm{e}^{\mathrm{j}(\frac{M\pi}{\Lambda}z + \omega_0 t)}}{2n_{\text{eff}}\omega_0}\sqrt{\left(\frac{\mu_0}{\varepsilon_0}\right)}\frac{\partial J_{\text{sp}x}}{\partial t}. \tag{2.77b}$$

And collecting all the radiation wave terms with factors $\mathrm{e}^{\mathrm{j}\frac{M-2m}{\Lambda}\pi z}$ yields

$$\left[\frac{\partial^2}{\partial x^2} + \frac{\partial^2}{\partial y^2} + k_0^2 n^2 - \left(\frac{M-2m}{\Lambda}\right)^2 \pi^2\right]\hat{e}_m = -k_0^2 n\phi(A_{-m}e^{\mathrm{f}} + A_{M-m}e^{\mathrm{b}}), \tag{2.77c}$$

where $m = 1, 2, 3, \ldots, M-1$.

Multiplying the optical mode $\phi(x, y)$ on both sides of equations (2.77a) and (2.77b), and integrating over the entire cross-sectional area yields

$$\left(\frac{1}{v_{\mathrm{g}}}\frac{\partial}{\partial t} + \frac{\partial}{\partial z}\right)e^{\mathrm{f}}(z,t) = \left[\mathrm{j}\delta + \mathrm{j}k_0\Gamma\Delta n(z,\omega_0) + \frac{1}{2}\Gamma g(z,\omega_0) - \frac{1}{2}\overline{\alpha}(z)\right]e^{\mathrm{f}}(z,t) + \mathrm{j}\kappa_M e^{\mathrm{b}}(z,t) + \frac{\mathrm{j}k_0}{2n_{\text{eff}}}\sum_{m=1}^{M-1}\int_{\Sigma} A_m(x,y)\hat{e}_m(x,y)\phi(x,y)\mathrm{d}x\,\mathrm{d}y + \widetilde{s}^{\mathrm{f}}(z,t), \tag{2.78a}$$

$$\left(\frac{1}{v_{\mathrm{g}}}\frac{\partial}{\partial t} - \frac{\partial}{\partial z}\right)e^{\mathrm{b}}(z,t) = \left[\mathrm{j}\delta + \mathrm{j}k_0\Gamma\Delta n(z,\omega_0) + \frac{1}{2}\Gamma g(z,\omega_0) - \frac{1}{2}\overline{\alpha}(z)\right]e^{\mathrm{b}}(z,t) + \mathrm{j}\kappa_{-M} e^{\mathrm{f}}(z,t) + \frac{\mathrm{j}k_0}{2n_{\text{eff}}}\sum_{m=1}^{M-1}\int_{\Sigma} A_{m-M}(x,y)\hat{e}_m(x,y)\phi(x,y)\mathrm{d}x\,\mathrm{d}y + \widetilde{s}^{\mathrm{b}}(z,t), \tag{2.78b}$$

where equations (2.41a), (2.41b), (2.35) and (2.37) have been used. Also, in deriving (2.78a) and (2.78b), we have used

$$k_0 \int_{\Sigma} n(x, y, \omega_0) A_M(x, y) \phi^2(x, y) \mathrm{d}x\,\mathrm{d}y$$
$$= k_0 \int_{\Sigma} n(x, y, \omega_0) \left[\Delta n_m(x, y) + \delta n_m - \frac{\mathrm{j}}{2k_0} \Delta g_m + \frac{\mathrm{j}}{2k_0} \Delta \alpha_m(x, y) \right] \phi^2(x, y) \mathrm{d}x\,\mathrm{d}y$$
$$= \overline{n} \left[\kappa_m^{\mathrm{IC}} + \kappa_m^{\mathrm{GCr}} - \mathrm{j}\kappa_m^{\mathrm{GCi}} + \mathrm{j}\kappa_m^{\mathrm{LC}} \right] \equiv \overline{n} \kappa_m, \tag{2.79}$$

where

$$k_0 \int_{\Sigma} n(x, y, \omega_0) \Delta n_m(x, y) \phi^2(x, y) \mathrm{d}x\,\mathrm{d}y \approx \overline{n} k_0 \int_{\Sigma} \Delta n_m(x, y) \phi^2(x, y) \mathrm{d}x\,\mathrm{d}y \equiv \overline{n} \kappa_m^{\mathrm{IC}}, \tag{2.80a}$$

$$k_0 \int_{\Sigma} n(x, y, \omega_0) \left(\delta n_m - \frac{\mathrm{j}}{2k_0} \Delta g_m \right) \phi^2(x, y) \mathrm{d}x\,\mathrm{d}y$$
$$\approx \overline{n} \left(k_0 \delta n_m - \frac{\mathrm{j}}{2} \Delta g_m \right) \equiv \overline{n} (\kappa_m^{\mathrm{GCr}} - \mathrm{j}\kappa_m^{\mathrm{GCi}}), \tag{2.80b}$$

$$\frac{\mathrm{j}}{2} \int_{\Sigma} n(x, y, \omega_0) \Delta \alpha_m(x, y) \phi^2(x, y) \mathrm{d}x\,\mathrm{d}y \approx \frac{\mathrm{j}\overline{n}}{2} \int_{\Sigma} \Delta \alpha_m(x, y) \phi^2(x, y) \mathrm{d}x\,\mathrm{d}y \equiv \mathrm{j}\overline{n} \kappa_m^{\mathrm{LC}}. \tag{2.80c}$$

(note that IC, GCr, GCi, and LC indicate the index coupling, the real part of the gain coupling, the imaginary part of the gain coupling, and the loss coupling, respectively.)

With its RHS viewed as an inhomogeneous driven source, equation (2.77c) can be formally solved by a Green's function approach. To do this, we need to solve [8, 9]

$$\left[\frac{\partial^2}{\partial x^2} + \frac{\partial^2}{\partial y^2} + k_0^2 n^2(x, y, \omega_0) - \left(\frac{M - 2m}{\Lambda} \right)^2 \pi^2 \right] G_m(x, y; x', y') = \delta(x - x', y - y'), \tag{2.81}$$

for $m = 1, 2, 3, \ldots, M - 1$ subject to the boundary conditions associated with the given waveguide structure. Once Green's functions are obtained, we can readily express the radiation waves in terms of the driven source, i.e., the forward and backward propagating waves

$$\hat{e}_m(x, y) = -k_0^2 \int_{\Sigma} n(x', y', \omega_0) \phi(x', y') [A_{-m}(x', y') e^{\mathrm{f}}(z, t)$$
$$+ A_{M-m}(x', y') e^{\mathrm{b}}(z, t)] G_m(x, y; x', y') \mathrm{d}x'\,\mathrm{d}y'$$
$$= h_m^{\mathrm{f}}(x, y) e^{\mathrm{f}}(z, t) + h_m^{\mathrm{b}}(x, y) e^{\mathrm{b}}(z, t), \tag{2.82}$$

where

$$h_m^{\mathrm{f}}(x,y) = -k_0^2 \int_\Sigma n(x',y',\omega_0)\phi(x',y')A_{-m}(x',y')G_m(x,y;x',y')\mathrm{d}x'\,\mathrm{d}y'$$
$$\approx -k_0^2\bar{n} \int_\Sigma \phi(x',y')A_{-m}(x',y')G_m(x,y;x',y')\mathrm{d}x'\,\mathrm{d}y'$$
$$h_m^{\mathrm{b}}(x,y) = -k_0^2 \int_\Sigma n(x',y',\omega_0)\phi(x',y')A_{M-m}(x',y')G_m(x,y;x',y')\mathrm{d}x'\,\mathrm{d}y'$$
$$\approx -k_0^2\bar{n} \int_\Sigma \phi(x',y')A_{M-m}(x',y')G_m(x,y;x',y')\mathrm{d}x'\,\mathrm{d}y', \tag{2.83}$$

with $m = 1, 2, 3, \ldots, M-1$. Finally, we plug equation (2.82) into equations (2.78a) and (2.78b) to obtain

$$\left(\frac{1}{v_{\mathrm{g}}}\frac{\partial}{\partial t} + \frac{\partial}{\partial z}\right)e^{\mathrm{f}}(z,t) = \left[\mathrm{j}\delta + \mathrm{j}k_0\Gamma\Delta n(z,\omega_0) + \frac{1}{2}\Gamma g(z,\omega_0) - \frac{1}{2}\overline{\alpha}(z) + \mathrm{j}\kappa_r^{\mathrm{ff}}\right]e^{\mathrm{f}}(z,t)$$
$$+\mathrm{j}(\kappa_M + \kappa_r^{\mathrm{fb}})e^{\mathrm{b}}(z,t) + \widetilde{s}^{\mathrm{f}}(z,t), \tag{2.84a}$$
$$\left(\frac{1}{v_{\mathrm{g}}}\frac{\partial}{\partial t} - \frac{\partial}{\partial z}\right)e^{\mathrm{b}}(z,t) = \left[\mathrm{j}\delta + \mathrm{j}k_0\Gamma\Delta n(z,\omega_0) + \frac{1}{2}\Gamma g(z,\omega_0) - \frac{1}{2}\overline{\alpha}(z) + \mathrm{j}\kappa_r^{\mathrm{bb}}\right]e^{\mathrm{b}}(z,t)$$
$$+\mathrm{j}(\kappa_{-M} + \kappa_r^{\mathrm{bf}})e^{\mathrm{f}}(z,t) + \widetilde{s}^{\mathrm{b}}(z,t), \tag{2.84b}$$

where

$$\kappa_r^{\mathrm{ff}} = \frac{k_0}{2n_{\mathrm{eff}}}\sum_{m=1}^{M-1}\int_\Sigma A_m(x,y)h_m^{\mathrm{f}}(x,y)\phi(x,y)\mathrm{d}x\,\mathrm{d}y$$
$$= \frac{-k_0^3}{2}\sum_{m=1}^{M-1}\int_\Sigma A_m(x,y)\left[\int_\Sigma \phi(x',y')A_{-m}(x',y')G_m(x,y;x',y')\mathrm{d}x'\,\mathrm{d}y'\right]\phi(x,y)\mathrm{d}x\,\mathrm{d}y, \tag{2.85a}$$

$$\kappa_r^{\mathrm{fb}} = \frac{k_0}{2n_{\mathrm{eff}}}\sum_{m=1}^{M-1}\int_\Sigma A_m(x,y)h_m^{\mathrm{b}}(x,y)\phi(x,y)\mathrm{d}x\,\mathrm{d}y$$
$$= \frac{-k_0^3}{2}\sum_{m=1}^{M-1}\int_\Sigma A_m(x,y)\left[\int_\Sigma \phi(x',y')A_{M-m}(x',y')G_m(x,y;x',y')\mathrm{d}x'\,\mathrm{d}y'\right]\phi(x,y)\mathrm{d}x\,\mathrm{d}y, \tag{2.85b}$$

$$\kappa_r^{\mathrm{bf}} = \frac{k_0}{2n_{\mathrm{eff}}}\sum_{m=1}^{M-1}\int_\Sigma A_{m-M}(x,y)h_m^{\mathrm{f}}(x,y)\phi(x,y)\mathrm{d}x\,\mathrm{d}y$$
$$= \frac{-k_0^3}{2}\sum_{m=1}^{M-1}\int_\Sigma A_{m-M}(x,y)\left[\int_\Sigma \phi(x',y')A_{-m}(x',y')G_m(x,y;x',y')\mathrm{d}x'\,\mathrm{d}y'\right]\phi(x,y)\mathrm{d}x\,\mathrm{d}y, \tag{2.85c}$$

$$\kappa_r^{\text{bb}} = \frac{k_0}{2n_{\text{eff}}} \sum_{m=1}^{M-1} \int_\Sigma A_{m-M}(x,y) h_m^{\text{b}}(x,y) \phi(x,y) \mathrm{d}x\,\mathrm{d}y$$

$$= \frac{-k_0^3}{2} \sum_{m=1}^{M-1} \int_\Sigma A_{m-M}(x,y) \left[\int_\Sigma \phi(x',y') A_{M-m}(x',y') G_m(x,y;x',y') \mathrm{d}x'\,\mathrm{d}y' \right] \phi(x,y) \mathrm{d}x\,\mathrm{d}y, \tag{2.85d}$$

with A_m given in equation (2.76b), ϕ and G_m the solutions to equations (2.34) and (2.81), respectively.

Equation (2.84), known as the coupled traveling wave equation, governs the slow-varying envelope functions of the forward and backward propagating waves. Comparing with equation (2.47), we find that the grating introduces distributed mutual coupling between the originally independently propagated forward and backward traveling waves along $\pm z$. The coupling coefficient κ_m can be taken as the ratio between the transmitted and reflected (due to the grating) waves per unit length, and is given by equation (2.79). It consists of four components as shown in equation (2.80a–c).

(1) κ_m^{IC} = (background) index coupling coefficient due to the periodically corrugated material background refractive index change; as shown in equation (2.80a), it is proportional to the mth Fourier expansion coefficient Δn_m, which is in turn determined by equation (2.75a), the corrugated material refractive index shape n_{p} in one period;

(2) κ_m^{GCr} = index change coupling coefficient or real part of the gain coupling coefficient due to the periodically corrugated material gain-induced index change; it is bias dependent and, as shown in equation (2.80b), is proportional to the mth Fourier expansion coefficient δn_m, which is in turn determined by equation (2.75b), the (bias induced) corrugated index change shape Δn_{p} in one period;

(3) κ_m^{GCi} = gain coupling coefficient or imaginary part of the gain coupling coefficient due to the periodically corrugated stimulated emission gain change; it is bias dependent and, as shown in equation (2.80b), is proportional to the mth Fourier expansion coefficient Δg_m, which is in turn determined by equation (2.75c), the (bias induced) corrugated stimulated emission gain shape g_{p} in one period;

(4) κ_m^{LC} = loss coupling coefficient due to the periodically corrugated optical loss change; as shown in equation (2.80c), it is proportional to the mth Fourier expansion coefficient $\Delta\alpha_m$, which is in turn determined by equation (2.75d), the corrugated optical loss shape α_{p} in one period.

Since $\kappa_m \neq \kappa_{-m}$ in general, from equation (2.84) we know that the coupling from the backward to forward and the coupling from the forward to backward propagating waves, measured by κ_m and κ_{-m}, respectively, are not symmetric. However, they do have restricted relations if the grating has a single coupling mechanism and has either symmetry or anti-symmetry. Actually, since the grating shape function must be real, from equation (2.75a–d) we have

$$\Delta n_{-m} = (\Delta n_m)^*, \quad \delta n_{-m} = (\delta n_m)^*,$$
$$\Delta g_{-m} = (\Delta g_m)^*, \text{ and } \Delta\alpha_{-m} = (\Delta\alpha_m)^*. \tag{2.86}$$

Consequently, from equation (2.80a–c) we find

$$\kappa_{-m}^{\rm IC} = (\kappa_m^{\rm IC})^*,$$
$$\kappa_{-m}^{\rm GCr} = (\kappa_m^{\rm GCr})^*,$$
$$\kappa_{-m}^{\rm GCi} = (\kappa_m^{\rm GCi})^*,$$
$$\kappa_{-m}^{\rm LC} = (\kappa_m^{\rm LC})^*, \tag{2.87}$$

if ϕ is a fully confined guided mode (hence ϕ is real). From equations (2.79) and (2.87), we know that for a purely index-coupled grating structure, $\kappa_m = \kappa_m^{\rm IC} + \kappa_m^{\rm GCr}$, hence $\kappa_{-m} = (\kappa_m)^*$. In particular, if the purely index-coupled grating structure is symmetric (e.g., the grating length is an integral number of half grating periods), κ_m is real and $\kappa_{-m} = \kappa_m$; whereas, if the purely index-coupled grating structure is anti-symmetric (e.g., the grating length is an integral number of grating periods), κ_m is imaginary and $\kappa_{-m} = -\kappa_m$. For a purely gain- or loss-coupled grating structure, $\kappa_m = -{\rm j}\kappa_m^{\rm GCi} + {\rm j}\kappa_m^{\rm LC}$, hence $\kappa_{-m} = -{\rm j}(\kappa_m^{\rm GCi})^* + {\rm j}(\kappa_m^{\rm LC})^*$. In particular, if the purely gain- or loss-coupled grating structure is symmetric, κ_m is imaginary and $\kappa_{-m} = \kappa_m$; whereas, if the purely gain- or loss-coupled grating structure is anti-symmetric, κ_m is real and $\kappa_{-m} = -\kappa_m$. For a complex-coupled grating structure, where both index and gain or loss coupling appear, or for a grating structure that is neither symmetric nor anti-symmetric, both κ_m and κ_{-m} are complex.

From equation (2.84), we know that a real κ_m (or κ_{-m}) brings in a phase shift, whereas an imaginary κ_m (or κ_{-m}) brings in an amplitude change, to the reflected wave. Therefore, different coupling mechanisms (i.e., index and gain or loss coupling) or grating symmetries (which can be controlled by the phase at either or both ends of the grating) may result in completely different "wave filtering" effects and give different lasing wavelength or side mode suppression ratio (SMSR) in DFB and DBR lasers or different reflection and transmission spectra in grating based devices. We will find the effect of different gratings in Chapter 10 through examples.

Also as seen in equation (2.84), from the backward to forward and from the forward to backward, the coupling coefficients between the propagating waves are given by κ_M and κ_{-M}, respectively, which is consistent with our original assumption that it is the Mth grating order that couples the forward to the backward propagating wave and vice versa.

The second difference between equations (2.84) and (2.47) is that there is an extra phase shift coefficient jδ associated with the propagating term that has nothing to do with the feedback. If we remove the grating by setting both κ_m and κ_{-m} to zero, this term still exists. This term naturally appears because we have shifted our reference propagation constants from the original $\pm\beta_0$ (when there is no grating) to $\pm M\pi/\Lambda$

(i.e., the grating Bragg wave number) for the forward and backward propagating waves, respectively. If we persist in using $\pm\beta_0$ as our reference propagation constants even for grating structures, we just need to replace the slow-varying envelope functions e^{f} and e^{b} by $e^{\mathrm{f}}\mathrm{e}^{\mathrm{j}\delta z}$ and $e^{\mathrm{b}}\mathrm{e}^{-\mathrm{j}\delta z}$ in equation (2.84), respectively, and we will find that the resulting equation remains the same except that the extra phase shift term $\mathrm{j}\delta$ disappears and the coupling coefficients are modified from κ_m and κ_{-m} to $\kappa_m \mathrm{e}^{-2\mathrm{j}\delta z}$ and $\kappa_{-m}\mathrm{e}^{2\mathrm{j}\delta z}$, respectively. Apparently, if we switch off the grating again, the resulting equation returns to equation (2.47).

Finally, if $M > 1$, the grating makes both forward and backward propagating waves couple themselves to the radiation waves ranking in order from 1 to $M-1$. Such coupling brings in two major effects as seen in equation (2.84). Firstly, it introduces additional phase shift and amplitude loss to the forward and backward propagating still waves, scaled by the real and imaginary parts of κ_r^{ff} and κ_r^{bb}, respectively. Secondly, it provides extra couplings from the backward to forward and from the forward to backward propagating waves, scaled by κ_r^{fb} and κ_r^{bf}, respectively. However, these effects are proportional to the second order perturbation of the grating, simply because any self-coupling of the forward or backward propagating wave and any mutual coupling between the forward and backward coupling waves through the radiating waves must experience the coupling between the propagating and the radiating waves twice: firstly from the propagating waves to the radiating waves and then back to the propagating waves. Otherwise, the radiating waves go away and nothing will be recaptured by the propagating waves. Since the coupling strength between the propagating and the radiating waves is in proportion to the grating strength (i.e., the periodically corrugated part or the grating Fourier expansion coefficients), the propagating to the radiating wave coupling and the recapture of the radiating wave by the propagating wave must have a coupling strength in proportion to the square of the grating strength. This can also be seen from equations (2.85a–d). Actually, taking the extreme case in equation (2.81) by assuming that the system does not give any spreading or damping to the impulse, as the response to the impulse, or the solution to equation (2.81), Green's functions will all become the δ-function. From equations (2.85a–d), we find

$$\kappa_r^{\mathrm{ff}} = \frac{-k_0^3}{2}\sum_{m=1}^{M-1}\int_\Sigma A_m(x,y)A_{-m}(x,y)\phi^2(x,y)\mathrm{d}x\,\mathrm{d}y, \tag{2.88a}$$

$$\kappa_r^{\mathrm{fb}} = \frac{-k_0^3}{2}\sum_{m=1}^{M-1}\int_\Sigma A_m(x,y)A_{M-m}(x,y)\phi^2(x,y)\mathrm{d}x\,\mathrm{d}y, \tag{2.88b}$$

$$\kappa_r^{\mathrm{bf}} = \frac{-k_0^3}{2}\sum_{m=1}^{M-1}\int_\Sigma A_{m-M}(x,y)A_{-m}(x,y)\phi^2(x,y)\mathrm{d}x\,\mathrm{d}y, \tag{2.88c}$$

$$\kappa_r^{\mathrm{bb}} = \frac{-k_0^3}{2}\sum_{m=1}^{M-1}\int_\Sigma A_{m-M}(x,y)A_{M-m}(x,y)\phi^2(x,y)\mathrm{d}x\,\mathrm{d}y. \tag{2.88d}$$

These self-coupling and mutual coupling coefficients bridged through the radiation wave are therefore given explicitly in the second order of the grating strength measured by A_m.

The above discussion suggests that, towards the first order approximation, the radiation effects are negligible in equation (2.84) even for higher order gratings with $M > 1$. Naturally, if the grating is of the first order with $M = 1$, no radiation can happen and those coefficients are zero according to equations (2.85a–d).

The coupled traveling wave equation (2.84) can be solved subject to certain initial and boundary conditions. Again, since the initial and boundary conditions are related to the operating conditions and structures of specific devices, we will find the effect of these conditions on device performance through examples shown in Chapter 10.

Finally, the real-world optical field is obtained by

$$\vec{E}(\vec{r},t) = \frac{1}{2}\vec{x}\phi(x,y)\left[e^{\mathrm{f}}(z,t)\mathrm{e}^{\mathrm{j}\frac{M\pi}{\Lambda}z} + e^{\mathrm{b}}(z,t)\mathrm{e}^{-\mathrm{j}\frac{M\pi}{\Lambda}z}\right]\mathrm{e}^{-\mathrm{j}\omega_0 t} + \frac{1}{2}\vec{x}\sum_{m=1}^{M-1}\hat{e}_m(x,y)\mathrm{e}^{\mathrm{j}\left(\frac{M-2m}{\Lambda}\pi z - \omega_0 t\right)} + \mathrm{c.c.} \tag{2.89}$$

2.5 Broadband optical traveling wave models

Unlike a single mode semiconductor laser that oscillates at a narrow wavelength range, some optoelectronic devices operate in a broad wavelength band, such as semiconductor optical amplifiers (SOA) in wavelength division multiplexing (WDM) systems where multiple wavelength channels must be processed, or gain-clamped (GC) SOAs where the lasing wavelength (for gain clamping) and the signal channel are far separated, or a superluminescent light-emitting diode (SLED) in which light emitted over a broadband is essential. For these broadband devices, the assumption in equation (2.23) is no longer valid as the medium dispersiveness is not negligible, i.e., the induced polarization of the host medium cannot instantly respond to the change of input field. Therefore, we have to use a different treatment from equation (2.23) in deriving the optical governing equations. Generally, there are three ways of dealing with this problem. The first approach is to retain the integration form of equation (2.9) in all of the derivations from Section 2.2 to 2.4. As a result, we will obtain differential–integral equations instead of differential equations. This is a natural consequence of a host medium with a limited response time: the induced polarization appears to have a memory as the host medium always responds to a fast-varying excitation field with a certain delay. To compute such delayed response, we have to count in the whole history by summing all the delayed components at any time instant, which gives us the convolution terms in the original differential equations. In principle, this approach can be numerically implemented through a digital filtering method in the time domain [10], yet its stability and convergence have not been proved, not to mention its poor computational efficiency. To show this general approach notwithstanding, we will follow a procedure similar to the one in Section 2.4.1 and designate it a direct convolution model. We will also introduce two other approaches based on some further approximations, which are stable, convergent, and more efficient in many applications [11, 12].

In this chapter, we take only the fully confined waveguide structure as an example. Similar derivations can readily be extended to partially confined or periodically corrugated waveguides.

2.5.1 The direct convolution model

If the optical field has a broadband, equation (2.23) is no longer valid because of the dispersiveness of the material susceptibility in the field band. However, we can view the susceptibility as a summation of a constant at the reference frequency and a frequency dependent part, or

$$\widetilde{\chi}(\vec{r}, \omega) = \widetilde{\chi}(\vec{r}, \omega_0) + \Delta\widetilde{\chi}(\vec{r}, \omega). \tag{2.90}$$

Equation (2.23) can then be written as

$$\begin{aligned}\vec{P}(\vec{r}, t) &= \varepsilon_0 F^{-1}\{[\widetilde{\chi}(\vec{r}, \omega_0) + \Delta\widetilde{\chi}(\vec{r}, \omega)]\widetilde{\vec{u}}(\vec{r}, \omega - \omega_0)\}\\ &= \varepsilon_0\widetilde{\chi}(\vec{r}, \omega_0)F^{-1}\left[\widetilde{\vec{u}}(\vec{r}, \omega - \omega_0)\right] + \varepsilon_0 F^{-1}\left[\Delta\widetilde{\chi}(\vec{r}, \omega)\widetilde{\vec{u}}(\vec{r}, \omega - \omega_0)\right]\\ &= \varepsilon_0\widetilde{\chi}(\vec{r}, \omega_0)\vec{u}(\vec{r}, t)\mathrm{e}^{-\mathrm{j}\omega_0 t} + \varepsilon_0\int_{-\infty}^{t}\Delta\chi(\vec{r}, t-\tau)\vec{u}(\vec{r}, \tau)\mathrm{e}^{-\mathrm{j}\omega_0\tau}\,\mathrm{d}\tau.\end{aligned} \tag{2.91}$$

By replacing the field in equation (2.91) with equation (2.32), we obtain

$$\begin{aligned}\vec{P}(\vec{r}, t) &= \vec{x}\varepsilon_0\phi(x, y)\mathrm{e}^{\mathrm{j}(\beta_0 z - \omega_0 t)}\left[\widetilde{\chi}(\vec{r}, \omega_0)e(z, t) + \int_{-\infty}^{t}\Delta\chi(\vec{r}, t-\tau)e(z, t)\mathrm{e}^{\mathrm{j}\omega_0(t-\tau)}\,\mathrm{d}\tau\right]\\ &= \vec{x}\phi(x, y)\mathrm{e}^{\mathrm{j}(\beta_0 z - \omega_0 t)}\left[\varepsilon_0\widetilde{\chi}(\vec{r}, \omega_0)e(z, t) + \Delta p(\vec{r}, t)\right],\end{aligned} \tag{2.92}$$

with the slow-varying envelope of the polarization defined as

$$\Delta p(\vec{r}, t) \equiv \varepsilon_0\int_{-\infty}^{t}\Delta\chi(\vec{r}, t-\tau)e(z, \tau)\mathrm{e}^{\mathrm{j}\omega_0(t-\tau)}\,\mathrm{d}\tau. \tag{2.93}$$

Insert equations (2.31) and (2.92) into the general wave equation (2.19), multiply the optical mode $\phi(x, y)$ on both sides of the equation obtained, and integrate over the entire cross-sectional area to yield

$$\begin{aligned}\frac{1}{v_g}\frac{\partial e(z, t)}{\partial t} + \frac{\partial e(z, t)}{\partial z} &= \left[\mathrm{j}k_0\Gamma\Delta n(z, \omega_0) + \frac{1}{2}\Gamma g(z, \omega_0) - \frac{1}{2}\overline{\alpha}(z)\right]e(z, t)\\ &\quad + \frac{\mathrm{j}k_0}{2n_{\mathrm{eff}}\varepsilon_0}\Delta p(z, t) + \widetilde{s}(z, t),\end{aligned} \tag{2.94}$$

where

$$\Delta p(z, t) \equiv \varepsilon_0\int_{-\infty}^{t}\Delta\chi(z, t-\tau)e(z, \tau)\mathrm{e}^{\mathrm{j}\omega_0(t-\tau)}\,\mathrm{d}\tau, \tag{2.95}$$

with

$$\Delta\chi(z,t) = \int_{\Sigma} \Delta\chi(\vec{r},t)\phi^2(x,y)\mathrm{d}x\,\mathrm{d}y. \tag{2.96}$$

Similarly to the derivation of equation (2.46), under the slow-varying envelope assumption ($|\partial^2 e/\partial z^2| \ll |\beta_0|^2|e|$ and $|\partial^2 e/\partial t^2| \ll |\omega_0|^2|e|$), $\partial^2 e/\partial z^2$ and $\partial^2 e/\partial t^2$ are dropped in deriving equation (2.94). Note that $\partial\Delta p/\partial t$ is also dropped for the reason explained in [13]. Equations (2.34), (2.35), (2.41a), and (2.41b) are also used with the spontaneous emission contribution given by equation (2.37).

Equation (2.94) governs the envelope function of the optical wave propagating along $+z$. For the optical wave that propagates along the opposite direction ($-z$), we simply need to use $-z$ to replace z in (2.94), as both the material property and the spontaneous emission contribution have bidirectional symmetry along $\pm z$. Therefore, we obtain

$$\left(\frac{1}{v_g}\frac{\partial}{\partial t} + \frac{\partial}{\partial z}\right)e^{\mathrm{f}}(z,t) = \left[\mathrm{j}k_0\Gamma\Delta n(z,\omega_0) + \frac{1}{2}\Gamma g(z,\omega_0) - \frac{1}{2}\overline{\alpha}(z)\right]e^{\mathrm{f}}(z,t) + \frac{\mathrm{j}k_0}{2n_{\mathrm{eff}}\varepsilon_0}\Delta p^{\mathrm{f}}(z,t) + \widetilde{s}^{\mathrm{f}}(z,t) \tag{2.97a}$$

$$\left(\frac{1}{v_g}\frac{\partial}{\partial t} - \frac{\partial}{\partial z}\right)e^{\mathrm{b}}(z,t) = \left[\mathrm{j}k_0\Gamma\Delta n(z,\omega_0) + \frac{1}{2}\Gamma g(z,\omega_0) - \frac{1}{2}\overline{\alpha}(z)\right]e^{\mathrm{b}}(z,t) + \frac{\mathrm{j}k_0}{2n_{\mathrm{eff}}\varepsilon_0}\Delta p^{\mathrm{b}}(z,t) + \widetilde{s}^{\mathrm{b}}(z,t), \tag{2.97b}$$

where

$$\Delta p^{\mathrm{f,b}}(z,t) \equiv \varepsilon_0 \int_{-\infty}^{t} \Delta\chi(z,t-\tau)e^{\mathrm{f,b}}(z,\tau)\mathrm{e}^{\mathrm{j}\omega_0(t-\tau)}\,\mathrm{d}\tau. \tag{2.98}$$

In equation (2.97), we have used the superscripts f and b to indicate the forward and backward propagating wave envelope functions, respectively. Also in equation (2.97), as the inhomogeneous spontaneous emission contributions to the forward and backward propagating waves, $\widetilde{s}^{\mathrm{f}}$ is given by equation (2.37) and $\widetilde{s}^{\mathrm{b}}(z,t) = \widetilde{s}^{\mathrm{f}}(-z,t)$, respectively.

Comparing equation (2.97) with equation (2.47), we find that the extra convolution term defined by equation (2.98) describes the broadband effect. Namely, because of the dispersiveness of the material susceptibility, as it passes through such material, the optical field depends not only on the instantaneous susceptibility, but also on the past susceptibility. If we view the material susceptibility as a system impulse response function, and the optical field as an input signal, then as the output signal the polarization carries the whole history of the system response to the input signal, which is represented by the convolution of the system impulse response and the input signal. Only if the system takes no time to respond to any fast-varying signal, i.e., the system frequency domain transfer function is a constant with unlimited bandwidth, expression (2.90) reduces to $\widetilde{\chi}(\vec{r},\omega) = \widetilde{\chi}(\vec{r},\omega_0)$ or $\Delta\widetilde{\chi}(\vec{r},\omega) = 0$, and we have $\Delta p^{\mathrm{f,b}}(z,t) = 0$ accordingly. Although in the real world there is no material (except a vacuum) bearing unlimited bandwidth, if the input signal has a narrow band, we can still approximate the system transfer function as a constant in the entire signal bandwidth, provided that the bandwidth of the system transfer function is much broader. Under such an approximation, the system still responds to the input signal without any delay in time. We can also understand that this is a situation in which the system carries no memory, so that its historical behavior has no effect on the current response. Therefore, we still have $\Delta p^{\mathrm{f,b}}(z,t)=0$, hence equation (2.47) when the optical field has a narrow band in comparison with the material susceptibility. In this sense, we can view equation (2.47) as a special case of equation (2.97).

As a general governing equation for the broadband optical field, equation (2.97) can be solved directly by exploring a digital filtering algorithm, which is discussed in Section 6.2.3.

2.5.2 The effective Bloch equation model

The frequency domain equivalent to equation (2.98) is

$$\Delta\tilde{p}^{\mathrm{f,b}}(z,\omega)=\varepsilon_0\Delta\tilde{\chi}(z,\omega-\omega_0)e^{\mathrm{f,b}}(z,\omega). \tag{2.99}$$

If we further write the frequency dependent part of the material susceptibility into a summation of the Lorentzian function [14]

$$\Delta\tilde{\chi}(z,\omega)=\sum_{i=1}^{M}\frac{A_i(z)}{\omega+\omega_0-\omega_{\mathrm{p}}(z)-\delta_i(z)+\mathrm{j}\Gamma_i(z)}, \tag{2.100}$$

equation (2.99) becomes

$$\Delta\tilde{p}^{\mathrm{f,b}}(z,\omega)=\varepsilon_0\sum_{i=1}^{M}\frac{A_i(z)e^{\mathrm{f,b}}(z,\omega)}{\omega-\omega_{\mathrm{p}}(z)-\delta_i(z)+\mathrm{j}\Gamma_i(z)}=\sum_{i=1}^{M}\Delta\tilde{p}_i^{\mathrm{f,b}}(z,\omega), \tag{2.101}$$

where

$$\Delta\tilde{p}_i^{\mathrm{f,b}}(z,\omega)\equiv\varepsilon_0\frac{A_i(z)e^{\mathrm{f,b}}(z,\omega)}{\omega-\omega_{\mathrm{p}}(z)-\delta_i(z)+\mathrm{j}\Gamma_i(z)}. \tag{2.102}$$

In the above expression, ω_{p} denotes the material gain peak frequency, A_i, δ_i, and Γ_i Lorentzian fitting parameters. They are all real numbers measured in rad/s. Actually, equation (2.100) assumes that the material gain and refractive index change profiles take the usual Lorentzian shape and its Kramers–Kronig transform in the form of

$$\Delta g(z,\omega)=\frac{k_0}{\bar{n}\Gamma}\sum_{i=1}^{M}\frac{A_i(z)\Gamma_i(z)}{[\omega+\omega_0-\omega_{\mathrm{p}}(z)-\delta_i(z)]^2+\Gamma_i^2(z)}, \tag{2.103a}$$

$$\Delta n(z,\omega)=\frac{1}{2\bar{n}\Gamma}\sum_{i=1}^{M}\frac{A_i(z)[\omega+\omega_0-\omega_{\mathrm{p}}(z)-\delta_i(z)]}{[\omega+\omega_0-\omega_{\mathrm{p}}(z)-\delta_i(z)]^2+\Gamma_i^2(z)}. \tag{2.103b}$$

This result becomes obvious if we first rewrite equation (2.40) in the form

$$\begin{aligned}\tilde{\chi}(\vec{r},\omega)&=n^2(x,y,\omega_0)-1+2n(x,y,\omega_0)\Delta n(z,\omega)\\&\quad-\frac{\mathrm{j}}{k_0}n(x,y,\omega_0)[g(z,\omega_0)+\Delta g(z,\omega)-\alpha(x,y,z)],\end{aligned} \tag{2.104}$$

then compare it with (2.90) to identify

$$\widetilde{\chi}(\vec{r}, \omega_0) = n^2(x, y, \omega_0) - 1 - \frac{\mathrm{j}}{k_0} n(x, y, \omega_0)[g(z, \omega_0) - \alpha(x, y, z)], \qquad (2.105a)$$

$$\Delta\widetilde{\chi}(\vec{r}, \omega) = 2n(x, y, \omega_0)\Delta n(z, \omega) - \frac{\mathrm{j}}{k_0} n(x, y, \omega_0)\Delta g(z, \omega). \qquad (2.105b)$$

In equation (2.104), we have split the material gain into a constant and a frequency dependent part, whereas the constant and frequency dependent components of the material refractive index are ascribed to the background refractive index and refractive index change, respectively. We have also assumed that the optical loss is dispersiveless. By taking the Fourier transform on both sides of equation (2.96) and replacing the frequency dependent material susceptibility in the integrand on the RHS with equation (2.105b), we find

$$\Delta\widetilde{\chi}(z, \omega) = \int_{\Sigma} \Delta\widetilde{\chi}(\vec{r}, \omega)\phi^2(x, y)\mathrm{d}x\,\mathrm{d}y \approx 2\overline{n}\Gamma\Delta n(z, \omega) - \frac{\mathrm{j}}{k_0}\overline{n}\Gamma\Delta g(z, \omega), \quad (2.106)$$

where approximations similar to equations (2.41b) and (2.42) are used with $\overline{n}$ and Γ given by equations (2.41a) and (2.44), respectively. Finally, equations (2.105a) and (2.105b) are obtained by setting the imaginary and real parts of equation (2.100) equal to the respective parts of equation (2.106).

We know from our physics-based gain model described in Chapter 4 that the semiconductor material gain is a summation over all contributions from transition pairs between the momentum matched conduction band electron and valence band hole. In bulk semiconductors, such summation in a three-dimensional (3D) momentum space is converted into a continuous integration over all possible electron-hole energy differences by the introduction of a 3D density of state (DOS) function. The integrand indeed takes the same form as equation (2.103a) if we let $M = 1$. Although we do not expect a Lorentzian function to retain the Lorentzian shape after being integrated in respect to one of its parameters, which means equation (2.103a) can never be strictly valid for bulk semiconductors, we can always look for the best fit between the rigorous physics-based gain model and the gain given in equation (2.103a) by choosing proper fitting parameters within the optical field bandwidth. That is to say, we first compute the physics-based gain numerically, then extract the fitting parameters in equation (2.103a) by minimizing the error between such obtained gain and the analytical gain calculated through equation (2.103a) over the entire optical field bandwidth. Once such a fitting is successful (i.e., the minimized error is less than a predetermined small value), we obtain a set of parameters that makes equation (2.103a) valid. In semiconductor quantum well (QW) structures, conduction band electrons and valence band holes have discrete energy levels along one direction. Hence the material gain is a summation over all contributions from the allowed transitions between these discrete energy levels. We should certainly set M in equation (2.103a) equal to the number of allowed (discrete) transitions. However, in computing the material gain, a summation is still required in the rest of the 2D momentum space.

This summation is converted into a continuous integration over all possible electron-hole energy differences by the introduction of a 2D DOS function. The integrand, again, takes the same form as the term inside the summation on the RHS of equation (2.103a). For the same reason as explained above, equation (2.103a) is not strictly valid for semiconductor QW structures either. However, we can still rely on the above mentioned fitting algorithm to obtain a set of parameters in equation (2.103a) to ensure its validation within the optical field bandwidth. Once parameters in equation (2.103a) are extracted, equation (2.103b) is uniquely determined through the Kramers–Kronig transformation where there are no more free parameters, hence no more unknowns. Before we return to explore the effective Bloch equation, it is worth mentioning that equation (2.103a) is strictly valid for semiconductor quantum dot (QD) structures, as the rigorous physics-based QD gain takes a form identical to equation (2.103a).

By rewriting equation (2.102) in the form

$$\mathrm{j}\omega\Delta\tilde{p}_i^{\mathrm{f,b}}(z,\omega) = \left[\mathrm{j}\omega_{\mathrm{p}}(z) + \mathrm{j}\delta_i(z) + \Gamma_i(z)\right]\Delta\tilde{p}_i^{\mathrm{f,b}}(z,\omega) + \mathrm{j}\varepsilon_0 A_i(z) e^{\mathrm{f,b}}(z,\omega), \quad (2.107)$$

we readily obtain the effective Bloch equation by further taking the inverse Fourier transform

$$\frac{\partial}{\partial t}\Delta p_i^{\mathrm{f,b}}(z,t) = [\mathrm{j}\omega_{\mathrm{p}}(z) + \mathrm{j}\delta_i(z) + \Gamma_i(z)]\Delta p_i^{\mathrm{f,b}}(z,t) + \mathrm{j}\varepsilon_0 A_i(z) e^{\mathrm{f,b}}(z,t), \quad (2.108)$$

where $i = 1, 2, 3, \ldots, M$. Also, taking the inverse Fourier transform of equation (2.101) and substituting the polarization contribution term (i.e., the extra convolution term) in equation (2.97) by the RHS of the resulting equation, we find

$$\left(\frac{1}{v_{\mathrm{g}}}\frac{\partial}{\partial t} + \frac{\partial}{\partial z}\right) e^{\mathrm{f}}(z,t) = \left[\mathrm{j}k_0\Gamma\Delta n(z,\omega_0) + \frac{1}{2}\Gamma g(z,\omega_0) - \frac{1}{2}\bar{\alpha}(z)\right] e^{\mathrm{f}}(z,t) + \frac{\mathrm{j}k_0}{2n_{\mathrm{eff}}\varepsilon_0}\sum_{i=1}^{M}\Delta p_i^{\mathrm{f}}(z,t) + \tilde{s}^{\mathrm{f}}(z,t), \quad (2.109\mathrm{a})$$

$$\left(\frac{1}{v_{\mathrm{g}}}\frac{\partial}{\partial t} - \frac{\partial}{\partial z}\right) e^{\mathrm{b}}(z,t) = \left[\mathrm{j}k_0\Gamma\Delta n(z,\omega_0) + \frac{1}{2}\Gamma g(z,\omega_0) - \frac{1}{2}\bar{\alpha}(z)\right] e^{\mathrm{b}}(z,t) + \frac{\mathrm{j}k_0}{2n_{\mathrm{eff}}\varepsilon_0}\sum_{i=1}^{M}\Delta p_i^{\mathrm{b}}(z,t) + \tilde{s}^{\mathrm{b}}(z,t). \quad (2.109\mathrm{b})$$

Equations (2.109) and (2.108) form a complete set of governing equations for dealing with the broadband optical field, where instead of computing the convolution equation (2.98) directly, we solve M effective Bloch equations in the form of equation (2.108) with their parameters obtained by a best matching of a rigorous physics-based gain to the one given in equation (2.103a). As a replacement of equation (2.98), equation (2.108) provides an alternative way of taking into account the history. Equation (2.98) reveals that the history can be computed through storing the historical responses and adding them together at the current step. In this case, we just need to memorize the history without much extra computational effort. Equation (2.108), however, indicates that the history can also be accounted for by computing the effect of the history at the current step. As opposed to the previous case, we now need extra computational effort in solving the effective Bloch equation without much extra memory requirement. In short, equation (2.98) memorizes the history while equation (2.108) computes the history. Therefore, we need fewer memory units but more CPU time to implement equation (2.108).

2.5.3 The wavelength slicing model

In some optoelectronic devices, the optical field has multiple discrete wavelength components spreading over a broad range, with each component changing with time because of modulation, such as in SOAs for in-line amplification in WDM systems or in directly modulated multi-mode Fabry–Perot (FP) laser diodes. The optical field can therefore be viewed as a multi-channel band limited modulation signal. If the modulation bandwidth is smaller than the channel frequency spacing, instead of equation (2.31), we can write an optical field with N wavelength channels in the form

$$\vec{E}(\vec{r}, t) = \vec{x}\phi(x, y)\sum_{k=1}^{N} e_k(z, t)\mathrm{e}^{\mathrm{j}(\beta_0 z-\omega_k t)}, \tag{2.110}$$

with ω_k indicating the kth channel reference frequency (carrier frequency), e_k the kth channel slow-varying longitudinal field envelope function, and $k = 1, 2, 3, \ldots, N$. Plugging equation (2.110) into equation (2.9) yields

$$\begin{aligned}\vec{P}(\vec{r}, t) &= \vec{x}\varepsilon_0\phi(x, y)\mathrm{e}^{\mathrm{j}\beta_0 z}\sum_{k=1}^{N}\int_{-\infty}^{t}\chi(\vec{r}, t-\tau)e_k(z, \tau)\mathrm{e}^{-\mathrm{j}\omega_k\tau}\,\mathrm{d}\tau\\ &= \vec{x}\varepsilon_0\phi(x, y)\mathrm{e}^{\mathrm{j}\beta_0 z}\sum_{k=1}^{N}F^{-1}[\tilde{\chi}(\vec{r}, \omega)\tilde{e}_k(z, \omega-\omega_k)]\\ &\approx \vec{x}\varepsilon_0\phi(x, y)\mathrm{e}^{\mathrm{j}\beta_0 z}\sum_{k=1}^{N}\tilde{\chi}(\vec{r}, \omega_k)F^{-1}[\tilde{e}_k(z, \omega-\omega_k)]\\ &= \vec{x}\varepsilon_0\phi(x, y)\mathrm{e}^{\mathrm{j}\beta_0 z}\sum_{k=1}^{N}\tilde{\chi}(\vec{r}, \omega_k)e_k(z, t)\mathrm{e}^{-\mathrm{j}\omega_k t}.\end{aligned} \tag{2.111}$$

In equation (2.111), we have utilized our presumption that e_k's are slow-varying functions of time hence every $\tilde{e}_k$ has a narrow band centralized at ω_k. As mentioned in Section 2.2, the material susceptibility can respond to very fast input function changes with time, which means that its frequency domain counterpart $\tilde{\chi}$ is relatively smooth. Therefore, in dealing with the product of a smooth ($\tilde{\chi}$) and a comb-like ($\sum_{k=1}^{N}\tilde{e}_k$) function, we can view the latter as a sampling function. The result of the product becomes the sampling function itself with its sharp peak amplitudes modulated by values of the smooth function sampled at these peaks. That is to say, the continuous smooth

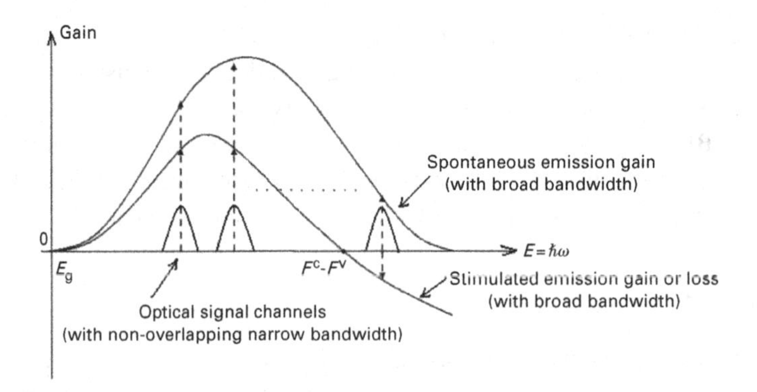

Fig. 2.4. Material stimulated and spontaneous emission gain profiles and narrow-banded, non-overlapping optical signal channels. When the signal passes through such material with an ultra-broad bandwidth, interferences between signal channels are negligible. Note that E_g is the gain medium bandgap energy, $F^c - F^v$ is the quasi-Fermi level separation due to external injection, and E is the photon energy variable.

function $\tilde{\chi}$ can be replaced by a set of discrete constants sampled at channel reference frequencies (ω_k's) corresponding to $\tilde{e}_k$'s, and consequently be taken out of the inverse Fourier transform. This is illustrated by Fig. 2.4.

By inserting equations (2.110) and (2.111) into equation (2.19), multiplying the optical mode $\phi(x, y)$ on both sides of the equation obtained, and integrating over the entire cross-sectional area, we obtain

$$\sum_{k=1}^{N} \mathrm{e}^{-\mathrm{j}\omega_k t} \frac{1}{v_{gk}} \frac{\partial e_k(z,t)}{\partial t} + \sum_{k=1}^{N} \mathrm{e}^{-\mathrm{j}\omega_k t} \frac{\partial e_k(z,t)}{\partial z}$$
$$= \sum_{k=1}^{N} \left[\mathrm{j}\frac{\Delta\omega_k}{\mathrm{c}}\bar{n} + \mathrm{j}k_0\Gamma\Delta n(z,\omega_k) + \frac{1}{2}\Gamma g(z,\omega_k) - \frac{1}{2}\bar{\alpha}(z) \right] e_k(z,t)\mathrm{e}^{-\mathrm{j}\omega_k t} + \mathrm{e}^{-\mathrm{j}\omega_0 t}\tilde{s}(z,t), \tag{2.112}$$

where the group velocity is rescaled by

$$v_{gk} = v_g\omega_0/\omega_k, \tag{2.113}$$

and $\omega_k + \omega_0$ is approximated by $2\omega_0$ and

$$\Delta\omega_k \equiv \omega_k - \omega_0. \tag{2.114}$$

In deriving equation (2.112), again we have dropped $\partial^2 e/\partial z^2$ and $\partial^2 e/\partial t^2$ because of the slow-varying envelope assumption ($|\partial^2 e/\partial z^2| \ll |\beta_0|^2|e|$ and $|\partial^2 e/\partial t^2| \ll |\omega_0|^2|e|$). Equations (2.34), (2.35), (2.41a), and (2.41b) are also used with the spontaneous emission contribution given by equation (2.37). In particular, we have assumed that the material background refractive index and optical loss are dispersiveless. Consequently, the following expressions hold along with equation (2.112) in a similar way to equations (2.38) to (2.40)

$$n^2(\vec{r}, \omega_k) = [n(x, y, \omega_k) + \Delta n(z, \omega_k)]^2 \approx [n(x, y, \omega_0) + \Delta n(z, \omega_k)]^2$$
$$\approx n^2(x, y, \omega_0) + 2n(x, y, \omega_0)\Delta n(z, \omega_k), \tag{2.115}$$

$$g(\vec{r}, \omega_k) = g(z, \omega_k) - \alpha(x, y, z), \tag{2.116}$$

$$1 + \widetilde{\chi}(\vec{r}, \omega_k) = [n(\vec{r}, \omega_k) - \frac{\mathrm{j}}{2k_0} g(\vec{r}, \omega_k)]^2$$
$$\approx n^2(x, y, \omega_0) + 2n(x, y, \omega_0)\Delta n(z, \omega_k)$$
$$- \frac{\mathrm{j}}{k_0} n(x, y, \omega_0)[g(z, \omega_k) - \alpha(x, y, z)]. \tag{2.117}$$

Multiplying by $\mathrm{e}^{\mathrm{j}\omega_l t}$ and taking a time average over a certain period T on both sides of equation (2.112) yields

$$\frac{1}{v_{gl}} \frac{\partial e_l(z, t)}{\partial t} + \frac{\partial e_l(z, t)}{\partial z} =$$
$$\left[\mathrm{j}\frac{\Delta\omega_l}{\mathrm{c}}\bar{n} + \mathrm{j}k_0\Gamma\Delta n(z, \omega_l) + \frac{1}{2}\Gamma g(z, \omega_l) - \frac{1}{2}\bar{\alpha}(z)\right] e_l(z, t) + \bar{s}_l(z, t), \tag{2.118}$$

where $l = 1, 2, 3, \ldots, N$ and the spontaneous emission contribution is a time-averaged result of equation (2.37)

$$\bar{s}_l(z, t) \equiv \frac{1}{T}\int_t^{t+T} \mathrm{e}^{\mathrm{j}(\omega_l - \omega_0)t}\widetilde{s}(z, t)\mathrm{d}t$$
$$= -\frac{\mathrm{j}}{2n_{\mathrm{eff}}\omega_0 T}\sqrt{\left(\frac{\mu_0}{\varepsilon_0}\right)}\int_t^{t+T} \mathrm{e}^{-\mathrm{j}(\beta_0 z - \omega_l t)}\left[\int_\Sigma \frac{\partial J_{\mathrm{sp}x}}{\partial t}\phi(x, y)\mathrm{d}x\,\mathrm{d}y\right]\mathrm{d}t. \tag{2.119}$$

In deriving equation (2.118), we have utilized the assumption that the modulation bandwidth is smaller than the channel frequency spacing, as such, every envelope function related term, i.e., $|\partial e_k/\partial t|$, $|\partial e_k/\partial z|$, and $|e_k|$ is a slow-varying function of time compared with the fast-varying factor $\mathrm{e}^{\mathrm{j}(\omega_l - \omega_k)t}$ when $l \neq k$. Therefore, these terms can be viewed as constants and taken out of the integrals. As a result, the time-average integral is simply reduced to

$$\frac{1}{T}\int_t^{t+T} \mathrm{e}^{\mathrm{j}(\omega_l - \omega_k)t}\mathrm{d}t = \frac{\mathrm{e}^{\mathrm{j}(\omega_l - \omega_k)(t+T)} - \mathrm{e}^{\mathrm{j}(\omega_l - \omega_k)t}}{\mathrm{j}(\omega_l - \omega_k)}$$
$$= \mathrm{e}^{\mathrm{j}(\omega_l - \omega_k)(t + \frac{T}{2})}\frac{\sin\left[(\omega_l - \omega_k)\frac{T}{2}\right]}{(\omega_l - \omega_k)\frac{T}{2}} \to \delta_{lk}, \tag{2.120}$$

as $(\omega_l - \omega_k)T/2 \to \infty$. Hence we obtain equation (2.118). Actually, this result is very similar to the phase matching condition. Because of our pre-assumption, there is no overlap of base-bands from different wavelength channels in the frequency domain. Therefore, there is no interference between different channels in the time domain in an average sense, because of the rapid phase change in time given by $e^{j\Delta\omega t}$, with $\Delta\omega$ indicating the channel frequency difference. Consequently, if and only if an equation in the form of equation (2.118) holds for every channel, equation (2.112) becomes possible.

If we consider wave propagations in both directions along $\pm z$ by following the approach introduced in Section 2.4.1 from equations (2.46) to (2.47), equation (2.118) can readily be expanded to

$$\left(\frac{1}{v_{gk}}\frac{\partial}{\partial t}+\frac{\partial}{\partial z}\right)e_k^{\rm f}(z,t)=\left[{\rm j}\frac{\Delta\omega_k}{\rm c}\bar{n}+{\rm j}k_0\Gamma\Delta n(z,\omega_k)+\frac{1}{2}\Gamma g(z,\omega_k)-\frac{1}{2}\bar{\alpha}(z)\right]e_k^{\rm f}(z,t)+\bar{s}_k^{\rm f}(z,t), \tag{2.121a}$$

$$\left(\frac{1}{v_{gk}}\frac{\partial}{\partial t}-\frac{\partial}{\partial z}\right)e_k^{\rm b}(z,t)=\left[{\rm j}\frac{\Delta\omega_k}{\rm c}\bar{n}+{\rm j}k_0\Gamma\Delta n(z,\omega_k)+\frac{1}{2}\Gamma g(z,\omega_k)-\frac{1}{2}\bar{\alpha}(z)\right]e_k^{\rm b}(z,t)+\bar{s}_k^{\rm b}(z,t), \tag{2.121b}$$

with $k = 1, 2, 3, \ldots, N$. Also in equation (2.121), as the averaged inhomogeneous spontaneous emission contributions to the forward and backward propagating waves, $\bar{s}_k^{\rm f}$ is given by equation (2.119) and $\bar{s}_k^{\rm b}(z,t)=\bar{s}_k^{\rm f}(-z,t)$, respectively.

Equation (2.121) is the governing equation for the broadband optical field with discrete wavelength channels and with each channel modulation bandwidth smaller than the channel spacing. If there is only one channel left, we can choose ω_0 as that channel frequency and equation (2.121) is reduced to equation (2.47) as we expected. In this sense, equation (2.121) is an extension of equation (2.47) with the interference contribution between different channels completely ignored, which is true when the base-bands of different channels do not overlap in the frequency domain. Once we cannot be certain that the broadband optical field comprises only discrete wavelength components, or we find that there is an overlap between the base-bands of any two different channels, we have to give up this wavelength slicing model and return to the direct convolution model or effective Bloch equation model.

2.6 Separation of spatial and temporal dependences – the standing wave model

Although the material susceptibility in equation (2.24) is given in its frequency domain value at the reference frequency, it can still be an implicit function of time because of the change of operating condition applied to the material. For example, changes of carrier injection or static electric field may introduce gain or absorption and refractive index changes in semiconductors. If these changes are introduced through the time-dependent external forward or backward bias, the material susceptibility will change with time accordingly through such bias. For this reason, we can write the material susceptibility in the form of

$$1+\widetilde{\chi}(\vec{r},\omega_0)\approx 1+\widetilde{\chi}_V(\vec{r},\omega_0)+\frac{\mathrm{d}\widetilde{\chi}_V(\vec{r},\omega_0)}{\mathrm{d}V}\Delta V(t). \tag{2.122}$$

In equation (2.122), $\widetilde{\chi}_V$ and $\mathrm{d}\widetilde{\chi}_V/\mathrm{d}V$ indicate the frequency domain material susceptibility and its derivative with respect to the external bias at an arbitrary static external bias V and the reference frequency ω_0, respectively; $\Delta V(t) = V(t+\Delta t) - V(t)$, the difference in the external bias between time $t+\Delta t$ and an arbitrary reference time t. To make equation (2.122) valid, we have to let $|\Delta V| \ll 1$. This is always possible for any continuous time-dependent function once we let $\Delta t \to 0$. Through equation (2.122), the material susceptibility is split into a time-invariant and a time-dependent part, with its time dependence given in an explicit form. As such, neither $\widetilde{\chi}_V$ nor $\mathrm{d}\widetilde{\chi}_V/\mathrm{d}V$ has time dependence anymore.

Using equation (2.122), we can rewrite equation (2.24) as

$$\begin{aligned}&\mathrm{j}\frac{2\omega_0}{\mathrm{c}^2}[1+\widetilde{\chi}(\vec{r},\omega_0)]\frac{\partial\vec{u}}{\partial t}\\&\quad=-\nabla^2\vec{u}-\frac{\omega_0^2}{\mathrm{c}^2}\left[1+\widetilde{\chi}_V(\vec{r},\omega_0)+\frac{\mathrm{d}\widetilde{\chi}_V(\vec{r},\omega_0)}{\mathrm{d}V}\Delta V(t)\right]\vec{u}+\mu_0\mathrm{e}^{\mathrm{j}\omega_0 t}\frac{\partial}{\partial t}\vec{J}_{\mathrm{sp}}.\end{aligned} \tag{2.123}$$

If we can solve the following eigenvalue problem

$$\nabla^2\vec{v}(\vec{r})+\frac{\omega_0^2}{\mathrm{c}^2}[1+\widetilde{\chi}_V(\vec{r},\omega_0)]\vec{v}(\vec{r})=\beta_V^2\vec{v}(\vec{r}), \tag{2.124}$$

subject to the boundary condition defined by the device structure and for any given static bias V, we will be able to separate the spatial and temporal variables in equation (2.123) by an integral transformation over the space domain, with the kernel function of the integral transform given as the eigenfunction obtained from equation (2.124). More specifically, we can expand our solution to equation (2.123) in the form

$$\vec{u}(\vec{r},t)=\sum_{l\in L}U_l(t)\vec{v}_l(\vec{r}), \tag{2.125a}$$

where integer l indicates the lth eigenvalue and eigenfunction, L the whole eigenvalue set, and $U_l(t)$ the time-dependent coefficient of the lth eigenfunction. Equation (2.125a) is valid since, as the solution to the eigenvalue problem in equation (2.124), $\vec{v}_l(\vec{r})$ $(l \in L)$ forms a complete and orthogonal set that can be used as a base to expand any continuous function. If the eigenvalues are continuous, l becomes a real vector in 3D and we should use a continuous integral instead of the discrete summation in equation (2.125a)

$$\vec{u}(\vec{r},t)=\int_L U_{\vec{l}}(t)\vec{v}_{\vec{l}}(\vec{r})\mathrm{d}\vec{l}. \tag{2.125b}$$

In the most general case, equation (2.124) gives both discrete and continuous eigensolutions so we should use the summation of the RHS of both equations (2.125a) and

(2.125b) to expand our solution. In this section, we will assume that there exist only discrete eigensolutions in equation (2.124), so equation (2.125a) will be used throughout. However, the generality of this approach will not be affected as we can achieve similar conclusions if equation (2.125b) or a combination of equations (2.125a) and (2.125b) is used.

Plugging equation (2.125a) into equation (2.123), multiplying by $\vec{v}_m(r)$ and integrating both sides yields

$$\frac{dU_m(t)}{dt} = \frac{v_g}{2}\left[j\frac{\beta_{Vm}^2}{\bar{n}k_0} + \Delta g_{Vmm}(t)\right]U_m(t) + \frac{v_g}{2}\sum_{l\in L, l\neq m}\Delta g_{lm}(t)U_l(t) + v_g\tilde{s}_m(t), \quad (2.126)$$

with $m \in L$, and

$$\Delta g_{Vlm}(t) \equiv j\frac{k_0}{\bar{n}}\frac{\int_\Omega \vec{v}_m(\vec{r})\frac{d\tilde{\chi}_V(\vec{r},\omega_0)}{dV}\Delta V(t)\vec{v}_l(\vec{r})d\vec{r}}{\int_\Omega \vec{v}_m^2(r)d\vec{r}}, \quad (2.127)$$

$$\tilde{s}_m(t) \equiv -\frac{je^{j\omega_0 t}}{2\bar{n}\omega_0}\sqrt{\left(\frac{\mu_0}{\varepsilon_0}\right)}\frac{\int_\Omega \vec{v}_m(\vec{r})\frac{\partial}{\partial t}\vec{J}_{sp}\, d\vec{r}}{\int_\Omega \vec{v}_m^2(r)d\vec{r}}. \quad (2.128)$$

In deriving equation (2.126), we have used equation (2.124) and the orthogonal condition among eigenfunctions

$$\int_\Omega \vec{v}_m(\vec{r})\vec{v}_l(\vec{r})d\vec{r} = \delta_{ml}\int_\Omega \vec{v}_m^2(\vec{r})d\vec{r}, \quad (2.129)$$

with an approximation made similarly to equation (2.41a)

$$\int_\Omega \vec{v}_m(\vec{r})[1+\tilde{\chi}(\vec{r},\omega_0)]\vec{v}_l(\vec{r})d\vec{r} \approx \int_\Omega \vec{v}_m(\vec{r})n^2(\vec{r},\omega_0)\vec{v}_l(\vec{r})d\vec{r}$$
$$\approx \bar{n}^2\int_\Omega \vec{v}_m(\vec{r})\vec{v}_l(\vec{r})d\vec{r} = \bar{n}^2\delta_{ml}\int_\Omega \vec{v}_m^2(r)d\vec{r}, \quad (2.130)$$

and with Ω in the above integrals defined as the entire space in which the optical field spreads.

Because of the homogeneity of the eigenequation (2.124), we can assign any dimensions to the eigenfunction $\vec{v}_m(\vec{r})$. The dimensions of the time-dependent coefficient $U_m(t)$ can therefore be assigned arbitrarily since equation (2.126) can always be normalized as long as the inhomogeneous spontaneous emission contribution is scaled in the same way. A convenient option is to assign $|U_m(t)|^2$ as the dimensionless photon number of the mth eigenfunction. As such, $\vec{v}_m(\vec{r})$ and $\tilde{s}_m(t)$ will take V/m and 1/m as their units, respectively.

Equation (2.126) gives the optical rate equation in its general form with the cross-coupling coefficients and inhomogeneous spontaneous emission contribution given by equations (2.127) and (2.128), respectively. Once we expand the optical field through time-invariant eigenfunctions, known as optical modes, their amplitudes are governed by equation (2.126). Following this approach, the spatial and temporal dependence of the optical field is separated through mode expansion, with optical modes carrying only the spatial dependence and their amplitudes describing only the temporal dependence. These optical modes are solutions to the eigenvalue problem equation (2.124) subject to the device boundary condition, whereas their amplitudes are governed by the optical rate equation (2.126). Apparently, equation (2.124) has no time dependence and is determined by the device structure under a static bias. Equation (2.126), on the other hand, is a set of ordinary differential equations (ODEs) without spatial dependence. Once equation (2.124) is solved, the parameters in equation (2.126), i.e., the self-coupling and cross-coupling coefficients, can be found from the solution of equation (2.124) (i.e., the eigenvalues) and from equation (2.127), respectively, hence equation (2.126) can be solved. It is also worth mentioning that the optical rate equation (2.126) is related to the device structure and operating condition (i.e., external bias) only through these self-coupling and cross-coupling parameters. Since parameter changes in an equation within a certain range can hardly change the nature of its solution, we know that the nature of the solution to equation (2.126) has little dependence on the detailed device structure and operating condition. Therefore, once we find a full solution of equations (2.124) and (2.126) for a certain "reference" device structure and "typical" operating condition, we do not expect any substantial change in the solution of equation (2.126) when the device structure and its operating condition deviate slightly from the "reference" structure and "typical" condition. In this sense, this model has its particular advantage in device structure refining and operation optimization.

A major drawback of this model, however, lies in the truncation of the modal expansion expressed in equation (2.125a). For any practical device, we need to extract power from at least one of its facets. More specifically, for an edge emitting device, we have to take power from the wave propagation direction along the cavity (z). Therefore, energy is not conserved inside the device cavity because of leakage at one or both facets. This fact indicates that, at least along the z direction, the optical field has a traveling wave component that cannot be expressed by a summation of a limited number of standing wave components. Hence the number of modes (i.e., the number of standing wave components) required in the expansion equation (2.125a) must be infinity. Fortunately, the number of eigenfunctions, or optical modes obtained from equation (2.124) is also infinity and their completeness is guaranteed [15]. Expansion in the form of equation (2.125a) is therefore theoretically valid for any case if we allow the summation to include an infinite number of modes. In practical computations, however, equation (2.125a) must be truncated, which brings the problem of how many terms must be retained to guarantee a certain accuracy. This unfortunately seems to be an open problem, although we have the following general guidance.

(1) For devices with more "closed" cavity or resonator structures with a higher quality (Q) factor, more power is stored inside the device and hence less energy leakage results. Fewer modes in equation (2.125a) will be required to ensure a given accuracy. Examples of such devices are various semiconductor lasers.

(2) For devices with more "open" cavity or resonator structures with a poor Q factor, less power is stored inside the device and hence high energy leakage results. More or even infinite modes will be required which makes equation (2.125a) computationally meaningless. Examples of such devices are SOAs, SLEDs and optoelectronic modulators.

To explore this approach further, we will again focus on the fully confined waveguide structure, as it is quite straightforward to expand this approach to other structures as mentioned in Section 2.4. Actually, comparing equation (2.122) with equation (2.40), we find

$$1+\tilde{\chi}_V(\vec{r},\omega_0) = n^2(x,y,\omega_0) + 2n(x,y,\omega_0)\Delta n_V(z,\omega_0) - \frac{\mathrm{j}}{k_0}n(x,y,\omega_0)[g_V(z,\omega_0) - \alpha(x,y,z)], \tag{2.131}$$

and

$$\frac{\mathrm{d}\tilde{\chi}_V(\vec{r},\omega_0)}{\mathrm{d}V} = 2n(x,y,\omega_0)\frac{\mathrm{d}\Delta n_V(z,\omega_0)}{\mathrm{d}V} - \frac{\mathrm{j}}{k_0}n(x,y,\omega_0)\frac{\mathrm{d}g_V(z,\omega_0)}{\mathrm{d}V}, \tag{2.132}$$

where we have assumed that the background refractive index and the optical loss are not external bias dependent, and have expanded the implicit time-dependent material gain and induced refractive index change in an explicit time-dependent form similar to equation (2.122). In the fully confined waveguide structure, we can write the 3D optical mode in equation (2.124) in a form similar to equation (2.32) with the time dependence removed

$$\vec{v}(\vec{r}) = \vec{x}\phi(x,y)e(z)\mathrm{e}^{\mathrm{j}\beta_0 z}, \tag{2.133}$$

where $e(z)$ is a slow-varying function along z and is known as the optical longitudinal mode distribution. Plugging equations (2.131) and (2.133) into equation (2.124), multiplying the transverse optical mode $\phi(x,y)$ on both sides of the equation obtained, and integrating over the entire cross-sectional area yields

$$\frac{\mathrm{d}e(z)}{\mathrm{d}z} + \left[-\mathrm{j}k_0\Gamma\Delta n_V(z,\omega_0) - \frac{1}{2}\Gamma g_V(z,\omega_0) + \frac{1}{2}\overline{\alpha}(z)\right]e(z) = -\mathrm{j}\frac{\beta_V^2}{2\beta_0}e(z), \tag{2.134}$$

where $\partial^2 e/\partial z^2$ is dropped under the slow-varying envelope assumption ($|\partial^2 e/\partial z^2| \ll |\beta_0|^2|e|$). In deriving equation (2.134), equations (2.34), (2.35), and (2.41b) to (2.43) are also used with the optical confinement factor Γ and optical modal loss $\overline{\alpha}$ defined as in equations (2.44) and (2.45), respectively.

To consider wave propagation in both directions along $\pm z$, we can readily expand equation (2.134) by following the approach introduced in Section 2.4.1

$$\frac{de^{\mathrm{f}}(z)}{dz}+\left[-jk_0\Gamma\Delta n_V(z,\omega_0)-\frac{1}{2}\Gamma g_V(z,\omega_0)+\frac{1}{2}\overline{\alpha}(z)\right]e^{\mathrm{f}}(z)=-j\frac{\beta_V^2}{2\beta_0}e^{\mathrm{f}}(z), \tag{2.135a}$$

$$-\frac{de^{\mathrm{b}}(z)}{dz}+\left[-jk_0\Gamma\Delta n_V(z,\omega_0)-\frac{1}{2}\Gamma g_V(z,\omega_0)+\frac{1}{2}\overline{\alpha}(z)\right]e^{\mathrm{b}}(z)=-j\frac{\beta_V^2}{2\beta_0}e^{\mathrm{b}}(z), \tag{2.135b}$$

with the 3D optical mode expression (2.133) modified to

$$\vec{v}(\vec{r})=\vec{x}\phi(x,y)[e^{\mathrm{f}}(z)e^{j\beta_0 z}+e^{\mathrm{b}}(z)e^{-j\beta_0 z}]. \tag{2.136}$$

Equation (2.135) forms an eigenvalue problem subject to the boundary condition given at the two facets of the device waveguide, which is normally in the form

$$\begin{aligned}e^{\mathrm{f}}(0)&=R_{\mathrm{l}}(\omega_0)e^{\mathrm{b}}(0)\\ e^{\mathrm{b}}(L)&=R_{\mathrm{r}}(\omega_0)e^{\mathrm{f}}(L),\end{aligned} \tag{2.137}$$

with R_{l} and R_{r} denoting the left and right facet amplitude reflectivity, respectively. We have also assumed in equation (2.137) that the device waveguide is between 0 and L. Also, under equation (2.136) the orthonormal condition (2.129) is reduced to

$$\begin{aligned}&\int_0^L [e_m^{\mathrm{f}}(z)e^{j\beta_0 z}+e_m^{\mathrm{b}}(z)e^{-j\beta_0 z}][e_l^{\mathrm{f}}(z)e^{j\beta_0 z}+e_l^{\mathrm{b}}(z)e^{-j\beta_0 z}]dz\\ &\quad=\delta_{ml}\int_0^L [e_m^{\mathrm{f}}(z)e^{j\beta_0 z}+e_m^{\mathrm{b}}(z)e^{-j\beta_0 z}]^2 dz,\end{aligned} \tag{2.138}$$

where equation (2.35) is used.

In a single mode waveguide, as the solution to equation (2.34) subject to the boundary condition of the cross-sectional structure, the transverse optical mode $\phi(x, y)$ is unique. As the solution to equation (2.135) subject to the boundary condition equation (2.137), however, the longitudinal optical mode pair $e^{\mathrm{f,b}}(z)$ is not unique. Actually, there must be an infinite number of longitudinal mode pairs because of the energy leakage. Therefore, the 3D mode index should be assigned to the longitudinal mode pair $e^{\mathrm{f,b}}(z)$. We further plug equations (2.132) and (2.136) into equation (2.127) to obtain

$$\Delta g_{Vlm}(t)=\frac{\Gamma\int_0^L\left[jk_0\frac{d\Delta n_V(z,\omega_0)}{dV}+\frac{1}{2}\frac{dg_V(z,\omega_0)}{dV}\right]\Delta V(t)\left[e_m^{\mathrm{f}}(z)e_l^{\mathrm{b}}(z)+e_m^{\mathrm{b}}(z)e_l^{\mathrm{f}}(z)\right]dz}{\int_0^L e_m^{\mathrm{f}}(z)e_m^{\mathrm{b}}(z)dz}, \tag{2.139}$$

where equation (2.42) is used and those fast oscillation terms with factors $e^{\pm 2j\beta_0 z}$ are all dropped.

Finally, by combining equations (2.22), (2.125a), and (2.136), we find the real-world optical field

$$\vec{E}(\vec{r},t) = \frac{1}{2}\vec{x}\phi(x,y)\sum_{l\in L} U_l(t)\left[e_l^{\mathrm{f}}(z)\mathrm{e}^{\mathrm{j}\beta_0 z} + e_l^{\mathrm{b}}(z)\mathrm{e}^{-\mathrm{j}\beta_0 z}\right]\mathrm{e}^{-\mathrm{j}\omega_0 t} + \mathrm{c.c.} \tag{2.140}$$

As opposed to the traveling wave model equation (2.47) described in Section 2.4.1, equations (2.135), (2.139), and (2.126) form the standing wave model where the longitudinal field envelope function has been split into a spatially dependent part, the longitudinal optical mode, and a time-dependent part, the mode amplitude, or the photon amplitude once it is normalized to a dimensionless quantity. Instead of solving PDEs in the longitudinal direction and time, we just need to solve a 1D eigenvalue problem (ODE) for the longitudinal mode and a set of rate equations (ODEs) for the mode amplitude, with the former accounting for the optical field longitudinal dependence and the latter describing the optical field time dependence, respectively. As for the optical field transverse dependence, both models describe it in the same way through the transverse optical mode governed by the eigenvalue problem equation (2.34) defined in the 2D cross-section. Published work [16–21] reveals that the standing wave model is far more efficient in static and small signal analysis for semiconductor lasers, especially for single mode lasers or when the number of lasing modes is limited to only a few. In large signal dynamic analysis, however, this method loses its computational efficiency very rapidly as the external bias changes with time more abruptly or as the number of modes required in expansion (2.125a) grows. Actually, to describe an abrupt change of the external bias in the time domain, we have to let longitudinal modes and associated eigenvalues float with the static bias to ensure the validity of equation (2.122) (i.e., to ensure $|\Delta V| \ll 1$). Therefore, the longitudinal eigenequation (2.135) needs to be solved at multiple static bias values, which increases the computational effort. Also, from equation (2.139) it is easy to conclude that the computational effort required on these coefficients in describing the cross-coupling terms in the optical rate equation (2.126) grows quadratically as the number of required longitudinal modes. Only for those structures under certain operation conditions where the longitudinal spatial hole burning (LSHB) effect is negligible, is equation (2.139) reduced to

$$\Delta g_{Vlm}(t) = \Gamma\left(2\mathrm{j}k_0\frac{\mathrm{d}\Delta n_V}{\mathrm{d}V} + \frac{\mathrm{d}g_V}{\mathrm{d}V}\right)\Delta V(t)\delta_{ml}. \tag{2.141}$$

Consequently, the optical rate equation (2.126) is greatly simplified to

$$\frac{\mathrm{d}U_m(t)}{\mathrm{d}t} = \frac{v_\mathrm{g}}{2}\left[\mathrm{j}\frac{\beta_{Vm}^2}{\bar{n}k_0} + 2\mathrm{j}k_0\Gamma\frac{\mathrm{d}\Delta n_V}{\mathrm{d}V}\Delta V(t) + \Gamma\frac{\mathrm{d}g_V}{\mathrm{d}V}\Delta V(t)\right]U_m(t) + v_\mathrm{g}\widetilde{s}_m(t), \tag{2.142}$$

where all the cross-coupling terms disappear. Obviously, the numerical computation will be extremely efficient in this situation. Other approaches have also been developed in an effort to reduce the computational cost of the cross-coupling coefficients described by equation (2.139) [22]. In situations where these approaches are applicable, the standing wave approach is still more efficient compared with the traveling wave method, even in large signal dynamic analysis.

In summary, the standing wave model serves as a complement of the traveling wave model in static and small signal analysis of devices with high Q resonant structure such as semiconductor lasers. It also has advantages in device structure refining and operation optimization. We will return to the numerical implementation and application of this model in Section 6.3.

2.7 Photon rate and phase equations – the behavior model

For some users, optoelectronic devices are viewed as symbolic nodes without physical dimensions. Such nodes process input signals and send them out as output signals. Therefore, only the terminal performance of such a device is of interest. A device model with only time dependence is thus desired and is referred to as the behavior model.

Actually, if we approximate the longitudinal optical field as uniformly distributed, equation (2.135) reduces to

$$\beta_V^2 = \beta_0[2k_0\Gamma\Delta n_V(z,\omega_0) - \mathrm{j}\Gamma g_V(z,\omega_0) + \mathrm{j}\bar{\alpha}(z)]. \tag{2.143}$$

Under uniform field distribution, the LSHB disappears, hence equation (2.142) is applicable. Plugging equation (2.143) into equation (2.142) yields

$$\frac{\mathrm{d}U(t)}{\mathrm{d}t} = \frac{v_g}{2}[2\mathrm{j}k_0\Gamma\Delta n(t) + \Gamma g(t) - \bar{\alpha}]U(t) + v_g\tilde{s}(t), \tag{2.144}$$

where we write $\Delta n_V + (\mathrm{d}\Delta n_V/\mathrm{d}V)\Delta V(t)$ and $g_V + (\mathrm{d}g_V/\mathrm{d}V)\Delta V(t)$ as $\Delta n(\omega_0)$ and $g(\omega_0)$, respectively. They still have implicit time dependence through the external bias.

By letting $U(t) = \surd(S(t))\mathrm{e}^{\mathrm{j}\varphi(t)}$, where $S(t) = |U(t)|^2$ and $\varphi(t)$ indicate the photon number and phase, respectively, we find from equation (2.144)

$$\frac{\mathrm{d}S(t)}{\mathrm{d}t} = v_g[\Gamma g(\omega_0) - \bar{\alpha}]S(t) + R_{sp}(t), \tag{2.145}$$

$$\frac{\mathrm{d}\varphi(t)}{\mathrm{d}t} = v_g k_0\Gamma\Delta n(\omega_0) + F_{sp}(t). \tag{2.146}$$

In equations (2.145) and (2.146), the spontaneous emission photon number and phase noises are defined as

$$R_{sp}(t) \equiv 2v_g Re[\tilde{s}(t)U^*(t)], \tag{2.147}$$

and

$$F_{sp}(t) \equiv v_g Im[\tilde{s}(t)U^*(t)]/S(t), \tag{2.148}$$

respectively.

Equations (2.145) and (2.146) are the photon rate and phase equations used for modeling the device behavior from an optical aspect. The photon number is related to the output optical power of the device, whereas the time derivative of the photon phase determines the optical frequency deviation from its reference frequency ω_0. We will skip further discussion on this model as this book mainly focuses on physics-based models.

2.8 The spontaneous emission noise treatment

Although the spontaneous emission contribution can be theoretically calculated through expressions given in above sections, we would rather take a phenomenological approach in evaluating the noise contribution to avoid unnecessary complication in implementations, because an exact treatment of noise without quantization of the optical field is not possible [23–27]. In this treatment, we will utilize the fact that, despite its random nature, the noise power can be evaluated directly. Fortunately, with respect to most of a device's characteristics, only the noise power matters. This approach is therefore justified.

Actually, by taking a small step d_z and integrating the steady state equation (2.46) along the wave propagation direction (z), we find

$$e(z + d_z) = e^{\left[jk_0\Gamma\Delta n(z,\omega_0)+\frac{1}{2}\Gamma g(z,\omega_0)-\frac{1}{2}\bar{\alpha}(z)\right]d_z} e(z) + \tilde{s}(z)d_z, \tag{2.149}$$

where the spontaneous emission contribution is added at the end of this step so that it is not yet amplified.

From the energy transmission point of view, propagation of the guided mode optical wave inside the waveguide is equivalent to plane wave propagation inside a free-space filled by the effective index, n_{eff}, of this mode. Since the absolute value of the Poynting vector of the latter (i.e., the plane wave) can be evaluated through $|S| = (\varepsilon_0 c/2)n_{\text{eff}}|\phi(x, y)|^2|e(z)|^2 = (n_{\text{eff}}\sqrt{(\varepsilon_0/\mu_0)}/2)|\phi(x, y)|^2|e(z)|^2$, and the absolute value of the Poynting vector is the power density flow along the wave propagation direction, we know that $\int_\Sigma |S| dx\, dy = (n_{\text{eff}}\sqrt{(\varepsilon_0/\mu_0)}/2)|e(z)|^2$ is the power carried by the optical wave $e(z)$ as it propagates along the waveguide, where we have utilized the fact that the optical wave is confined in the cross section Σ (as the guided mode) and is normalized.

Multiplying equation (2.149) by its complex conjugate and a factor of $n_{\text{eff}}\sqrt{(\varepsilon_0/\mu_0)}/2$ in $1/\Omega$ on both sides yields

$$P(z + d_z) = e^{[\Gamma g(z,\omega_0)-\bar{\alpha}(z)]d_z} P(z) + \frac{n_{\text{eff}}\, d_z^2}{2}\sqrt{\left(\frac{\varepsilon_0}{\mu_0}\right)}|\tilde{s}(z)|^2. \tag{2.150}$$

In equation (2.150), we have ignored cross terms as the noise has a random phase, which makes the averaged contribution zero. Equation (2.150) clearly shows a power balance with the last term on its RHS indicating the spontaneous emission noise power contribution within the small step d_z. On the other hand, the spontaneous emission noise power contribution in a small section d_z can be phenomenologically evaluated as [28, 29]

$$\frac{n_{\text{eff}}\, d_z^2}{2}\sqrt{\left(\frac{\varepsilon_0}{\mu_0}\right)}|\tilde{s}(z)|^2 = \gamma\Gamma v_g g_{sp}(z, \omega_0)\hbar\omega_0, \tag{2.151}$$

with γ denoting the dimensionless coupling coefficient of the spontaneous emission over the entire spatial sphere and spread over the entire frequency spectrum to the waveguide mode at the reference frequency, $\hbar = h/2\pi$ the reduced Planck's constant in J s, and g_{sp} the spontaneous emission gain in 1/m. The phenomenological parameter γ must be introduced since spontaneously emitted photons go in every direction in a spatial sphere whereas the waveguide is built in one direction. Therefore, the waveguide cannot capture all these photons, and hence we need a parameter to describe the percentage of photons coupled to the guided mode [30, 31]. Moreover, spontaneously emitted photons take every possible frequency over the entire emission spectrum, and a coupling factor therefore needs to be introduced to represent the proportion of the photons emitted in the neighborhood of the reference frequency, as only these photons will be coupled to the operating frequency. It is also worth mentioning that the material spontaneous emission gain introduced here and the material stimulated emission gain (simply called material gain) introduced in Section 2.4 are not the same unless the ground state is completely empty. The stimulated emission gain, triggered by incoming photons, is the net contribution of the downward transition (which emits photons) minus the upward transition (which absorbs photons) between the excited and ground state, whereas the spontaneous emission gain is the contribution of the downward transition only. We will further discuss the calculation of these gains in Chapter 7 on the implementation of our material model.

As a Langevin noise source, the amplitude and phase of the spontaneous emission noise $\tilde{s}(z, t)$ given in equation (2.37) can be approximately modeled by the Gaussian and uniformly distributed random processes, respectively. The Gaussian distributed random process takes zero mean with the autocorrelation function given as [32–35]

$$\langle|\tilde{s}(z,t)||\tilde{s}(z',t')|\rangle = 2\sqrt{\left(\frac{\mu_0}{\varepsilon_0}\right)}\frac{\gamma\Gamma g_{sp}(z,\omega_0)\hbar\omega_0}{n_{eff}}\delta(z-z')\delta(t-t'), \qquad (2.152)$$

where δ denotes the Dirac function. At the steady state as $\mathrm{d}_z \to 0$, $z \to z'$, $t \to t'$, $\delta(z - z') \to 1/\mathrm{d}_z$, and $\delta(t - t') \to v_g/\mathrm{d}_z$, the LHS and RHS of (2.152) approach $< |\tilde{s}(z)|^2 >$ and $2\sqrt{(\mu_0/\varepsilon_0)}[\gamma\Gamma v_g g_{sp}(z,\omega_0)\hbar\omega_0/(n_{eff}\mathrm{d}_z^2)]$, respectively, which is consistent with (2.151). The uniformly distributed random process for the phase of $\tilde{s}(z, t)$ is over $[0, 2\pi]$. In numerical calculations, the spontaneous emission contribution given in the form of equation (2.37) is therefore modeled by two independent random number generators (RNGs): the RNG that takes the Gaussian distribution with zero mean and autocorrelation in the form of equation (2.152) describes the amplitude, the other RNG that takes the uniform distribution over $[0, 2\pi]$ represents the phase of $\tilde{s}(z, t)$.

The spontaneous emission noise $\tilde{s}(x, z, t)$ given in equation (2.57) can be treated by a similar approach, with its amplitude autocorrelation function modified to

$$< |\tilde{s}(x,z,t)||\tilde{s}(x',z',t')| >= 2\sqrt{\left(\frac{\mu_0}{\varepsilon_0}\right)}\frac{\gamma\Gamma g_{sp}(z,\omega_0)\hbar\omega_0}{n_{eff}}\delta(x-x')\delta(z-z')\delta(t-t'), \qquad (2.153)$$

where as $x \to x'$, $\delta(x-x') \to \mathrm{d}_x$, with d_x denoting the length of a small section along x.

Still following a similar approach, we obtain the amplitude autocorrelation function of the spontaneous emission noise $\bar{s}_l(z,t)$ given in equation (2.119) as

$$<|\bar{s}_l(z,t)||\bar{s}_l(z',t')|>=2\sqrt{\left(\frac{\mu_0}{\varepsilon_0}\right)}\frac{\gamma_s\Gamma g_{sp}(z,\omega_l)\hbar\omega_0}{n_{eff}}\delta(z-z')\delta(t-t'), \quad (2.154)$$

with γ_s only counting the spatial coupling factor.

We again apply this approach to deal with the spontaneous emission noise in the standing wave model. Actually, by ignoring those cross-coupling terms in the optical rate equation (2.126) and integrating it over a small time step Δt, we find

$$U_m(t+\Delta t)=e^{\frac{v_g}{2}\left[j\frac{\beta^2_{Vm}}{\bar{n}k_0}+\Delta g_{Vmm}(t)\right]\Delta t}U_m(t)+v_g\Delta t\tilde{s}_m(t), \quad (2.155)$$

where the spontaneous emission contribution is added at the end of this step. Multiplying equation (2.155) by its complex conjugate and $\hbar\omega_0$ on both sides yields

$$P_m(t+\Delta t)=e^{v_g\left[\frac{Im(\beta^2_{Vm})}{\bar{n}k_0}+\Delta g_{Vmm}(t)\right]\Delta t}P_m(t)+(v_g\Delta t)^2|\tilde{s}_m(t)|^2. \quad (2.156)$$

In equation (2.156), we have ignored all cross terms due to the random phase of the noise, which leads to zero averaged contribution. Equation (2.156) clearly shows a power balance with the last term on its RHS indicating the spontaneous emission noise power contribution within the small time step Δt. Therefore, following equation (2.151), we find

$$(v_g\Delta t)^2|\tilde{s}_m(t)|^2=\gamma\Gamma v_g\bar{g}_{sp}(\omega_0)\hbar\omega_0, \quad (2.157)$$

with

$$\bar{g}_{sp}(\omega_0)=\frac{\int_0^L g_{sp}(z,\omega_0)e^f_m(z)e^b_m(z)dz}{\int_0^L e^f_m(z)e^b_m(z)dz}. \quad (2.158)$$

We again assume that the amplitude and phase of the spontaneous emission noise $\tilde{s}_m(t)$ given in equation (2.128) follow a Gaussian and a uniform distributed random process, respectively. The Gaussian distributed random process takes zero mean with its autocorrelation function given as

$$<|\tilde{s}_m(t)||\tilde{s}_m(t')|>=\frac{\gamma\Gamma\bar{g}_{sp}(\omega_0)\hbar\omega_0}{v_g\Delta t}\delta(t-t'). \quad (2.159)$$

The uniform distributed random process for the phase of $\tilde{s}_m(t)$ is over $[0, 2\pi]$.

In the behavior model, $\hbar\omega_0 dS(t)/dt$ has the dimensions of optical power; according to equation (2.145), $\hbar\omega_0 R_{sp}(t)$ must be the spontaneous emission noise power. Its mean value can therefore be estimated as

$$<\hbar\omega_0 R_{sp}(t)>=\gamma\Gamma v_g g_{sp}(\omega_0)\hbar\omega_0. \quad (2.160)$$

Regarding the spontaneous emission contribution to the photon phase, since it is inversely proportional to the photon number according to equation (2.148), we can take it as a random process with zero mean.

By taking out their mean values, we define

$$\widetilde{R}_{sp}(t) \equiv R_{sp}(t) - \gamma\Gamma v_g g_{sp}(\omega_0) \tag{2.161}$$

and

$$\widetilde{F}_{sp}(t) \equiv F_{sp}(t) \tag{2.162}$$

Hence the photon rate and phase equation can be rewritten as

$$\frac{dS(t)}{dt} = v_g[\Gamma g(\omega_0) - \overline{\alpha}]S(t) + \gamma\Gamma v_g g_{sp}(\omega_0) + \widetilde{R}_{sp}(t), \tag{2.163}$$

$$\frac{d\varphi(t)}{dt} = v_g k_0 \Gamma \Delta n(\omega_0) + \widetilde{F}_{sp}(t), \tag{2.164}$$

where $\widetilde{R}_{sp}(t)$ and $\widetilde{F}_{sp}(t)$ are independent Gaussian distributed random processes with zero means and autocorrelation functions given by [36, 37]

$$< \widetilde{R}_{sp}(t)\widetilde{R}_{sp}(t') > = 2\gamma\Gamma v_g g_{sp}(\omega_0)S(t)\delta(t-t'), \tag{2.165}$$

$$< \widetilde{F}_{sp}(t)\widetilde{F}_{sp}(t') > = \frac{\gamma\Gamma v_g g_{sp}(\omega_0)}{2S(t)}\delta(t-t'). \tag{2.166}$$

Different random distributions (such as the Poisson distribution) can also be assigned to the spontaneous emission noise. We find, however, that no appreciable difference can be found in modeling those deterministic device characteristics, as long as the noise autocorrelation function follows equations (2.152) to (2.154), (2.159), (2.165), and (2.166) [12] [38].

Finally, it is worth mentioning that in the above expressions for spontaneous emission noise powers and autocorrelation functions, we have absorbed Petermann's factor [39] into the spontaneous emission noise coupling coefficient γ or γ_s, instead of showing it explicitly. This is because we have followed a phenomenological approach instead of using the original noise amplitude expressions given in equations (2.37), (2.57), (2.119), (2.128), (2.147) and (2.148). Since the focus of this book is on numerical modeling and solution techniques, a rigorous treatment [40–44] of the spontaneous emission noise is beyond our scope.

References

[1] K. S. Yee, Numerical solution of initial boundary value problems involving Maxwell's equations in isotropic media. *IEEE Trans. Antennas and Propagation*, **14** (1966), 302–7.

[2] J. C. Strikwerda, *Finite Difference Schemes and Partial Differential Equations*, 1st edn (New York: Chapman & Hall, 1989).

[3] A. V. Chelnokov, J.-M. Lourtioz, and P. Gavrilovic, Numerical modeling of the spatial and spectro-temporal behavior of wide aperture unstable resonator semiconductor lasers. *IEEE Photon. Tech. Lett.*, **7**:8 (1995), 863–5.

[4] S. F. Yu, A quasi-three-dimensional large-signal dynamic model of distributed feedback lasers. *IEEE Journal of Quantum Electron.*, **QE-32**:3 (1996), 424–31.

[5] C. Elachi, Waves in active and passive periodic structures: a review. *Proc. IEEE*, **64** (1976), 1666–98.

[6] T. K. Gaylord, Analysis and applications of optical diffraction by gratings. *Proc. IEEE*, **73** (1985), 894–937.

[7] W. Streifer, Coupled wave analysis of DFB and DBR lasers. *IEEE Journal of Quantum Electron.*, **QE-13**:4 (1977), 134–41.

[8] R. F. Kazarinov and C. H. Henry, Second-order distributed feedback lasers with mode selection provided by first-order radiation losses. *IEEE Journal of Quantum Electron.*, **QE-21**:2 (1985), 144–50.

[9] A. Shams, X. Li, and W.-P. Huang, Second and higher-order resonant gratings with gain or loss: Part 1, Green's function analysis. *IEEE Journal of Quantum Electron.*, **QE-36**:12 (2000), 1421–30.

[10] J. G. Proakis and D. G. Manolakis, *Digital Signal Processing, Principles, Algorithms, and Applications*, 1st edn (Upper Saddle River, New Jersey: Prentice-Hall, 1996).

[11] X. Li and J.-W. Park, Time-domain simulation of channel crosstalk and inter-modulation distortion in gain clamped semiconductor optical amplifiers. *Optics Comm.*, **263** (2006), 219–28.

[12] J.-W. Park, X. Li, and W.-P. Huang, Performance simulation and design optimization of gain-clamped semiconductor optical amplifiers based on distributed Bragg reflectors. *IEEE Journal of Quantum Electron.*, **QE-39**:11 (2003), 1415–23.

[13] A. E. Siegman, *Lasers*, 1st edn (Mill Valley, CA: University Science Books, 1986).

[14] C. Z. Ning, R. A. Indik, and J. V. Moloney, Effective Bloch equations for semiconductor lasers and amplifiers. *IEEE Journal of Quantum Electron.*, **QE-33**:9 (1997), 1543–50.

[15] P. Blanchard and E. Bruning, *Mathematical Methods in Physics*, 1st edn (Boston: Birkhauser, 2003).

[16] B. Tromborg, H. Olesen, and X. Pan, Theory of linewidth for multielectrode laser diodes with spatially distributed noise sources. *IEEE Journal of Quantum Electron.*, **QE-27**:2 (1991), 178–92.

[17] H. Olesen, B. Tromborg, X. Pan, and H. E. Lassen, Stability and dynamic properties of multielectrode laser diodes using a Green's function approach. *IEEE Journal of Quantum Electron.*, **QE-29**:8 (1993), 2282–301.

[18] H. Wenzel and H.-J. Wunsche, An equation for the amplitude of the modes in semiconductor lasers. *IEEE Journal of Quantum Electron.*, **QE-30**:9 (1994), 2073–80.

[19] J. Hong, W.-P. Huang, and T. Makino, Static and dynamic simulation for ridge-waveguide MQW DFB lasers. *IEEE Journal of Quantum Electron.*, **QE-31**:1 (1995), 49–59.

[20] X. Li, A. D. Sadovnikov, W.-P. Huang, and T. Makino, A physics-based three-dimensional model for distributed feedback (DFB) laser diodes. *IEEE Journal of Quantum Electron.*, **QE-34**:9 (1998), 1545–53.

[21] Y.-P. Xi, X. Li, and W.-P. Huang, Time-domain standing-wave approach based on cold cavity modes for simulation of DFB lasers. *IEEE Journal of Quantum Electron.*, **QE-44**: 10 (2008), 931–7.

[22] Y.-P. Xi, W.-P. Huang, and X. Li, An efficient solution to the standing-wave model based on cold cavity modes for simulation of DFB lasers. To appear in *IEEE/OSA Journal of Lightwave Tech.*

[23] L. D. Landau and E. M. Lifshitz, *Electrodynamics of Continuous Media*, 1st edn (Reading, Massachusetts: Addison-Wesley, 1960).

[24] M. Lax, Quantum noise IX: quantum theory of noise sources, *Phys. Rev.*, **145** (1965), 110–29.

[25] M. Lax, Classical noise V: noise in self-sustained oscillators. *Phys. Rev.*, **160** (1967), 290–307.

[26] M. Lax, Quantum noise VII: the rate equations and amplitude noise in lasers, *IEEE Journal of Quantum Electron.*, **QE-3** (1967), 37–46.

[27] M. Lax and W. H. Louisell, Quantum noise IX: quantum Fokker–Planck solution for laser noise. *IEEE Journal of Quantum Electron.*, **QE-3** (1967), 47–58.

[28] A. Yariv, *Quantum Electronics*, 4th edn (New York: Wiley, 1989).

[29] D. D. Marcenac and J. E. Carroll, Quantum-mechanical model for realistic Fabry–Perot lasers. *IEE Proc. J*, **140**:3 (1993), 157–71.

[30] M. Yamada, Advanced theory of semiconductor lasers. In *Handbook of Semiconductor Lasers and Photonic Integrated Circuits*, ed. Y. Suematsu and A. R. Adams. (London: Chapman & Hall, 1994).

[31] C. F. Janz and J. N. McMullin, Spontaneous emission coupling to radiation and guided modes of planar waveguides structures. *IEEE Journal of Quantum Electron.*, **QE-31**:7 (1995), 1344–53.

[32] C. H. Henry, Theory of the linewidth of semiconductor lasers. *IEEE Journal of Quantum Electron.*, **QE-18**:2 (1982), 259–64.

[33] C. H. Henry, Phase noise in semiconductor lasers. *IEEE/OSA Journal of Lightwave Tech.*, **LT-4**:3 (1986), 298–311.

[34] K. Vahala and A. Yariv, Semiclassical theory of noise in semiconductor lasers: Part I. *IEEE Journal of Quantum Electron.*, **QE-19**:6 (1983), 1096–101.

[35] K. Vahala and A. Yariv, Semiclassical theory of noise in semiconductor lasers: Part II. *IEEE Journal of Quantum Electron.*, **QE-19**:6 (1983), 1102–9.

[36] M. Lax, Classical noise IV: Langevin methods. *Rev. Mod. Phys.*, **38** (1966), 541–66.

[37] C. H. Henry, Theory of spontaneous emission noise in open resonators and its application to lasers and optical amplifiers. *IEEE/OSA Journal of Lightwave Tech.*, **LT-4**:3 (1986), 288–97.

[38] J.-W. Park, W.-P. Huang, and X. Li, Investigation of semiconductor optical amplifier integrated with DBR laser for high saturation power and fast gain dynamics. *IEEE Journal of Quantum Electron.*, **QE-40**:11 (2004), 1540–7.

[39] K. Petermann, Calculated spontaneous emission factor for double-heterostructure injection lasers with gain-induced waveguiding. *IEEE Journal of Quantum Electron.*, **QE-15**:7 (1979), 566–70.

[40] H. A. Haus and S. Kawakami, On the "excess spontaneous emission factor" in gain-guided laser amplifiers. *IEEE Journal of Quantum Electron.*, **QE-21**:1 (1985), 63–9.

[41] A. E. Siegman, Excess spontaneous emission in non-Hermitian optical systems: I. laser amplifiers. *Phys. Rev. A*, **39**:3 (1989), 1253–63.

[42] A. E. Siegman, Excess spontaneous emission in non-Hermitian optical systems: II. laser oscillators. *Phys. Rev. A*, **39**:3, (1989), 1264–8.

[43] W. A. Hamel and J. P. Woerdman, Nonorthogonality of the longitudinal eigenmodes of a laser. *Phys. Rev. A*, **40**:5 (1989), 2785–7.

[44] D. Lenstra, Linewidth modifications due to nonorthogonality, inhomogeneity, and directionality of longitudinal laser modes. *IEEE Journal of Quantum Electron.*, **QE-29**:3 (1993), 954–60.

3 Material model I: Semiconductor band structures

To be able to model material optical properties such as the gain and refractive index change, also known as material excitations, we need first to solve the single electron band structure in semiconductors. That is to say, we will solve the material eigenstates before we consider their excitations. Our discussion in this chapter is limited to compound semiconductors with direct bandgaps where k–p theory is applicable.

3.1 Single electron in bulk semiconductors

3.1.1 The Schrödinger equation and Hamiltonian operator

For a system with K nuclei with charge Z_k and N electrons, we can write the time-dependent Schrödinger equation in the form

$$\mathrm{j}\hbar\frac{\partial}{\partial t}\Psi(\vec{r},t) = \vec{H}\Psi(\vec{r},t), \tag{3.1}$$

with its Hamiltonian operator given as [1]

$$\begin{aligned}\vec{H} = &-\frac{\hbar^2}{2\mathrm{m}_0}\sum_{n=1}^{N}\nabla_n^2 - \frac{\hbar^2}{2}\sum_{k=1}^{K}\frac{\nabla_k^2}{M_k} + \frac{1}{2}\frac{\mathrm{e}^2}{4\pi\varepsilon_0}\sum_{n,m=1,n\neq m}^{N}\frac{1}{|\vec{r}_n - \vec{r}_m|}\\ &-\frac{\mathrm{e}^2}{4\pi\varepsilon_0}\sum_{k=1}^{K}\sum_{n=1}^{N}\frac{Z_k}{|\vec{r}_n - \vec{R}_k|} + \frac{1}{2}\frac{\mathrm{e}^2}{4\pi\varepsilon_0}\sum_{k,l=1,k\neq l}^{K}\frac{Z_k Z_l}{|\vec{R}_k - \vec{R}_l|}.\end{aligned} \tag{3.2}$$

In equations (3.1) and (3.2), ψ indicates the dimensionless wave function of the N electron system, m_0 the electron mass in kg, M_k the kth nucleus mass in kg, e the elementary charge in C. Since the Hamiltonian operator has no explicit time dependence, by introducing

$$\Psi(\vec{r},t) = \mathrm{e}^{-\mathrm{j}\frac{\varepsilon_\mathrm{h}}{\hbar}t}\psi(\vec{r}), \tag{3.3}$$

we obtain the steady-state Schrödinger equation

$$\vec{H}\psi(\vec{r}) = \varepsilon_\mathrm{h}\psi(\vec{r}). \tag{3.4}$$

In equation (3.4), $\psi(\vec{r})$ and $\varepsilon_{\rm h}$ denote the eigenfunction and eigenvalue of the given Hamiltonian operator, known as the system wave function and energy, respectively. The first and second terms in equation (3.2) are the summations of the kinetic energy of the electrons and nuclei, respectively, the third and last terms the summations of the Coulomb interactions (repulsion) between the electrons and nuclei, respectively, and the fourth term the summation of the Coulomb interactions (attraction) between the nuclei and the electrons. The extra factor of $1/2$ in the third and last terms rectifies the double counting in repulsion energy summation.

Without any approximation of the Hamiltonian in the form of equation (3.2), the static Schrödinger equation (3.4) can only be solved exactly for a system of one nucleus with one electron, e.g., a hydrogen atom, under the mass–center coordinates.

Because the nuclei move much slower than the electrons, pure nuclei contributions can be ignored according to the Born–Oppenheimer approximation. Hence the Hamiltonian is reduced to

$$\vec{H}=-\frac{\hbar^2}{2\mathrm{m}_0}\sum_{n=1}^{N}\nabla_n^2+\frac{\mathrm{e}^2}{8\pi\varepsilon_0}\sum_{n,m=1,n\neq m}^{N}\frac{1}{|\vec{r}_n-\vec{r}_m|}-\frac{\mathrm{e}^2}{4\pi\varepsilon_0}\sum_{k=1}^{K}\sum_{n=1}^{N}\frac{Z_k}{|\vec{r}_n-\vec{R}_k|}. \tag{3.5}$$

Under such a Hamiltonian, the static Schrödinger equation (3.4) can still be solved exactly only for a few simple systems with no more than two electrons, e.g., a helium atom.

In considering the Hamiltonian for a single electron (e.g., the nth electron) in this system, we can employ the Hartree self-consistent model by taking the interaction between all the electron pairs as an average effect on the nth electron. Hence, the Hamiltonian for the nth electron becomes

$$\begin{aligned}\vec{H}_n = &-\frac{\hbar^2}{2\mathrm{m}_0}\nabla_n^2-\frac{\mathrm{e}^2}{4\pi\varepsilon_0}\sum_{k=1}^{K}\frac{Z_k}{|\vec{r}_n-\vec{R}_k|}\\ &+\frac{\mathrm{e}^2}{8\pi\varepsilon_0}\sum_{m=1,m\neq n}^{N}\int\phi_m^*(\vec{r}_m)\frac{1}{|\vec{r}_n-\vec{r}_m|}\phi_m(\vec{r}_m)\mathrm{d}\vec{r}_m.\end{aligned} \tag{3.6}$$

Once the wave function for the nth electron $\phi_n(\vec{r}_n)$ is solved by using equation (3.6) in equation (3.4), we obtain the electron wave function for the whole system as

$$\psi(\vec{r})=\prod_{n=1}^{N}\phi_n(\vec{r}_n). \tag{3.7a}$$

Unfortunately, this model is proven not to be accurate in many cases. The reason is that the spin of the electron is not considered. With spin, the electron wave function of the whole system must be anti-symmetric under the (two electron) swapping operation. However, equation (3.7a) is obviously a symmetric function under such an operation. To solve this problem, the wave function should be fitted into the Slater determinant in the following form rather than equation (3.7a)

$$\psi(\vec{r}) = \frac{1}{\sqrt{N!}} \begin{vmatrix} \phi^1_{\vec{r}1} & \phi^2_{\vec{r}1} & \cdots & \phi^N_{\vec{r}1} \\ \phi^1_{\vec{r}2} & \phi^2_{\vec{r}2} & \cdots & \phi^N_{\vec{r}2} \\ \cdots & \cdots & \cdots & \cdots \\ \phi^1_{\vec{r}N} & \phi^2_{\vec{r}N} & \cdots & \phi^N_{\vec{r}N} \end{vmatrix}. \tag{3.7b}$$

Consequently, the Hamiltonian in equation (3.6) needs to be modified to the following Hartree–Fock form

$$\vec{H}_n \phi_n(\vec{r}_n) = \left[-\frac{\hbar^2}{2\mathrm{m}_0}\nabla_n^2 - \frac{\mathrm{e}^2}{4\pi\varepsilon_0}\sum_{k=1}^{K}\frac{Z_k}{|\vec{r}_n - \vec{R}_k|}\right]\phi_n(\vec{r}_n)$$

$$+ \frac{\mathrm{e}^2}{8\pi\varepsilon_0}\sum_{m=1}^{N}\int \phi_m^*(\vec{r}_m)\frac{1}{|\vec{r}_n - \vec{r}_m|}[\phi_m(\vec{r}_m)\phi_n(\vec{r}_n) + \phi_n(\vec{r}_m)\phi_m(\vec{r}_n)]\,\mathrm{d}\vec{r}_m. \tag{3.8a}$$

In looking for the single electron eigenstates, we assume that the effect of all the other electrons, i.e., the Coulomb and the exchange operator as the last two terms in equation (3.8a), can be approximated by grouped repulsion centers that coincide with the nuclei in the form of

$$\frac{\mathrm{e}^2}{4\pi\varepsilon_0}\sum_{k=1}^{K}\frac{\sigma_n}{|\vec{r}_n - \vec{R}_k|}. \tag{3.8b}$$

We will consider the inter-electron many-body effect, in its exact form only when we calculate the optical gain and refractive index change in chapter 4. In this chapter, therefore, the Hamiltonian is reduced to

$$\vec{H}_n = -\frac{\hbar^2}{2\mathrm{m}_0}\nabla_n^2 - \frac{\mathrm{e}^2}{4\pi\varepsilon_0}\sum_{k=1}^{K}\frac{Z_k - \sigma_n}{|\vec{r}_n - \vec{R}_k|}. \tag{3.9}$$

In bulk semiconductors with perfect crystalline structures, all the nuclei rank in order in the 3D space and form a periodic structure. We can further write the Hamiltonian in equation (3.9) in the form

$$\vec{H}_0 = -\frac{\hbar^2}{2\mathrm{m}_0}\nabla^2 + V_0(\vec{r}) = \frac{\vec{p}^2}{2\mathrm{m}_0} + V_0(\vec{r}), \tag{3.10}$$

with $\vec{p} \equiv \hbar\nabla/\mathrm{j}$ denoting the momentum operator, $\vec{p}^2/(2\mathrm{m}_0) = -\hbar^2\nabla^2/(2\mathrm{m}_0)$ the electron kinetic energy operator, and

$$V_0(\vec{r}) = -\frac{\mathrm{e}^2}{4\pi\varepsilon r}, \tag{3.11}$$

the Coulomb potential energy, which is periodic

$$V_0(\vec{r}) = V_0[\vec{r} + \vec{R}(\vec{n})], \tag{3.12}$$

over the entire bulk semiconductor. In equation (3.11), ε denotes the bulk semiconductor background frequency domain permittivity. The lattice vector $\vec{R}$ in equation (3.12) is defined as

$$\vec{R}(\vec{n}) \equiv \sum_{i=1}^{3} n_i \vec{a}_i, \tag{3.13}$$

with n_i as an integer defined on $[-N_i/2, N_i/2]$ for $i = 1, 2, 3$. In equation (3.13) $\vec{a}_i$ denotes the primitive vector of the lattice and N_i the total number of primitive cells along dimension L_i with $i = 1, 2, 3$, respectively. L_i's (for $i = 1, 2, 3$) define a 3D bulk semiconductor with

$$L_i = |\vec{a}_i| N_i. \tag{3.14}$$

Equation (3.10) reveals that the Hamiltonian comprises the kinetic energy and the Coulomb potential energy. It is clear that the kinetic energy is the energy of a free electron moving in a vacuum box that occupies the same space as the bulk semiconductor, whereas the Coulomb potential energy describes the additional effect on this electron if we fill the vacuum box with a semiconductor lattice.

3.1.2 Bloch's theorem and band structure

The idea behind the Bloch theorem is to find a simple operator $\vec{P}$, which commutes with the Hamiltonian operator $\vec{H}_0$. As such, the eigenstates of $\vec{H}_0$ can be chosen as the eigenstates of $\vec{P}$ simultaneously. If the eigenstates of $\vec{P}$ can be readily determined, we can use them to block-diagonalize the Hamiltonian, hence reducing the solution domain. In this particular case, $\vec{P}$ is selected as the lattice translation operator

$$\vec{P}\psi(\vec{r}) = \psi[\vec{r} + \vec{R}(\vec{n})]. \tag{3.15}$$

By taking p as the eigenvalue of the lattice translation operator $\vec{P}$, we have $\vec{P}\psi(\vec{r}) = p\psi(\vec{r})$, which means

$$p\psi(\vec{r}) = \psi[\vec{r} + \vec{R}(\vec{n})]. \tag{3.16}$$

It is apparent that we must have $|p| = 1$ as otherwise the wave function ψ would be unbounded (i.e., $\psi \to \infty$ as $|p| > 1$ and $\psi \to 0$ as $|p| < 1$) if we repeatedly apply equation (3.15). We can therefore select

$$p = \mathrm{e}^{\mathrm{j}\vec{k}\cdot\vec{R}(\vec{n})}, \tag{3.17}$$

with $\vec{k}$ introduced as the wave vector in 1/m. Hence equation (3.16) becomes

$$\psi[\vec{r} + \vec{R}(\vec{n})] = \mathrm{e}^{\mathrm{j}\vec{k}\cdot\vec{R}(\vec{n})}\psi(\vec{r}). \tag{3.18}$$

By introducing the periodic boundary condition at each facet of the 3D bulk semiconductor, i.e., $\psi[\vec{r} + \vec{R}(\vec{N}/2)] = \psi[\vec{r} + \vec{R}(-\vec{N}/2)]$, we find

$$\mathrm{e}^{\mathrm{j}\vec{k}\cdot\sum_{i=1}^{3} N_i \vec{a}_i} = 1. \tag{3.19}$$

This is possible only if we let

$$\vec{k} = \sum_{i=1}^{3} \frac{n_i}{N_i} \vec{b}_i, \tag{3.20}$$

with $\vec{b}_i$ defined such that $\vec{b}_i \cdot \vec{a}_l = 2\pi\delta_{il}$. $\vec{b}_i$'s are known as the primitive vectors of the reciprocal lattice and can be explicitly expressed in the form of primitive vectors through

$$\vec{b}_1 = 2\pi \frac{\vec{a}_2 \times \vec{a}_3}{|\vec{a}_1 \cdot (\vec{a}_2 \times \vec{a}_3)|}, \vec{b}_2 = 2\pi \frac{\vec{a}_3 \times \vec{a}_1}{|\vec{a}_2 \cdot (\vec{a}_3 \times \vec{a}_1)|}, \vec{b}_3 = 2\pi \frac{\vec{a}_1 \times \vec{a}_2}{|\vec{a}_3 \cdot (\vec{a}_1 \times \vec{a}_2)|}. \tag{3.21}$$

Since the Hamiltonian operator is invariant under the lattice translation according to equations (3.10) and (3.12), $\vec{P}$ commutes with $\vec{H}_0$. By applying $\vec{P}$ on both sides of the single electron Schrödinger equation in bulk semiconductors

$$\vec{H}_0 \Psi(\vec{r}) = \varepsilon_{\mathrm{h}} \Psi(\vec{r}), \tag{3.22}$$

we find the LHS and RHS are

$$\vec{P}\vec{H}_0 \psi(\vec{r}) = \vec{H}_0 \vec{P} \psi(\vec{r}) = \vec{H}_0 \psi[\vec{r} + \vec{R}(\vec{n})],$$

and

$$\vec{P}\varepsilon_{\mathrm{h}} \psi(\vec{r}) = \varepsilon_{\mathrm{h}} \vec{P} \psi(\vec{r}) = \varepsilon_{\mathrm{h}} \psi[\vec{r} + \vec{R}(\vec{n})],$$

respectively. Hence $\psi[\vec{r} + \vec{R}(\vec{n})]$ also serves as a solution of equation (3.22). We can therefore write the general solution to equation (3.22) as a linear combination of all possible $\psi[\vec{r} + \vec{R}(\vec{n})]$'s with the lattice vector $\vec{R}$ given in equation (3.13), giving

$$\psi(\vec{r}) = \sum_{i=1}^{3} \sum_{n_i=-N_i/2}^{N_i/2} A_{n_i} \psi[\vec{r} + \vec{R}(\vec{n})], \tag{3.23}$$

where A_n is the coefficient of the nth term in the linear combination.

Plugging equation (3.18) into equation (3.23) yields

$$\psi(\vec{r}) = \sum_{i=1}^{3} \sum_{n_i=-N_i/2}^{N_i/2} A_{n_i} \mathrm{e}^{\mathrm{j}\vec{k} \cdot \vec{R}(\vec{n})} \psi(\vec{r}) \quad \text{or} \quad \sum_{i=1}^{3} \sum_{n_i=-N_i/2}^{N_i/2} A_{n_i} \mathrm{e}^{\mathrm{j}\vec{k} \cdot \vec{R}(\vec{n})} = 1.$$

This must be valid for all $\vec{k}$'s given in the form of equation (3.20), hence

$$A_{n_i} = \frac{1}{N_1 N_2 N_3} \mathrm{e}^{-\mathrm{j}\vec{k} \cdot \vec{R}(\vec{n})}. \tag{3.24}$$

Equation (3.23) can therefore be written as

$$\begin{aligned}\psi(\vec{r}) &= \frac{1}{N_1N_2N_3}\sum_{i=1}^{3}\sum_{n_i=-N_i/2}^{N_i/2} \mathrm{e}^{-\mathrm{j}\vec{k}\cdot\vec{R}(\vec{n})}\psi[\vec{r}+\vec{R}(\vec{n})] \\ &= \mathrm{e}^{\mathrm{j}\vec{k}\cdot\vec{r}}\frac{1}{N_1N_2N_3}\sum_{i=1}^{3}\sum_{n_i=-N_i/2}^{N_i/2} \mathrm{e}^{-\mathrm{j}\vec{k}\cdot[\vec{r}+\vec{R}(\vec{n})]}\psi[\vec{r}+\vec{R}(\vec{n})] \\ &= \mathrm{e}^{\mathrm{j}\vec{k}\cdot\vec{r}}u_{\vec{k}}(\vec{r}),\end{aligned} \tag{3.25}$$

with the lattice wave function defined as

$$u_{\vec{k}}(\vec{r}) \equiv \frac{1}{N_1N_2N_3}\sum_{i=1}^{3}\sum_{n_i=1}^{N_i} \mathrm{e}^{-\mathrm{j}\vec{k}\cdot[\vec{r}+\vec{R}(\vec{n})]}\psi[\vec{r}+\vec{R}(\vec{n})]. \tag{3.26}$$

We can easily prove that $u_{\vec{k}}[\vec{r}+\vec{R}(\vec{n})] = u_{\vec{k}}(\vec{r})$ since the summation on the RHS of equation (3.26) goes through every possible lattice, hence the lattice wave function is a periodic function in terms of the lattice vector $\vec{R}$. Equation (3.25) is known as Bloch's theorem. It does not provide a final solution but effectively reduces the solution domain from $\psi(\vec{r})$ in the entire bulk semiconductor that comprises $N_1N_2N_3$ primitive cells to $u_{\vec{k}}(\vec{r})$ in one primitive cell. The latter obeys an equation obtained by plugging the Bloch equation (3.25) back into the static Schrödinger equation (3.22) with the Hamiltonian operator given in equation (3.10), which is in the form

$$\left[-\frac{\hbar^2}{2\mathrm{m}_0}(\nabla+\mathrm{j}k)^2 + V_0(\vec{r})\right]u_{\vec{k}}(\vec{r}) = \varepsilon_{\mathrm{h}}u_{\vec{k}}(\vec{r}). \tag{3.27}$$

Once the new eigenvalue problem equation (3.27) is solved in a primitive cell, the single electron wave function in the entire bulk semiconductor can be found from equation (3.25). In dealing with equation (3.27), however, we have to take the wave vector $\vec{k}$ as a varying parameter and solve the eigenvalue problem for all the possible values of $\vec{r}$ restricted by equation (3.20), i.e., in the first Brillouin zone (BZ). Therefore, strictly speaking we have to solve equation (3.27) $N_1N_2N_3$ times. In this sense, equation (3.27) does not appear easier to solve than equation (3.22). Fortunately, there are many efficient approaches developed for solving equation (3.27) under various assumptions and for different applications, which drastically reduce its computational cost [2–4]. In our applications, the material optical property is the major concern. Therefore, there is no need to solve equation (3.27) accurately for the $\vec{k}$'s that are not in the neighborhood of their extremes, as states having such $\vec{k}$ values are almost empty anyway in the time scale in which we are interested. Particularly for those semiconductors with direct bandgaps, the extremes line up at a single $\vec{k}$. Therefore, we need only to solve equation (3.27) once at this extreme $\vec{k}$, and then expand our solution to its neighborhood through the perturbation approach. Known as the k–p theory, this model is well justified for our purpose, and hence we will use it throughout this chapter.

Before moving on to solve equation (3.27), we will further extract the in-depth information given by Bloch's theorem.

First, for each wave vector $\vec{k}$ in the first BZ there is a discrete infinite set of eigenvalues ε_h. This conclusion becomes clearer if we utilize the knowledge that the lattice periodicity lies in both V_0 and $u_{\vec{k}}$ to transform equation (3.27) into an algebraic equation. Actually, by introducing the reciprocal lattice vector in 1/m

$$\vec{G}(\vec{m}) \equiv \sum_{i=1}^{3} m_i \vec{b}_i, \tag{3.28}$$

with $m_i = 0, \pm1, \pm2, \pm3\ldots$, taking any integer for $i = 1, 2, 3$, we can expand V_0 and $u_{\vec{k}}$ into Fourier series

$$V_0(\vec{r}) = \sum_{i=1}^{3} \sum_{m_i=-\infty}^{\infty} \widetilde{V}_{\vec{G}} e^{j\vec{G}(\vec{m})\cdot\vec{r}}, \tag{3.29}$$

$$u_{\vec{k}}(\vec{r}) = \sum_{i=1}^{3} \sum_{m_i=-\infty}^{\infty} \widetilde{u}_{\vec{k}\vec{g}} e^{j\vec{G}(\vec{m})\cdot\vec{r}}, \tag{3.30}$$

where

$$\widetilde{V}_{\vec{G}} = \frac{1}{\Omega} \int_{\Omega} V_0(\vec{r}) e^{-j\vec{G}(\vec{m})\cdot\vec{r}} d\vec{r}, \tag{3.31}$$

$$\widetilde{u}_{\vec{k}\vec{G}} = \frac{1}{\Omega} \int_{\Omega} u_{\vec{k}}(\vec{r}) e^{-j\vec{G}(\vec{m})\cdot\vec{r}} d\vec{r}, \tag{3.32}$$

with Ω denoted as the whole volume of the bulk semiconductor. (If the bulk is rectangular in shape and the primitive cell is a simple cubic, we simply have $\Omega = L_1L_2L_3$.)

Substituting equation (3.29) and (3.30) into equation (3.27) and equating the various Fourier components yields

$$\frac{\hbar^2}{2m_0}[\vec{k} + \vec{G}(\vec{m})]^2 \widetilde{u}_{\vec{k}\vec{G}} + \sum_{i=1}^{3} \sum_{m'_i=-\infty}^{\infty} \widetilde{V}_{\vec{G}'} \widetilde{u}_{\vec{k}(\vec{G}-\vec{G}')} = \varepsilon_h \widetilde{u}_{\vec{k}\vec{G}}. \tag{3.33}$$

Equation (3.33) is an infinite set of linear algebraic equations. For non-trivial solutions, for each $\vec{k}$ in the first BZ, we must let its determinate be zero, which results in an algebraic equation of an infinite order for the unknown eigenvalue ε_h. In principle we can solve this equation to obtain ε_h. Hence we have an infinite number of discrete ε_h's for each $\vec{k}$ in the first BZ. Therefore, we can explicitly record ε_h as $\varepsilon_{n\vec{k}}$. The positive integer n runs through an ascending order to index these discrete ε_h's with $n = 1$ marking the smallest ε_h. As illustrated by Fig. 3.1, since $\Delta\vec{G} = \sum_{i=1}^{3} \vec{b}_i \gg \Delta\vec{k} = \sum_{i=1}^{3} \vec{b}_i/N_i$, or the spacing between the adjacent reciprocal lattice vectors is much larger than the spacing between the adjacent wave vectors, we find from equation (3.33) that the change in the eigenvalue

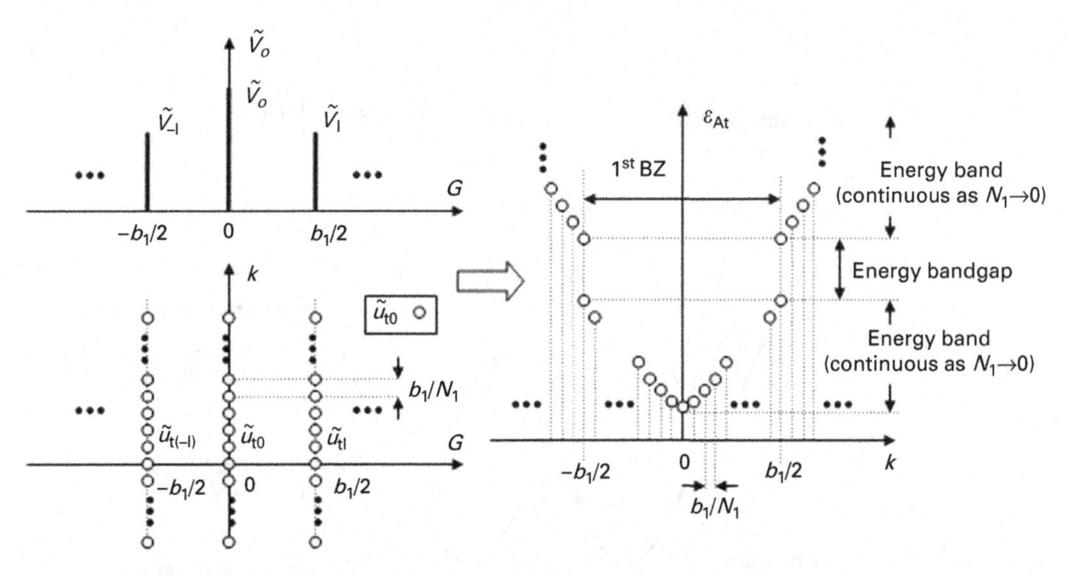

Fig. 3.1. The creation of energy bands in a one-dimensional lattice where energy bandgaps appear because of the disturbance of the periodic potential that is non-vanishing only at the reciprocal lattice in the wave vector domain.

ε_h caused by the change in the reciprocal lattice vector $\vec{G}$ must be larger than that caused by the change in the wave vector $\vec{k}$. And this conclusion holds for any periodic potential energy V_0 as it has only non-vanishing Fourier components on the reciprocal lattice and has no $\vec{k}$ dependence. Therefore, $\varepsilon_{n\vec{k}}$ is densely distributed in $\vec{k}$ but sparsely distributed in n since $\Delta\varepsilon|_{\Delta n} = \varepsilon_{n\vec{k}} - \varepsilon_{(n-1)\vec{k}} > \Delta\varepsilon|_{\Delta\vec{k}} = |\varepsilon_{n\vec{k}} - \varepsilon_{n(\vec{k}-\Delta\vec{k})}|$. Normally the bulk volume is huge in relation to the primitive cell, or $N_i \gg 1$ for $i = 1, 2, 3$; also we have $\Delta\vec{k} = \sum_{i=1}^{3} \vec{b}_i/N_i \to 0$ and hence $\vec{k}$ becomes continuous as well as $\varepsilon_{n\vec{k}}$ because of the one to one correspondence between $\varepsilon_{n\vec{k}}$ and $\vec{k}$ as revealed by equation (3.33). However, $\varepsilon_{n\vec{k}}$ still changes discretely with n. This explains the well-known inherent energy band structure associated with the periodic potential.

Noticing that the BZ boundaries are the bisecting planes where Bragg scattering occurs, whereas inside the BZ there are no such boundaries, we may conclude that $\varepsilon_{n\vec{k}}$ must be an analytical function of $\vec{k}$ inside the BZ whereas its non-analytical dependence on $\vec{k}$ is allowed only at the boundaries. Actually, for crystal structures with certain symmetry, the solution to equation (3.27) can be real. One such example is for structures with inversion symmetry $V_0(\vec{r}) = V_0(-\vec{r})$, the solution to equation (3.27) does not change if $\vec{r}$ and $\vec{k}$ are inverted simultaneously. Hence it is always possible to select the solution to be real at all $\vec{k}$. According to equation (3.11), $V_0(\vec{r}) = -\text{e}^2/(4\pi\varepsilon r) = -\text{e}^2/(4\pi\varepsilon\sqrt{(x^2+y^2+z^2)}) = V_0(-\vec{r})$, there is inversion symmetry for the Coulomb potential and hence the periodic function $u_{\vec{k}}$ can be real. Further following Bloch's theorem equation (3.25), we find that if $\vec{k}$ does not lie on a BZ boundary, as a complex variable, the wave function ψ is not the same as ψ^*. Hence the probability flux density

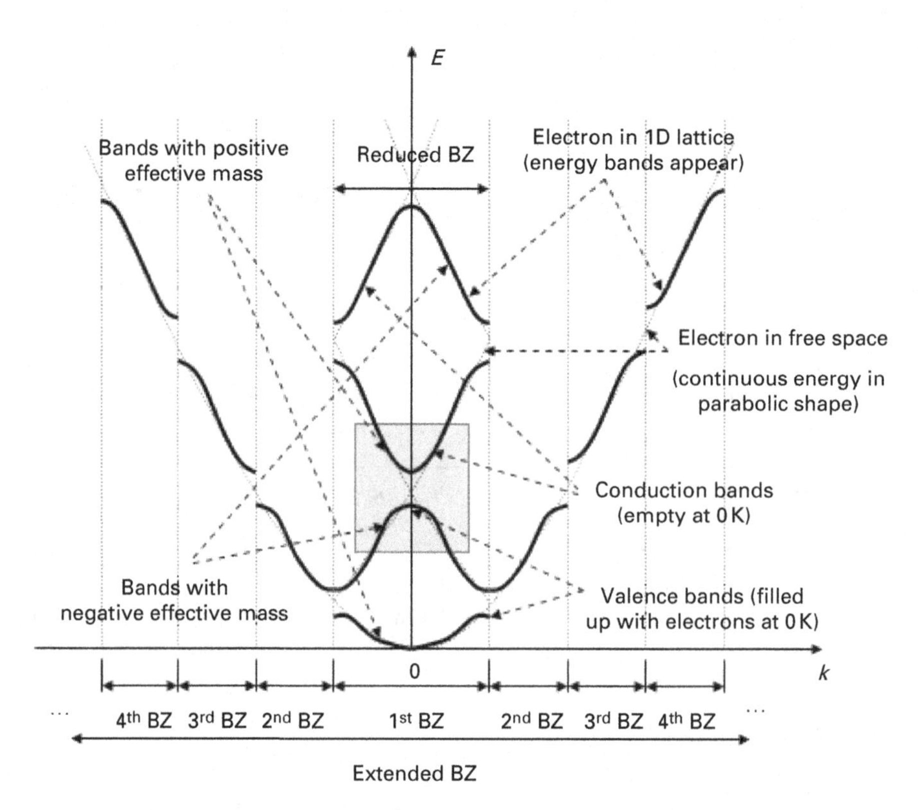

Fig. 3.2. The single electron band structure in a one-dimensional lattice and Brillouin zones. If we assume that the electrons fill up to the second band, the grey area that comprises the top valence and bottom conduction bands is the area of our interest, as the two major processes in optoelectronic devices, i.e., the electron transition and transport all happen inside this area.

defined by $\psi^*\nabla\psi - \psi\nabla\psi^*$ does not vanish. This implies that for a small perturbation (e.g., an externally applied electrical field) the single electron eigenstate and its energy will change accordingly. This change is made by mixing of the original state with states having a neighboring wave vector whose energies are arbitrarily close to the original energy. That is to say, $\varepsilon_{n\vec{k}}$ is a continuous function of $\vec{k}$. However, once $\vec{k}$ lies on a BZ boundary in a certain direction, ψ itself becomes a periodic function in the same direction according to equation (3.18) and hence it satisfies equation (3.27). Therefore, ψ can be a real function and there is no flux carried by such states in this direction. As such, a small perturbation may not be sufficient to change its original state and energy, or $\mathrm{d}\varepsilon_{n\vec{k}}/\mathrm{d}\vec{k} = 0$ in the direction along which those states have their wave vectors $\vec{k}$ on the BZ boundary. These states correspond to the extremes on the $\varepsilon_{n\vec{k}} \sim \vec{k}$ curves at which bandgaps appear once $\Delta\varepsilon|_{\Delta n} \neq 0$, as illustrated by Fig. 3.2 for the 1D case.

Although in the above derivations we have restricted $\vec{k}$ by equation (3.20), i.e., in the first BZ, Bloch's theorem can readily be expanded to any $\vec{k}$ at the reciprocal lattice. Actually, equation (3.25) can also be written as

$$\psi(\vec{r}) = \mathrm{e}^{\mathrm{j}[\vec{k}+\vec{G}(\vec{m})]\cdot\vec{r}}[\mathrm{e}^{-\mathrm{j}[\vec{G}(\vec{m})]\cdot\vec{r}}u_{\vec{k}}(\vec{r})] = \mathrm{e}^{\mathrm{j}[\vec{k}+\vec{G}(\vec{m})]\cdot\vec{r}}v_{\vec{k}+\vec{G}(\vec{m})}(\vec{r}), \tag{3.34}$$

where $v_{\vec{k}+\vec{G}(\vec{m})}(\vec{r}) \equiv \mathrm{e}^{-\mathrm{j}[\vec{G}(\vec{m})]\cdot\vec{r}} u_{\vec{k}}(\vec{r})$. From

$$v_{\vec{k}+\vec{G}(\vec{m})}[\vec{r}+\vec{R}(\vec{n})] = \mathrm{e}^{-\mathrm{j}[\vec{G}(\vec{m})]\cdot[\vec{r}+\vec{R}(\vec{n})]} u_{\vec{k}}[\vec{r}+\vec{R}(\vec{n})]$$
$$= \mathrm{e}^{-\mathrm{j}[\vec{G}(\vec{m})]\cdot\vec{r}} u_{\vec{k}}(\vec{r}) = v_{\vec{k}+\vec{G}(\vec{m})}(\vec{r}),$$

we can tell that, similarly to $u_{\vec{k}}$, $v_{\vec{k}+\vec{G}(\vec{m})}$ is still a periodic function of the lattice. Hence equation (3.34) indicates that Bloch's theorem holds for any $\vec{k}+\vec{G}(\vec{m})$. Similarly to equation (3.27), we obtain the following governing equation for $v_{\vec{k}+\vec{G}(\vec{m})}$

$$\left\{-\frac{\hbar^2}{2m_0}[\nabla + \mathrm{j}k + \mathrm{j}\vec{G}(\vec{m})]^2 + V_0(\vec{r})\right\} v_{\vec{k}+\vec{G}(\vec{m})}(\vec{r}) = \varepsilon_{\mathrm{h}} v_{\vec{k}+\vec{G}(\vec{m})}(\vec{r}). \tag{3.35}$$

By expanding periodic functions V_0 and $v_{\vec{k}+\vec{G}(\vec{m})}$ into Fourier series and substituting them into equation (3.35), we find that the Fourier components of $v_{\vec{k}+\vec{G}(\vec{m})}$ obey exactly the same equation as equation (3.33). Therefore, by tracing the smallest eigenvalue as $\vec{k}$ varies over the entire reciprocal space, we will be able to recover all the eigenvalues as $\vec{k}$ varies in the first BZ. For this reason, there are two alternative schemes to depict the band structure ε_{h}. In the first scheme (reduced BZ scheme), it is recorded as $\varepsilon_{n\vec{k}}$ where $\vec{k}$ is restricted to the first BZ as we described above. In the second scheme (extended BZ scheme), however, it is just recorded as $\varepsilon_{\vec{k}}$ but $\vec{k}$ is allowed to take on any value in the entire reciprocal space, i.e., $\vec{k} = \sum_{i=1}^{3}[m_i + (n_i/N_i)]\vec{b}_i$ with n_i as an integer defined on $[-N_i/2, N_i/2]$ and m_i as any integer for $i = 1, 2, 3$.

Although we will use the reduced BZ scheme throughout this chapter, the second scheme is still introduced here because we are attempting to establish a link between single electron behavior in a vacuum and in the bulk semiconductor. Actually, if we take away the semiconductor lattice in real space by turning off the potential energy in the Hamiltonian operator, we will obtain a well-known result, i.e., a single electron in a vacuum behaves like a free electron with a static wave function specified by $\mathrm{e}^{\mathrm{j}\vec{k}\cdot\vec{r}}/\sqrt{\Omega}$, which actually represents a plane wave if we restore the time-dependent factor $\mathrm{e}^{-\mathrm{j}\frac{\varepsilon_{\mathrm{h}}}{\hbar}t}$ which has been taken out in equation (3.3). The plane wave vector is defined as $\vec{k} = \sum_{i=1}^{3}(2\pi l_i/L_i)(\vec{b}_i/|\vec{b}_i|)$ with l_i indicating any integer and $\vec{b}_i/|\vec{b}_i|$ the unit vector in the reciprocal space for $i = 1, 2, 3$. It is obvious that we can also write the wave vector as $\vec{k} = \sum_{i=1}^{3}(l_i/N_i)\vec{b}_i = \sum_{i=1}^{3}[m_i + (n_i/N_i)]\vec{b}_i$ with n_i as an integer defined on $[-N_i/2, N_i/2]$ and m_i as any integer for $i = 1, 2, 3$. For this free electron, its energy–wave vector dependence is found as $\varepsilon_{\vec{k}} = \hbar^2|\vec{k}|^2/(2m_0)$. We can take this picture as a description into the second scheme with the wave vector $\vec{k}$ defined in the extended BZ. With the periodic potential energy turned on, the single electron wave function in the bulk semiconductor is modified to the Bloch function in equation (3.34), which is still defined in the extended BZ. The wave function is now factorized by a "global" plane wave function and a periodic "local" function. The former is just the same plane wave: a solution for a free electron in a vacuum taking the same space with the periodic potential (i.e., the semiconductor lattice structure) removed. The latter reflects the effect of the periodic potential of the semiconductor lattice structure on a single electron. In particular, at the BZ centers, i.e., $n_i = 0$ for $i = 1, 2, 3$, $\vec{k} = \vec{G}(\vec{m})$, this periodic function obeys a reduced equation having the same form as the one that governs a fully bound electron. In this sense, the first factor in the Bloch function describes the global behavior of the electron in the entire bulk semiconductor, whereas the second factor depicts the local behavior of the electron in a primitive cell perturbed by its neighboring cells (because of its $\vec{k}$ dependence). Therefore, if the periodic potential is taken away, the second factor disappears and we obtain a pure plane wave function which is indeed the solution of a free electron in a vacuum occupying the same space. On the other hand, at the BZ centers or reciprocal lattice points, the first factor disappears and we obtain a pure localized wave function which is the solution of a fully bound electron around the lattice in real space. More specifically, for Coulomb potential these bound states are similar to electron states in a hydrogen atom. This solution structure is dictated by the boundary conditions in the bulk semiconductor. The periodic boundary assumption at each facet of the bulk gives only a "loose" constraint on the allowed values taken by the wave vector $\vec{k}$. Once we allow the bulk dimensions to go to infinity this constraint is virtually removed as $\vec{k}$ can take any real value as $\Delta\vec{k} \rightarrow 0$. This condition imposes no constraint on the electron states at all, as for any given wave vector $\vec{k}$, we can always find a corresponding state with an energy given as $\varepsilon_{\vec{k}} = \hbar^2|\vec{k}|^2/(2\text{m}_0)$. The condition at the BZ boundaries because of the Bragg scattering of the periodic lattice in real space, however, gives a "tight" constraint on the allowed states. Such constraint, however, is imposed only on the states at the BZ boundaries, with no effect on the states inside the BZ. This means that $\varepsilon_{\vec{k}}$ can possibly become discrete once $\vec{k}$ takes a value at the BZ boundary, and is not affected otherwise. Hence we have the well-known bulk semiconductor band structure. Since the wave function of a single electron in a bulk semiconductor can be viewed as a periodic local function under the influence of the semiconductor lattice modulated by a global free electron wave function in a vacuum space, this electron is partially bound and partially free, as opposed to its two extremes, i.e., a fully bound electron around a lattice and a fully free electron in a vacuum. In the classical model, a single electron in a bulk semiconductor moves around a lattice and in the whole bulk space simultaneously. Consequently its motion can be seen as an orbit around a lattice (with momentum $\vec{p} = \hbar\nabla/j$) and free motion in the entire empty space (with momentum $\hbar\vec{k}$), as revealed by equation (3.27).

In bulk semiconductors, the wave vector $\vec{k}$ must be conserved modulo any reciprocal lattice vector $\vec{G}$ i.e., $k_1 + k_2 \equiv 0(\text{mod } G)$. It is analogous to ordinary momentum in free space but with the extra feature that it is conserved only within one primitive reciprocal cell, or one BZ. For an optical process, because of the negligible photon wave vector, we must have the two excitations at vectors $\vec{k}_1$ and $\vec{k}_2$ to satisfy $\vec{k}_1 + \vec{k}_2 = 0$, or $\vec{k}_1 = -\vec{k}_2$, which means that in bulk semiconductors we have to let $\vec{k}_1 = \vec{G}(\vec{m}) - \vec{k}_2$ where $\vec{G}(\vec{m}) - \vec{k}_2$ and $\vec{k}_1$ are in the same BZ. In the reduced BZ scheme, this condition is conveniently interpreted as meaning that only vertical transitions between energy bands with different index n at the same $\vec{k}$ are allowed. This is the major reason that we choose the reduced BZ scheme to model the material optical property.

To facilitate following-up derivations in k–p theory, we adopt the Dirac notation to rewrite the static Schrödinger equation (3.22) as $\vec{H}_0|\phi^b_{n\vec{k}}\rangle = \varepsilon_{n\vec{k}}|\phi^b_{n\vec{k}}\rangle$, and the Bloch equation (3.25) as $\langle\vec{r}\mid\phi^b_{n\vec{k}}\rangle = \mathrm{e}^{\mathrm{j}\vec{k}\cdot\vec{r}}\langle\vec{r}\mid n\vec{k}\rangle$, with $|\phi^b_{n\vec{k}}\rangle$ and $|n\vec{k}\rangle$ representing a single electron state in a bulk semiconductor and a periodic lattice state, respectively. The latter obeys equation (3.27), which can also be expressed as

$$\left\{-\frac{\hbar^2}{2\mathrm{m}_0}[\nabla+\mathrm{j}k]^2+V_0(\vec{r})\right\}\left|n\vec{k}\right\rangle = \varepsilon_{n\vec{k}}\left|n\vec{k}\right\rangle,$$

or

$$(\vec{H}_0+\frac{\hbar}{\mathrm{m}_0}\vec{k}\cdot\vec{p})\left|n\vec{k}\right\rangle = (\varepsilon_{n\vec{k}}-\frac{\hbar^2k^2}{2\mathrm{m}_0})\left|n\vec{k}\right\rangle. \tag{3.36}$$

Finally, we have following orthonormal conditions

$$\left\langle\phi^b_{m\vec{k}}\middle|\phi^b_{n\vec{k}'}\right\rangle = \delta_{\vec{k}\vec{k}'}\delta_{mn} \quad \text{and} \quad \left\langle m\vec{k}\middle|n\vec{k}\right\rangle = \delta_{mn}. \tag{3.37}$$

3.1.3 Solution at $\vec{k}=0$: Kane's model

At $\vec{k}=0$, equation (3.36) becomes

$$\vec{H}_0\left|n\right\rangle = (-\frac{\hbar^2\nabla^2}{2\mathrm{m}_0}-\frac{\mathrm{e}^2}{4\pi\varepsilon r})\left|n\right\rangle = \varepsilon_{n\vec{k}}\left|n\right\rangle. \tag{3.38}$$

Equation (3.38) has the same form as the equation that describes a single electron in a hydrogen atom with $\vec{r}$ denoting the difference coordinate. The solution to equation (3.38) is therefore in the form [5]

$$\left|n\right\rangle = \left|R_rY_{\theta\varphi}\right\rangle. \tag{3.39}$$

The angular dependent component in equation (3.39) follows the spherical harmonic function

$$Y^{lm}_{\theta\varphi} = \surd\left(\frac{2l+1}{4\pi}\frac{(l-|m|)!}{(l+|m|)!}\right)(-1)^{\frac{m+|m|}{2}}\mathrm{e}^{\mathrm{j}m\varphi}P_{l|m|}(\cos\theta),\ l\subset I(\text{integer}),\ m=0,\pm1,\ldots,\pm l \tag{3.40a}$$

with

$$P_{lm}(x) = \frac{1}{2^l l!}(1-x^2)^{\frac{m}{2}}\frac{\mathrm{d}^{l+m}}{\mathrm{d}x^{l+m}}(x^2-1)^l, \tag{3.40b}$$

denoting the Legendre function. The radial dependence can be expressed in the form of Laguerre polynomials [5].

The first few spherical harmonic functions are

$$l=0, \quad m=0, \quad Y_{\theta\varphi}^{00} = \frac{1}{\sqrt{(4\pi)}} \to |0,0\rangle \equiv j\,|S\rangle\,, \tag{3.41a}$$

$$l=1, \quad m=0, \quad Y_{\theta\varphi}^{10} = \sqrt{\left(\frac{3}{4\pi}\right)}\cos\theta = \sqrt{\left(\frac{3}{4\pi}\right)}\frac{z}{r} \to |1,0\rangle \equiv |Z\rangle\,, \tag{3.41b}$$

$$\begin{aligned} l=1, \quad m=\pm 1, \quad Y_{\theta\varphi}^{1\pm 1} &= \mp\sqrt{\left(\frac{3}{8\pi}\right)}\sin\theta \mathrm{e}^{\pm \mathrm{j}\varphi} \\ &= \mp\sqrt{\left(\frac{3}{8\pi}\right)}\frac{x\pm \mathrm{j}y}{r} \to |1,\pm 1\rangle \equiv \mp\frac{|X\rangle \pm \mathrm{j}\,|Y\rangle}{\sqrt{2}}, \end{aligned} \tag{3.41c}$$

with $l = 0$ and $l = 1$ denoting the single s-state and triple degenerated p-states, respectively.

According to equations (3.41a) to (3.41c), in the coordinate representation we can record the angular-dependent component in the eigenstates of equation (3.38) (i.e., the orbital angular momentum eigenstates $|l, m\rangle$) as

$$\begin{aligned} \begin{bmatrix} |0,0\rangle \\ |1,1\rangle \\ |1,-1\rangle \\ |1,0\rangle \end{bmatrix} &= \begin{bmatrix} \mathrm{j} & 0 & 0 & 0 \\ 0 & -\frac{1}{\sqrt{2}} & -\frac{\mathrm{j}}{\sqrt{2}} & 0 \\ 0 & \frac{1}{\sqrt{2}} & -\frac{\mathrm{j}}{\sqrt{2}} & 0 \\ 0 & 0 & 0 & 1 \end{bmatrix} \begin{bmatrix} |S\rangle \\ |X\rangle \\ |Y\rangle \\ |Z\rangle \end{bmatrix} \quad \text{or} \quad \begin{bmatrix} |S\rangle \\ |X\rangle \\ |Y\rangle \\ |Z\rangle \end{bmatrix} \\ &= \begin{bmatrix} -\mathrm{j} & 0 & 0 & 0 \\ 0 & -\frac{1}{\sqrt{2}} & \frac{1}{\sqrt{2}} & 0 \\ 0 & \frac{\mathrm{j}}{\sqrt{2}} & \frac{\mathrm{j}}{\sqrt{2}} & 0 \\ 0 & 0 & 0 & 1 \end{bmatrix} \begin{bmatrix} |0,0\rangle \\ |1,1\rangle \\ |1,-1\rangle \\ |1,0\rangle \end{bmatrix}. \end{aligned} \tag{3.42}$$

The eigenvalues related to the single s-state (conduction band) and triple degenerated p-states (valence band) are E_{s} and E_{p}, respectively. Under the orbital angular momentum base $[|0,0\rangle\,, |1,1\rangle\,, |1,-1\rangle\,, |1,0\rangle]^{\mathrm{T}}$ (where $[\ldots]^{\mathrm{T}}$ means taking the transpose operation), Hamiltonian $\vec{H}_0$ is diagonalized

$$\vec{H}_0 \to \begin{bmatrix} E_{\mathrm{s}} & 0 & 0 & 0 \\ 0 & E_{\mathrm{p}} & 0 & 0 \\ 0 & 0 & E_{\mathrm{p}} & 0 \\ 0 & 0 & 0 & E_{\mathrm{p}} \end{bmatrix} \begin{matrix} |0,0\rangle \\ |1,1\rangle \\ |1,-1\rangle \\ |1,0\rangle \end{matrix} \equiv \overline{H}_0. \tag{3.43}$$

When we consider the electron spin effect, however, each state will further split into two degenerated states. Therefore, under the extended base

$$[|0,0\uparrow\rangle\,, |1,1\uparrow\rangle\,, |1,-1\uparrow\rangle\,, |1,0\uparrow\rangle\,, |0,0\downarrow\rangle\,, |1,1\downarrow\rangle\,, |1,-1\downarrow\rangle\,, |1,0\downarrow\rangle]^{\mathrm{T}}, \tag{3.44}$$

the 8×8 Hamiltonian is diagonalized as

$$\vec{H}_0 \to \begin{bmatrix} \overline{H}_0 & 0 \\ 0 & \overline{H}_0 \end{bmatrix}_{8\times 8},$$

with the diagonalized 4×4 block $\overline{H}_0$ given in equation (3.43).

However, $\vec{H}_0$ is not sufficient to describe the spin effect. We must include the spin–orbit coupling in the Hamiltonian, which changes the Hamiltonian from equation (3.10) to [6]

$$\vec{H}_0 + \frac{\hbar}{4\mathrm{m}_0^2\mathrm{c}^2}(\nabla V \times \vec{p}) \cdot \vec{\sigma}, \tag{3.45}$$

where

$$\vec{\sigma} = \begin{bmatrix} 0 & 1 \\ 1 & 0 \end{bmatrix}\hat{x} + \begin{bmatrix} 0 & -\mathrm{j} \\ \mathrm{j} & 0 \end{bmatrix}\hat{y} + \begin{bmatrix} 1 & 0 \\ 0 & -1 \end{bmatrix}\hat{z}, \tag{3.46}$$

is the Pauli spin matrix.

Consequently, equation (3.36) becomes

$$[\vec{H}_0 + \frac{\hbar}{\mathrm{m}_0}\vec{k} \cdot \vec{p} + \frac{\hbar}{4\mathrm{m}_0^2\mathrm{c}^2}(\nabla V \times \vec{p}) \cdot \vec{\sigma} + \frac{\hbar}{4\mathrm{m}_0^2\mathrm{c}^2}(\nabla V \times \hbar\vec{k}) \cdot \vec{\sigma}]\left|n\vec{k}\right\rangle$$
$$= \left(\varepsilon_{n\vec{k}} - \frac{\hbar^2 k^2}{2\mathrm{m}_0}\right)\left|n\vec{k}\right\rangle. \tag{3.47}$$

The last (fourth) term in the modified Hamiltonian is relatively small in comparison with the second last (third) term because the electron global free moving momentum is much smaller than its local orbiting momentum, i.e., $\hbar\vec{k} \ll \vec{p}$. By neglecting the last term on the LHS of equation (3.47), we find

$$\left[\vec{H}_0 + \frac{\hbar}{\mathrm{m}_0}\vec{k} \cdot \vec{p} + \frac{\hbar}{4\mathrm{m}_0^2\mathrm{c}^2}(\nabla V \times \vec{p}) \cdot \vec{\sigma}\right]\left|n\vec{k}\right\rangle = \left(\varepsilon_{n\vec{k}} - \frac{\hbar^2 k^2}{2\mathrm{m}_0}\right)\left|n\vec{k}\right\rangle. \tag{3.48}$$

Because of the extra term introduced by the spin–orbit coupling, the new Hamiltonian will not be diagonalized under equation (3.44). However, under the re-ordered base

$$[|0, 0 \downarrow\rangle, |1, -1 \uparrow\rangle, |1, 0 \downarrow\rangle, |1, 1 \uparrow\rangle, |0, 0 \uparrow\rangle, |1, 1 \downarrow\rangle, |1, 0 \uparrow\rangle, |1, -1 \downarrow\rangle]^{\mathrm{T}}, \tag{3.49}$$

the 8×8 Hamiltonian can still be block-diagonalized as

$$\begin{bmatrix} \overline{H} & 0 \\ 0 & \overline{H} \end{bmatrix}_{8\times 8},$$

with the 4×4 block given as [7, 8]

$$\overline{H} = \begin{bmatrix} E_{\mathrm{s}} & 0 & kP & 0 \\ 0 & E_{\mathrm{p}} - \frac{\Delta}{3} & \frac{\sqrt{(2)}\Delta}{3} & 0 \\ kP & \frac{\sqrt{(2)}\Delta}{3} & E_{\mathrm{p}} & 0 \\ 0 & 0 & 0 & E_{\mathrm{p}} + \frac{\Delta}{3} \end{bmatrix}, \tag{3.50}$$

where $\vec{k} = k\hat{z}$ is assumed with the following definitions

$$P \equiv -\mathrm{j}\frac{\hbar}{\mathrm{m}_0}\langle S|\, p_z\, |Z\rangle\,, \tag{3.51a}$$

$$\Delta \equiv \mathrm{j}\frac{3\hbar}{4\mathrm{m}_0^2\mathrm{c}^2}\left\langle X|\frac{\partial V}{\partial x}p_y - \frac{\partial V}{\partial y}p_x|Y\right\rangle. \tag{3.51b}$$

Equation (3.50) can readily be diagonalized again. Actually, taking the reference energy such that $E_\mathrm{p} + \Delta/3 = 0$ and $E_\mathrm{s} = E_\mathrm{g}$, we find equation (3.50) to be

$$\overline{H} = \begin{bmatrix} E_\mathrm{g} & 0 & kP & 0 \\ 0 & -\frac{2\Delta}{3} & \frac{\sqrt{(2)}\Delta}{3} & 0 \\ kP & \frac{\sqrt{(2)}\Delta}{3} & -\frac{\Delta}{3} & 0 \\ 0 & 0 & 0 & 0 \end{bmatrix}. \tag{3.52}$$

The associated determinantal equation becomes $\det|\overline{H} - E_x\overline{I}| = 0$ or

$$E_x = 0, \quad \text{and} \quad E_x(E_x - E_\mathrm{g})(E_x + \Delta) - k^2P^2(E_x + 2\Delta/3) = 0. \tag{3.53}$$

At $\vec{k} = 0$, we find the solutions are $E_x = 0$, $E_x = E_\mathrm{g}$, $E_x = -\Delta$. In the neighborhood of $\vec{k} = 0$, by following a normal perturbation approach, we can find the solution to equation (3.53). Actually, by letting $E_x = 0 + \varepsilon(k^2)$ in equation (3.53), we obtain

$$\varepsilon(k^2)(-E_\mathrm{g})(\Delta) - k^2P^2(2\Delta/3) = 0, \quad \text{or} \quad \varepsilon(k^2) = -2k^2P^2/(3E_\mathrm{g}).$$

Then by letting $E_x = E_\mathrm{g} + \varepsilon(k^2)$ in equation (3.53), we obtain

$$E_\mathrm{g}\varepsilon(k^2)(E_\mathrm{g}+\Delta) - k^2P^2(E_\mathrm{g}+2\Delta/3) = 0, \text{ or } \varepsilon(k^2) = k^2P^2(E_\mathrm{g}+2\Delta/3)/[E_\mathrm{g}(E_\mathrm{g}+\Delta)].$$

Finally by letting $E_x = -\Delta + \varepsilon(k^2)$ in equation (3.53), we obtain

$$(-\Delta)(-\Delta - E_\mathrm{g})\varepsilon(k^2) - k^2P^2(-\Delta/3) = 0, \text{ or } \varepsilon(k^2) = -k^2P^2/[3(E_\mathrm{g} + \Delta)].$$

From equation (3.48), $E_x = \varepsilon_{n\vec{k}} - \hbar^2k^2/(2\mathrm{m}_0)$, hence we find:

- conduction band $\varepsilon^{\mathrm{c}}_{\vec{k}} = E_\mathrm{g} + \hbar^2k^2/(2\mathrm{m}_0) + k^2P^2(E_\mathrm{g} + 2\Delta/3)/[E_\mathrm{g}(E_\mathrm{g} + \Delta)]$, (3.54a)
- heavy-hole band $\varepsilon^{\mathrm{hh}}_{\vec{k}} = \hbar^2k^2/(2\mathrm{m}_0)$, (3.54b)
- light-hole band $\varepsilon^{\mathrm{lh}}_{\vec{k}} = \hbar^2k^2/(2\mathrm{m}_0) - 2k^2P^2/(3E_\mathrm{g})$, (3.54c)
- spin–orbit split band $\varepsilon^{\mathrm{so}}_{\vec{k}} = -\Delta + \hbar^2k^2/(2\mathrm{m}_0) - k^2P^2/[3(E_\mathrm{g} + \Delta)]$. (3.54d)

To obtain the associated eigenvectors, by using the upper Hamiltonian block in equation (3.52), we have

$$\begin{bmatrix} E_\mathrm{g} - E_x & 0 & kP & 0 \\ 0 & -\frac{2\Delta}{3} - E_x & \frac{\sqrt{(2)}\Delta}{3} & 0 \\ kP & \frac{\sqrt{(2)}\Delta}{3} & -\frac{\Delta}{3} - E_x & 0 \\ 0 & 0 & 0 & -E_x \end{bmatrix}\begin{bmatrix} a_n \\ b_n \\ c_n \\ d_n \end{bmatrix} = 0. \tag{3.55}$$

From equation (3.55), we find that the heavy-hole band (i.e., $|1, 1 \uparrow\rangle$, with its energy eigenvalue 0 at $\vec{k} = 0$) is still decoupled from the other bands. However, the conduction band (i.e., $|0, 0 \downarrow\rangle$, with its energy eigenvalue E_g at $\vec{k} = 0$), the light-hole band (i.e., $|1, -1 \uparrow\rangle$, with its energy eigenvalue 0 at $\vec{k} = 0$), and the spin–orbit split band (i.e., $|1, 0 \downarrow\rangle$, with its energy eigenvalue $-\Delta$ at $\vec{k} = 0$) are coupled under the spin–orbit interaction. Therefore, equation (3.55) is reduced to

$$\begin{bmatrix} E_g - E_x & 0 & kP \\ 0 & -\frac{2\Delta}{3} - E_x & \frac{\sqrt{(2)}\Delta}{3} \\ kP & \frac{\sqrt{(2)}\Delta}{3} & -\frac{\Delta}{3} - E_x \end{bmatrix} \begin{bmatrix} a_n \\ b_n \\ c_n \end{bmatrix} = 0. \tag{3.56}$$

In the neighborhood of $\vec{k} = 0$ and under the normalization condition $a_n^2 + b_n^2 + c_n^2 = 1$, we find:

- Conduction band $\begin{bmatrix} 0 & 0 & 0 \\ 0 & -\frac{2\Delta}{3} - E_g & \frac{\sqrt{(2)}\Delta}{3} \\ 0 & \frac{\sqrt{(2)}\Delta}{3} & -\frac{\Delta}{3} - E_g \end{bmatrix} \begin{bmatrix} a_c \\ b_c \\ c_c \end{bmatrix} = 0,$
or $a_c = 1, b_c = c_c = 0$.
- Light-hole band $\begin{bmatrix} E_g & 0 & 0 \\ 0 & -\frac{2\Delta}{3} & \frac{\sqrt{(2)}\Delta}{3} \\ 0 & \frac{\sqrt{(2)}\Delta}{3} & -\frac{\Delta}{3} \end{bmatrix} \begin{bmatrix} a_{lh} \\ b_{lh} \\ c_{lh} \end{bmatrix} = 0,$
or $a_{lh} = 0, b_{lh} = \frac{1}{\sqrt{3}}, c_{lh} = \sqrt{\frac{2}{3}}$.
- Spin–orbit split band $\begin{bmatrix} E_g + \Delta & 0 & 0 \\ 0 & \frac{\Delta}{3} & \frac{\sqrt{(2)}\Delta}{3} \\ 0 & \frac{\sqrt{(2)}\Delta}{3} & \frac{2\Delta}{3} \end{bmatrix} \begin{bmatrix} a_{so} \\ b_{so} \\ c_{so} \end{bmatrix} = 0,$
or $a_{so} = 0, b_{so} = \sqrt{\frac{2}{3}}, c_{so} = -\frac{1}{\sqrt{3}}$.

Therefore, the Hamiltonian block equation (3.52) is diagonalized to

$$\begin{bmatrix} E_g + \frac{\hbar^2 k^2}{2m_0} + \frac{k^2 P^2 (E_g + 2\Delta/3)}{E_g(E_g + \Delta)} & 0 & 0 & 0 \\ 0 & -\Delta + \frac{\hbar^2 k^2}{2m_0} - \frac{k^2 P^2}{3(E_g + \Delta)} & 0 & 0 \\ 0 & 0 & \frac{\hbar^2 k^2}{2m_0} - \frac{2k^2 P^2}{3E_g} & 0 \\ 0 & 0 & 0 & \frac{\hbar^2 k^2}{2m_0} \end{bmatrix}$$

under the new base subset

$$\left[|0, 0 \downarrow\rangle, \frac{1}{\sqrt{3}}|1, -1 \uparrow\rangle + \sqrt{\left(\frac{2}{3}\right)}|1, 0 \downarrow\rangle, \sqrt{\left(\frac{2}{3}\right)}|1, -1 \uparrow\rangle - \frac{1}{\sqrt{3}}|1, 0 \downarrow\rangle, |1, 1 \uparrow\rangle\right]^T. \tag{3.57a}$$

A similar approach applies to the lower Hamiltonian block in equation (3.52) as well. Hence we obtain the same diagonalized Hamiltonian under following base subset

$$\left[|0, 0 \uparrow\rangle, \frac{1}{\sqrt{3}}|1, 1 \downarrow\rangle + \sqrt{\left(\frac{2}{3}\right)}|1, 0 \uparrow\rangle, \sqrt{\left(\frac{2}{3}\right)}|1, 1 \downarrow\rangle - \frac{1}{\sqrt{3}}|1, 0 \uparrow\rangle, |1, -1 \downarrow\rangle\right]^T. \tag{3.57b}$$

We name the new base as:

- conduction band energy $\varepsilon^{c}_{\vec{k}=0} = E_g$, and state

$$|0, 0 \downarrow\uparrow\rangle = \mathrm{j}\,|S \downarrow\uparrow\rangle\,, \tag{3.58a}$$

- heavy-hole band energy $\varepsilon^{hh}_{\vec{k}=0} = 0$, and state

$$\begin{aligned}\left|\frac{3}{2}, \frac{3}{2}\right\rangle &\equiv |1, 1 \uparrow\rangle = -\frac{1}{\sqrt{2}}(|X \uparrow\rangle + \mathrm{j}\,|Y \uparrow\rangle)\\ \left|\frac{3}{2}, -\frac{3}{2}\right\rangle &\equiv |1, -1 \downarrow\rangle = \frac{1}{\sqrt{2}}(|X \downarrow\rangle - \mathrm{j}\,|Y \downarrow\rangle),\end{aligned} \tag{3.58b}$$

- light-hole band energy $\varepsilon^{lh}_{\vec{k}=0} = 0$, and state

$$\begin{aligned}\left|\frac{3}{2}, -\frac{1}{2}\right\rangle &\equiv \frac{1}{\sqrt{3}}|1, -1 \uparrow\rangle + \sqrt{\left(\frac{2}{3}\right)}|1, 0 \downarrow\rangle = \frac{1}{\sqrt{6}}(|X \uparrow\rangle - \mathrm{j}\,|Y \uparrow\rangle) + \sqrt{\left(\frac{2}{3}\right)}|Z \downarrow\rangle\\ \left|\frac{3}{2}, \frac{1}{2}\right\rangle &\equiv \frac{1}{\sqrt{3}}|1, 1 \downarrow\rangle + \sqrt{\left(\frac{2}{3}\right)}|1, 0 \uparrow\rangle = -\frac{1}{\sqrt{6}}(|X \downarrow\rangle + \mathrm{j}\,|Y \downarrow\rangle) + \sqrt{\left(\frac{2}{3}\right)}|Z \uparrow\rangle\,,\end{aligned} \tag{3.58c}$$

- spin–orbit split band energy $\varepsilon^{so}_{\vec{k}=0} = -\Delta$, and state

$$\begin{aligned}\left|\frac{1}{2}, -\frac{1}{2}\right\rangle &\equiv \sqrt{\left(\frac{2}{3}\right)}|1, -1 \uparrow\rangle - \frac{1}{\sqrt{3}}|1, 0 \downarrow\rangle = \frac{1}{\sqrt{3}}(|X \uparrow\rangle - \mathrm{j}\,|Y \uparrow\rangle) - \frac{1}{\sqrt{3}}|Z \downarrow\rangle\\ \left|\frac{1}{2}, \frac{1}{2}\right\rangle &\equiv \sqrt{\left(\frac{2}{3}\right)}|1, 1 \downarrow\rangle - \frac{1}{\sqrt{3}}|1, 0 \uparrow\rangle = -\frac{1}{\sqrt{3}}(|X \downarrow\rangle + \mathrm{j}\,|Y \downarrow\rangle) - \frac{1}{\sqrt{3}}|Z \uparrow\rangle\,.\end{aligned} \tag{3.58d}$$

Unfortunately, this approach is oversimplified and hence its result is not accurate. For example, equation (3.54b) obviously gives the wrong sign of the heavy-hole mass. The inaccuracy comes from the truncation of the eigenstates: there are an infinite number of eigenstates in the angular momentum base $|l, m\rangle$, but we have considered only the first four (or eight if we have the electron spin effect included) in dealing with equation (3.38). Although such a truncation has little effect on the solution at $\vec{k} = 0$, it spoils the perturbation solution obtained in the neighborhood of $\vec{k} = 0$, as the rest of the eigenstates that have been ignored in this analysis indeed have significant effect on those retained states at $\vec{k} \neq 0$.

3.1.4 Solution at $\vec{k} \neq 0$: Luttinger–Kohn's model

Since the above solution is correct only at $\vec{k} = 0$, for $\vec{k} \neq 0$, we attempt to modify the perturbation approach by including the effect of the rest of the bands in the full solution to equation (3.48) other than the conduction, heavy-hole, light-hole, and spin–orbit split bands. For this purpose, Lowdin's renormalization method [9] is employed where we assign the strongly coupled bands to one group (group A) and leave the weakly coupled bands in another (group B). The bands in group A will be treated simultaneously as the fundamental (unperturbed) component based on the result obtained at $\vec{k} = 0$ from Kane's model, while the bands in group B will be treated as a perturbation to the former. An iteration algorithm can be established to refine the result to the required accuracy.

In k–p theory, we attempt to obtain the solution in the neighborhood of $\vec{k} = 0$, hence we identify

$$\vec{H}' \equiv (\hbar/\mathrm{m}_0)\vec{k} \cdot \vec{p} \tag{3.59}$$

in (3.48) as a perturbation to the Hamiltonian operator at $\vec{k} = 0$. As such, equation (3.48) can be rewritten as

$$\left[\vec{H}_0 + \frac{\hbar}{4\mathrm{m}_0^2\mathrm{c}^2}(\nabla V \times \vec{p}) \cdot \vec{\sigma} + \frac{\hbar^2 k^2}{2\mathrm{m}_0} + \vec{H}'\right] \left|n\vec{k}\right\rangle = \varepsilon_{n\vec{k}} \left|n\vec{k}\right\rangle. \tag{3.60}$$

We expand the solution to equation (3.60) in terms of the solution obtained at $\vec{k} = 0$, i.e., the states given in equation (3.58a–d) plus those that have been truncated, and separate them into two groups [10, 11]

$$\left|n\vec{k}\right\rangle = \sum_{\mathrm{j}}^{\mathrm{A}} a_{\mathrm{j}}^{\mathrm{A}}(\vec{k}) \left|n_{\mathrm{j}}^{\mathrm{A}}\right\rangle + \sum_{v}^{\mathrm{B}} a_{v}^{\mathrm{B}}(\vec{k}) \left|n_{v}^{\mathrm{B}}\right\rangle, \tag{3.61}$$

with group A containing the following six strongly coupled valence bands, i.e., the two degenerated (at $\vec{k} = 0$) heavy-hole and two degenerated light-hole bands, and the two spin–orbit split bands

$$\begin{aligned} &\left|n_1^{\mathrm{A}}\right\rangle \equiv \left|\frac{3}{2}, \frac{3}{2}\right\rangle \quad \left|n_2^{\mathrm{A}}\right\rangle \equiv \left|\frac{3}{2}, \frac{1}{2}\right\rangle \quad \left|n_3^{\mathrm{A}}\right\rangle \equiv \left|\frac{3}{2}, -\frac{1}{2}\right\rangle \\ &\left|n_4^{\mathrm{A}}\right\rangle \equiv \left|\frac{3}{2}, -\frac{3}{2}\right\rangle \quad \left|n_5^{\mathrm{A}}\right\rangle \equiv \left|\frac{1}{2}, \frac{1}{2}\right\rangle \quad \left|n_6^{\mathrm{A}}\right\rangle \equiv \left|\frac{1}{2}, -\frac{1}{2}\right\rangle, \end{aligned} \tag{3.62}$$

and group B containing all remaining bands, respectively.

From Section 3.1.3, we know that

$$\vec{H}_{\vec{k}=0} \left|n_j^{\mathrm{A}}\right\rangle \equiv \left[\vec{H}_0 + \frac{\hbar}{4\mathrm{m}_0^2\mathrm{c}^2}(\nabla V \times \vec{p}) \cdot \vec{\sigma}\right] \left|n_j^{\mathrm{A}}\right\rangle = \varepsilon_j^{\mathrm{A}} \left|n_j^{\mathrm{A}}\right\rangle, \tag{3.63}$$

$$\text{with subscript } j = 1, 2, 3, 4, 5, 6, \; \varepsilon_1^{\mathrm{A}} = \varepsilon_2^{\mathrm{A}} = \varepsilon_3^{\mathrm{A}} = \varepsilon_4^{\mathrm{A}} = 0, \text{ and } \varepsilon_5^{\mathrm{A}} = \varepsilon_6^{\mathrm{A}} = -\Delta. \tag{3.64}$$

The bands in group B ($|n_v^{\mathrm{B}}\rangle$) are unknown, so are the expanding coefficients in both groups ($a_{\mathrm{j}}^{\mathrm{A}}$ and (a_v^{B}).

From Lowdin's theory in Appendix A, coefficients in group A are obtained by solving the following linear algebraic equations

$$\sum_{j=1}^{6} (U_{ij}^{\mathrm{A}} - E\delta_{ij}) a_j^{\mathrm{A}}(\vec{k}) = 0, \tag{3.65}$$

with $i = 1, 2, 3, 4, 5, 6$ belonging to group A. To first order accuracy we have

$$U_{ij}^{\mathrm{A}} \approx (\vec{H}_1)_{ij} + \sum_{v \neq i,j}^{\mathrm{B}} \frac{(\vec{H}_1)_{iv}(\vec{H}_1)_{vj}}{E - (\vec{H}_1)_{vv}}, \tag{3.66}$$

where

$$\vec{H}_1 \equiv \vec{H}_{\vec{k}=0} + \frac{\hbar^2 k^2}{2\mathrm{m}_0} + \vec{H}'. \tag{3.67}$$

Noting that group A bands are either degenerate or almost degenerate (since $\Delta \sim 0$) at $\vec{k} = 0$, as $\vec{k} \neq 0$, E in expression (3.66) can be replaced by the averaged eigenvalue of those group A bands, i.e., $-\Delta/3$.

For states in group A, $i, j \subset A$, we have

$$(\vec{H}_1)_{ij} = (\vec{H}_{\vec{k}=0})_{ij} + \left(\frac{\hbar^2 k^2}{2\mathrm{m}_0}\right)_{ij} + (\vec{H}')_{ij} = (\varepsilon_j^{\mathrm{A}} + \frac{\hbar^2 k^2}{2\mathrm{m}_0})\delta_{ij}, \tag{3.68}$$

since $(\vec{H}')_{ij} = \left(\frac{\hbar}{\mathrm{m}_0}\vec{k} \cdot \vec{p}\right)_{ij} = 0.$.

For states in group B, $v \subset B \neq i, j$, we also have

$$(\vec{H}_{\vec{k}=0})_{iv} = (\vec{H}_{\vec{k}=0})_{vj} = \left[\frac{\hbar^2 k^2}{2\mathrm{m}_0}\right]_{iv} = \left[\frac{\hbar^2 k^2}{2\mathrm{m}_0}\right]_{vi} = 0, \tag{3.69}$$

because of the orthogonality among different eigenstates.

Moreover, we find

$$(\vec{H}')_{iv} = \left(\frac{\hbar}{\mathrm{m}_0}\vec{k} \cdot \vec{p}\right)_{iv} = \sum_{\alpha=x,y,z} \frac{\hbar k_\alpha}{\mathrm{m}_0} p_{iv}^{\alpha}$$

$$(\vec{H}')_{vj} = \left(\frac{\hbar}{\mathrm{m}_0}\vec{k} \cdot \vec{p}\right)_{vj} = \sum_{\alpha=x,y,z} \frac{\hbar k_\alpha}{\mathrm{m}_0} p_{vj}^{\alpha}, \tag{3.70}$$

for $i, j \subset A, v \subset B \neq i, j$, and $p_{iv}^{\alpha} \equiv \langle n_i^{\mathrm{A}}| \, p_\alpha \, |n_v^{\mathrm{B}}\rangle$, $p_{vj}^{\alpha} \equiv \langle n_v^{\mathrm{B}}| \, p_\alpha \, |n_j^{\mathrm{A}}\rangle$, with $\alpha = x, y, z$ and

$$\vec{p} \equiv \sum_{\alpha=x,y,z} \vec{p}_\alpha = p_x\hat{x} + p_y\hat{y} + p_z\hat{z}. \tag{3.71}$$

If we further define

$$\varepsilon_v^{\mathrm{B}} \equiv (\vec{H}_1)_{vv}, \tag{3.72}$$

equation (3.66) is reduced to

$$\begin{aligned} U_{ij}^{\mathrm{A}} &\approx \left(\varepsilon_j^{\mathrm{A}} + \frac{\hbar^2 k^2}{2\mathrm{m}_0}\right)\delta_{ij} + \sum_{v\neq i,j}^{\mathrm{B}} \frac{(\vec{H}')_{iv}(\vec{H}')_{vj}}{(-\Delta/3) - \varepsilon_v^{\mathrm{B}}} \\ &= \left(\varepsilon_j^{\mathrm{A}} + \frac{\hbar^2 k^2}{2\mathrm{m}_0}\right)\delta_{ij} + \frac{\hbar^2}{\mathrm{m}_0^2}\sum_{v}^{\mathrm{B}} \frac{1}{(-\Delta/3) - \varepsilon_v^{\mathrm{B}}} \sum_{\alpha,\beta=x,y,z} k_\alpha k_\beta p_{iv}^{a} p_{vj}^{\beta} \\ &= \left(\varepsilon_j^{\mathrm{A}} + \frac{\hbar^2 k^2}{2\mathrm{m}_0}\right)\delta_{ij} + \sum_{\alpha,\beta=x,y,z}\left[\frac{\hbar^2}{\mathrm{m}_0^2}\sum_{v}^{\mathrm{B}} \frac{p_{iv}^{a} p_{vj}^{\beta}}{(-\Delta/3) - \varepsilon_v^{\mathrm{B}}}\right] k_\alpha k_\beta, \end{aligned} \tag{3.73}$$

where equations (3.68) to (3.72) are used.

Under the following definitions

$$A \equiv \frac{\hbar^2}{2\mathrm{m}_0} + \frac{\hbar^2}{\mathrm{m}_0^2}\sum_{v}^{\mathrm{B}} \frac{\langle X|\, p_x\, |v\rangle \langle v|\, p_x\, |X\rangle}{(-\Delta/3) - \varepsilon_v^{\mathrm{B}}}, \tag{3.74a}$$

$$B \equiv \frac{\hbar^2}{2\mathrm{m}_0} + \frac{\hbar^2}{\mathrm{m}_0^2}\sum_{v}^{\mathrm{B}} \frac{\langle X|\, p_y\, |v\rangle \langle v|\, p_y\, |X\rangle}{(-\Delta/3) - \varepsilon_v^{\mathrm{B}}}, \tag{3.74b}$$

$$C \equiv \frac{\hbar^2}{\mathrm{m}_0^2}\sum_{v}^{\mathrm{B}} \frac{\langle X|\, p_x\, |v\rangle \langle v|\, p_y\, |Y\rangle + \langle X|\, p_y\, |v\rangle \langle v|\, p_x\, |Y\rangle}{(-\Delta/3) - \varepsilon_v^{\mathrm{B}}}, \tag{3.74c}$$

and

$$-\frac{\hbar^2}{2\mathrm{m}_0}\gamma_1 \equiv \frac{1}{3}(A + 2B), \tag{3.75a}$$

$$-\frac{\hbar^2}{2\mathrm{m}_0}\gamma_2 \equiv \frac{1}{6}(A - B), \tag{3.75b}$$

$$-\frac{\hbar^2}{2\mathrm{m}_0}\gamma_3 \equiv \frac{1}{6}C, \tag{3.75c}$$

Equation (3.65) can be recorded in matrix form [12]

$$-\begin{bmatrix} P+Q & -S & R & 0 & -\frac{S}{\sqrt{2}} & \sqrt{(2)}R \\ -S^* & P-Q & 0 & R & -\sqrt{(2)}Q & \sqrt{\left(\frac{3}{2}\right)}S \\ R^* & 0 & P-Q & S & \sqrt{\left(\frac{3}{2}\right)}S^* & \sqrt{(2)}Q \\ 0 & R^* & S^* & P+Q & -\sqrt{(2)}R^* & -\frac{S^*}{\sqrt{2}} \\ -\frac{S^*}{\sqrt{2}} & -\sqrt{(2)}Q^* & \sqrt{\left(\frac{3}{2}\right)}S & -\sqrt{(2)}R & P+\Delta & 0 \\ \sqrt{(2)}R^* & \sqrt{\left(\frac{3}{2}\right)}S^* & \sqrt{(2)}Q^* & -\frac{S}{\sqrt{2}} & 0 & P+\Delta \end{bmatrix} \begin{bmatrix} a_1^{\mathrm{A}} \\ a_2^{\mathrm{A}} \\ a_3^{\mathrm{A}} \\ a_4^{\mathrm{A}} \\ a_5^{\mathrm{A}} \\ a_6^{\mathrm{A}} \end{bmatrix} = E \begin{bmatrix} a_1^{\mathrm{A}} \\ a_2^{\mathrm{A}} \\ a_3^{\mathrm{A}} \\ a_4^{\mathrm{A}} \\ a_5^{\mathrm{A}} \\ a_6^{\mathrm{A}} \end{bmatrix}, \tag{3.76}$$

where we have introduced

$$P \equiv \frac{\hbar^2 \gamma_1}{2m_0}(k_x^2 + k_y^2 + k_z^2), \tag{3.77a}$$

$$Q \equiv \frac{\hbar^2 \gamma_2}{2m_0}(k_x^2 + k_y^2 - 2k_z^2), \tag{3.77b}$$

$$R \equiv \frac{\sqrt{(3)}\hbar^2}{2m_0}[-\gamma_2(k_x^2 - k_y^2) + j2\gamma_3 k_x k_y], \tag{3.77c}$$

$$S \equiv \frac{\sqrt{3}\hbar^2 \gamma_3}{m_0}(k_x - jk_y)k_z, \tag{3.77d}$$

as the energies that compose the Luttinger–Kohn Hamiltonian matrix elements U_{ij}^{A} in the wave vector (momentum) domain.

In these expressions, parameters $\gamma_{1,2,3}$ are known as the material Luttinger band structure constants. They are normally found by experiment rather than through equations (3.75a–c) and (3.74a–c). In numerical implementations, these parameters can be found from a semiconductor material database such as [13, 14].

By solving equation (3.76), we will be able to obtain the eigenvalues E_n as $\varepsilon_{n\vec{k}}$ in equation (3.60) and the eigenstate coefficients a_{jn}^{A} with $j = 1, 2, 3, 4, 5, 6$ and $n = 1, 2, 3, 4, 5, 6$ (due to the double degeneracy, there are only three different eigenvalues) under the following normalization condition

$$\sum_{j=1}^{6} |a_{jn}^{\mathrm{A}}(\vec{k})|^2 = 1, \tag{3.78}$$

with $n = 1, 2, 3, 4, 5, 6$.

Therefore, we obtain the valence band energies ($\varepsilon_{n\vec{k}}$) and the corresponding lattice states in the neighborhood of $\vec{k} = 0$ in the form of

$$\left|n\vec{k}\right\rangle \approx \sum_{j=1}^{6} a_{jn}^{\mathrm{A}}(\vec{k}) \left|n_j^{\mathrm{A}}\right\rangle, \tag{3.79a}$$

with the base $\left|n_j^{\mathrm{A}}\right\rangle$ for $j = 1, 2, 3, 4, 5, 6$ given in equations (3.62) and (3.58b–d) as the eigensolution at $\vec{k} = 0$. Finally, following Bloch's theorem, the normalized bulk semiconductor valence band electron (hole) wave function corresponding to $\varepsilon_{n\vec{k}}$ is given as

$$\left\langle \vec{r}/\phi_{n\vec{k}}^{\mathrm{bv}} \right\rangle = \frac{e^{j\vec{k}\cdot\vec{r}}}{\sqrt{\Omega}} \left\langle \vec{r}/n\vec{k} \right\rangle = \frac{e^{j\vec{k}\cdot\vec{r}}}{\sqrt{\Omega}} \sum_{j=1}^{6} a_{jn}^{\mathrm{A}}(\vec{k}) \left\langle \vec{r}/n_j^{\mathrm{A}} \right\rangle, \tag{3.79b}$$

where $n = 1, 2, 3, 4, 5, 6$.

If we let group A contain only a single band $\left|n_0^{\mathrm{A}}\right\rangle \equiv |0, 0\downarrow\rangle$ or $\left|n_0^{\mathrm{A}}\right\rangle \equiv |0, 0\uparrow\rangle$, we still have equation (3.63) with $\varepsilon_0^{\mathrm{A}} = E_g$. Again from Lowdin's theory, the group A coefficient is obtained from

$$(U_{00}^{\mathrm{A}} - E\delta_{00})a_0^{\mathrm{A}}(\vec{k}) = 0, \tag{3.80}$$

where to first order accuracy we have

$$\begin{aligned} E = U_{00}^{\mathrm{A}} &\approx (\vec{H}_1)_{00} + \sum_{v\neq 0}^{\mathrm{B}} \frac{(\vec{H}_1)_{0v}(\vec{H}_1)_{v0}}{E - (\vec{H}_1)_{vv}}, \\ &= E_{\mathrm{g}} + \frac{\hbar^2 k^2}{2\mathrm{m}_0} + \frac{\hbar^2}{\mathrm{m}_0^2} \sum_{v}^{\mathrm{B}} \frac{\sum_{\alpha,\beta=x,y,z} k_\alpha k_\beta p_{0v}^{a} p_{v0}^{\beta}}{E_{\mathrm{g}} - \varepsilon_v^{\mathrm{B}}}, \\ &= E_{\mathrm{g}} + \frac{\hbar^2 k^2}{2m_e}. \end{aligned} \tag{3.81}$$

In equation (3.81), we have defined the conduction band effective mass as

$$\frac{1}{m_e} \equiv \frac{1}{\mathrm{m}_0}\left(1 + \frac{2}{\mathrm{m}_0 k^2} \sum_{v}^{\mathrm{B}} \frac{\sum_{\alpha,\beta=x,y,z} k_\alpha k_\beta p_{0v}^{a} p_{v0}^{\beta}}{E_{\mathrm{g}} - \varepsilon_v^{\mathrm{B}}}\right). \tag{3.82}$$

From the normalization condition we find $a_0^{\mathrm{A}}(\vec{k}) = 1$. Therefore, we obtain the conduction band energy E, i.e., $\varepsilon_{n\vec{k}}$ in equation (3.60), as given in equation (3.81) and the double degenerated lattice state in the neighborhood of $\vec{k} = 0$ in the form

$$\left|n\vec{k}\right\rangle \approx \left|n_0^{\mathrm{A}}\right\rangle = |0, 0\downarrow\rangle \text{ and } |0, 0\uparrow\rangle, \tag{3.83a}$$

with the base $\left|n_0^{\mathrm{A}}\right\rangle$ given in equation (3.58a) as the eigensolution at $\vec{k} = 0$. Finally, following Bloch's theorem, the normalized bulk semiconductor conduction band electron wave function is given as

$$\left\langle \vec{r} \mid \phi_{n\vec{k}}^{\mathrm{bc}} \right\rangle = \frac{\mathrm{e}^{\mathrm{j}\vec{k}\cdot\vec{r}}}{\sqrt{\Omega}} \langle \vec{r} \mid n\vec{k} \rangle = \frac{\mathrm{e}^{\mathrm{j}\vec{k}\cdot\vec{r}}}{\sqrt{\Omega}} \left\langle \vec{r} \mid n_0^{\mathrm{A}} \right\rangle = \frac{\mathrm{e}^{\mathrm{j}\vec{k}\cdot\vec{r}}}{\sqrt{\Omega}} \langle \vec{r} \mid 0, 0\downarrow\uparrow\rangle. \tag{3.83b}$$

In summary, for bulk material, the conduction band (double degenerated sub-bands with opposite spin) energy is given in equation (3.81) with the effective mass defined by equation (3.82). The valence band (three pairs of double degenerated sub-bands at $\vec{k} = 0$, total six sub-bands) energies are obtained by solving the eigenvalue equation (3.76).

The conduction band electron lattice state is given by equation (3.83a), which is the same as the eigenstate at $\vec{k} = 0$, and is also the same as the eigenstate when there is no spin–orbit interaction. This eigenstate is given in equations (3.41a) through (3.58a), with its space domain angular distribution taking the same form as an s-orbiting electron in a hydrogen atom, as shown in Fig. 3.3a.

The valence band electron (hole) lattice states are given by equation (3.79a), which are all linear combinations of eigenstates at $\vec{k} = 0$.The coefficients are obtained by

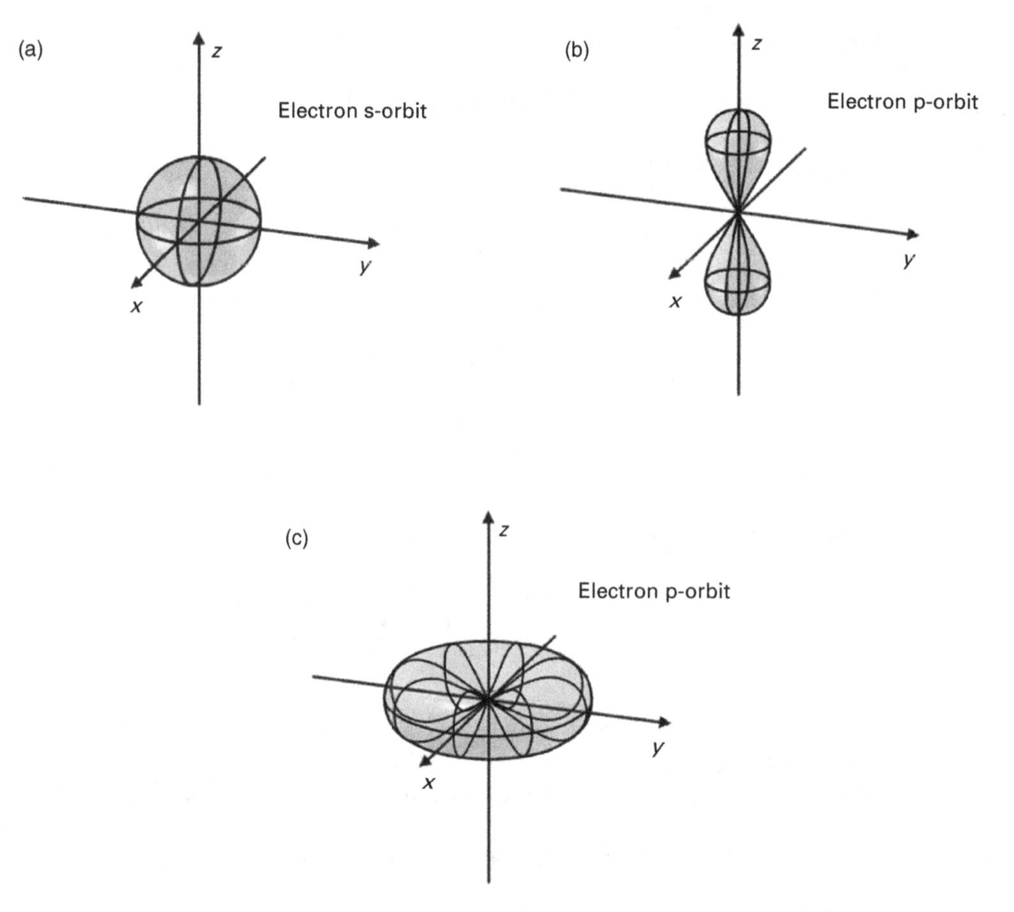

Fig. 3.3. Single electron space domain angular distributions in centralized Coulomb attractive field. (a) s-orbit. (b) and (c) p-orbit.

solving the linear algebraic equations (3.76) once the eigenvalues are found, subject to the normalization condition of equation (3.78). Those eigenstates at $\vec{k} = 0$ are given as equations (3.58b–d) through (3.62). They are again the linear superposition of eigenstates at $\vec{k} = 0$ when there is no spin–orbit interaction. The latter are expressed as equations (3.41b&c), with their space domain angular distributions taking the same form as a triple degenerated p-orbiting electron in a hydrogen atom, as shown in Fig. 3.3b&c.

3.1.5 Solution under 4 × 4 Hamiltonian and axial approximation

The above k–p theory under Luttinger–Kohn's implementation (6 × 6 Hamiltonian) can readily be expanded to have more bands included in group A, to form, e.g., an 8 × 8 Hamiltonian in dealing with semiconductors with a narrow bandgap where the coupling between the conduction and valence bands is strong, or even a 10 × 10 Hamiltonian in dealing with diluted nitrides where an extra nitrogen resonant band appears in the neighborhood of the conduction band edge [15]. For GaAs and InP based semiconductors, the spin–orbit split bands have an appreciably different energy from the degenerated heavy- and light-hole bands at $\vec{k} = 0$, i.e., Δ is significant. Therefore, we need just to deal with a reduced 4×4 Hamiltonian with the two spin–orbit split bands excluded. We can further block diagonalize the 4×4 Hamiltonian to obtain two decoupled 2×2 Hamiltonians, through which an analytical solution is obtained.

Actually, by dropping the two spin–orbit split bands, we find the following 4×4 Luttinger–Kohn Hamiltonian from equation (3.76)

$$\overline{H}^{\mathrm{LK}} \begin{bmatrix} a_1^{\mathrm{A}} \\ a_2^{\mathrm{A}} \\ a_3^{\mathrm{A}} \\ a_4^{\mathrm{A}} \end{bmatrix} \equiv - \begin{bmatrix} P+Q & -S & R & 0 \\ -S^* & P-Q & 0 & R \\ R^* & 0 & P-Q & S \\ 0 & R^* & S^* & P+Q \end{bmatrix} \begin{bmatrix} a_1^{\mathrm{A}} \\ a_2^{\mathrm{A}} \\ a_3^{\mathrm{A}} \\ a_4^{\mathrm{A}} \end{bmatrix} = E \begin{bmatrix} a_1^{\mathrm{A}} \\ a_2^{\mathrm{A}} \\ a_3^{\mathrm{A}} \\ a_4^{\mathrm{A}} \end{bmatrix}. \tag{3.84}$$

The 4×4 Hamiltonian $\overline{H}^{\mathrm{LK}}$ in equation (3.84) can be transformed into

$$\overline{H} = \overline{U}\,\overline{H}^{\mathrm{LK}}\,\overline{U}^{+} = - \begin{bmatrix} P+Q & R' & 0 & 0 \\ R'^{*} & P-Q & 0 & 0 \\ 0 & 0 & P-Q & R' \\ 0 & 0 & R'^{*} & P+Q \end{bmatrix}, \tag{3.85}$$

where

$$R' \equiv |R| - \mathrm{j}|S|, \quad R'^{*} \equiv |R| + \mathrm{j}|S|, \quad R \equiv |R|\mathrm{e}^{\mathrm{j}\theta_R}, \quad S \equiv |S|\mathrm{e}^{\mathrm{j}\theta_S}. \tag{3.86}$$

Under this transformation, the old base is linked to the new base through

$$\begin{bmatrix} |n_1^{\mathrm{A}}\rangle \\ |n_2^{\mathrm{A}}\rangle \\ |n_3^{\mathrm{A}}\rangle \\ |n_4^{\mathrm{A}}\rangle \end{bmatrix} \equiv \overline{U}^{+} \begin{bmatrix} |1\rangle \\ |2\rangle \\ |3\rangle \\ |4\rangle \end{bmatrix} = \begin{bmatrix} \alpha & 0 & 0 & \alpha \\ 0 & -\beta^* & \beta^* & 0 \\ 0 & \beta & \beta & 0 \\ -\alpha^* & 0 & 0 & \alpha^* \end{bmatrix} \begin{bmatrix} |1\rangle \\ |2\rangle \\ |3\rangle \\ |4\rangle \end{bmatrix}, \tag{3.87}$$

where

$$\begin{aligned} \alpha &\equiv \mathrm{e}^{\mathrm{j}[(\theta_S+\theta_R)/2+\pi/4]}/\sqrt{2}, \\ \beta &\equiv \mathrm{e}^{\mathrm{j}[(\theta_S-\theta_R)/2+\pi/4]}/\sqrt{2}, \end{aligned} \tag{3.88}$$

with the unitary transformation matrix in equation (3.85) given by

$$\overline{U} \equiv \begin{bmatrix} \alpha^* & 0 & 0 & -\alpha \\ 0 & -\beta & \beta^* & 0 \\ 0 & \beta & \beta^* & 0 \\ \alpha^* & 0 & 0 & \alpha \end{bmatrix}. \tag{3.89}$$

It is easy to prove that we indeed have $\overline{U}\,\overline{U}^{+} = \overline{U}^{+}\overline{U} = \overline{U}\,\overline{U}^{-1} = \overline{U}^{-1}\overline{U} = 1$, where $\overline{U}^{+}$ and $\overline{U}^{-1}$ are the Hermitian conjugate and inverse matrix of $\overline{U}$, respectively.

Therefore, by diagonalizing the two decoupled 2×2 Hamiltonians in equation (3.85) individually, we obtain the following eigenvalues

$$E^{\mathrm{b}}_{\mathrm{hh}} = -P + \surd(Q^2 + |R|^2 + |S|^2), \tag{3.90a}$$

$$E^{\mathrm{b}}_{\mathrm{lh}} = -P - \surd(Q^2 + |R|^2 + |S|^2). \tag{3.90b}$$

Since the eigenvalues of the two 2×2 Hamiltonians are identical, we find that the energies in equation (3.90a&b) are double degenerated. Therefore, the valence band structure comprises two double degenerated sub-bands, known as the double degenerated heavy-hole and light-hole bands, respectively; the reason will soon be revealed.

By substituting eigenvalues in equations (3.90a&b) into the following equations

$$\begin{bmatrix} P+Q+E & R' \\ R'^* & P-Q+E \end{bmatrix}\begin{bmatrix} c_1 \\ c_2 \end{bmatrix} = 0, \quad \begin{bmatrix} P-Q+E & R' \\ R'^* & P+Q+E \end{bmatrix}\begin{bmatrix} c_3 \\ c_4 \end{bmatrix} = 0, \tag{3.91}$$

we can find the coefficients under the new base

$$c_1(E) = \frac{-R'}{\surd((P+Q+E)^2 + |R|^2 + |S|^2)} \quad c_2(E) = \frac{P+Q+E}{\surd((P+Q+E)^2 + |R|^2 + |S|^2)}$$

$$c_3(E) = \frac{P+Q+E}{\surd((P+Q+E)^2 + |R|^2 + |S|^2)} \quad c_4(E) = \frac{-R'^*}{\surd((P+Q+E)^2 + |R|^2 + |S|^2)}, \tag{3.92}$$

with the normalization condition $|c_1(E)|^2 + |c_2(E)|^2 = |c_3(E)|^2 + |c_4(E)|^2 = 1$ satisfied.

The double degenerated eigenstates are given as

$$\begin{bmatrix} |\mathrm{hh}1\rangle \\ |\mathrm{hh}2\rangle \end{bmatrix} = \overline{C}(E^{\mathrm{b}}_{\mathrm{hh}})\begin{bmatrix} |1\rangle \\ |2\rangle \\ |3\rangle \\ |4\rangle \end{bmatrix} = \overline{C}(E^{\mathrm{b}}_{\mathrm{hh}})\overline{U}\begin{bmatrix} |n^{\mathrm{A}}_1\rangle \\ |n^{\mathrm{A}}_2\rangle \\ |n^{\mathrm{A}}_3\rangle \\ |n^{\mathrm{A}}_4\rangle \end{bmatrix}, \tag{3.93a}$$

$$\begin{bmatrix} |\mathrm{lh}1\rangle \\ |\mathrm{lh}2\rangle \end{bmatrix} = \overline{C}(E^{\mathrm{b}}_{\mathrm{lh}})\begin{bmatrix} |1\rangle \\ |2\rangle \\ |3\rangle \\ |4\rangle \end{bmatrix} = \overline{C}(E^{\mathrm{b}}_{\mathrm{lh}})\overline{U}\begin{bmatrix} |n^{\mathrm{A}}_1\rangle \\ |n^{\mathrm{A}}_2\rangle \\ |n^{\mathrm{A}}_3\rangle \\ |n^{\mathrm{A}}_4\rangle \end{bmatrix}, \tag{3.93b}$$

with the coefficient matrix defined as

$$\overline{C}(E) = \begin{bmatrix} c_1(E) & c_2(E) & 0 & 0 \\ 0 & 0 & c_3(E) & c_4(E) \end{bmatrix}. \tag{3.94}$$

We can also write these lattice states in a general form similar to equation (3.79a)

$$\begin{bmatrix} |1\vec{k}\rangle \\ |2\vec{k}\rangle \\ |3\vec{k}\rangle \\ |4\vec{k}\rangle \end{bmatrix} = \begin{bmatrix} |\mathrm{hh}1\rangle \\ |\mathrm{hh}2\rangle \\ |\mathrm{lh}1\rangle \\ |\mathrm{lh}2\rangle \end{bmatrix} = [a^{\mathrm{A}}_{jn}(\vec{k})]_{4\times 4} \begin{bmatrix} |n^{\mathrm{A}}_1\rangle \\ |n^{\mathrm{A}}_2\rangle \\ |n^{\mathrm{A}}_3\rangle \\ |n^{\mathrm{A}}_4\rangle \end{bmatrix}, \tag{3.95}$$

with

$$[a^{\mathrm{A}}_{jn}(\vec{k})]_{4\times 4} \equiv \begin{bmatrix} c_1(E^{\mathrm{b}}_{\mathrm{hh}}) & c_2(E^{\mathrm{b}}_{\mathrm{hh}}) & 0 & 0 \\ 0 & 0 & c_3(E^{\mathrm{b}}_{\mathrm{hh}}) & c_4(E^{\mathrm{b}}_{\mathrm{hh}}) \\ c_1(E^{\mathrm{b}}_{\mathrm{lh}}) & c_2(E^{\mathrm{b}}_{\mathrm{lh}}) & 0 & 0 \\ 0 & 0 & c_3(E^{\mathrm{b}}_{\mathrm{lh}}) & c_4(E^{\mathrm{b}}_{\mathrm{lh}}) \end{bmatrix} \overline{U}. \tag{3.96}$$

For a large variety of group III-V compound semiconductors, we have

$$\gamma_1 > \gamma_2 \approx \gamma_3. \tag{3.97}$$

Therefore, we can introduce an axial approximation by introducing an effective Luttinger parameter

$$\overline{\gamma} = (\gamma_2 + \gamma_3)/2, \tag{3.98}$$

to replace both γ_2 and γ_3 in the formula for R in equation (3.77c). Since the crystalline structures of these group III-V compounds are not simple cubic, there is no inherent isotropy in the $k_x k_y$ plane. The axial approximation, however, removes the slight anisotropy of the band structure along different directions in the $k_x k_y$ plane, resulting in cylindrical symmetry around the k_z axis.

Using equations (3.97) and (3.98) we find

$$R \approx -\frac{\sqrt{(3)}\hbar^2\overline{\gamma}}{2\mathrm{m}_0}(k_x - \mathrm{j}k_y)^2 = \frac{\sqrt{(3)}\hbar^2\overline{\gamma}}{2\mathrm{m}_0}(k_x^2 + k_y^2)\mathrm{e}^{\mathrm{j}\left[\pi - 2\tan\left(\frac{k_y}{k_x}\right)\right]},$$

$$S = \frac{\sqrt{(3)}\hbar^2\gamma_3}{\mathrm{m}_0}(k_x - \mathrm{j}k_y)k_z = \frac{\sqrt{(3)}\hbar^2\gamma_3}{\mathrm{m}_0}k_z\sqrt{(k_x^2 + k_y^2)}\mathrm{e}^{-\mathrm{j}\tan\left(\frac{k_y}{k_x}\right)}.$$

Plugging these expressions into equation (3.86) yields

$$R' = \frac{\sqrt{(3)}\hbar^2\overline{\gamma}}{2\mathrm{m}_0}(k_x^2 + k_y^2)\left(1 - \mathrm{j}\frac{2\gamma_3}{\overline{\gamma}}\frac{k_z}{\sqrt{(k_x^2 + k_y^2)}}\right). \tag{3.99}$$

Equation (3.88) becomes

$$\alpha \equiv \frac{1}{\sqrt{2}}\mathrm{e}^{\mathrm{j}\left[\frac{3}{4}\pi - \frac{3}{2}\tan\left(\frac{k_y}{k_x}\right)\right]}, \quad \beta \equiv \frac{1}{\sqrt{2}}\mathrm{e}^{\mathrm{j}\left[-\frac{1}{4}\pi + \frac{1}{2}\tan\left(\frac{k_y}{k_x}\right)\right]}. \tag{3.100}$$

According to equation (3.90a&b), we obtain

$$E_{\mathrm{hh}}^{\mathrm{b}} \approx -\frac{\hbar^2 k^2}{2\mathrm{m}_0}(\gamma_1 - 2\gamma_2) = -\frac{\hbar^2 k^2}{2m_{\mathrm{hh}}}, \tag{3.101a}$$

$$E_{\mathrm{lh}}^{\mathrm{b}} \approx -\frac{\hbar^2 k^2}{2\mathrm{m}_0}(\gamma_1 + 2\gamma_2) = -\frac{\hbar^2 k^2}{2m_{\mathrm{lh}}}, \tag{3.101b}$$

with

$$m_{\mathrm{hh}} \equiv \frac{\mathrm{m}_0}{\gamma_1 - 2\gamma_2}, \quad m_{\mathrm{lh}} \equiv \frac{\mathrm{m}_0}{\gamma_1 + 2\gamma_2}, \tag{3.102}$$

defined as the valence band heavy hole and light-hole effective mass, respectively. As $m_{\mathrm{hh}} > m_{\mathrm{lh}}$ always holds, we can easily identify equations (3.101a&b) as the heavy-hole and the light-hole energy band, respectively.

3.1.6 Hamiltonians for different semiconductors

As a summary of this section, we give the following examples on Hamiltonian selection in k–p theory based on Luttinger–Kohn's model.

- Two bands in group A (heavy-hole and light-hole) with 4×4 Hamiltonian: InGaAs-AlGaAs-GaAs, InGaP-AlInGaP-GaAs, InGaAsP-InP and AlGaInAs-InP [12].
- Three bands in group A (heavy-hole, light-hole, and spin–orbit split) with 6×6 Hamiltonian: InGaAsP-InP, AlGaInAs-InP and group III nitrides with a wurtzite structure such as InGaN-AlGaN [16].
- Four bands (e.g., conduction, heavy-hole, light-hole, and spin–orbit split) with 8×8 Hamiltonian: wide bandgap group II-VI compounds [17], group III nitrides with wurtzite structure such as InGaN-AlGaN [18], group III antimonides, and narrow bandgap group II-VI compounds [19].
- Five bands (e.g., N-resonant, conduction, heavy-hole, light-hole, and spin–orbit split) with 10×10 Hamiltonian: diluted nitrides such as GaInNAs-AlGaAs-GaAs [20].

3.2 Single electron in semiconductor quantum well structures

3.2.1 The effective mass theory and governing equation

As shown in Fig. 3.4, a semiconductor quantum well (QW) has dimensions on an intermediate scale which is comparatively larger than a single lattice but much smaller than the dimensions of a bulk semiconductor. As such, the QW can be viewed as a local disturbance in the semiconductor lattices in the background, and consequently a single electron will experience not only the periodic Coulomb potential, but also a potential disturbance on an intermediate scale introduced by the QW.

Therefore, a single electron in a semiconductor QW structure follows the steady-state Schrödinger equation with a modified Hamiltonian for bulk semiconductors to take into

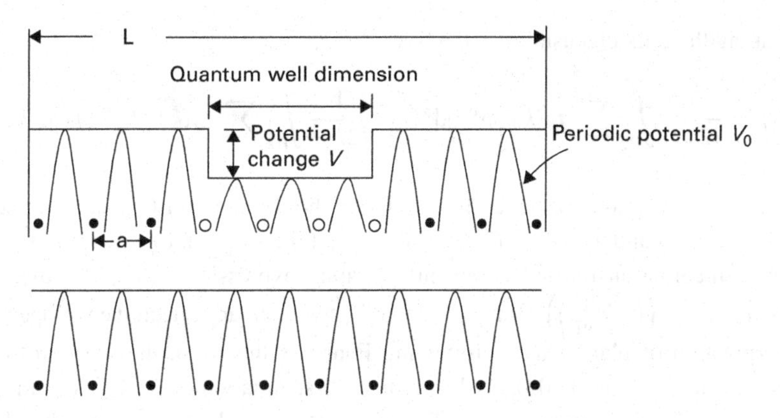

Fig. 3.4. The one-dimensional semiconductor quantum well structure.

account the potential change inside the QW

$$[\vec{H}_{\mathrm{b}}+V(\vec{r})]\,|\psi\rangle=\left[\vec{H}_0+\frac{\hbar}{4\mathrm{m}_0^2\mathrm{c}^2}(\nabla V\times\vec{p})\cdot\vec{\sigma}+V(\vec{r})\right]|\psi\rangle=\varepsilon_{\mathrm{h}}\,|\psi\rangle\,, \tag{3.103}$$

where we have defined

$$\vec{H}_{\mathrm{b}}\equiv\vec{H}_0+\frac{\hbar}{4\mathrm{m}_0^2\mathrm{c}^2}(\nabla V\times\vec{p})\cdot\vec{\sigma}=\frac{\vec{p}^2}{2\mathrm{m}_0}+V_0+\frac{\hbar}{4\mathrm{m}_0^2\mathrm{c}^2}(\nabla V\times\vec{p})\cdot\vec{\sigma}, \tag{3.104}$$

as the Hamiltonian for bulk semiconductors with spin–orbit interaction. In this expression, $\vec{H}_{\mathrm{b}}$ is periodic with respect to the lattice. However, $\vec{H}_{\mathrm{b}}+V(\vec{r})$ is no longer periodic because of the local disturbance on an intermediate scale described by $V(\vec{r})$. Therefore, Bloch's theorem is not directly applicable.

In Section 3.1, we have already obtained the solution for bulk semiconductors, i.e., $\vec{H}_{\mathrm{b}}|\phi^{\mathrm{b}}_{n\vec{k}}\rangle=\varepsilon_{n\vec{k}}|\phi^{\mathrm{b}}_{n\vec{k}}\rangle$. Following Bloch's theorem, we have $\langle\vec{r}|\phi^{\mathrm{b}}_{n\vec{k}}\rangle=\mathrm{e}^{\mathrm{j}\vec{k}\cdot\vec{r}}\langle\vec{r}|n\vec{k}\rangle$ with $|n\vec{k}\rangle\approx|n_0^{\mathrm{A}}\rangle=|0,0\downarrow\rangle,\,|0,0\uparrow\rangle$ and $|n\vec{k}\rangle\approx\sum_{j=1}^{6}a^{\mathrm{A}}_{jn}(\vec{k})|n^{\mathrm{A}}_j\rangle$, where $n=1,2,3,4,5,6$ as the periodic lattice states for the double degenerated conduction band electron and the valence band electron (hole), respectively. To solve equation (3.103) further with an extra potential contribution (V) in the Hamiltonian, at first sight, we might consider using Lowdin's renormalization theorem again. However, this extra potential contribution is by no means a perturbation, hence the expansion in the perturbation approach will never converge.

The effective mass theory is therefore introduced to solve equation (3.103) [10].

Actually, to best utilize what we have obtained from the bulk "background" material, we take a generalized integral transform on equation (3.103) anyway by utilizing the bulk eigenstates as the core function. That is to say, we expand the solution to equation (3.103) by taking the bulk eigenstates as the base

$$|\psi\rangle = \frac{1}{(2\pi)^3}\int_{\mathrm{BZ}}\sum_n f_n(\vec{k})\left|\phi^{\mathrm{b}}_{n\vec{k}}\right\rangle \mathrm{d}^3\vec{k} = \frac{1}{(2\pi)^3}\int_{\mathrm{BZ}}\sum_n f_n(\vec{k})\mathrm{e}^{\mathrm{j}\vec{k}\cdot\vec{r}}\left|n\vec{k}\right\rangle \mathrm{d}^3\vec{k}, \quad (3.105)$$

where the integration goes over the first BZ in the wave vector (momentum) space.

On the one hand, equation (3.105) shows that the solution to equation (3.103) is the inverse integral transform of a momentum domain expression (f_n) under a discrete set of core functions $\left(\left\langle\vec{r}\,/\phi^{\mathrm{b}}_{n\vec{k}}\right\rangle\right)$. Since a "global" wave with any specific envelope is a linear superposition of plane waves (following Fourier's theorem), the solution to the QW structure in equation (3.103), as shown in the first equation of (3.105), can therefore be viewed as a linear superposition of the eigenstates of the bulk background with the QW removed. On the other hand, this solution can also be viewed as a superposition of the periodic lattice wave functions of the bulk background. In bulk semiconductors, a single electron wave function can be factorized as a global plane wave and a periodic local lattice wave function. In QW structures, however, a single electron wave function is factorized as a global wave with a specific envelope and the original periodic local lattice wave function, as given in the second equation of (3.105). By removing the semiconductor lattices, in bulk material we obtain a vacuum space, hence the single electron eigenstate is described by a plane wave; in a QW structure, however, we obtain a space filled by a non-uniform potential distribution $V(\vec{r})$, hence the single electron eigenstate must be described by a wave having its envelope with a specific shape determined by the given potential. Since an arbitrary wave can be viewed as a linear superposition of those plane waves, again we can expand such a wave in a QW structure in terms of the plane waves. Finally, we consider the lattice effect by introducing the periodic lattice wave function as a factor in such an expansion to obtain equation (3.105).

As bulk semiconductor eigenstates and periodic lattice states, $\left|\phi^{\mathrm{b}}_{n\vec{k}}\right\rangle$ and $n\vec{k}$ are solved in Section 3.1, our task is then to find those coefficients $f_n(\vec{k})$ in expansion (3.105). For this purpose, we substitute equation (3.105) into equation (3.103) to yield

$$\frac{1}{(2\pi)^3}\int_{\mathrm{BZ}}\sum_n (\varepsilon_{n\vec{k}}-\varepsilon_{\mathrm{h}})f_n(\vec{k})\left|\phi^{\mathrm{b}}_{n\vec{k}}\right\rangle \mathrm{d}^3\vec{k} + \frac{1}{(2\pi)^3}\int_{\mathrm{BZ}}\sum_n f_n(\vec{k})V(\vec{r})\left|\phi^{\mathrm{b}}_{n\vec{k}}\right\rangle \mathrm{d}^3\vec{k} = 0. \quad (3.106)$$

Multiply $\left\langle\phi^{\mathrm{b}}_{m\vec{k}'}\right|$ on both sides of equation (3.106) to obtain

$$(\varepsilon_{m\vec{k}'}-\varepsilon_{\mathrm{h}})f_m(\vec{k}') + \frac{1}{(2\pi)^3}\int_{\mathrm{BZ}}\sum_n f_n(\vec{k})\left\langle\phi_{m\vec{k}'}\right|V(\vec{r})\left|\phi_{n\vec{k}}\right\rangle \mathrm{d}^3\vec{k} = 0,$$

or

$$(\varepsilon_{n\vec{k}}-\varepsilon_{\mathrm{h}})f_n(\vec{k}) + \frac{1}{(2\pi)^3}\int_{\mathrm{BZ}}\sum_m f_m(\vec{k}')\left\langle\phi_{n\vec{k}}\right|V(\vec{r})\left|\phi_{m\vec{k}'}\right\rangle \mathrm{d}^3\vec{k}' = 0, \quad (3.107)$$

where orthonormal condition (3.37) has been used.

Noting $|n\vec{k}\rangle$'s periodicity on the lattice, we find

$$\begin{aligned}
\left\langle \phi_{n\vec{k}} \middle| V(\vec{r}) \middle| \phi_{m\vec{k}'} \right\rangle &= \int_\Omega V(\vec{r})\phi^*_{n\vec{k}}(\vec{r})\phi_{m\vec{k}'}(\vec{r})\mathrm{d}^3\vec{r} \\
&= \int_\Omega V(\vec{r})u^*_{n\vec{k}}(\vec{r})u_{m\vec{k}'}(\vec{r})\mathrm{e}^{-\mathrm{j}(\vec{k}-\vec{k}')\cdot\vec{r}}\mathrm{d}^3\vec{r} \\
&= \int_\Omega V(\vec{r})\left[\sum_{i=1}^{3}\sum_{l_i=-\infty}^{\infty}\widetilde{C}_{\vec{G}}\mathrm{e}^{\mathrm{j}\vec{G}(\vec{l})\cdot\vec{r}}\right]\mathrm{e}^{-\mathrm{j}(\vec{k}-\vec{k}')\cdot\vec{r}}\mathrm{d}^3\vec{r} \\
&= \sum_{i=1}^{3}\sum_{l_i=-\infty}^{\infty}\widetilde{C}_{\vec{G}}\int_\Omega V(\vec{r})\mathrm{e}^{-\mathrm{j}[\vec{k}-\vec{k}'-\vec{G}(\vec{l})]\cdot\vec{r}}\mathrm{d}^3\vec{r} \\
&= \sum_{i=1}^{3}\sum_{l_i=-\infty}^{\infty}\widetilde{C}_{\vec{G}}\widetilde{V}[\vec{k}-\vec{k}'-\vec{G}(\vec{l})],
\end{aligned} \tag{3.108}$$

with $\widetilde{V}$ denoting the inverse Fourier transform of the potential energy $V(\vec{r})$. Also, in equation (3.108) the Fourier expansion of $u^*_{n\vec{k}}(\vec{r})u_{m\vec{k}'}(\vec{r})$ is used since the product of periodic functions on the lattice is still a periodic function on the lattice.

If we assume that $V(\vec{r})$ is a slow-varying function in space in terms of the lattice, the DC component in $\widetilde{V}$ will be dominant. Therefore, $\widetilde{V}[\vec{k}-\vec{k}'-\vec{G}(\vec{l})] \neq 0$ only if $\vec{k}-\vec{k}'-\vec{G}(\vec{l}) = 0$. Since both $\vec{k}$ and $\vec{k}'$ are within the first BZ, $\vec{k}-\vec{k}'-\vec{G}(\vec{l}) = 0$ will be possible only if $\vec{G}(\vec{l}) = 0$ and $\vec{k} = \vec{k}'$. The associated Fourier coefficient can be evaluated through

$$\widetilde{C}_{\vec{G}=0} = \frac{1}{\Omega}\int_\Omega u^*_{n\vec{k}}(\vec{r})u_{m\vec{k}'}(\vec{r})\mathrm{d}^3\vec{r} = \delta_{nm}. \tag{3.109}$$

Equation (3.108) can therefore be simplified to

$$\left\langle \phi_{n\vec{k}} \middle| V(\vec{r}) \middle| \phi_{m\vec{k}'} \right\rangle \approx \widetilde{V}(\vec{k}-\vec{k}')\delta_{nm}. \tag{3.110}$$

Plugging (3.110) into (3.107) yields

$$(\varepsilon_{n\vec{k}} - \varepsilon_\mathrm{h})\, f_n(\vec{k}) + \frac{1}{(2\pi)^3}\int_\mathrm{BZ} f_n(\vec{k}')\widetilde{V}(\vec{k}-\vec{k}')\mathrm{d}^3\vec{k}' = 0. \tag{3.111}$$

By introducing an envelope function in the space domain defined as the inverse Fourier transform of the coefficient $f_n(\vec{k})$ in the momentum domain, we have

$$F_n(\vec{r}) = \frac{1}{(2\pi)^3}\int_\mathrm{BZ} f_n(\vec{k})\mathrm{e}^{\mathrm{j}\vec{k}\cdot\vec{r}}\mathrm{d}^3\vec{k}. \tag{3.112}$$

By taking the inverse Fourier transform of equation (3.111), we finally obtain

$$[\varepsilon_{n(\nabla/j)} + V(\vec{r})]F_n(\vec{r}) = \varepsilon_\mathrm{h} F_n(\vec{r}). \tag{3.113}$$

Once the eigenvalue problem equation (3.113) is solved for $\varepsilon_{\rm h}$ and F_n, from equation (3.105), we find the single electron state

$$|\psi\rangle = \sum_n \frac{1}{(2\pi)^3} \int_{\rm BZ} |n\vec{k}\rangle f_n(\vec{k}) {\rm e}^{{\rm j}\vec{k}\cdot\vec{r}} {\rm d}^3\vec{k}$$

$$\approx \sum_n |n\rangle \frac{1}{(2\pi)^3} \int_{\rm BZ} f_n(\vec{k}) {\rm e}^{{\rm j}\vec{k}\cdot\vec{r}} {\rm d}^3\vec{k} = \sum_n F_n(\vec{r})\,|n\rangle, \tag{3.114}$$

where $|n\vec{k}\rangle \approx |n\rangle$ has been assumed in the neighborhood of $\vec{k} = 0$, i.e., the weak dependence of $|n\vec{k}\rangle$ on $\vec{k}$ has been ignored, which is a valid assumption in k–p theory. Actually, we can further expand the lattice state $|n\vec{k}\rangle$ in terms of the base $|n\rangle$ (i.e., the lattice state at $\vec{k} = 0$, or the BZ center) in equation (3.105). Those $\vec{k}$ dependent coefficients will merge with f_n while equation (3.109) still holds for $\vec{k} = \vec{k}' = 0$, therefore, we will reach the same effective mass equation in the form of equation (3.113), and equation (3.114) will be strictly valid. In addition, this approximation is not necessary if computational efficiency is not a concern, as we can always solve equation (3.113) to obtain $\varepsilon_{\rm h}$ and F_n, take the inverse of the Fourier transform to find f_n, and use the first equation (3.114) to compute the electron state.

From equation (3.114), we find that the solution to equation (3.103) is simply a periodic local lattice wave function at the BZ center ($\langle\vec{r}/n\rangle$) modulated by an envelope function. The former is purely determined by the periodic semiconductor lattice background, whereas the latter is determined by the QW induced potential disturbance with the lattice background removed: the lattice has only an aggregate effect through its own eigenvalue, i.e., the kinetic energy $\varepsilon_{n(\nabla/j)}$, as in equation (3.113). More specifically, for a given bulk semiconductor energy dependence on $\vec{k}$ (e.g., the parabolic dependence known as the parabolic band in most cases), the lattice effect is parameterized in equation (3.113) (e.g., in a parabolic band, the lattice effect is brought into one parameter, i.e., the effective mass, only). This indicates that the main property of the envelope function depends on the QW structure only. The lattice background can only affect the parameter but not the shape of the envelope function. Therefore, to some extent, we can separate the design of the QW structure from the selection of its lattice background, i.e., the bulk semiconductor as its host medium.

3.2.2 Conduction band (without degeneracy)

For the conduction band, the above scheme is straightforward as we know that the bulk semiconductor eigenvalue is explicitly given as $E_{\rm g} + \hbar^2k^2/(2m_e)$. Therefore, we have

$$\varepsilon_{0(\nabla/j)} = E_{\rm g} - \frac{\hbar^2\nabla^2}{2m_e}. \tag{3.115}$$

Hence equation (3.113) becomes

$$\left[-\frac{\hbar^2\nabla^2}{2m_e} + V(\vec{r})\right] F_0^{\rm c}(\vec{r}) = (\varepsilon_0^{\rm c} - E_{\rm g}) F_0^{\rm c}(\vec{r}), \tag{3.116}$$

where $\varepsilon_{\rm h}$ is further denoted as $\varepsilon_0^{\rm c}$ to indicate explicitly that this eigenvalue belongs to the conduction band s-state and superscript c is also appended to the envelope function for the same reason.

As an eigenvalue problem, equation (3.116) may have multiple discrete and/or continuous eigenvalues depending on the given potential disturbance $V(\vec{r})$ and boundary conditions. Once equation (3.116) is solved for $\varepsilon_{0j}^{\rm c}$ and $F_{0j}^{\rm c}(\vec{r})$, following equation (3.114), we find the corresponding conduction band electron states as

$$\left|\phi_j^{\rm qc}\right\rangle = F_{0j}^{\rm c}(\vec{r})\left|n_0^{\rm A}\right\rangle = F_{0j}^{\rm c}(\vec{r})\left|0,0\uparrow\downarrow\right\rangle. \tag{3.117}$$

3.2.3 Valence band (with degeneracy)

For the valence band, the above scheme is not directly applicable as the bulk semiconductor eigenvalues are not given in the explicit form. The reason is that the Hamiltonian is not diagonalized under base $\left|n\vec{k}\right\rangle$ because of the degeneracy between some of the states at $\vec{k} = 0$. However, such degeneracy can be eliminated by selecting a proper linear combination of eigenstates $\left|n\vec{k}\right\rangle$ at $\vec{k} = 0$ as the new base. That is to say, under a transformed base constructed by a proper combination of $\left|n\vec{k}\right\rangle$ at $\vec{k} = 0$, such a problem will disappear. Actually, taking the same expansion as equation (3.61), with the base selected as equation (3.62) and with those group B states ignored, we find from equation (3.105)

$$|\psi\rangle = \frac{1}{(2\pi)^3}\int_{\rm BZ}\sum_{n=1}^{6} f_n(\vec{k})e^{j\vec{k}\cdot\vec{r}}\left|n_n^{\rm A}\right\rangle d^3\vec{k}. \tag{3.118}$$

By following the approach set out in Section 3.2.1, we obtain equation (3.111) again with $n = 1, 2, 3, 4, 5, 6$. Since $\varepsilon_{n\vec{k}}$ in equation (3.111) is the valence band energy of the bulk semiconductor, according to equation (3.65), we have

$$\varepsilon_{n\vec{k}}\, f_n(\vec{k}) = \sum_{m=1}^{6} U_{nm}^{\rm A}\, f_m(\vec{k}), \tag{3.119}$$

with $n = 1, 2, 3, 4, 5, 6$. Substituting (3.119) into (3.111) yields

$$\sum_{m=1}^{6} U_{nm}^{\rm A}\, f_m(\vec{k}) - \varepsilon_{\rm h}\, f_n(\vec{k}) + \frac{1}{(2\pi)^3}\int_{\rm BZ} f_n(\vec{k}')\widetilde{V}(\vec{k}-\vec{k}')d^3\vec{k}' = 0, \tag{3.120}$$

with $n = 1, 2, 3, 4, 5, 6$. By taking the inverse Fourier transform of equation (3.120), we obtain

$$\sum_{m=1}^{6}\left[\widetilde{U}_{nm}^{\rm A} + V(\vec{r})\delta_{nm}\right]F_m^{\rm v}(\vec{r}) = \varepsilon_n^{\rm v}F_n^{\rm v}(\vec{r}), \tag{3.121}$$

with $n = 1, 2, 3, 4, 5, 6$ and $\varepsilon_{\rm h}$ in equation (3.120) recorded as $\varepsilon_n^{\rm v}$ in equation (3.121) to indicate explicitly that this eigenvalue belongs to those valence band states. The superscript v is also appended to the envelope function for the same reason. It forms a Fourier transform pair with f_n as defined in equation (3.112).

By applying the inverse Fourier transform on equation (3.73), we find those Luttinger–Kohn matrix elements as

$$\widetilde{U}^{\mathrm{A}}_{nm} = \left(\varepsilon^{\mathrm{A}}_{m} - \frac{\hbar^2\nabla^2}{2\mathrm{m}_0}\right)\delta_{nm} + \sum_{\alpha,\beta=x,y,z}\left[\frac{\hbar^2}{\mathrm{m}_0^2}\sum_{\nu}^{\mathrm{B}}\frac{p^{a}_{n\upsilon}p^{\beta}_{\upsilon m}}{(-\Delta/3)-\varepsilon^{\mathrm{B}}_{\upsilon}}\right]\left(-\mathrm{j}\frac{\partial}{\partial\alpha}\right)\left(-\mathrm{j}\frac{\partial}{\partial\beta}\right), \tag{3.122}$$

with $n, m = 1, 2, 3, 4, 5, 6$. They all become operators in the space domain.

Equation (3.122) can still be written in the Luttinger–Kohn Hamiltonian matrix form

$$\widetilde{H}^{\mathrm{LK}} = -\begin{bmatrix} \widetilde{P}+\widetilde{Q} & -\widetilde{S} & \widetilde{R} & 0 & -\frac{\widetilde{S}}{\sqrt{2}} & \sqrt{(2)}\widetilde{R} \\ -\widetilde{S}^* & \widetilde{P}-\widetilde{Q} & 0 & \widetilde{R} & -\sqrt{(2)}\widetilde{Q} & \sqrt{\left(\frac{3}{2}\right)}\widetilde{S} \\ \widetilde{R}^* & 0 & \widetilde{P}-\widetilde{Q} & \widetilde{S} & \sqrt{\left(\frac{3}{2}\right)}\widetilde{S}^* & \sqrt{(2)}\widetilde{Q} \\ 0 & \widetilde{R}^* & \widetilde{S}^* & \widetilde{P}+\widetilde{Q} & -\sqrt{(2)}\widetilde{R}^* & -\frac{\widetilde{S}^*}{\sqrt{2}} \\ -\frac{\widetilde{S}^*}{\sqrt{2}} & -\sqrt{(2)}\widetilde{Q}^* & \sqrt{\left(\frac{3}{2}\right)}\widetilde{S} & -\sqrt{(2)}\widetilde{R} & \widetilde{P}+\Delta & 0 \\ \sqrt{(2)}\widetilde{R}^* & \sqrt{\left(\frac{3}{2}\right)}\widetilde{S}^* & \sqrt{(2)}\widetilde{Q}^* & -\frac{\widetilde{S}}{\sqrt{2}} & 0 & \widetilde{P}+\Delta \end{bmatrix}, \tag{3.123}$$

with the operators in equation (3.123) defined as

$$\widetilde{P} \equiv -\frac{\hbar^2\gamma_1}{2\mathrm{m}_0}\nabla^2 = -\frac{\hbar^2\gamma_1}{2\mathrm{m}_0}\left(\frac{\partial^2}{\partial x^2}+\frac{\partial^2}{\partial y^2}+\frac{\partial^2}{\partial z^2}\right), \tag{3.124a}$$

$$\widetilde{Q} \equiv -\frac{\hbar^2\gamma_2}{2\mathrm{m}_0}\left(\frac{\partial^2}{\partial x^2}+\frac{\partial^2}{\partial y^2}-2\frac{\partial^2}{\partial z^2}\right), \tag{3.124b}$$

$$\widetilde{R} \equiv -\frac{\sqrt{(3)}\hbar^2}{2\mathrm{m}_0}\left[-\gamma_2\left(\frac{\partial^2}{\partial x^2}-\frac{\partial^2}{\partial y^2}\right)+\mathrm{j}2\gamma_3\frac{\partial^2}{\partial x\partial y}\right], \tag{3.124c}$$

$$\widetilde{S} \equiv -\frac{\sqrt{(3)}\hbar^2\gamma_3}{\mathrm{m}_0}\left(\frac{\partial^2}{\partial x\partial z}-\mathrm{j}\frac{\partial^2}{\partial y\partial z}\right). \tag{3.124d}$$

As an eigenvalue problem, equation (3.121) may have multiple discrete and/or continuous eigenvalues depending on the given potential disturbance $V(\vec{r})$ and boundary conditions. Once equation (3.121) is solved for $\varepsilon^{\mathrm{v}}_{nj}$ and $F^{\mathrm{v}}_{nj}(\vec{r})$, $n = 1, 2, 3, 4, 5, 6$, with the operators given as equation (3.124a–d) in the Luttinger–Kohn Hamiltonian equation (3.123), following equation (3.118), we find the valence band electron (hole) states as

$$\left|\phi^{\mathrm{qv}}_{j}\right\rangle = \sum_{n=1}^{6} F^{\mathrm{v}}_{nj}(\vec{r})\left|n^{\mathrm{A}}_{n}\right\rangle, \tag{3.125}$$

with $\left|n^{\mathrm{A}}_{n}\right\rangle$, $n = 1, 2, 3, 4, 5, 6$ given in equations (3.58b–d) through (3.62).

3.2.4 Quantum well band structures

The above effective mass theory is derived for a 3D potential disturbance $V(\vec{r})$. Therefore, the governing equations are applicable not only to QW structures, but also to quantum wire and quantum dot structures. For 1D QW structure along the z direction, the spatial dependence of the potential disturbance is reduced to 1D, i.e., we have $V(z)$ in equation (3.103). The envelope function defined in equation (3.112) becomes a mixed space and wave vector domain quantity, with its 1D space dependence along z and 2D wave vector dependence in the $k_x k_y$ plane

$$F_n(k_x, k_y, z) = \frac{\mathrm{e}^{\mathrm{j}(k_x x + k_y y)}}{2\pi\sqrt{\Sigma}} \int_{\mathrm{BZ}} f_n(\vec{k}) \mathrm{e}^{\mathrm{j}k_z z} \mathrm{d}k_z, \tag{3.126}$$

with Σ defined as the cross-sectional area of the bulk semiconductor.

For those non-degenerated bands such as the conduction band, $f_n(\vec{k}) = f_0(k_z)$, which makes the space and wave vector dependent parts in equation (3.126) separable

$$F_n(k_x, k_y, z) = \frac{\mathrm{e}^{\mathrm{j}(k_x x + k_y y)}}{2\pi\sqrt{\Sigma}} \int_{\mathrm{BZ}} f_0(k_z) \mathrm{e}^{\mathrm{j}k_z z} \mathrm{d}k_z = \frac{\mathrm{e}^{\mathrm{j}(k_x x + k_y y)}}{\sqrt{\Sigma}} F_0^{\mathrm{c}}(z), \tag{3.127}$$

where the 1D space domain envelope function is defined as

$$F_0^{\mathrm{c}}(z) \equiv \frac{1}{2\pi} \int_{\mathrm{BZ}} f_0(k_z) \mathrm{e}^{\mathrm{j}k_z z} \mathrm{d}k_z. \tag{3.128}$$

Therefore, in the cross-section (i.e., the xy plane), we still have the bulk semiconductor solution, i.e., the plane wave in the form of $\mathrm{e}^{\mathrm{j}(k_x x + k_y y)}/\sqrt{\Sigma}$. As such, in the effective mass equation (3.113), we should only replace k_z with $-\mathrm{j}\partial/\partial z$ but leave k_x and k_y as they are since the inverse Fourier transform is performed only on k_z. Hence we have

$$\left[-\frac{\hbar^2}{2} \frac{\mathrm{d}}{\mathrm{d}z} \left(\frac{1}{m_e} \frac{\mathrm{d}}{\mathrm{d}z} \right) + \frac{\hbar^2 k_t^2}{2m_e} + V(z) \right] F_0^{\mathrm{c}}(z) = (\varepsilon_0^{\mathrm{c}} - E_{\mathrm{g}}) F_0^{\mathrm{c}}(z), \tag{3.129}$$

where $k_t^2 \equiv k_x^2 + k_y^2$. The effective mass m_e in equation (3.129) has been taken inside the first derivative operator because of its z dependence, which guarantees the existence of the second order derivative of the envelope function (i.e., the continuity of the wave function and its first order derivative that correspond to the probability and the probability flux, respectively) hence the Hermitian of the Hamiltonian is assured [21].

For a given potential disturbance in the QW structure, equation (3.129) can be solved for $\varepsilon_{0j}^{\mathrm{c}}$ and $F_{0j}^{\mathrm{c}}(z)$. We then find the conduction band electron states as

$$\left|\phi_j^{\mathrm{qc}}\right\rangle = \frac{\mathrm{e}^{\mathrm{j}(k_x x + k_y y)}}{\sqrt{\Sigma}} F_{0j}^{\mathrm{c}}(z) \left|n_0^{\mathrm{A}}\right\rangle = \frac{\mathrm{e}^{\mathrm{j}(k_x x + k_y y)}}{\sqrt{\Sigma}} F_{0j}^{\mathrm{c}}(z) \left|0, 0 \uparrow\downarrow\right\rangle. \tag{3.130a}$$

Therefore, in a semiconductor QW structure the conduction band electron wave functions are given as

$$\left\langle \vec{r} \mid \phi_j^{\mathrm{qc}} \right\rangle = \frac{\mathrm{e}^{\mathrm{j}(k_x x + k_y y)}}{\sqrt{\Sigma}} F_{0j}^{\mathrm{c}}(z) \left\langle \vec{r} \mid n_0^{\mathrm{A}} \right\rangle = \frac{\mathrm{e}^{\mathrm{j}(k_x x + k_y y)}}{\sqrt{\Sigma}} F_{0j}^{\mathrm{c}}(z) \left\langle \vec{r} \mid 0, 0 \uparrow\downarrow \right\rangle. \quad (3.130\mathrm{b})$$

For degenerated bands such as the valence band, in the cross-section we do not have a simple plane wave solution because of the inseparable operator dependence along z and in the xy plane, which is a result of the coupling between the eigenstates (a linear combination of the orbital angular momentum eigenstates $|l, m\rangle$) in coordinate representation. This can be understood as the state having orbital angular symmetry in the bulk semiconductor losing its symmetry in the QW structure. The only exception is when the state has the highest orbital angular symmetry, i.e., a spherical shape, as we have discussed above for the conduction band electron.

Following equations (3.118) and (3.126), we can expand the valence band electron (hole) state as

$$|\psi\rangle = \frac{\mathrm{e}^{\mathrm{j}(k_x x + k_y y)}}{2\pi\sqrt{\Sigma}} \int_{\mathrm{BZ}} \sum_{n=1}^{6} f_n(\vec{k}) \mathrm{e}^{\mathrm{j}k_z z} \left| n_n^A \right\rangle \mathrm{d}k_z. \quad (3.131)$$

By following the approach described in Section 3.2.1, we find that equation (3.111) is modified to

$$(\varepsilon_{n\vec{k}} - \varepsilon_{\mathrm{h}}) f_n(\vec{k}) + \frac{1}{2\pi} \int_{\mathrm{BZ}} f_n(\vec{k}') \widetilde{V}(k_z - k_z') \mathrm{d}k_z' = 0, \quad (3.132)$$

with $n = 1, 2, 3, 4, 5, 6$. Since $\varepsilon_{n\vec{k}}$ in equation (3.132) is the valence band energy of the bulk semiconductor, equation (3.119) still holds with $n = 1, 2, 3, 4, 5, 6$. Substituting equation (3.119) into (3.132) yields

$$\sum_{m=1}^{6} U_{nm}^{\mathrm{A}} f_m(\vec{k}) - \varepsilon_{\mathrm{h}} f_n(\vec{k}) + \frac{1}{2\pi} \int_{\mathrm{BZ}} f_n(\vec{k}') \widetilde{V}(k_z - k_z') \mathrm{d}k_z' = 0, \quad (3.133)$$

with $n = 1, 2, 3, 4, 5, 6$. Multiplying $\mathrm{e}^{\mathrm{j}(k_x x + k_y y)}/\sqrt{\Sigma}$ on both sides of equation (3.133) and taking the inverse Fourier transform, we obtain

$$\sum_{m=1}^{6} [\widetilde{U}_{nm}^{\mathrm{AZ}} + V(z)\delta_{nm}] F_m^{\mathrm{v}}(k_x, k_y, z) = \varepsilon_n^{\mathrm{v}} F_n^{\mathrm{v}}(k_x, k_y, z), \quad (3.134)$$

with $n = 1, 2, 3, 4, 5, 6$ and ε_{h} in equation (3.133) recorded as $\varepsilon_n^{\mathrm{v}}$ in equation (3.134) to indicate explicitly that this eigenvalue belongs to those valence band states. The superscript v is also appended to the mixed space and wave vector domain envelope function for the same reason. It forms a 1D Fourier transform pair in z with f_n as defined in equation (3.126).

By applying the 1D inverse Fourier transform in terms of z to equation (3.73), we find the Luttinger–Kohn matrix elements

$$\widetilde{U}_{nm}^{\mathrm{AZ}} = \left[\varepsilon_m^{\mathrm{A}} + \frac{\hbar^2 k_t^2}{2\mathrm{m}_0} - \frac{\hbar^2}{2\mathrm{m}_0}\frac{\mathrm{d}^2}{\mathrm{d}z^2}\right]\delta_{nm} + \sum_{\alpha,\beta=x,y,z}\left[\frac{\hbar^2}{\mathrm{m}_0^2}\sum_{v}^{\mathrm{B}}\frac{p_{nv}^{a}p_{vm}^{\beta}}{(-\Delta/3)-\varepsilon_v^{\mathrm{B}}}\right]\Delta_\alpha\Delta_\beta, \tag{3.135}$$

with $\Delta_\alpha = -\mathrm{j}\mathrm{d}/\mathrm{d}\alpha$ if $\alpha = z$, and $\Delta_\alpha = k_\alpha$ if $\alpha = x, y$, $\Delta_\beta = -\mathrm{j}\mathrm{d}/\mathrm{d}\beta$ if $\beta = z$, and $\Delta_\beta = k_\beta$ if $\beta = x, y$, where $n, m = 1, 2, 3, 4, 5, 6$. They are operators in z and normal functions of k_x and k_y.

Equation (3.135) can still be written as the Luttinger–Kohn Hamiltonian matrix that takes the same form as equation (3.123) but with the "partial" operators modified to

$$\widetilde{P}^{\mathrm{Z}} \equiv \frac{\hbar^2\gamma_1}{2\mathrm{m}_0}(k_t^2 - \frac{\mathrm{d}^2}{\mathrm{d}z^2}), \tag{3.136a}$$

$$\widetilde{Q}^{\mathrm{Z}} \equiv \frac{\hbar^2\gamma_2}{2\mathrm{m}_0}(k_t^2 + 2\frac{\mathrm{d}^2}{\mathrm{d}z^2}), \tag{3.136b}$$

$$\widetilde{R}^{\mathrm{Z}} \equiv \frac{\sqrt{(3)}\hbar^2}{2\mathrm{m}_0}\left[-\gamma_2(k_x^2 - k_y^2) + \mathrm{j}2\gamma_3 k_x k_y\right], \tag{3.136c}$$

$$\widetilde{S}^{\mathrm{Z}} \equiv -\mathrm{j}\frac{\sqrt{(3)}\hbar^2\gamma_3}{\mathrm{m}_0}(k_x - jk_y)\frac{\mathrm{d}}{\mathrm{d}z}. \tag{3.136d}$$

As an eigenvalue problem, equation (3.134) may have multiple discrete and/or continuous eigenvalues depending on the given potential disturbance $V(z)$ and boundary conditions. Once equation (3.134) is solved for $\varepsilon_{nj}^{\mathrm{v}}$ and $F_{nj}^{\mathrm{v}}(k_x, k_y, z), n = 1, 2, 3, 4, 5, 6$, with the "partial" operators given as equations (3.136a–d) in the Luttinger–Kohn Hamiltonian matrix elements having the same form as equation (3.123), by following equations (3.131) and (3.126), we find the valence band electron (hole) states

$$\left|\phi_j^{\mathrm{qv}}\right\rangle = \sum_{n=1}^{6} F_{nj}^{\mathrm{v}}(k_x, k_y, z)\left|n_n^{\mathrm{A}}\right\rangle, \tag{3.137a}$$

with $\left|n_n^{\mathrm{A}}\right\rangle, n = 1, 2, 3, 4, 5, 6$ given in equations (3.58b–d) through (3.62). Finally, in the semiconductor QW structure the valence band electron (hole) wave functions are

$$\left\langle\vec{r}/\phi_j^{\mathrm{qv}}\right\rangle = \sum_{n=1}^{6} F_{nj}^{\mathrm{v}}(k_x, k_y, z)\left\langle\vec{r}/n_n^{\mathrm{A}}\right\rangle. \tag{3.137b}$$

If we drop the two spin–orbit split bands, the left 4×4 can readily be diagonalized into two decoupled 2×2 blocks under a new base $[|1\rangle \quad |2\rangle \quad |3\rangle \quad |4\rangle]^{\mathrm{T}}$ [22, 23]. The new base is linked to the original base through a unitary transformation matrix shown in equations (3.87) and (3.89). The eigenvalue problem, equation (3.134), is therefore reduced to

$$\left[\frac{\hbar^2 k_t^2(\gamma_1+\gamma_2)}{2m_0} - \frac{\hbar^2(\gamma_1-2\gamma_2)}{2m_0}\frac{d^2}{dz^2} + V(z)\right]F_1^v + \frac{\sqrt{(3)}\hbar^2 k_t}{2m_0}\left(\overline{\gamma}k_t - j2\gamma_3\left|\frac{d}{dz}\right|\right)F_2^v$$
$$= -\varepsilon^v F_1^v$$
$$\frac{\sqrt{(3)}\hbar^2 k_t}{2m_0}\left(\overline{\gamma}k_t + j2\gamma_3\left|\frac{d}{dz}\right|\right)F_1^v + \left[\frac{\hbar^2 k_t^2(\gamma_1-\gamma_2)}{2m_0} - \frac{\hbar^2(\gamma_1+2\gamma_2)}{2m_0}\frac{d^2}{dz^2} + V(z)\right]F_2^v$$
$$= -\varepsilon^v F_2^v,$$

and

$$\left[\frac{\hbar^2 k_t^2(\gamma_1-\gamma_2)}{2m_0} - \frac{\hbar^2(\gamma_1+2\gamma_2)}{2m_0}\frac{d^2}{dz^2} + V(z)\right]F_3^v + \frac{\sqrt{(3)}\hbar^2 k_t}{2m_0}\left(\overline{\gamma}k_t - j2\gamma_3\left|\frac{d}{dz}\right|\right)F_4^v$$
$$= -\varepsilon^v F_3^v$$
$$\frac{\sqrt{(3)}\hbar^2 k_t}{2m_0}\left(\overline{\gamma}k_t + j2\gamma_3\left|\frac{d}{dz}\right|\right)F_3^v + \left[\frac{\hbar^2 k_t^2(\gamma_1+\gamma_2)}{2m_0} - \frac{\hbar^2(\gamma_1-2\gamma_2)}{2m_0}\frac{d^2}{dz^2} + V(z)\right]F_4^v$$
$$= -\varepsilon^v F_4^v,$$

where equations (3.85), (3.86), and (3.136a–d) have been used. Axial approximation is also used for simplification of equation (3.136c). As such, it is reduced to

$$\widetilde{R}^Z \equiv -\frac{\sqrt{(3)}\hbar^2\overline{\gamma}}{2m_0}(k_x - jk_y)^2.$$

Following the convention of using the terms "heavy" and "light" holes according to their masses in the z direction, we define

$$m_{hhz} = \frac{m_0}{\gamma_1 - 2\gamma_2},\ m_{hht} = \frac{m_0}{\gamma_1+\gamma_2},\ m_{lhz} = \frac{m_0}{\gamma_1+2\gamma_2},\ m_{lht} = \frac{m_0}{\gamma_1-\gamma_2}. \tag{3.138}$$

Therefore, the governing equation for the mixed space and wave vector domain envelope function becomes

$$\left[\frac{\hbar^2 k_t^2}{2m_{hht}} - \frac{\hbar^2}{2}\frac{d}{dz}\left(\frac{1}{m_{hhz}}\frac{d}{dz}\right) + V(z)\right]F_1^v + \frac{\sqrt{(3)}\hbar^2 k_t}{2m_0}\left(\overline{\gamma}k_t - j2\gamma_3\left|\frac{d}{dz}\right|\right)F_2^v = -\varepsilon^v F_1^v$$
$$\frac{\sqrt{(3)}\hbar^2 k_t}{2m_0}\left(\overline{\gamma}k_t + j2\gamma_3\left|\frac{d}{dz}\right|\right)F_1^v + \left[\frac{\hbar^2 k_t^2}{2m_{lht}} - \frac{\hbar^2}{2}\frac{d}{dz}\left(\frac{1}{m_{lhz}}\frac{d}{dz}\right) + V(z)\right]F_2^v = -\varepsilon^v F_2^v, \tag{3.139a}$$

$$\left[\frac{\hbar^2 k_t^2}{2m_{\mathrm{lht}}} - \frac{\hbar^2}{2}\frac{\mathrm{d}}{\mathrm{d}z}\left(\frac{1}{m_{\mathrm{lhz}}}\frac{\mathrm{d}}{\mathrm{d}z}\right) + V(z)\right] F_3^{\mathrm{v}} + \frac{\sqrt{(3)}\hbar^2 k_t}{2\mathrm{m}_0}\left(\overline{\gamma} k_t - \mathrm{j}2\gamma_3 \left|\frac{\mathrm{d}}{\mathrm{d}z}\right|\right) F_4^{\mathrm{v}} = -\varepsilon^{\mathrm{v}} F_3^{\mathrm{v}}$$

$$\frac{\sqrt{(3)}\hbar^2 k_t}{2\mathrm{m}_0}\left(\overline{\gamma} k_t + \mathrm{j}2\gamma_3 \left|\frac{\mathrm{d}}{\mathrm{d}z}\right|\right) F_3^{\mathrm{v}} + \left[\frac{\hbar^2 k_t^2}{2m_{\mathrm{hht}}} - \frac{\hbar^2}{2}\frac{\mathrm{d}}{\mathrm{d}z}\left(\frac{1}{m_{\mathrm{hhz}}}\frac{\mathrm{d}}{\mathrm{d}z}\right) + V(z)\right] F_4^{\mathrm{v}} = -\varepsilon^{\mathrm{v}} F_4^{\mathrm{v}}, \tag{3.139b}$$

where the effective masses in the z direction have been taken inside the first derivative operator for the reason given after equation (3.129).

Equation (3.138) clearly shows that the valence band effective mass becomes anisotropic and mass reversal occurs since the heavy hole has a smaller in-plane mass than the light hole. From equation (3.139a&b), we find that in the neighborhood of $\vec{k}_t = 0$ the two sets of equations degenerate with the off-diagonal coupling terms all disappearing. These equations thus reduce to

$$\left[-\frac{\hbar^2}{2}\frac{\mathrm{d}}{\mathrm{d}z}\left(\frac{1}{m_{\mathrm{hhz}}}\frac{\mathrm{d}}{\mathrm{d}z}\right) + V(z)\right] F_{1,4}^{\mathrm{v}} = -\varepsilon^{\mathrm{v}} F_{1,4}^{\mathrm{v}}$$

$$\left[-\frac{\hbar^2}{2}\frac{\mathrm{d}}{\mathrm{d}z}\left(\frac{1}{m_{\mathrm{lhz}}}\frac{\mathrm{d}}{\mathrm{d}z}\right) + V(z)\right] F_{2,3}^{\mathrm{v}} = -\varepsilon^{\mathrm{v}} F_{2,3}^{\mathrm{v}}. \tag{3.140}$$

Since $m_{\mathrm{hhz}} > m_{\mathrm{lhz}}$, we must have $|\varepsilon_{\mathrm{hh}}^{\mathrm{v}}| < |\varepsilon_{\mathrm{lh}}^{\mathrm{v}}|$ according to equation (3.140), and therefore in contrast to the bulk semiconductor, the heavy-hole and light-hole bands are no longer degenerate at $\vec{k}_t = 0$ in a QW structure and the heavy-hole band will always be on top (since a hole has negative energy in our reference). For this reason, one would imagine also that the heavy-hole band and light-hole band will have crossing points since the heavy-hole band edge (at $\vec{k}_t = 0$) is above the light-hole band edge, yet the heavy-hole band drops more rapidly as $\vec{k}_t$ increases because of its smaller in-plane mass ($m_{\mathrm{hht}} < m_{\mathrm{lht}}$) [21]. However, the band crossing can never happen physically, which means that our solution does not any longer allow the bands to take parabolic shapes. This is known as band mixing, a well-known effect in semiconductor QW structures [12].

3.3 Single electron in strained layer structures

3.3.1 A general approach

The strain tensor induced in a crystal lattice can generally be expressed as

$$\overline{e} \equiv \left[e_{\alpha\beta}\right]_{\alpha,\beta=x,y,z}, \tag{3.141}$$

where $e_{\alpha\beta}$ is a dimensionless relative shift in the space domain because of the lattice deformation.

In strained materials, all governing equations such as equations (3.48) for bulk semiconductors and (3.103) for semiconductor QW structures still hold with every coordinate in the space domain in the strained system. However, material parameters are all given in the unstrained system. Therefore, the strain effect will be reflected by transferring the wave functions back from the strained coordinate system to the unstrained coordinate system, so as to make the description for the strained structure consistent [24, 25].

Assuming that $\vec{r}$ and $\vec{r}'$ are the space vectors in the unstrained and strained systems, respectively, we can link the two coordinate systems through

$$r'_\alpha = r_\alpha + \sum_{\beta=x,y,z} e_{\alpha\beta} r_\beta. \tag{3.142}$$

Since the strain has always to be small to avoid lattice relaxation through dislocation, its effect is a typical perturbation of the unstrained system. By ignoring all of the terms of the order of $O(\overline{e}^2)$ or higher, we have

$$\begin{aligned} r_\alpha &= r'_\alpha - \sum_{\beta=x,y,z} e_{\alpha\beta} r_\beta = r'_\alpha - \sum_{\beta=x,y,z} e_{\alpha\beta}\left[r'_\beta - \sum_\gamma e_{\beta\gamma} r_\gamma\right] \\ &= r'_\alpha - \sum_{\beta=x,y,x} e_{\alpha\beta} r'_\beta + \sum_{\beta=x,y,x} e_{\alpha\beta} \sum_{\gamma=x,y,z} e_{\beta\gamma} r_\gamma \approx r'_\alpha - \sum_{\beta=x,y,z} e_{\alpha\beta} r'_\beta \\ &= \sum_{\beta=x,y,z} (\delta_{\alpha\beta} - e_{\alpha\beta}) r'_\beta, \end{aligned}$$

or

$$r_\beta \approx \sum_{\alpha=x,y,z} (\delta_{\beta\alpha} - e_{\beta\alpha}) r'_\alpha. \tag{3.143}$$

We also find

$$\frac{\partial}{\partial r'_\alpha} \equiv \sum_{\beta=x,y,z} \frac{\partial r_\beta}{\partial r'_\alpha}\frac{\partial}{\partial r_\beta} \approx \sum_{\beta=x,y,z} (\delta_{\alpha\beta} - e_{\alpha\beta})\frac{\partial}{\partial r_\beta} = \frac{\partial}{\partial r_\alpha} - \sum_{\beta=x,y,z} e_{\alpha\beta}\frac{\partial}{\partial r_\beta}, \tag{3.144a}$$

$$\begin{aligned} \frac{\partial^2}{\partial r'^2_\alpha} &\approx \left[\sum_{\beta=x,y,z} (\delta_{\alpha\beta} - e_{\alpha\beta})\frac{\partial}{\partial r_\beta}\right]\left[\sum_{\beta=x,y,z} (\delta_{\alpha\beta} - e_{\alpha\beta})\frac{\partial}{\partial r_\beta}\right] \\ &\approx \sum_{\beta=x,y,z} (\delta_{\alpha\beta} - 2e_{\alpha\beta})\frac{\partial^2}{\partial r_\alpha \partial r_\beta} = \frac{\partial^2}{\partial r^2_\alpha} - 2\sum_{\beta=x,y,z} e_{\alpha\beta}\frac{\partial^2}{\partial r_\alpha \partial r_\beta}. \end{aligned} \tag{3.144b}$$

Therefore, we have

$$k_\alpha p^{\alpha'} \approx k_\alpha p^\alpha - \sum_{\beta=x,y,z} k_\alpha e_{\alpha\beta} p^\beta, \tag{3.145}$$

$$V_0(\vec{r}') \approx V_0(\vec{r}) + \sum_{\alpha,\beta=x,y,z} V_0^{\alpha\beta} e_{\alpha\beta}, \tag{3.146}$$

where $V_0^{\alpha\beta}$ is the derivative of the lattice periodic potential $V_0(\vec{r}')$ with respect to $e_{\alpha\beta}$.

For bulk materials, the most complicated Hamiltonian in equation (3.48) can therefore be converted back from the deformed lattice (because of the strain) to the normal lattice where there is no strain [26]

$$\begin{aligned}&\frac{\vec{p}^2}{2\mathrm{m}_0} + V_0(\vec{r}\,') + \frac{\hbar}{\mathrm{m}_0}\vec{k}\cdot\vec{p} + \frac{\hbar}{4\mathrm{m}_0^2\mathrm{c}^2}(\nabla V \times \vec{p})\cdot\vec{\sigma}\\ &\quad\approx -\frac{\hbar^2\nabla^2}{2\mathrm{m}_0} + V_0(\vec{r}) + \frac{\hbar}{\mathrm{m}_0}\vec{k}\cdot\vec{p} + \frac{\hbar}{4\mathrm{m}_0^2\mathrm{c}^2}(\nabla V \times \vec{p})\cdot\vec{\sigma} + \sum_{\alpha,\beta=x,y,x} S^{\alpha\beta}e_{\alpha\beta},\end{aligned} \tag{3.147}$$

where

$$S^{\alpha\beta} = \frac{\hbar^2}{\mathrm{m}_0}\frac{\partial^2}{\partial r_\alpha \partial r_\beta} + V_0^{\alpha\beta} - \frac{\hbar}{\mathrm{m}_0}k_\alpha p^\beta. \tag{3.148}$$

The contribution of the lattice conversion related to the spin–orbit interaction has been dropped as it is of a higher order.

Taking the last term in equation (3.147) as a perturbation and by following the same procedures as in Section 3.1.4, we find that equation (3.73) should be modified to

$$U_{ij}^{\mathrm{AS}} = \left(\varepsilon_j^{\mathrm{A}} + \frac{\hbar^2k^2}{2\mathrm{m}_0}\right)\delta_{ij} + \sum_{\alpha,\beta=x,y,z}\left[\frac{\hbar^2}{\mathrm{m}_0^2}\sum_{v}^{\mathrm{B}}\frac{p_{iv}^{a}p_{vj}^{\beta}}{(-\Delta/3)-\varepsilon_v^{\mathrm{B}}}\right]k_a k_\beta + \sum_{\alpha,\beta=x,y,z} S_{ij}^{\alpha\beta}e_{\alpha\beta}, \tag{3.149}$$

with

$$S_{ij}^{\alpha\beta} = \left\langle n_i^{\mathrm{A}}\right| S^{\alpha\beta}\left|n_j^{\mathrm{A}}\right\rangle, \tag{3.150}$$

for strained bulk semiconductors. From equation (3.149), we find that a mapping from $e_{\alpha\beta}$ to $k_\alpha k_\beta$ can be established. Therefore, we can skip all derivations to append our final results directly from bulk semiconductor and QW structures to strained bulk semiconductor and strained layer QW structures, respectively, by following an analogous approach.

3.3.2 Strained bulk semiconductors

In bulk semiconductors, for the conduction band, following the mapping $e_{\alpha\beta} \leftrightarrow k_\alpha k_\beta$, we find

$$k_x^2 \leftrightarrow e_{xx}, \quad k_y^2 \leftrightarrow e_{yy}, \quad k_z^2 \leftrightarrow e_{zz}, \quad \frac{\hbar^2}{2m_e} \leftrightarrow \mathrm{a_c}.$$

As such, we further obtain

$$\frac{\hbar^2k^2}{2m_e} \leftrightarrow \mathrm{a_c}(e_{xx} + e_{yy} + e_{zz}).$$

By defining the conduction band hydrostatic energy as

$$P_e^{\mathrm{c}} \equiv \mathrm{a_c}(e_{xx} + e_{yy} + e_{zz}), \tag{3.151}$$

we find that the only effect introduced by the strain is that the eigenvalue has a shift at $\vec{k} = 0$ in the amount of conduction band hydrostatic energy according to equation (3.149). Therefore, we find that equation (3.81) should be modified to

$$E = E_g + \frac{\hbar^2 k^2}{2m_e} + P_e^c. \tag{3.152}$$

The conduction band electron lattice state $\left|n_0^A\right\rangle$ is still the same, as given in equation (3.83a).

For the valence band, following the mapping $e_{\alpha\beta} \leftrightarrow k_\alpha k_\beta$, we find

$$k_x^2 \leftrightarrow e_{xx}, \quad k_y^2 \leftrightarrow e_{yy}, \quad k_z^2 \leftrightarrow e_{zz}, \quad k_x k_y \leftrightarrow e_{xy}, \quad k_x k_z \leftrightarrow e_{xz}, \quad k_y k_z \leftrightarrow e_{yz},$$

and

$$-\frac{\hbar^2}{2\mathrm{m}_0}\gamma_1 \leftrightarrow -S_\mathrm{v}^\mathrm{d} = \mathrm{a_v}, \quad -\frac{\hbar^2}{2\mathrm{m}_0}\gamma_2 \leftrightarrow -\frac{S_\mathrm{u}}{3} = \frac{\mathrm{b}}{2}, \quad -\frac{\hbar^2}{2\mathrm{m}_0}\gamma_3 \leftrightarrow -\frac{S_\mathrm{u}'}{3} = \frac{\mathrm{d}}{2\sqrt{3}},$$

with S_v^d, S_u, and S_u' defined as different components of $S_{ij}^{\alpha\beta}$ in equation (3.150). As such, the following mappings can be established

$$\begin{aligned}
\frac{\hbar^2\gamma_1}{2\mathrm{m}_0}(k_x^2 + k_y^2 + k_z^2) &\leftrightarrow -\mathrm{a_v}(e_{xx} + e_{yy} + e_{zz}), \\
\frac{\hbar^2\gamma_2}{2\mathrm{m}_0}(k_x^2 + k_y^2 - 2k_z^2) &\leftrightarrow -\frac{\mathrm{b}}{2}(e_{xx} + e_{yy} - 2e_{zz}), \\
\frac{\sqrt{(3)}\hbar^2}{2\mathrm{m}_0}[-\gamma_2(k_x^2 - k_y^2) + \mathrm{j}2\gamma_3 k_x k_y] &\leftrightarrow \frac{\sqrt{(3)}\mathrm{b}}{2}(e_{xx} - e_{yy}) - \mathrm{jd}e_{xy}, \\
\frac{\sqrt{(3)}\hbar^2\gamma_3}{\mathrm{m}_0}(k_x - \mathrm{j}k_y)k_z &\leftrightarrow -\mathrm{d}(e_{xz} - \mathrm{j}e_{yz}).
\end{aligned}$$

By introducing the following definitions:

- valence band hydrostatic energy $P_e^\mathrm{v} \equiv -\mathrm{a_v}(e_{xx} + e_{yy} + e_{zz})$, (3.153a)
- valence band shear energy $Q_e \equiv -\frac{\mathrm{b}}{2}(e_{xx} + e_{yy} - 2e_{zz})$, (3.153b)
- $R_e \equiv \frac{\sqrt{(3)}\mathrm{b}}{2}(e_{xx} - e_{yy}) - \mathrm{jd}e_{xy}$ (3.153c)
- $S_e \equiv -\mathrm{d}(e_{xz} - \mathrm{j}e_{yz})$, (3.153d)

we find that the resulting Luttinger–Kohn Hamiltonian matrix still takes the same form as given in equation (3.76), but with the energies in its elements given in equation (3.77a–d) modified to

$$\widetilde{P} \equiv \frac{\hbar^2\gamma_1}{2m_0}(k_x^2 + k_y^2 + k_z^2) + P_e^v, \tag{3.154a}$$

$$\widetilde{Q} \equiv \frac{\hbar^2\gamma_2}{2m_0}(k_x^2 + k_y^2 - 2k_z^2) + Q_e, \tag{3.154b}$$

$$\widetilde{R} \equiv \frac{\sqrt{(3)}\hbar^2}{2m_0}[-\gamma_2(k_x^2 - k_y^2) + j2\gamma_3 k_x k_y] + R_e, \tag{3.154c}$$

$$\widetilde{S} \equiv \frac{\sqrt{(3)}\hbar^2\gamma_3}{m_0}(k_x - jk_y)k_z + S_e, \tag{3.154d}$$

according to equation (3.149). The parameters a_c, a_v, b, and d used in this section are the material deformation potential constants. They can be found in a semiconductor material database such as [13, 14].

Once equation (3.76) is solved with such modified energies in the Luttinger–Kohn Hamiltonian matrix elements subject to the normalization condition (3.78), we can obtain the valence band electron (hole) states through equation (3.79a) with the unchanged base $|n_j^A\rangle$, $j = 1, 2, 3, 4, 5, 6$ given in equations (3.58b–d) through (3.62).

3.3.3 Strained layer quantum well structures

By using the mapping approach introduced in dealing with strained bulk semiconductors in Section 3.3.2 and the effective mass theory introduced in Section 3.2, we can readily obtain governing equations for the band structure of a single electron in strained layer QW structures.

Actually, for the conduction band electron, equation (3.116) should be modified to

$$\left[-\frac{\hbar^2\nabla^2}{2m_e} + V(\vec{r})\right] F_0^c(\vec{r}) = (\varepsilon_0^c - E_g - P_e^c) F_0^c(\vec{r}). \tag{3.155}$$

Once equation (3.155) is solved for ε_{0j}^c and $F_{0j}^c(\vec{r})$, the electron state is obtained through equation (3.117).

For the valence band electron (hole), equation (3.121) should still be solved with the Hamiltonian given in equation (3.123). However, in replacing equations (3.124a–d), we should use the following operators to construct the Luttinger–Kohn Hamiltonian matrix elements in equation (3.123)

$$\widetilde{P} = -\frac{\hbar^2\gamma_1}{2m_0}\left(\frac{\partial^2}{\partial x^2} + \frac{\partial^2}{\partial y^2} + \frac{\partial^2}{\partial z^2}\right) + P_e^v, \tag{3.156a}$$

$$\widetilde{Q} = -\frac{\hbar^2\gamma_2}{2\mathrm{m}_0}\left(\frac{\partial^2}{\partial x^2}+\frac{\partial^2}{\partial y^2}-2\frac{\partial^2}{\partial z^2}\right)+Q_e, \tag{3.156b}$$

$$\widetilde{R} \equiv -\frac{\sqrt{(3)}\hbar^2}{2\mathrm{m}_0}\left[-\gamma_2\left(\frac{\partial^2}{\partial x^2}-\frac{\partial^2}{\partial y^2}\right)+\mathrm{j}2\gamma_3\frac{\partial^2}{\partial x\partial y}\right]+R_e, \tag{3.156c}$$

$$\widetilde{S} \equiv -\frac{\sqrt{(3)}\hbar^2\gamma_3}{\mathrm{m}_0}\left(\frac{\partial^2}{\partial x\partial z}-\mathrm{j}\frac{\partial^2}{\partial y\partial z}\right)+S_e. \tag{3.156d}$$

Once equation (3.121) is solved for $\varepsilon^{\mathrm{v}}_{nj}$ and $F^{\mathrm{v}}_{nj}(\vec{r})$, $n = 1, 2, 3, 4, 5, 6$, through the operators given in equations (3.156a–d) through (3.123), the valence band electron (hole) states are obtained through equation (3.125).

3.3.4 Semiconductors with the zinc blende structure

For zinc blende structures with in-plane (i.e., in the xy plane) biaxial strains, equation (3.141) is reduced to

$$e_{xx} = e_{yy} \neq e_{zz}, \quad e_{xy} = e_{yx} = e_{xz} = e_{zx} = e_{yz} = e_{zy} = 0. \tag{3.157}$$

When the strain is introduced by a lattice mismatch between two layers, we have

$$e_{xx} = e_{yy} = \frac{\mathrm{a}_0 - a}{a} \equiv e_0, \text{ and } e_{zz} = -2\frac{C_{12}}{C_{11}}e_0, \tag{3.158}$$

with a_0 and a denoted as the lattice constant of the unstrained (substrate) layer and the strained layer, respectively. C_{ij}'s are the material elastic stiffness constants, which can be found from a semiconductor material database [13, 14].

As such, for semiconductors with a zinc blende structure, we find

$$P^{\mathrm{c}}_e = 2\mathrm{a}_{\mathrm{c}}e_0\left(1-\frac{C_{12}}{C_{11}}\right), \quad P^{\mathrm{v}}_e = -2\mathrm{a}_{\mathrm{v}}e_0\left(1-\frac{C_{12}}{C_{11}}\right),$$
$$Q_e = -\mathrm{b}e_0\left(1+2\frac{C_{12}}{C_{11}}\right), \quad R_e = S_e = 0. \tag{3.159}$$

In particular, in strained bulk semiconductors, when considering the 4×4 Luttinger–Kohn Hamiltonian with the spin–orbit split bands excluded from group A, we find that the strain changes only the diagonal elements in equation (3.84). Therefore, the same diagonalizing technique introduced in Section 3.1.5 is still applicable and hence we can find analytical solutions. Actually, by noticing equations (3.154a–d) and (3.159), we can write the Luttinger–Kohn Hamiltonian in equation (3.84) as

$$-\begin{bmatrix} P+P_e^{\mathrm{v}}+Q+Q_e & -S & R & 0 \\ -S^* & P+P_e^{\mathrm{v}}-Q-Q_e & 0 & R \\ R^* & 0 & P+P_e^{\mathrm{v}}-Q-Q_e & S \\ 0 & R^* & S^* & P+P_e^{\mathrm{v}}+Q+Q_e \end{bmatrix}, \tag{3.160}$$

to show the strain effect explicitly, where the matrix elements R and S are still given by equation (3.77a–d). Adopting the same approach by which we obtain equation (3.90a&b) from equation (3.84), we find the following analytical expressions for the valence band energies in the strained bulk semiconductors

$$E_{\mathrm{hh}}^{\mathrm{bs}} = -(P+P_e^{\mathrm{v}}) + \sqrt{\left((Q+Q_e)^2 + |R|^2 + |S|^2\right)}, \tag{3.161a}$$

$$E_{\mathrm{lh}}^{\mathrm{bs}} = -(P+P_e^{\mathrm{v}}) - \sqrt{\left((Q+Q_e)^2 + |R|^2 + |S|^2\right)}. \tag{3.161b}$$

Under biaxial compressive strain, $a > a_0$, $e_0 < 0$, according to equation (3.159), we have $P_e^{\mathrm{c}} < 0$, $P_e^{\mathrm{v}} > 0$, and $Q_e > 0$. Comparing with the unconstrained case, we find that the conduction band will shift down by the conduction band hydrostatic energy (P_e^{c}), the heavy-hole band will shift down by the valence band hydrostatic energy (P_e^{v}) plus a shift up by almost the shear energy (Q_e, once $|Q+Q_e| \gg \sqrt{(|R|^2+|S|^2)}$, e.g., at $\vec{k}=0$), and the light-hole band will shift down by the valence band hydrostatic energy (P_e^{v}) plus another shift down by almost the shear energy (Q_e, once $|Q+Q_e| \gg \sqrt{(|R|^2+|S|^2)}$, e.g., at $\vec{k}=0$). As a result, the degenerated heavy-hole and light-hole band at $\vec{k}=0$ will split by twice the shear energy ($2Q_e$), with the heavy-hole band on top.

Under biaxial tensile strain, $a < a_0$, $e_0 > 0$, according to equation (3.140), we have $P_e^{\mathrm{c}} > 0$, $P_e^{\mathrm{v}} < 0$, and $Q_e < 0$. Comparing with the unconstrained case, we find that the conduction band will shift up by the conduction band hydrostatic energy (P_e^{c}), the heavy-hole band will shift up by the valence band hydrostatic energy (P_e^{v}) plus a shift down by almost the shear energy (Q_e, once $|Q+Q_e| \gg \sqrt{(|R|^2+|S|^2)}$, e.g., at $\vec{k}=0$), and the light-hole band will shift up by the valence band hydrostatic energy (P_e^{v}) plus another shift up by almost the shear energy (Q_e, once $|Q+Q_e| \gg \sqrt{(|R|^2+|S|^2)}$, e.g., at $\vec{k}=0$). As a result, the degenerated heavy-hole and light-hole bands at $\vec{k}=0$ will also split by twice the shear energy ($2Q_e$), but with the light-hole band on top.

In strained layer QW structures, strain produces similar effects. However, in a QW structure there is no degeneracy at $\vec{k}=0$ between the heavy-hole and light-hole bands, with the heavy-hole band always on top. In this case, the heavy-hole and light-hole bands will be further apart at $\vec{k}=0$ by the compressive strain. On the contrary, the heavy-hole and light-hole bands will be brought closer at $\vec{k}=0$ by a tensile strain. Under a considerable tensile strain, reversal may happen with the light-hole band on top.

The strain effect on both a bulk semiconductor and a QW structure is illustrated in Figs. 3.5(a) and (b), respectively.

The strain has other effects on the band structure, such as enhancing the material anisotropy and changing the effective mass [12, 21]. These effects are more complicated

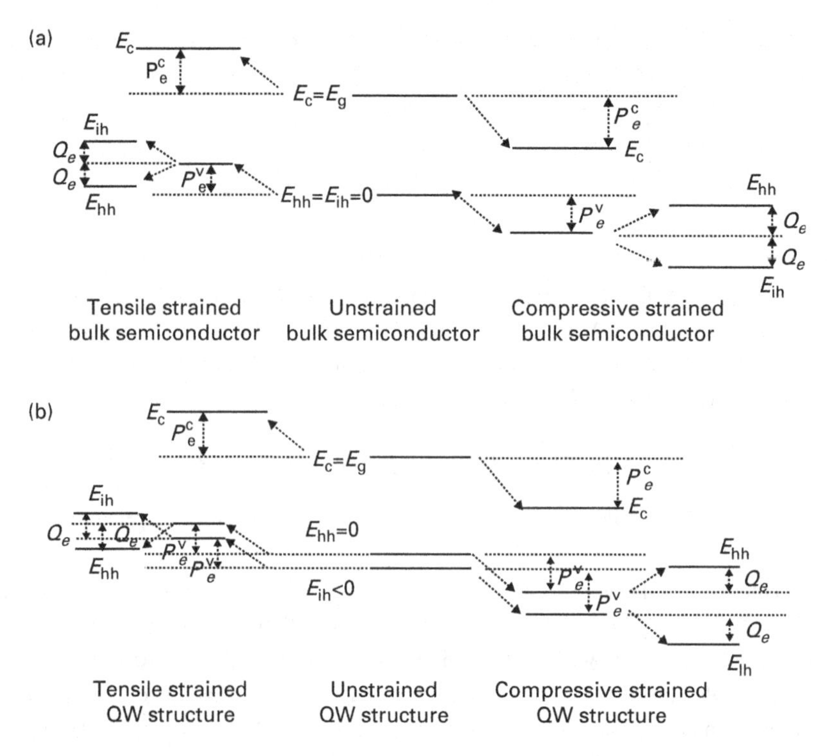

Fig. 3.5. The strain effect on band edge energies. (a) In bulk semiconductors. (b) In QW structures.

in QW structures where mass reversal and band mixing occur even without strain [21]. Discussion on these effects, however, is beyond the scope of this book.

3.4 Summary of the k–p theory

To summarize this chapter following the k–p approach, a step by step procedure is set out below showing how the band structure is calculated for a single electron in bulk semiconductors and in QW structures, with or without strain applied.

Step 1 uniform lattice at $k = 0$.
Hamiltonian: kinetic energy + periodic lattice potential.
Solution technique: Bloch's theorem.

(1) Factorize the solution as a global plane wave and a periodic local function.
(2) Map to a single hydrogen atom system in solving the local function.

Drawbacks:

(1) valid only for $k = 0$;
(2) spin–orbit coupling effect is missing.

Step 2 uniform lattice at k = 0 with spin–orbit coupling.
Hamiltonian: kinetic energy + periodic lattice potential + s–o energy.
Solution technique: Kane's approach.

(1) Treat the s–o energy as a perturbation term.
(2) Ignore the existence of other bands except for the conduction band, the heavy-hole band, the light-hole band and the s–o split band.

Drawback:

(1) poor accuracy.

Step 3 uniform lattice in the neighborhood of k = 0 with spin–orbit coupling.
Hamiltonian: kinetic energy + periodic lattice potential + s–o energy + k–p energy.
Solution technique: Luttinger–Kohn's approach.

(1) Treat the k–p energy as a perturbation term.
(2) Effect of the rest of the bands is included in considering the bands of interest.
(3) Degeneracy eliminated through proper combination of the base functions.

This is the final step for bulk semiconductors.

Step 4 uniform lattice in the neighborhood of k = 0 with spin–orbit coupling + QW.
Hamiltonian: kinetic energy + periodic lattice potential + s–o energy + k–p energy + non-periodic QW potential energy.
Solution technique: effective mass theory.

(1) Factorize the solution as a global wave envelope with an unknown shape and a specific set of periodic local functions, i.e., the L–K base function obtained for bulk, which is equivalent to an integral transform taking the L–K base as the core function.
(2) Under this transform, the unknown shape satisfies an integral equation; the only integral term involves the newly added QW potential energy, which can be approximated to a convolution, and provided that this extra energy does not change drastically in the lattice scale, under the inverse Fourier transform, a set of PDEs is obtained for the unknown shape in the space domain.
(3) Solve the set of PDEs to obtain the unknown shape in the space domain, and construct the final solution by taking a combination of L–K base functions modulated by the PDE solution in the space domain.

This is the final step for QW structures without strains applied.

Step 5 uniform lattice in the neighborhood of k = 0 with spin-orbit coupling + QW + strain.
Hamiltonian: kinetic energy + periodic lattice potential + s–o energy + k–p energy + non-periodic QW potential energy + lattice deformation energy.
Solution technique: perturbation method.

(1) Treat the lattice deformation energy as a perturbation.
(2) Since the lattice deformation energy is still periodic, the treatment is similar to the L–K approach.
(3) Mapping is used to skip the tedious derivations for the effective Hamiltonian matrix.

This is the final step for QW structures or bulk semiconductors with strains applied.

References

[1] R. M. Martin, *Electronic Structure*, 1st edn (Cambridge, UK: Cambridge University Press, 2004).

[2] J. M. Thijssen, *Computational Physics*, 1st edn (Cambridge, UK: Cambridge University Press, 1999).

[3] J. Singh, *Electronic and Optoelectronic Properties of Semiconductor Structures*, 1st edn (London, UK: Cambridge University Press, 2003).

[4] P. Y. Yu and M. Cardona, *Fundamentals of Semiconductors: Physics and Materials Properties*, 3rd edn (New York: Springer-Verlag, 2001).

[5] L. D. Landau and E. M. Lifshitz, *Quantum Mechanics*, 3rd edn (Oxford, UK: Pergamon Press, 1977).

[6] L. Schiff, *Quantum Mechanics*, 1st edn (New York: McGraw-Hill, 1968).

[7] E. O. Kane, Band structure of indium antimonide. *Journal of Phys. Chem. Solids*, **1** (1957), 249–61.

[8] E. O. Kane, The k·p method. In *Semiconductor and Semimetals, Vol. 1*, ed. R. K. Willardson and A. C. Beer. (New York: Academic, 1966).

[9] P. Lowdin, A note on the quantum-mechanical perturbation theory. *Journal of Chem. Phys.*, **19** (1951), 1396–401.

[10] J. M. Luttinger and W. Kohn, Motion of electrons and holes in perturbed periodic fields. *Phys. Rev.*, **97** (1955), 869–83.

[11] J. M. Luttinger, Quantum theory of cyclotron resonance in semiconductors: general theory. *Phys. Rev.*, **102** (1956), 1030–41.

[12] S. L. Chuang, *Physics of Optoelectronic Devices*, 1st edn (New York: John Wiley & Sons, 1995).

[13] Landolt-Bornstein, Vol. 17 Semiconductors, ed. O. Madelung, M. Schulz, and H. Weiss. In *Numerical Data and Fundamental Relationships in Science and Technology*, ed. K. H. Hellwege. (Berlin: Springer-Verlag, 1982).

[14] S. Adachi, *Physical Properties of III-V Semiconductor Compounds, InP, InAs, GaAs, GaP, InGaAs, and InGaAsP*, 1st edn (New York: John Wiley & Sons, 1992).

[15] W. Shan, W. Walukiewicz, J. W. Ager III, *et al.*, Band anticrossing in GaInNAs alloys. *Phys. Rev. Lett.*, **82**:6 (1999), 1221–4.

[16] C. Y.-P. Chao and S. L. Chuang, Spin-orbit-coupling effects on the valence-band structure of strained semiconductor quantum wells. *Phys. Rev. B*, **46**:7 (1992), 4110–22.

[17] Y. Rajakarunanayake, R. H. Miles, G. Y. Wu, and T. C. McGill, Band structure of ZnSe-ZnTe superlattices. *Phys. Rev. B*, **37**:17 (1988), 10212–5.

[18] S. L. Chuang and C. S. Chang, K·p method for strained wurtzite semiconductors. *Phys. Rev. B*, **54**:4 (1996), 2491–504.

[19] X. C. Zhang, A. Pfeuffer-Jeschke, K. Ortner, *et al.*, Rashba splitting in n-type modulation-doped HgTe quantum wells with an inverted band structure. *Phys. Rev. B*, **63** (2001), 245305-1–8.

[20] S. Tomic and E. P. O'Reilly, Optimization of material parameters in 1.3-μm InGaAsN-GaAs lasers. *IEEE Photon. Tech. Lett.*, **15**:1 (2003), 6-8.
[21] W. W. Chow and S. W. Koch, *Semiconductor Laser Fundamentals: Physics of the Gain Materials*, 1st edn (Berlin: Springer-Verlag, 1999).
[22] D. Ahn and S. L. Chuang, Optical gain in a strained-layer quantum-well laser. *IEEE Journal of Quantum Electron.*, **QE-24**:12 (1988), 2400–6.
[23] S. L. Chuang, Efficient band-structure calculations of strained quantum wells using a two-by-two Hamiltonian. *Phys. Rev. B*, **43** (1991), 9649–61.
[24] C. Kittel, *Introduction to Solid State Physics*, 1st edn (New York: Wiley & Sons, 1971).
[25] C. Kittel, *Quantum Theory of Solids*, 1st edn (New York: Wiley & Sons, 1967).
[26] G. L. Bir and G. E. Pikus, *Symmetry and Strain-Induced Effects in Semiconductors*, 1st edn (New York: Wiley & Sons, 1974).

4 Material model II: Optical gain

Based on the single electron band structures that have been solved in Chapter 3, we are ready to derive models for calculation of material optical properties.

4.1 A comprehensive model with many-body effect

4.1.1 Introduction

As summarized in Table 4.1, a real world physics process is described in the time–space domain with time and space vectors as arguments. Through Fourier transform, we can also analyze this process in the frequency (energy) – wave vector (momentum) domain with the frequency and wave vector as arguments. The original arguments and their alternatives form conjugate pairs which satisfy Heisenberg's uncertainty relation.

Table 4.2 shows all possible combinations for describing a physics process in different domains.

Selecting a suitable domain to describe a physics process may bring the following advantages:

- a derivative operation in the original domain (time or space) will become an algebraic operation in the alternative domain (frequency or momentum). Therefore, under an integration transform with the core function obtained from solving an eigenvalue problem, a linear differential equation can be converted to an algebraic equation. In particular, if the linear differential equation contains only invariant parameters, this integration transform is the Fourier transform where the plane wave function serves as the core function, as it is the eigenfunction of any linear and parameter invariant system.
- any periodic function in one domain will be a discrete function after being transformed into its conjugate domain. For example, a periodic time domain function must have a discrete frequency or energy spectrum, a periodic space domain function must have a discrete wave vector or momentum dependence, and vice versa.

As will be seen in this chapter, many-body interactions among polaritons (through which photons are emitted), electrons, holes, and phonons (lattice wave) can be described

Table 4.1. The original and alternative arguments

Original argument	Relation	Alternative argument
Time (t)	single-side Fourier transform	frequency (ω) or energy ($E = \hbar\omega$)
Space vector ($\vec{r}$)	Fourier transform	wave vector ($\vec{k}$) or momentum ($p = \hbar\vec{k}$)

Table 4.2. The description domains for a physical process

Combination	Eigenstate	Application
Time–space domain	$\vec{r} = a(t)$	classic matter trajectory
Time–momentum domain	$\vec{k} = b(t)$	many-body interaction
Frequency–space domain	$\omega = c(\vec{r})$	band diagram
Frequency–momentum domain	$\omega = d(\vec{k})$	classic wave dispersion curve

in the time–momentum domain through the Heisenberg equation, through which the material optical properties such as gain and refractive index change can also be extracted. As opposed to conventional time–space domain modeling, we need to deal only with ODEs rather than PDEs in this approach, at the cost of having to solve a large number of equations rather than a few.

4.1.2 The Heisenberg equation

If we stay in the eigenstates that have been solved in Chapter 3, and attribute the time evolution of any physical quantity to its associated operator ($\vec{O}$) dependence on time, we will obtain the Heisenberg equation in the form of [1]

$$\mathrm{j}\hbar\frac{\mathrm{d}\vec{O}}{\mathrm{d}t} = [\vec{O}, \vec{H}] = \vec{O}\vec{H} - \vec{H}\vec{O}, \tag{4.1}$$

with $\vec{H}$ denoting the system Hamiltonian.

There are several advantages of using this picture to describe many-body interactions in the time–momentum domain:

- there is no need to introduce or compute any new eigenstates;
- the governing equation is readily obtained in the form of equation (4.1), once we find the system Hamiltonian and the density distribution operators for particles involved in the interaction;
- bookkeeping is easier in counting those interactions.

4.1.3 A comprehensive model

In the many-body microscopic model for semiconductors, interaction between conduction band electrons and valence band holes, which generates polaritons contributing to the electromagnetic wave at optical wavelength, and other interactions between electrons, holes and phonons, i.e., Coulomb repulsion between electrons, Coulomb repulsion between holes, Coulomb attraction between electrons and holes, and scattering of electrons and holes by phonons, should be considered.

By introducing the annihilation and creation operators for conduction band electrons, for valence band holes, and for phonons, respectively, we can write the system Hamiltonian in the wave vector domain as [2, 3]

$$\vec{H} = \vec{H}_{\text{kin}} + \vec{H}_{\text{pol}} + \vec{H}_{\text{C}} + \vec{H}_{\text{sct}}. \tag{4.2}$$

The terms on the RHS of equation (4.2) are the kinetic energy, the carrier and field (i.e., the electron–hole pair and the optical field) interaction energy, the carrier Coulomb interaction energy, and the carrier–phonon scattering energy Hamiltonians. The first term is given by

$$\begin{aligned}
\vec{H}_{\text{kin}} &= \int_{\Omega} \mathrm{d}\vec{r} \left[\vec{\psi}^{+}(\vec{r})(-\frac{\hbar^2\nabla^2}{2m_n})\vec{\psi}(\vec{r}) \right] \\
&= \sum_{n=\text{c,v}} \sum_{\vec{k}} \varepsilon_{n\vec{k}} \hat{a}^{+}_{n\vec{k}} \hat{a}_{n\vec{k}} \\
&= \sum_{\vec{k}} (\varepsilon_{\text{c}\vec{k}} \hat{a}^{+}_{\text{c}\vec{k}} \hat{a}_{\text{c}\vec{k}} + \varepsilon_{\text{v}\vec{k}} \hat{a}^{+}_{\text{v}\vec{k}} \hat{a}_{\text{v}\vec{k}}) \\
&= \sum_{\vec{k}} [(E_{\text{g}} + \varepsilon_{\text{e}\vec{k}}) \hat{a}^{+}_{\vec{k}} \hat{a}_{\vec{k}} + \varepsilon_{\text{h}\vec{k}} \hat{b}^{+}_{-\vec{k}} \hat{b}_{-\vec{k}}].
\end{aligned} \tag{4.3}$$

Throughout this chapter, we use the hat symbol ($\hat{o}$) and the vector symbol ($\vec{O}$) to present the microscopic and macroscopic operators, respectively, with symbols $\hat{o}^{+}$ and $\vec{O}^{+}$ indicating the Hermitian conjugate operators of $\hat{o}$ and $\vec{O}$, respectively. In equation (4.3), we have the field operator of all electrons (i.e., conduction band electrons and valence band holes) defined as

$$\vec{\psi}(\vec{r}, t) \equiv |\psi\rangle\,\langle\psi| = \sum_{n=\text{c,v}} \sum_{\vec{k}} \hat{a}_{n\vec{k}}(t) \langle \vec{r} \mid \phi_{n\vec{k}} \rangle, \tag{4.4}$$

with $|\psi\rangle$ defined as the state for all electrons, $|\phi_{n\vec{k}}\rangle$ and $\langle\vec{r} \mid \phi_{n\vec{k}}\rangle$ the single electron state and wave function, respectively. Also in equations (4.3) and (4.4), n denotes the band index integer, m_n the nth band electron effective mass, $\varepsilon_{n\vec{k}} = \hbar^2 k^2/(2m_n)$ the nth band electron energy, $\hat{a}_{n\vec{k}}$ and $\hat{a}^{+}_{n\vec{k}}$ the nth band electron annihilation and creation operator, respectively. In this model, we have considered only two bands, i.e., the conduction band and valence band, and have defined

$$\varepsilon_{c\vec{k}} = \hbar^2k^2/(2m_e), \quad \varepsilon_{h\vec{k}} = \hbar^2k^2/(2m_h) = -\hbar^2k^2/(2m_v) = -\varepsilon_{v\vec{k}}, \tag{4.5}$$

as the conduction and valence band energies, respectively. For convenience, we have further introduced $\hat{a}_{\vec{k}}$ and $\hat{a}^+_{\vec{k}}$ as the conduction band electron, $\hat{b}_{\vec{k}}$ and $\hat{b}^+_{\vec{k}}$ as the valence band hole annihilation and creation operators, respectively. In deriving the last part of equation (4.3), we have used relations $\hat{b}^+_{-\vec{k}} = \hat{a}_{v\vec{k}}$ (the annihilation of a valence band electron is the creation of a valence band hole), and $\hat{b}_{-\vec{k}} = \hat{a}^+_{v\vec{k}}$ (the creation of a valence band electron is the annihilation of a valence band hole). As such, we find $\hat{a}^+_{v\vec{k}}\hat{a}_{v\vec{k}} = \hat{b}_{-\vec{k}}\hat{b}^+_{-\vec{k}} = 1 - \hat{b}^+_{-\vec{k}}\hat{b}_{-\vec{k}}$ by utilizing the anticommutation relation for the Fermion operator. The reference energy is chosen such that the constant energy term $\sum_{\vec{k}} \varepsilon_{v\vec{k}}$ is zero. The summation over the entire $\vec{k}$ space also includes different spin states.

The second term on the RHS of equation (4.2) is the carrier (electron and hole) and field (electromagnetic wave at optical wavelength) interaction Hamiltonian, which is given in the form

$$\begin{aligned}\vec{H}_{\text{pol}} &= -\Omega\vec{P}\cdot\vec{E} \\ &= -\sum_{\vec{k}}(\mu_{\vec{k}}\hat{a}^+_{\vec{k}}\hat{b}^+_{-\vec{k}} + \mu^*_{\vec{k}}\hat{b}_{-\vec{k}}\hat{a}_{\vec{k}})\cdot\vec{E}.\end{aligned} \tag{4.6}$$

In equation (4.6), $\vec{E}$ is the electrical field of the optical wave propagating inside the medium that we are analyzing, $\vec{P}$ the polarization operator of the medium, and $\mu_{\vec{k}}$ the dipole matrix element between the conduction and valence bands given in the form

$$\vec{\mu}_{\vec{k}} = \mathrm{e}\langle\phi^{\mathrm{c}}_{\vec{k}}|\vec{r}\,|\phi^{\mathrm{v}}_{\vec{k}}\rangle, \tag{4.7}$$

with $|\phi^{\mathrm{c}}_{\vec{k}}\rangle$ and $|\phi^{\mathrm{v}}_{\vec{k}}\rangle$ representing the single conduction band electron and single valence band hole state, respectively. Once the band structure for a single electron is solved, following this definition, the dipole matrix elements are readily obtained. The electron and hole eigenstates (i.e., the energy bands or discrete energy levels and wave functions) can be calculated for a given structure through the method introduced in Chapter 3.

Considering either a direct or an exchange electron–electron collision process as shown in Fig. 4.1, we can write the carrier Coulomb interaction Hamiltonian as the third term on the RHS of equation (4.2) in the form

$$\begin{aligned}\vec{H}_{\mathrm{C}} &= \int_\Omega \mathrm{d}\vec{r}_1 \int_\Omega \mathrm{d}\vec{r}_2 \left[\vec{\psi}^+(\vec{r}_1)\vec{\psi}^+(\vec{r}_2)\frac{\mathrm{e}^2}{4\pi\varepsilon_0|\vec{r}_1-\vec{r}_2|}\vec{\psi}(\vec{r}_2)\vec{\psi}(\vec{r}_1)\right] \\ &= \frac{1}{2}\sum_{\vec{k},\vec{k}'}\sum_{\vec{q}\neq 0} V_{|\vec{q}|}\left[\hat{a}^+_{\mathrm{c}(\vec{k}+\vec{q})}\hat{a}^+_{\mathrm{c}(\vec{k}'-\vec{q})}\hat{a}_{\mathrm{c}\vec{k}'}\hat{a}_{\mathrm{c}\vec{k}}\right. \\ &\quad \left. + \hat{a}^+_{\mathrm{v}(\vec{k}+\vec{q})}\hat{a}^+_{\mathrm{v}(\vec{k}'-\vec{q})}\hat{a}_{\mathrm{v}\vec{k}'}\hat{a}_{\mathrm{v}\vec{k}} + 2\hat{a}^+_{\mathrm{c}(\vec{k}+\vec{q})}\hat{a}^+_{\mathrm{v}(\vec{k}'-\vec{q})}\hat{a}_{\mathrm{v}\vec{k}'}\hat{a}_{\mathrm{c}\vec{k}}\right]\end{aligned}$$

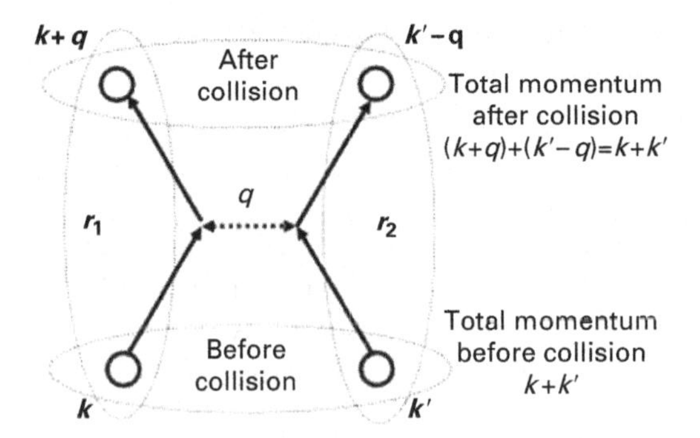

Fig. 4.1. The electron–electron collision process, in which total momentum is conserved.

$$= \frac{1}{2} \sum_{\vec{k},\vec{k}'} \sum_{\vec{q} \neq 0} V_{|\vec{q}|} \left(\hat{a}^+_{\vec{k}+\vec{q}} \hat{a}^+_{\vec{k}'-\vec{q}} \hat{a}_{\vec{k}'} \hat{a}_{\vec{k}} + \hat{b}^+_{\vec{k}+\vec{q}} \hat{b}^+_{\vec{k}'-\vec{q}} \hat{b}_{\vec{k}'} \hat{b}_{\vec{k}} \right.$$
$$\left. -2\hat{a}^+_{\vec{k}+\vec{q}} \hat{b}^+_{\vec{k}'-\vec{q}} \hat{b}_{\vec{k}'} \hat{a}_{\vec{k}} \right) + \sum_{\vec{k}} \sum_{\vec{q} \neq 0} V_{|\vec{q}|} \hat{b}^+_{-\vec{k}} \hat{b}_{-\vec{k}}, \tag{4.8}$$

with the first two terms within parenthesis on the RHS describing the intraband carrier interaction, i.e., the conduction band electron and valence band hole Coulomb repulsion, and the third term the interband carrier interaction, i.e., the conduction band and valence band hole Coulomb attraction. In deriving equation (4.8), we have used the fact that the Coulomb scattering does not change the spin orientations of the electrons involved. The contribution from $\vec{q} = 0$ must be excluded since it has to be cancelled by the corresponding terms of the electron–ion and ion–ion Coulomb interaction. We have also dropped the terms that do not conserve the number of electrons in each band since such terms do not conserve energy. In equation (4.8), the unscreened Coulomb potential in the wave vector domain is obtained from

$$V_{\vec{k},\vec{k}',\vec{k}'-\vec{q},\vec{k}+\vec{q}} = \int_\Omega \mathrm{d}\vec{r}_1 \int_\Omega \mathrm{d}\vec{r}_2 \left[\phi^*_{\vec{k}+\vec{q}}(\vec{r}_1) \phi^*_{\vec{k}'-\vec{q}}(\vec{r}_2) \frac{\mathrm{e}^2}{4\pi\varepsilon_0 |\vec{r}_1 - \vec{r}_2|} \phi_{\vec{k}'}(\vec{r}_2) \phi_{\vec{k}}(\vec{r}_1) \right]. \tag{4.9a}$$

For bulk semiconductors, we approximate all carrier wave functions involved in equation (4.9a) to plane waves. Therefore, equation (4.9a) reduces to

$$V_{\vec{k},\vec{k}',\vec{k}'-\vec{q},\vec{k}+\vec{q}} = \frac{1}{\Omega^2} \int_\Omega \mathrm{d}\vec{r}_1 \int_\Omega \mathrm{d}\vec{r}_2$$
$$\left[\mathrm{e}^{-\mathrm{j}(\vec{k}+\vec{q})\cdot\vec{r}_1} \mathrm{e}^{-\mathrm{j}(\vec{k}'-\vec{q})\cdot\vec{r}_2} \frac{\mathrm{e}^2}{4\pi\varepsilon_0 |\vec{r}_1 - \vec{r}_2|} \mathrm{e}^{\mathrm{j}\vec{k}'\cdot\vec{r}_2} \mathrm{e}^{\mathrm{j}\vec{k}\cdot\vec{r}_1} \right]$$

$$= \frac{e^2}{4\pi\varepsilon_0\Omega^2}\int_\Omega d\vec{R}_1 \int_\Omega d\vec{r}_2 \left[\frac{e^{-j\vec{q}\cdot(\vec{r}_1-\vec{r}_2)}}{|\vec{r}_1-\vec{r}_2|}\right]$$

$$= \frac{e^2}{4\pi\varepsilon_0\Omega}\int_\Omega d\vec{r}\left(\frac{e^{-j\vec{q}\cdot\vec{r}}}{r}\right).$$

To make this integration converge, we further multiply the integrand by an extra decay factor e^{-ar} in the radial direction and then take the limit as $a \to 0$. As such, we obtain

$$\begin{aligned} V_{|\vec{q}|} &\equiv \frac{e^2}{4\pi\varepsilon_0\Omega}\lim_{a\to 0}\int_\Omega d\vec{r}\left(\frac{e^{-j\vec{q}\cdot\vec{r}}e^{-ar}}{r}\right) \\ &= \frac{e^2}{2\varepsilon_0\Omega}\lim_{a\to 0}\int_0^\infty dr\int_0^\pi r\sin\theta d\theta(e^{-jqr\cos\theta}e^{-ar}) \\ &= \frac{e^2}{\varepsilon_0\Omega q}\lim_{a\to 0}\int_0^\infty e^{-ar}\sin(qr)dr \\ &= \frac{e^2}{\varepsilon_0\Omega q^2}\lim_{a\to 0}\frac{1}{1+(a/q)^2} = \frac{e^2}{\varepsilon_0\Omega q^2}, \end{aligned} \tag{4.9b}$$

with the wave vector $\vec{q}$ defined in the 3D wave vector domain.

For an idealized quantum well (i.e., a square well with infinite barrier), equation (4.9a) becomes

$$\begin{aligned} V_{\vec{k},\vec{k}',\vec{k}'-\vec{q},\vec{k}+\vec{q}} &= \frac{1}{\Sigma^2}\int_\Sigma d\vec{r}_{t1}\int_{Z_1} dz_1\int_\Sigma d\vec{r}_{t2} \\ &\times\int_{Z_2} dz_2\left[e^{-j(\vec{k}_t+\vec{q}_t)\cdot\vec{r}_{t1}}F^*(z_1)e^{-j(\vec{k}_{t'}-\vec{q}_t)\cdot\vec{r}_{t2}}F^*(z_2)\right. \\ &\left.\times\frac{e^2}{4\pi\varepsilon_0|\vec{r}_1-\vec{r}_2|}e^{j\vec{k}'_t\cdot\vec{r}_{t2}}F(z_2)e^{j\vec{k}_t\cdot\vec{r}_{t1}}F(z_1)\right] \\ &= \frac{e^2}{4\pi\varepsilon_0\Sigma^2}\int_\Sigma d\vec{r}_{t1}\int_\Sigma d\vec{r}_{t2}\left[\frac{e^{-j\vec{q}_t\cdot(\vec{r}_{t1}-\vec{r}_{t2})}}{|\vec{r}_{t1}-\vec{r}_{t2}|}\right] \\ &= \frac{e^2}{4\pi\varepsilon_0\Sigma}\int_\Sigma d\vec{r}_t\left(\frac{e^{-j\vec{q}_t\cdot\vec{r}_t}}{r_t}\right). \end{aligned}$$

By following the same method in obtaining equation (4.9b), we find

$$\begin{aligned} V_{|\vec{q}|} &\equiv \frac{e^2}{4\pi\varepsilon_0\Sigma}\lim_{a\to 0}\int_\Sigma d\vec{r}_t\left(\frac{e^{-j\vec{q}_t\cdot\vec{r}_t}e^{-ar_t}}{r_t}\right) \\ &= \frac{e^2}{4\pi\varepsilon_0\Sigma}\lim_{a\to 0}\int_0^{2\pi}\int_0^\infty e^{-(a+jq\cos\varphi)r}dr\, d\varphi \end{aligned}$$

$$= \frac{e^2}{2\pi\varepsilon_0\Sigma}\lim_{a\to 0}\int_0^{\pi}\frac{1}{a+jq\cos\varphi}\,d\varphi$$
$$= \frac{e^2}{2\pi\varepsilon_0\Sigma}\lim_{a\to 0}\frac{\pi}{\sqrt{(a^2+q^2)}} = \frac{e^2}{2\varepsilon_0\Sigma q}, \tag{4.9c}$$

with the wave vector $\vec{q}$ defined in the 2D wave vector domain.

Similarly to arbitrary quantum wells, the Coulomb potential depends on wave functions of the involved carrier states

$$V_{\vec{k},\vec{k}',\vec{k}'-\vec{q},\vec{k}+\vec{q}} = \frac{1}{\Sigma^2}\int_{\Sigma}d\vec{r}_{t1}\int_{Z_1}dz_1\int_{\Sigma}d\vec{r}_{t2}\int_{Z_2}dz_2$$
$$\Big[e^{-j(\vec{k}_t+\vec{q}_t)\cdot\vec{r}_{t1}}F^*_{\vec{k}_t+\vec{q}_t}(z_1)e^{-j(\vec{k}'_t-\vec{q}_t)\cdot\vec{r}_{t2}}F^*_{\vec{k}'_t-\vec{q}_t}(z_2)$$
$$\times\frac{e^2}{4\pi\varepsilon_0|\vec{r}_1-\vec{r}_2|}e^{j\vec{k}'_t\cdot\vec{r}_{t2}}F_{\vec{k}'_t}(z_2)e^{j\vec{k}_t\cdot\vec{r}_{t1}}F_{\vec{k}_t}(z_1)\Big]$$
$$= \frac{e^2}{4\pi\varepsilon_0\Sigma^2}\int_{Z_1}dz_1\int_{Z_2}dz_2\Big\{F^*_{\vec{k}_t+\vec{q}_t}(z_1)F^*_{\vec{k}'_t-\vec{q}_t}(z_2)$$
$$\times\Big[\int_{\Sigma}d\vec{r}_{t1}\int_{\Sigma}d\vec{r}_{t2}\frac{e^{-j\vec{q}_t\cdot(\vec{r}_{t1}-\vec{r}_{t2})}}{|\vec{r}_1-\vec{r}_2|}\Big]F_{\vec{k}'_t}(z_2)F_{\vec{k}_t}(z_1)\Big\}.$$

According to [4], this integral is evaluated through

$$V_{|\vec{k}-\vec{k}'|} \equiv \frac{e^2}{2\varepsilon_0\Sigma|\vec{k}-\vec{k}'|}\int_{Z_1}dz_1\int_{Z_2}dz_2\Big[F^*_{\vec{k}+\vec{q}}(z_1)F^*_{\vec{k}'-\vec{q}}(z_2)$$
$$\times e^{-|\vec{k}-\vec{k}'||z_1-z_2|}F_{\vec{k}'}(z_2)F_{\vec{k}}(z_1)\Big], \tag{4.9d}$$

with wave vectors $\vec{k}$, $\vec{k}'$, and $\vec{q}$ all defined in the 2D wave vector domain.

The last term on the RHS of equation (4.2) is the carrier scattering Hamiltonian, which is given as

$$\vec{H}_{\rm sct} = \sum_{\vec{k},\vec{q}}\hbar G_{\vec{q}}\Big[\hat{a}^+_{\vec{k}+\vec{q}}\hat{a}_{\vec{k}}(\hat{b}^{\rm LO}_{\vec{q}}+\hat{b}^{\rm LO+}_{-\vec{q}})+\hat{b}^+_{\vec{k}+\vec{q}}\hat{b}_{\vec{k}}(\hat{b}^{\rm LO}_{\vec{q}}+\hat{b}^{\rm LO+}_{-\vec{q}})\Big], \tag{4.10}$$

with $\hat{b}^{\rm LO}_{\vec{q}}$ and $\hat{b}^{\rm LO+}_{\vec{q}}$ defined as the annihilation and creation operators of the longitudinal optical (LO) phonons, respectively. Equation (4.10) describes scattering of electrons or holes inside the conduction or valence band by emitting or absorbing LO phonons. Note that $G_{\vec{q}}$ is the Fröhlich electron – LO phonon coupling matrix element that gives the linear interaction coefficient of an electron or hole with lattice polarization [2, 5]

$$G_{|\vec{k}|} = \sqrt{\Big[\frac{\omega_{\rm LO}\widetilde{V}_{|\vec{k}|}}{2\hbar}\Big(\frac{1}{\varepsilon_\infty}-\frac{1}{\varepsilon}\Big)\Big]}, \tag{4.11}$$

with ω_{LO} denoted as the LO phonon angular frequency, ε_∞ the frequency domain permittivity of the medium at high frequency. In equation (4.11), $\widetilde{V}_{|\vec{k}|}$ is the screened wave vector domain Coulomb potential which will be derived in Section 4.3.2.

By introducing

$$\hat{f}^{\mathrm{e}}_{\vec{k}} \equiv \hat{a}^{+}_{\vec{k}}\hat{a}_{\vec{k}}, \quad \hat{f}^{\mathrm{h}}_{\vec{k}} \equiv \hat{b}^{+}_{-\vec{k}}\hat{b}_{-\vec{k}}, \quad \hat{p}_{\vec{k}} \equiv \hat{b}_{-\vec{k}}\hat{a}_{\vec{k}}, \tag{4.12}$$

as the conduction band electron, the valence band hole and the polariton number operator, respectively, we find the governing equations for these operators through the Heisenberg equation (4.1)

$$\frac{\mathrm{d}\hat{f}^{\mathrm{e}}_{\vec{k}}}{\mathrm{d}t} = \frac{\mathrm{j}}{\hbar}[\hat{H}, \hat{f}^{\mathrm{e}}_{\vec{k}}], \quad \frac{\mathrm{d}\hat{f}^{\mathrm{h}}_{\vec{k}}}{\mathrm{d}t} = \frac{\mathrm{j}}{\hbar}[\hat{H}, \hat{f}^{\mathrm{h}}_{\vec{k}}], \quad \frac{\mathrm{d}\hat{p}_{\vec{k}}}{\mathrm{d}t} = \frac{\mathrm{j}}{\hbar}[\hat{H}, \hat{p}_{\vec{k}}], \tag{4.13}$$

with the Hamiltonian given by equation (4.2).

Equation (4.13) is a set of ODEs in the time–wave vector domain. Once they are solved, we obtain the corresponding macroscopic conduction band electron and valence band hole densities as

$$N_{\mathrm{e}}(t) = \frac{1}{\Omega}\sum_{\vec{k}} f^{\mathrm{e}}_{\vec{k}}, \quad N_{\mathrm{h}}(t) = \frac{1}{\Omega}\sum_{\vec{k}} f^{\mathrm{h}}_{\vec{k}}, \tag{4.14}$$

respectively. According to equation (4.6), the macroscopic polarization operator is linked to the polariton number operator by

$$\hat{\vec{P}} = \frac{1}{\Omega}\sum_{\vec{k}} (\vec{\mu}_{\vec{k}}\hat{p}^{+}_{\vec{k}} + \vec{\mu}^{*}_{\vec{k}}\hat{p}_{\vec{k}}). \tag{4.15}$$

Therefore, the macroscopic polarization is given as

$$\vec{P}(t) \equiv <\hat{\vec{P}}> = \frac{1}{\Omega}\sum_{\vec{k}} (\vec{\mu}_{\vec{k}} p^{*}_{\vec{k}} + \vec{\mu}^{*}_{\vec{k}} p_{\vec{k}}). \tag{4.16}$$

In equations (4.14) and (4.16), by taking away the hat, we use symbol $o \equiv <\hat{o}>$ to represent the average value (i.e., the expectation) of operator $\hat{o}$.

4.1.4 General governing equations

Equation (4.13) can be simplified by utilizing the properties of the bilinear product of Fermion operators in the form of $\hat{\alpha}_{\vec{k}}\hat{\beta}_{\vec{k}}$. Actually, we have [5]

$$[\hat{\alpha}_{\vec{k}}\hat{\beta}_{\vec{k}}, \hat{\gamma}_{\vec{k}'}\hat{\delta}_{\vec{k}'}] = 0, \quad \text{for } \vec{k} \neq \vec{k}', \tag{4.17a}$$

$$[\hat{\alpha}^{+}_{\vec{k}}\hat{\alpha}_{\vec{k}}, \beta_{\vec{k}'}] = [\hat{\alpha}_{\vec{k}}\hat{\alpha}^{+}_{\vec{k}}, \beta_{\vec{k}'}] = 0, \quad \text{for } \vec{k} \neq \vec{k}', \tag{4.17b}$$

$$\hat{\alpha}^+_{\vec{k}}\hat{\alpha}_{\vec{k}} + \hat{\alpha}_{\vec{k}}\hat{\alpha}^+_{\vec{k}} = 1, \tag{4.17}$$

$$\hat{\alpha}_{\vec{k}}\hat{\alpha}_{\vec{k}} = \hat{\alpha}^+_{\vec{k}}\hat{\alpha}^+_{\vec{k}} = 0. \tag{4.17a}$$

By rewriting equation (4.2) as

$$\begin{aligned}\vec{H} &= \vec{H}_{\text{kin}} + \vec{H}_{\text{pol}} + \vec{H}_{\text{C}} + \vec{H}_{\text{sct}}\\ &= \sum_{\vec{k}}[(E_{\text{g}} + \varepsilon_{\text{e}\vec{k}})\hat{f}^{\text{e}}_{\vec{k}} + \varepsilon_{\text{h}\vec{k}}\hat{f}^{\text{h}}_{\vec{k}}] - \sum_{\vec{k}}(\mu_{\vec{k}}\hat{p}^+_{\vec{k}} + \mu^*_{\vec{k}}\hat{p}_{\vec{k}})\cdot\vec{E} + \vec{H}_{\text{C}} + \vec{H}_{\text{sct}},\end{aligned} \tag{4.18}$$

and by utilizing equation (4.17a) we find from equation (4.13)

$$\frac{\mathrm{d}\hat{f}^{\text{e}}_{\vec{k}}}{\mathrm{d}t} = \frac{\text{j}}{\hbar}[(E_{\text{g}} + \varepsilon_{\text{e}\vec{k}})\hat{f}^{\text{e}}_{\vec{k}} + \varepsilon_{\text{h}\vec{k}}\hat{f}^{\text{h}}_{\vec{k}} - (\vec{\mu}_{\vec{k}}\hat{p}^+_{\vec{k}} + \vec{\mu}^*_{\vec{k}}\hat{p}_{\vec{k}})\cdot\vec{E}, \hat{f}^{\text{e}}_{\vec{k}}] + \frac{\text{j}}{\hbar}[\vec{H}_{\text{C}} + \vec{H}_{\text{sct}}, \hat{f}^{\text{e}}_{\vec{k}}], \tag{4.19a}$$

$$\frac{\mathrm{d}\hat{f}^{\text{h}}_{\vec{k}}}{\mathrm{d}t} = \frac{\text{j}}{\hbar}[(E_{\text{g}} + \varepsilon_{\text{e}\vec{k}})\hat{f}^{\text{e}}_{\vec{k}} + \varepsilon_{\text{h}\vec{k}}\hat{f}^{\text{h}}_{\vec{k}} - (\vec{\mu}_{\vec{k}}\hat{p}^+_{\vec{k}} + \vec{\mu}^*_{\vec{k}}\hat{p}_{\vec{k}})\cdot\vec{E}, \hat{f}^{\text{h}}_{\vec{k}}] + \frac{\text{j}}{\hbar}[\vec{H}_{\text{C}} + \vec{H}_{\text{sct}}, \hat{f}^{\text{h}}_{\vec{k}}], \tag{4.19b}$$

$$\frac{\mathrm{d}\hat{p}_{\vec{k}}}{\mathrm{d}t} = \frac{\text{j}}{\hbar}[(E_{\text{g}} + \varepsilon_{\text{e}\vec{k}})\hat{f}^{\text{e}}_{\vec{k}} + \varepsilon_{\text{h}\vec{k}}\hat{f}^{\text{h}}_{\vec{k}} - (\vec{\mu}_{\vec{k}}\hat{p}^+_{\vec{k}} + \vec{\mu}^*_{\vec{k}}\hat{p}_{\vec{k}})\cdot\vec{E}, \hat{p}_{\vec{k}}] + \frac{\text{j}}{\hbar}[\vec{H}_{\text{C}} + \vec{H}_{\text{sct}}, \hat{p}_{\vec{k}}]. \tag{4.19c}$$

The following commuting relations can also be found from equation (4.17b–d)

$$[\hat{f}^{\text{e}}_{\vec{k}}, \hat{f}^{\text{e}}_{\vec{k}}] = [\hat{f}^{\text{h}}_{\vec{k}}, \hat{f}^{\text{h}}_{\vec{k}}] = [\hat{p}_{\vec{k}}, \hat{p}_{\vec{k}}] = [\hat{p}^+_{\vec{k}}, \hat{p}^+_{\vec{k}}] = 0,$$

$$[\hat{f}^{\text{e}}_{\vec{k}}, \hat{f}^{\text{h}}_{\vec{k}}] = \hat{a}^+_{\vec{k}}\hat{a}_{\vec{k}}\hat{b}^+_{-\vec{k}}\hat{b}_{-\vec{k}} - \hat{b}^+_{-\vec{k}}\hat{b}_{-\vec{k}}\hat{a}^+_{\vec{k}}\hat{a}_{\vec{k}} = 0,$$

$$\begin{aligned}[\hat{f}^{\text{e}}_{\vec{k}}, \hat{p}_{\vec{k}}] &= \hat{a}^+_{\vec{k}}\hat{a}_{\vec{k}}\hat{b}_{-\vec{k}}\hat{a}_{\vec{k}} - \hat{b}_{-\vec{k}}\hat{a}_{\vec{k}}\hat{a}^+_{\vec{k}}\hat{a}_{\vec{k}} = \hat{b}_{-\vec{k}}\hat{a}^+_{\vec{k}}\hat{a}_{\vec{k}}\hat{a}_{\vec{k}} - \hat{b}_{-\vec{k}}\hat{a}_{\vec{k}}(1 - \hat{a}_{\vec{k}}\hat{a}^+_{\vec{k}})\\ &= -\hat{b}_{-\vec{k}}\hat{a}_{\vec{k}} + \hat{b}_{-\vec{k}}\hat{a}_{\vec{k}}\hat{a}_{\vec{k}}\hat{a}^+_{\vec{k}} = -\hat{b}_{-\vec{k}}\hat{a}_{\vec{k}} = -\hat{p}_{\vec{k}},\end{aligned}$$

$$\begin{aligned}[\hat{f}^{\text{e}}_{\vec{k}}, \hat{p}^+_{\vec{k}}] &= \hat{a}^+_{\vec{k}}\hat{a}_{\vec{k}}\hat{a}^+_{\vec{k}}\hat{b}^+_{-\vec{k}} - \hat{a}^+_{\vec{k}}\hat{b}^+_{-\vec{k}}\hat{a}^+_{\vec{k}}\hat{a}_{\vec{k}} = (1 - \hat{a}_{\vec{k}}\hat{a}^+_{\vec{k}})\hat{a}^+_{\vec{k}}\hat{b}^+_{-\vec{k}} - \hat{a}^+_{\vec{k}}\hat{a}^+_{\vec{k}}\hat{a}_{\vec{k}}\hat{b}^+_{-\vec{k}}\\ &= \hat{a}^+_{\vec{k}}\hat{b}^+_{-\vec{k}} - \hat{a}_{\vec{k}}\hat{a}^+_{\vec{k}}\hat{a}^+_{\vec{k}}\hat{b}^+_{-\vec{k}} = \hat{a}^+_{\vec{k}}\hat{b}^+_{-\vec{k}} = \hat{p}^+_{\vec{k}}, [\hat{f}^{\text{h}}_{\vec{k}}, \hat{p}_{\vec{k}}]\\ &= \hat{b}^+_{-\vec{k}}\hat{b}_{-\vec{k}}\hat{b}_{-\vec{k}}\hat{a}_{\vec{k}} - \hat{b}_{-\vec{k}}\hat{a}_{\vec{k}}\hat{b}^+_{-\vec{k}}\hat{b}_{-\vec{k}} = -\hat{b}_{-\vec{k}}\hat{b}^+_{-\vec{k}}\hat{b}_{-\vec{k}}\hat{a}_{\vec{k}}\\ &= -(1 - \hat{b}^+_{-\vec{k}}\hat{b}_{-\vec{k}})\hat{b}_{-\vec{k}}\hat{a}_{\vec{k}} = -\hat{b}_{-\vec{k}}\hat{a}_{\vec{k}} + \hat{b}^+_{-\vec{k}}\hat{b}_{-\vec{k}}\hat{b}_{-\vec{k}}\hat{a}_{\vec{k}}\\ &= -\hat{b}_{-\vec{k}}\hat{a}_{\vec{k}} = -\hat{p}_{\vec{k}},\end{aligned}$$

$$
\begin{aligned}
[\hat{f}^{\mathrm{h}}_{\vec{k}}, \hat{p}^{+}_{\vec{k}}] &= \hat{b}^{+}_{-\vec{k}}\hat{b}_{-\vec{k}}\hat{a}^{+}_{\vec{k}}\hat{b}^{+}_{-\vec{k}} - \hat{a}^{+}_{\vec{k}}\hat{b}^{+}_{-\vec{k}}\hat{b}^{+}_{-\vec{k}}\hat{b}_{-\vec{k}} = \hat{a}^{+}_{\vec{k}}\hat{b}^{+}_{-\vec{k}}\hat{b}_{-\vec{k}}\hat{b}^{+}_{-\vec{k}} \\
&= \hat{a}^{+}_{\vec{k}}\hat{b}^{+}_{-\vec{k}}(1 - \hat{b}^{+}_{-\vec{k}}\hat{b}_{-\vec{k}}) = \hat{a}^{+}_{\vec{k}}\hat{b}^{+}_{-\vec{k}} - \hat{a}^{+}_{\vec{k}}\hat{b}^{+}_{-\vec{k}}\hat{b}^{+}_{-\vec{k}}\hat{b}_{-\vec{k}} = \hat{a}^{+}_{\vec{k}}\hat{b}^{+}_{-\vec{k}} = \hat{p}^{+}_{\vec{k}}, \\
[\hat{p}_{\vec{k}}, \hat{p}^{+}_{\vec{k}}] &= \hat{b}_{-\vec{k}}\hat{a}_{\vec{k}}\hat{a}^{+}_{\vec{k}}\hat{b}^{+}_{-\vec{k}} - \hat{a}^{+}_{\vec{k}}\hat{b}^{+}_{-\vec{k}}\hat{b}_{-\vec{k}}\hat{a}_{\vec{k}} = \hat{b}_{-\vec{k}}\hat{b}^{+}_{-\vec{k}}\hat{a}_{\vec{k}}\hat{a}^{+}_{\vec{k}} - \hat{b}^{+}_{-\vec{k}}\hat{b}_{-\vec{k}}\hat{a}^{+}_{\vec{k}}\hat{a}_{\vec{k}} \\
&= (1 - \hat{b}^{+}_{-\vec{k}}\hat{b}_{-\vec{k}})(1 - \hat{a}^{+}_{\vec{k}}\hat{a}_{\vec{k}}) - \hat{b}^{+}_{-\vec{k}}\hat{b}_{-\vec{k}}\hat{a}^{+}_{\vec{k}}\hat{a}_{\vec{k}} = 1 - \hat{b}^{+}_{-\vec{k}}\hat{b}_{-\vec{k}} - \hat{a}^{+}_{\vec{k}}\hat{a}_{\vec{k}} \\
&= 1 - \hat{f}^{\mathrm{h}}_{\vec{k}} - \hat{f}^{\mathrm{e}}_{\vec{k}}.
\end{aligned}
$$

By utilizing these commuting relations, we can further simplify equation (4.19a–c) to obtain

$$
\frac{\mathrm{d}\hat{f}^{\mathrm{e}}_{\vec{k}}}{\mathrm{d}t} = \frac{\mathrm{j}}{\hbar}(\vec{\mu}_{\vec{k}}\hat{p}^{+}_{\vec{k}} - \vec{\mu}^{*}_{\vec{k}}\hat{p}_{\vec{k}}) \cdot \vec{E} + \frac{\mathrm{j}}{\hbar}[\vec{H}_{\mathrm{C}}, \hat{f}^{\mathrm{e}}_{\vec{k}}] + \frac{\mathrm{j}}{\hbar}[\vec{H}_{\mathrm{sct}}, \hat{f}^{\mathrm{e}}_{\vec{k}}], \tag{4.20a}
$$

$$
\frac{\mathrm{d}\hat{f}^{\mathrm{h}}_{\vec{k}}}{\mathrm{d}t} = \frac{\mathrm{j}}{\hbar}(\vec{\mu}_{\vec{k}}\hat{p}^{+}_{\vec{k}} - \vec{\mu}^{*}_{\vec{k}}\hat{p}_{\vec{k}}) \cdot \vec{E} + \frac{\mathrm{j}}{\hbar}[\vec{H}_{\mathrm{C}}, \hat{f}^{\mathrm{h}}_{\vec{k}}] + \frac{\mathrm{j}}{\hbar}[\vec{H}_{\mathrm{sct}}, \hat{f}^{\mathrm{h}}_{\vec{k}}], \tag{4.20b}
$$

$$
\begin{aligned}
\frac{\mathrm{d}\hat{p}_{\vec{k}}}{\mathrm{d}t} &= -\frac{\mathrm{j}}{\hbar}(E_{\mathrm{g}} + \varepsilon_{\mathrm{e}\vec{k}} + \varepsilon_{\mathrm{h}\vec{k}})\hat{p}_{\vec{k}} - \frac{\mathrm{j}}{\hbar}\vec{\mu}_{\vec{k}}(\hat{f}^{\mathrm{e}}_{\vec{k}} + \hat{f}^{\mathrm{h}}_{\vec{k}} - 1) \cdot \vec{E} \\
&\quad + \frac{\mathrm{j}}{\hbar}[\vec{H}_{\mathrm{C}}, \hat{p}_{\vec{k}}] + \frac{\mathrm{j}}{\hbar}[\vec{H}_{\mathrm{sct}}, \hat{p}_{\vec{k}}].
\end{aligned} \tag{4.20c}
$$

Taking the average value (i.e., the expectation) of the operator equations (4.20a–c), we find

$$
\frac{\mathrm{d}f^{\mathrm{e}}_{\vec{k}}}{\mathrm{d}t} = \frac{\mathrm{j}}{\hbar}[(\vec{\mu}_{\vec{k}} \cdot \vec{E})p^{*}_{\vec{k}} - (\vec{\mu}^{*}_{\vec{k}} \cdot \vec{E})p_{\vec{k}}] + \frac{\mathrm{j}}{\hbar} < [\vec{H}_{\mathrm{C}}, \hat{f}^{\mathrm{e}}_{\vec{k}}] > + \frac{\mathrm{j}}{\hbar} < [\vec{H}_{\mathrm{sct}}, \hat{f}^{\mathrm{e}}_{\vec{k}}] >, \tag{4.21a}
$$

$$
\frac{\mathrm{d}f^{\mathrm{h}}_{\vec{k}}}{\mathrm{d}t} = \frac{\mathrm{j}}{\hbar}[(\vec{\mu}_{\vec{k}} \cdot \vec{E})p^{*}_{\vec{k}} - (\vec{\mu}^{*}_{\vec{k}} \cdot \vec{E})p_{\vec{k}}] + \frac{\mathrm{j}}{\hbar} < [\vec{H}_{\mathrm{C}}, \hat{f}^{\mathrm{h}}_{\vec{k}}] > + \frac{\mathrm{j}}{\hbar} < [\vec{H}_{\mathrm{sct}}, \hat{f}^{\mathrm{h}}_{\vec{k}}] >, \tag{4.21b}
$$

$$
\begin{aligned}
\frac{\mathrm{d}p_{\vec{k}}}{\mathrm{d}t} &= -\frac{\mathrm{j}}{\hbar}(E_{\mathrm{g}} + \varepsilon_{\mathrm{e}\vec{k}} + \varepsilon_{\mathrm{h}\vec{k}})p_{\vec{k}} - \frac{\mathrm{j}}{\hbar}(\vec{\mu}_{\vec{k}} \cdot \vec{E})(f^{\mathrm{e}}_{\vec{k}} + f^{\mathrm{h}}_{\vec{k}} - 1) \\
&\quad + \frac{\mathrm{j}}{\hbar} < [\vec{H}_{\mathrm{C}}, \hat{p}_{\vec{k}}] > + \frac{\mathrm{j}}{\hbar} < [\vec{H}_{\mathrm{sct}}, \hat{p}_{\vec{k}}] > .
\end{aligned} \tag{4.21c}
$$

Dealing with the carrier (i.e., the conduction band electron and valence band hole) and polariton number operators commuting with the kinetic energy and carrier-field interaction Hamiltonians is different from dealing with number operators commuting with the Coulomb Hamiltonian. In the former, closed forms can still be reached in terms of the carrier and polariton number operators, but in the latter we do not have closed forms due to the virtually countless number operators involved in the microscopic domain. Actually, the Coulomb Hamiltonian not only couples every single number operator to every other single operator, it also couples every single operator to every other pair, every pair to every other pair, and so on. As a result, the commutation between the Coulomb Hamiltonian and the number operator must have an almost infinite number of expansion terms in order to describe these interactions in every possible combination. To make computation of such a commutation possible, we have to truncate the expansion. Therefore, we follow the Hartree–Fock approach to rank the expansion in order of number operators, and view the higher order term as the one with more number operators involved in a product form. In such an expansion ordered from the lowest to the highest, we will only keep:

(1) zeroth order terms, i.e., those terms proportional to the polariton number operator, which bring in a Coulomb potential dependent energy shift to the valence band and a Coulomb field renormalization term to the dipole matrix element;
(2) first order terms, i.e., those terms proportional to the product of the carrier number and polariton operators and the product of two different polariton operators, which bring in an effect known as the exchange energy shift and also Coulomb field renormalization on the dipole matrix element;
(3) second order terms, i.e., the terms proportional to the product of three and four different carrier number operators, and the product of three different carrier number operators and one polariton operator, which describe the carrier–carrier collision and bring in a high order correction to the Coulomb field renormalization on the dipole matrix element due to such a collision.

The rest of the terms involved in the Coulomb interaction are all neglected. The zeroth and first order terms are proportional to the Coulomb potential, whereas the second order term is proportional to the product of the Coulomb potentials. Using equations (4.8), (4.12), (4.17a–d), we find [3, 6, 7]

$$
\begin{aligned}
<[\vec{H}_{\mathrm{C}}, \hat{f}^{\mathrm{e}}_{\vec{k}}]> = & \sum_{\vec{k}',\vec{q}\neq 0} V_{|\vec{q}|} < \hat{a}^{+}_{\vec{k}}\hat{a}^{+}_{\vec{k}'-\vec{q}}\hat{a}_{\vec{k}-\vec{q}}\hat{a}_{\vec{k}'} - \hat{a}^{+}_{\vec{k}+\vec{q}}\hat{a}^{+}_{\vec{k}'-\vec{q}}\hat{a}_{\vec{k}}\hat{a}_{\vec{k}'} \\
& + \hat{a}^{+}_{\vec{k}}\hat{a}_{\vec{k}-\vec{q}}\hat{b}^{+}_{\vec{k}'-\vec{q}}\hat{b}_{\vec{k}'} - \hat{a}^{+}_{\vec{k}+\vec{q}}\hat{a}_{\vec{k}}\hat{b}^{+}_{\vec{k}'-\vec{q}}\hat{b}_{\vec{k}'} > \\
\approx & \sum_{\vec{k}'\neq\vec{k}} V_{|\vec{k}'-\vec{k}|} p_{\vec{k}'} p^{*}_{\vec{k}} - \sum_{\vec{k}'\neq\vec{k}} V_{|\vec{k}'-\vec{k}|} p^{*}_{\vec{k}'} p_{\vec{k}} + \mathrm{j}(\Sigma^{\mathrm{e-out}}_{\vec{k}} \\
& + \Sigma^{\mathrm{e-in}}_{\vec{k}}) f^{\mathrm{e}}_{\vec{k}} - j\Sigma^{\mathrm{e-in}}_{\vec{k}},
\end{aligned} \tag{4.22a}
$$

$$
\begin{aligned}
<[\vec{H}_{\mathrm{C}}, \hat{f}^{\mathrm{h}}_{\vec{k}}]> = & \sum_{\vec{k}',\vec{q}\neq 0} V_{|\vec{q}|} < \hat{b}^{+}_{-\vec{k}}\hat{b}^{+}_{\vec{k}'-\vec{q}}\hat{b}_{-\vec{k}-\vec{q}}\hat{b}_{\vec{k}'} - \hat{b}^{+}_{-\vec{k}+\vec{q}}\hat{b}^{+}_{\vec{k}'-\vec{q}}\hat{b}_{-\vec{k}}\hat{b}_{\vec{k}'} \\
& + \hat{a}^{+}_{\vec{k}'-\vec{q}}\hat{a}_{\vec{k}'}\hat{b}^{+}_{-\vec{k}}\hat{b}_{-\vec{k}-\vec{q}} - \hat{a}^{+}_{\vec{k}'-\vec{q}}\hat{a}_{\vec{k}'}\hat{b}^{+}_{-\vec{k}+\vec{q}}\hat{b}_{-\vec{k}} > \\
\approx & \sum_{\vec{k}'\neq\vec{k}} V_{|\vec{k}'-\vec{k}|} p_{\vec{k}'} p^{*}_{\vec{k}} - \sum_{\vec{k}'\neq\vec{k}} V_{|\vec{k}'-\vec{k}|} p^{*}_{\vec{k}'} p_{\vec{k}} + \mathrm{j}(\Sigma^{\mathrm{h-out}}_{\vec{k}} \\
& + \Sigma^{\mathrm{h-in}}_{\vec{k}}) f^{\mathrm{h}}_{\vec{k}} - \mathrm{j}\Sigma^{\mathrm{h-in}}_{\vec{k}},
\end{aligned} \tag{4.22b}
$$

$$
\begin{aligned}
<[\vec{H}_{\mathrm{C}}, \hat{p}_{\vec{k}}]> = & \sum_{\vec{k}', \vec{q}\neq 0} V_{|\vec{q}|} < \hat{a}^{+}_{\vec{k}'+\vec{q}} \hat{b}_{-\vec{k}} \hat{a}_{\vec{k}'} \hat{a}_{\vec{k}+\vec{q}} + \hat{b}^{+}_{\vec{k}'-\vec{q}} \hat{b}_{\vec{k}'} \hat{a}_{\vec{k}} \hat{b}_{-\vec{k}-\vec{q}} \\
& - \hat{a}^{+}_{\vec{k}'+\vec{q}} \hat{b}_{-\vec{k}+\vec{q}} \hat{a}_{\vec{k}'} \hat{a}_{\vec{k}} - \hat{b}^{+}_{\vec{k}'-\vec{q}} \hat{b}_{\vec{k}'} \hat{a}_{\vec{k}-\vec{q}} \hat{b}_{-\vec{k}} > \\
& - \sum_{\vec{q}\neq 0} V_{|\vec{q}|} <\hat{p}_{\vec{k}}> + \sum_{\vec{q}\neq 0} V_{|\vec{q}|} < \hat{b}_{-\vec{k}+\vec{q}} \hat{a}_{\vec{k}-\vec{q}} > \\
\approx & - \sum_{\vec{q}\neq 0} V_{|\vec{q}|} p_{\vec{k}} + \sum_{\vec{k}'\neq\vec{k}} V_{|\vec{k}'-\vec{k}|} p_{\vec{k}'} + \sum_{\vec{k}'\neq\vec{k}} V_{|\vec{k}'-\vec{k}|} (f^{\mathrm{e}}_{\vec{k}'} + f^{\mathrm{h}}_{\vec{k}'}) p_{\vec{k}} \\
& - \sum_{\vec{k}'\neq\vec{k}} V_{|\vec{k}'-\vec{k}|} p_{\vec{k}'} (f^{\mathrm{e}}_{\vec{k}} + f^{\mathrm{h}}_{\vec{k}}) + \mathrm{j} \sum_{\vec{k}'} \Lambda_{\vec{k}\vec{k}'} p_{\vec{k}'},
\end{aligned}
\tag{4.22c}
$$

where we have defined

$$
\Sigma^{\mathrm{e/h-out}}_{\vec{k}} \equiv \pi \sum_{\alpha=\mathrm{e,h}} \sum_{\vec{l}\neq 0} \sum_{\vec{k}'} (2\widetilde{V}^{2}_{|\vec{l}|} - \delta_{(\mathrm{e/h})\alpha} \widetilde{V}_{|\vec{l}|} \widetilde{V}_{|\vec{k}'-\vec{l}-\vec{k}|})
$$
$$
\times \delta[\varepsilon_{(\mathrm{e/h})\vec{k}} + \varepsilon_{\alpha\vec{k}'} - \varepsilon_{(\mathrm{e/h})(\vec{k}+\vec{l})} - \varepsilon_{\alpha(\vec{k}'-\vec{l})}](1 - f^{\mathrm{e/h}}_{\vec{k}+\vec{l}}) f^{\alpha}_{\vec{k}'} (1 - f^{\alpha}_{\vec{k}'-\vec{l}}),
\tag{4.23a}
$$

$$
\Sigma^{\mathrm{e/h-in}}_{\vec{k}} \equiv \pi \sum_{\alpha=\mathrm{e,h}} \sum_{\vec{l}\neq 0} \sum_{\vec{k}'} (2\widetilde{V}^{2}_{|\vec{l}|} - \delta_{(\mathrm{e/h})\alpha} \widetilde{V}_{|\vec{l}|} \widetilde{V}_{|\vec{k}'-\vec{l}-\vec{k}|})
$$
$$
\times \delta[\varepsilon_{(\mathrm{e/h})\vec{k}} + \varepsilon_{\alpha\vec{k}'} - \varepsilon_{(\mathrm{e/h})(\vec{k}+\vec{l})} - \varepsilon_{\alpha(\vec{k}'-\vec{l})}] f^{\mathrm{e/h}}_{\vec{k}+\vec{l}} (1 - f^{\alpha}_{\vec{k}'}) f^{\alpha}_{\vec{k}'-\vec{l}},
\tag{4.23b}
$$

$$
\Lambda_{\vec{k}\vec{k}'} \equiv \begin{cases}
\sum_{\alpha,\beta=\mathrm{e,h}} \sum_{\vec{k}''} \sum_{\vec{l}\neq 0} \left(2\widetilde{V}^{2}_{|\vec{l}|} - \delta_{\alpha\beta} \widetilde{V}_{|\vec{l}|} \widetilde{V}_{|\vec{k}''-\vec{l}-\vec{k}|}\right) g\left[\varepsilon_{\alpha\vec{k}} + \varepsilon_{\beta\vec{k}''} - \varepsilon_{\alpha(\vec{k}+\vec{l})} \right. \\
\left. -\varepsilon_{\beta(\vec{k}''-\vec{l})}\right] \left[f^{\alpha}_{\vec{k}+\vec{l}} \left(1 - f^{\beta}_{\vec{k}''}\right) f^{\beta}_{\vec{k}''-\vec{l}} + \left(1 - f^{\alpha}_{\vec{k}+\vec{l}}\right) f^{\beta}_{\vec{k}''} \left(1 - f^{\beta}_{\vec{k}''-\vec{l}}\right)\right] \\
\hfill \vec{k}' = \vec{k} \\
\sum_{\alpha,\beta=\mathrm{e,h}} \sum_{\vec{k}''} \left(2\widetilde{V}^{2}_{|\vec{k}'-\vec{k}|} - \delta_{\alpha\beta} \widetilde{V}_{|\vec{k}'-\vec{k}|} \widetilde{V}_{|\vec{k}''-\vec{k}'|}\right) g\left[-\varepsilon_{\alpha\vec{k}} - \varepsilon_{\beta\vec{k}''} + \varepsilon_{\alpha\vec{k}'} \right. \\
\left. + \varepsilon_{\beta(\vec{k}''-\vec{k}'+\vec{k})}\right] \left[\left(1 - f^{\alpha}_{\vec{k}}\right)\left(1 - f^{\beta}_{\vec{k}''}\right) f^{\beta}_{\vec{k}''-\vec{k}'+\vec{k}} + f^{\alpha}_{\vec{k}} f^{\beta}_{\vec{k}''} \left(1 - f^{\beta}_{\vec{k}''-\vec{k}'+\vec{k}}\right)\right] \\
\hfill \vec{k}' \neq \vec{k}
\end{cases}
\tag{4.23c}
$$

In equation (4.23c), the generalized Dirac function is defined as [6]

$$
g(x) \equiv \lim_{\gamma\to 0} \left[\frac{\mathrm{j}}{x + \mathrm{j}\gamma}\right] = \pi\delta(x) + \mathrm{j}PV\left(\frac{1}{x}\right),
\tag{4.24}
$$

with $PV(1/x)$ indicating the principal value integral of $1/x$. In the last equations of (4.22a) and (4.22b) there is no zeroth order term. The first two terms on the RHS are both first order, which involves the Coulomb interaction between two different polaritons, whereas the last two terms are both second order, which describes the carrier–carrier collision. In the last equation of (4.22c), the first two terms on the RHS are both zeroth order, where the first term gives a Coulomb potential dependent energy shift to the valence band through the self-coupling, and the second term contributes to the Coulomb field renormalization on the dipole matrix element through the cross-coupling. The third and fourth terms are both first order, which describes Coulomb interactions between a single carrier and a single polariton, where the former introduces an exchange energy shift and the latter still contributes to the Coulomb field renormalization. Finally, the last term is second order, which gives the carrier–carrier collision effect on the polariton. It brings in a high order correction to the Coulomb field renormalization. As shown in equations (4.23a–c), for those second order terms in equation (4.22a–c), the Coulomb potential is replaced by its screened counterpart, for reasons we explain later in Section 4.3.1.

In dealing with carrier–phonon scattering, i.e., the last commutator in equation (4.21a–c), we will assume that the LO phonons move much faster than carriers, so that in the carrier–phonon scattering process, the LO phonons are at their equilibrium status described by the Bose–Einstein distribution in the form

$$f^{\mathrm{p}}_{\vec{q}} = \langle \hat{f}^{\mathrm{p}}_{\vec{q}} \rangle = \langle \hat{b}^{\mathrm{LO}+}_{\vec{q}} \hat{b}^{\mathrm{LO}}_{\vec{q}} \rangle = \bar{f}^{\mathrm{p}}_{\vec{q}} \equiv \frac{1}{\mathrm{e}^{\hbar\omega_{\mathrm{LO}}/k_{\mathrm{B}}T} - 1}, \tag{4.25}$$

with $\hat{f}^{\mathrm{p}}_{\vec{q}}$ defined as the LO phonon number operator, $f^{\mathrm{p}}_{\vec{q}}$ the expectation of $\hat{f}^{\mathrm{p}}_{\vec{q}}$, $\bar{f}^{\mathrm{p}}_{\vec{q}}$ the steady state solution of $f^{\mathrm{p}}_{\vec{q}}$, k_{B} the Boltzmann constant in J/K, and T the semiconductor lattice (ambient) temperature in K. Under this assumption we can solve the carrier–phonon scattering commutator through a second-order perturbation approach, i.e., formally integrate the last commutator in equations (4.21a–c) and iterate twice, to obtain

$$< [\vec{H}_{\mathrm{sct}}, \hat{f}^{\mathrm{e}}_{\vec{k}}] > = \mathrm{j} \sum_{\vec{q} \neq 0, \pm} \Sigma^{\mathrm{e-p}}_{\vec{q},\pm} f^{\mathrm{p}}_{\vec{q}} + \mathrm{j} \sum_{\vec{q} \neq 0, \pm} \Sigma^{\mathrm{e}}_{\vec{q},\pm}, \tag{4.26a}$$

$$< [\vec{H}_{\mathrm{sct}}, \hat{f}^{\mathrm{h}}_{\vec{k}}] > = \mathrm{j} \sum_{\vec{q} \neq 0, \pm} \Sigma^{\mathrm{h-p}}_{\vec{q},\pm} f^{\mathrm{p}}_{\vec{q}} + \mathrm{j} \sum_{\vec{q} \neq 0, \pm} \Sigma^{\mathrm{h}}_{\vec{q},\pm}, \tag{4.26b}$$

$$< [\vec{H}_{\mathrm{sct}}, \hat{p}_{\vec{k}}] > = \mathrm{j} \sum_{\vec{q}} \Lambda^{\mathrm{p}}_{\vec{k}\vec{q}} p_{\vec{k}+\vec{q}}, \tag{4.26c}$$

where we have defined

$$\begin{aligned} \Sigma^{\mathrm{e/h-p}}_{\vec{q},\pm} &\equiv 2\pi\hbar^2 G^2_{|\vec{q}|} \delta[\varepsilon_{(\mathrm{e/h})\vec{k}} - \varepsilon_{(\mathrm{e/h})(\vec{k}-\vec{q})} \mp \hbar\omega_{\mathrm{LO}}] \\ &\quad \times [f^{\mathrm{e/h}}_{\vec{k}}(1 - f^{\mathrm{e/h}}_{\vec{k}-\vec{q}}) - (1 - f^{\mathrm{e/h}}_{\vec{k}}) f^{\mathrm{e/h}}_{\vec{k}-\vec{q}}], \end{aligned} \tag{4.27a}$$

$$\begin{aligned} \Sigma^{\mathrm{e/h}}_{\vec{q},\pm} &\equiv 2\pi\hbar^2 G^2_{|\vec{q}|} \delta[\varepsilon_{(\mathrm{e/h})\vec{k}} - \varepsilon_{(\mathrm{e/h})(\vec{k}-\vec{q})} \mp \hbar\omega_{\mathrm{LO}}] \\ &\quad \times \left[f^{\mathrm{e/h}}_{\vec{k}}(1 - f^{\mathrm{e/h}}_{\vec{k}-\vec{q}}) \left(\frac{1}{2} \pm \frac{1}{2}\right) - (1 - f^{\mathrm{e/h}}_{\vec{k}}) f^{\mathrm{e/h}}_{\vec{k}-\vec{q}} \left(\frac{1}{2} \mp \frac{1}{2}\right) \right], \end{aligned} \tag{4.27b}$$

$$\Lambda^{\mathrm{p}}_{\vec{k}\vec{q}} \equiv \begin{cases} \hbar^2 \sum\limits_{\alpha=\mathrm{e,h}} \sum\limits_{\vec{l}\neq 0} G^2_{|\vec{l}|} \Big\{ g\Big[\varepsilon_{\alpha\vec{k}} - \varepsilon_{\alpha(\vec{k}-\vec{l})} - \hbar\omega_{\mathrm{LO}}\Big]\Big[\Big(1 - f^{\alpha}_{\vec{k}-\vec{l}}\Big) f^{\mathrm{p}}_{\vec{l}} \\ + f^{\alpha}_{\vec{k}-\vec{l}}\Big(1 + f^{\mathrm{p}}_{\vec{l}}\Big)\Big] + g\Big[\varepsilon_{\alpha\vec{k}} - \varepsilon_{\alpha(\vec{k}-\vec{l})} + \hbar\omega_{\mathrm{LO}}\Big] \\ \Big[\Big(1 - f^{\alpha}_{\vec{k}-\vec{l}}\Big)\Big(1 + f^{\mathrm{p}}_{\vec{l}}\Big) + f^{\alpha}_{\vec{k}-\vec{l}} f^{\mathrm{p}}_{\vec{l}}\Big]\Big\} & \vec{q} = 0 \\ \hbar^2 \sum\limits_{\alpha=\mathrm{e,h}} G^2_{|\vec{q}|} \Big\{ g\Big[\varepsilon_{\alpha\vec{k}} - \varepsilon_{\alpha(\vec{k}-\vec{q})} - \hbar\omega_{\mathrm{LO}}\Big]\Big[\Big(1 - f^{\alpha}_{\vec{k}}\Big) f^{\mathrm{p}}_{\vec{q}} \\ + f^{\alpha}_{\vec{k}}\Big(1 + f^{\mathrm{p}}_{\vec{q}}\Big)\Big] + g\Big[\varepsilon_{\alpha\vec{k}} - \varepsilon_{\alpha(\vec{k}-\vec{q})} + \hbar\omega_{\mathrm{LO}}\Big] \\ \Big[\Big(1 - f^{\alpha}_{\vec{k}}\Big)\Big(1 + f^{\mathrm{p}}_{\vec{q}}\Big) + f^{\alpha}_{\vec{k}} f^{\mathrm{p}}_{\vec{q}}\Big]\Big\}. & \vec{q} \neq 0 \end{cases} \tag{4.27c}$$

Finally, substituting equations (4.22a–c) and (4.26a–c) into equations (4.21a–c) we obtain the governing equations for the expectations of the microscopic polariton number, conduction band electron number, and valence band hole number

$$\frac{\mathrm{d}p_{\vec{k}}}{\mathrm{d}t} = -\frac{\mathrm{j}}{\hbar}\sum_{\vec{k}'} \Theta_{\vec{k}\vec{k}'} p_{\vec{k}'} - \frac{\mathrm{j}}{\hbar}(\vec{\mu}_{\vec{k}} \cdot \vec{E})(f^{\mathrm{e}}_{\vec{k}} + f^{\mathrm{h}}_{\vec{k}} - 1) - \frac{1}{\hbar}\sum_{\vec{k}'} \Lambda_{\vec{k}\vec{k}'} p_{\vec{k}'} - \frac{1}{\hbar}\sum_{\vec{q}} \Lambda^{\mathrm{p}}_{\vec{k}\vec{q}} p_{\vec{k}+\vec{q}}, \tag{4.28a}$$

$$\frac{\mathrm{d}f^{\mathrm{e}}_{\vec{k}}}{\mathrm{d}t} = \mathrm{j}\Omega_{\vec{k}} p^*_{\vec{k}} - \mathrm{j}\Omega^*_{\vec{k}} p_{\vec{k}} - \frac{1}{\hbar}(\Sigma^{\mathrm{e-out}}_{\vec{k}} + \Sigma^{\mathrm{e-in}}_{\vec{k}}) f^{\mathrm{e}}_{\vec{k}} + \frac{1}{\hbar}\Sigma^{\mathrm{e-in}}_{\vec{k}} - \frac{1}{\hbar}\sum_{\vec{q}\neq 0,\pm} \Sigma^{\mathrm{e-p}}_{\vec{q},\pm} f^{\mathrm{p}}_{\vec{q}} - \frac{1}{\hbar}\sum_{\vec{q}\neq 0,\pm} \Sigma^{\mathrm{e}}_{\vec{q},\pm}, \tag{4.28b}$$

$$\frac{\mathrm{d}f^{\mathrm{h}}_{\vec{k}}}{\mathrm{d}t} = \mathrm{j}\Omega_{\vec{k}} p^*_{\vec{k}} - \mathrm{j}\Omega^*_{\vec{k}} p_{\vec{k}} - \frac{1}{\hbar}(\Sigma^{\mathrm{h-out}}_{\vec{k}} + \Sigma^{\mathrm{h-in}}_{\vec{k}}) f^{\mathrm{h}}_{\vec{k}} + \frac{1}{\hbar}\Sigma^{\mathrm{h-in}}_{\vec{k}} - \frac{1}{\hbar}\sum_{\vec{q}\neq 0,\pm} \Sigma^{\mathrm{h-p}}_{\vec{q},\pm} f^{\mathrm{p}}_{\vec{q}} - \frac{1}{\hbar}\sum_{\vec{q}\neq 0,\pm} \Sigma^{\mathrm{h}}_{\vec{q},\pm}, \tag{4.28c}$$

where we have defined

$$\Theta_{\vec{k}\vec{k}'} \equiv \begin{cases} \hbar\bar{\omega}_{\vec{k}} - \sum\limits_{\vec{l}\neq\vec{k}} V_{|\vec{l}-\vec{k}|}(f^{\mathrm{e}}_{\vec{l}} + f^{\mathrm{h}}_{\vec{l}} - 1) & \vec{k}' = \vec{k} \\ V_{|\vec{k}'-\vec{k}|}(f^{\mathrm{e}}_{\vec{k}} + f^{\mathrm{h}}_{\vec{k}} - 1) & \vec{k}' \neq \vec{k}, \end{cases} \tag{4.29}$$

$$\Omega_{\vec{k}} \equiv \frac{1}{\hbar}(\vec{\mu}_{\vec{k}} \cdot \vec{E} + \sum_{\vec{l}\neq\vec{k}} V_{|\vec{l}-\vec{k}|} p_{\vec{l}}), \tag{4.30}$$

with

$$\bar{\omega}_{\vec{k}} \equiv \frac{1}{\hbar}(E_g + \varepsilon_{e\vec{k}} + \varepsilon_{h\vec{k}}). \tag{4.31}$$

More specifically, for bulk semiconductors, we have

$$\bar{\omega}_{\vec{k}} = \frac{1}{\hbar}\left(E_g + \frac{\hbar^2 k^2}{2m_e} + \frac{\hbar^2 k^2}{2m_h}\right) = \frac{1}{\hbar}\left(E_g + \frac{\hbar^2 k^2}{2m_r}\right), \tag{4.32}$$

where $1/m_r \equiv 1/m_e + 1/m_h$ with m_r known as the reduced electron–hole effective mass.

Equations (4.28a–c) are coupled first order non-linear ODEs. Since the solution of a first order linear ODE in the form $\mathrm{d}f(t)/\mathrm{d}t = -\mathrm{j}A(t)f(t) + B(t)$ can generally be written as $f(t) = [\int B(t)e^{\mathrm{j}\int A(t)\mathrm{d}t}dt]e^{-\mathrm{j}\int A(t)\mathrm{d}t} + C$, we find that the coefficient $A(t)$ and the inhomogeneous driving term $B(t)$ serve as the exponential growth rate and the "seed," respectively, in the evolution of the solution with time. More specifically, if $A(t)$ can be viewed as a constant in the time scale of our interest, the real and imaginary part of $A(t)$ becomes the harmonic frequency and gain of the solution, respectively.

Therefore, in the governing equation for the polariton number, equation (4.28a), the first term on the RHS describes the "coherence" level of the polaritons, as the transition energy matrix given in equation (4.29) is always real. If we remove the Coulomb potential, this matrix becomes diagonalized, which means there is no interference between different polaritons. Under this assumption, a polariton with momentum $\hbar\vec{k}$ will oscillate at its own harmonic frequency $\bar{\omega}_{\vec{k}}$ that is determined by the transition energy as given in equation (4.31). Different polaritons with different momentum will oscillate at different frequencies without mixing. With the Coulomb potential included, however, the non-zero off-diagonal matrix element couples and mixes polaritons with different momentum. Such a coupling brings in a dephasing effect to polaritons so the oscillation frequency of a polariton with momentum $\hbar\vec{k}$ may be pulled away from its own harmonic frequency $\bar{\omega}_{\vec{k}}$, and other polaritons with their momentum different from $\hbar\vec{k}$ may take the harmonic frequency $\bar{\omega}_{\vec{k}}$. It is also interesting to note that $\sum_{\vec{k}} \Theta_{\vec{k}\vec{k}'} = \hbar\bar{\omega}_{\vec{k}'}$, that is to say, the contribution of the Coulomb interaction to the summation of the off-diagonal column elements cancels out with that of the diagonal elements. Moreover, for a given $\vec{k}$ (i.e., along the same row of matrix $\bar{\Phi} = [\Phi_{\vec{k}\vec{k}}]$), the contributions of the Coulomb interaction to the diagonal element and off-diagonal elements have opposite signs. This leads to a compensation effect on the Coulomb interaction to some extent. This observation also suggests that, in solving equation (4.28a), we either ignore the Coulomb potential completely, or consider its effect on both diagonal and off-diagonal elements. Considering the Coulomb effect only on the diagonal elements with the off-diagonal elements ignored (i.e., with the polariton coupling ignored) would be likely to lead to a large error.If the coupling between different polaritons is a weak effect, we can treat the off-diagonal terms as an inhomogeneous driving force. Actually, by splitting the diagonal and off-diagonal elements in $\bar{\Phi} = [\Phi_{\vec{k}\vec{k}}]$ into two separate terms, we obtain from equation (4.28a)

$$\frac{\mathrm{d}p_{\vec{k}}}{\mathrm{d}t} = -\frac{\mathrm{j}}{\hbar}[\hbar\bar{\omega}_{\vec{k}} - \sum_{\vec{l}\neq\vec{k}} V_{|\vec{l}-\vec{k}|}(f^{\mathrm{e}}_{\vec{l}} + f^{\mathrm{h}}_{\vec{l}} - 1)]p_{\vec{k}}$$
$$-\mathrm{j}\Omega_{\vec{k}}(f^{\mathrm{e}}_{\vec{k}} + f^{\mathrm{h}}_{\vec{k}} - 1) - \frac{1}{\hbar}\sum_{\vec{k}'} \Lambda_{\vec{k}\vec{k}'}p_{\vec{k}'} - \frac{1}{\hbar}\sum_{\vec{q}} \Lambda^{\mathrm{p}}_{\vec{k}\vec{q}}p_{\vec{k}+\vec{q}}. \quad (4.33)$$

The second term on the RHS of equation (4.33) provides the "seed" to the polariton evolution, which is in proportion to the population inversion $\hat{f}^{\mathrm{e}}_{\vec{k}} + \hat{f}^{\mathrm{h}}_{\vec{k}} - 1$ with the coefficient known as the renormalized Rabi frequency given as equation (4.30). It is obvious that the Coulomb interaction brings in two major effects; one is an exchange shift added to the transition energy, which tunes the oscillation frequency of the polaritons and hence shifts the material gain or absorption profiles around in the frequency domain. The other is a Coulomb field renormalization term added to the dipole matrix element energy ($\vec{\mu}_{\vec{k}} \cdot \vec{E}$), which changes the "seed" intensity and hence either enhances or reduces the strength of the material gain or absorption. Since the Coulomb interaction brings in contributions to the first and second terms on the RHS of equation (4.33) in the opposite sign, depending on the sign of the population inversion, it has a different impact on the material gain and absorption profiles. Actually, once the population is inverted, i.e., $\hat{f}^{\mathrm{e}}_{\vec{k}} + \hat{f}^{\mathrm{h}}_{\vec{k}} - 1 > 0$, the Coulomb interaction introduces an energy shift that will effectively decrease the transition energy from $\hbar\bar{\omega}_{\vec{k}}$, and will effectively enhance the dipole matrix element. On the contrary, if the population is not inverted, i.e., $\hat{f}^{\mathrm{e}}_{\vec{k}} + \hat{f}^{\mathrm{h}}_{\vec{k}} - 1 < 0$, the Coulomb interaction brings in an energy shift that will effectively increase the transition energy from $\hbar\bar{\omega}_{\vec{k}}$, and will effectively reduce the dipole matrix element. Therefore, models without the many-body Coulomb interaction effect included (e.g., the free carrier model) will mistakenly position the material gain profile at the shorter wavelength side and will underestimate the material gain strength. It will affect the absorption calculation in the opposite way, i.e., without the inclusion of Coulomb interaction in the model, the material absorption profile will be mistakenly positioned at the longer wavelength side with its absorption strength overestimated. This effect is illustrated by Fig. 4.2.

As for the last two terms on the RHS of equation (4.28a), they provide higher order corrections to the first term. Unlike matrix elements $\Phi_{\vec{k}\vec{k}}$ that are all real, matrix elements $\Lambda_{\vec{k}\vec{k}'}$ and $\Lambda^{\mathrm{p}}_{\vec{k}\vec{q}}$ defined as equations (4.23c) and (4.27c) are generally complex since the generalized Dirac function given in equation (4.24) is complex. Therefore, the imaginary part of $\Lambda_{\vec{k}\vec{k}'}$ and $\Lambda^{\mathrm{p}}_{\vec{k}\vec{q}}$ provide direct correction on $\Theta_{\vec{k}\vec{k}'}$, which means the carrier–carrier collision and carrier–phonon scattering also introduce transition energy shifts through the diagonal elements $\Lambda_{\vec{k}\vec{k}}$ and $\Lambda^{\mathrm{p}}_{\vec{k}0}$ and dephasings through the off-diagonal elements $\Lambda_{\vec{k}(\vec{k}'\neq\vec{k})}$ and $\Lambda^{\mathrm{p}}_{\vec{k}(\vec{q}\neq 0)}$, because of the change of carrier energy in such processes. The real parts of $\Lambda_{\vec{k}\vec{k}'}$ and $\Lambda^{\mathrm{p}}_{\vec{k}\vec{q}}$ introduce attenuations (i.e., damping of the oscillation) to

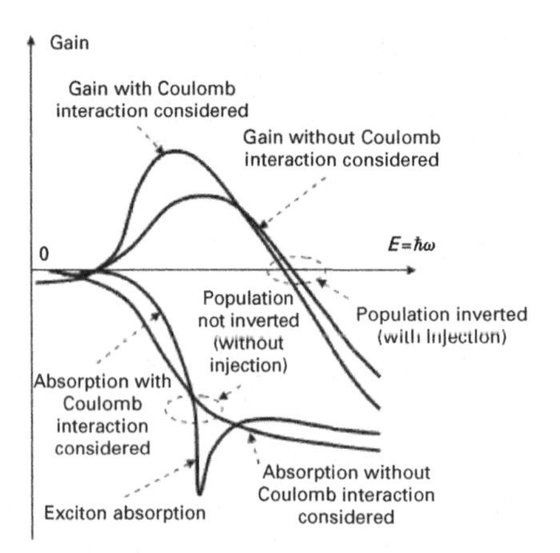

Fig. 4.2. The many-body Coulomb interaction effect on gain and absorption profiles.

the polariton through their diagonal elements and damping to other coupled polaritons through their off-diagonal elements. The contribution of the off-diagonal elements can also be viewed as being brought to the dipole matrix element. The imaginary parts of the off-diagonal elements $\mathrm{Im}[\Lambda_{\vec{k}(\vec{k}'\neq\vec{k})}]$ and $\mathrm{Im}[\Lambda^{\mathrm{p}}_{\vec{k}(\vec{q}\neq 0)}]$ may effectively enhance or reduce the dipole matrix element, whereas the real parts of the off-diagonal elements $\mathrm{Re}[\Lambda_{\vec{k}(\vec{k}'\neq\vec{k})}]$ and $\mathrm{Re}[\Lambda^{\mathrm{p}}_{\vec{k}(\vec{q}\neq 0)}]$ will introduce a dephasing effect to the dipole matrix element as they turn the equivalent Rabi frequency complex. Again, it is easy to prove that $\sum_{\vec{k}}\sum_{\vec{k}'}\Lambda_{\vec{k}\vec{k}'}p_{\vec{k}'}=0$, which means the carrier–carrier collision induced second-order Coulomb interaction brings in a purely interference effect to the polaritons because of conservation of the total kinetic energy in such a collision process. The off-diagonal elements have the effect of partially canceling out the influence of the diagonal elements [6]. Therefore, it is crucial to consider the contribution of the higher order many-body Coulomb interaction to both diagonal and off-diagonal terms in the polariton equation simultaneously. This means that if the Coulomb interaction is included up to a certain order in the diagonal term, the cross-coupling (off-diagonal) terms with the Coulomb interaction of the same order must be included simultaneously.

In the governing equations for the carrier (electron and hole) number equations (4.28b&c), the first two terms on the RHS give the net carrier change when interacting with the polariton. The Coulomb interaction modifies the Rabi frequency, which can be viewed as a change on the dipole matrix element. After renormalization of the transition energy and the dipole matrix element, the set of governing equations (4.33), as a replacement of equations (4.28a–c) with the zeroth and first order Coulomb interactions included, resemble two-level Bloch equations. In this sense, equations (4.28b&c) are still in the form of carrier rate equations, with a modified Rabi frequency that includes an extra term describing the correction on the carrier recombination and creation rates. This extra term comes from the cross-coupling between different polaritons because of the Coulomb interaction.

If we consider only many-body carrier transport with the carrier–phonon scattering process ignored, only the third and forth terms on the RHS of equations (4.28b&c) will remain and we obtain the carrier–carrier Boltzmann equation

$$\frac{\mathrm{d}f_{\vec{k}}^{\mathrm{e/h}}}{\mathrm{d}t} = -\frac{1}{\hbar}(\Sigma_{\vec{k}}^{\mathrm{e/h-out}} + \Sigma_{\vec{k}}^{\mathrm{e/h-in}})f_{\vec{k}}^{\mathrm{e/h}} + \frac{1}{\hbar}\Sigma_{\vec{k}}^{\mathrm{e/h-in}}. \tag{4.34}$$

As given in equations (4.23a&b), both $\Sigma_{\vec{k}}^{\mathrm{e/h-out}}/\hbar$ and $\Sigma_{\vec{k}}^{\mathrm{e/h-in}}/\hbar$ are carrier $f_{\vec{k}}^{\mathrm{e/h}}$ dependent. By letting $\mathrm{d}f_{\vec{k}}^{\mathrm{e/h}}/\mathrm{d}t = 0$, we find the steady state solution $\bar{f}_{\vec{k}}^{\mathrm{e/h}}$ of equation (4.34) by solving

$$\Sigma_{\vec{k}}^{\mathrm{e/h-out}}(\bar{f}_{\vec{k}}^{\mathrm{e/h}})\bar{f}_{\vec{k}}^{\mathrm{e/h}} = \Sigma_{\vec{k}}^{\mathrm{e/h-in}}(\bar{f}_{\vec{k}}^{\mathrm{e/h}})(1 - \bar{f}_{\vec{k}}^{\mathrm{e/h}}), \tag{4.35}$$

known as the detailed balance equation. Actually, the solution to equation (4.35) follows the well-known quasi-Fermi distribution in the form

$$\bar{f}_{\vec{k}}^{\mathrm{e/h}} \equiv \frac{1}{\mathrm{e}^{\frac{\varepsilon_{(\mathrm{e/h})\vec{k}} - F^{\mathrm{c/v}}}{k_{\mathrm{B}}T}} + 1}, \tag{4.36}$$

with $F^{\mathrm{c/v}}$ defined as the quasi-Fermi level in the conduction band and valence band, respectively. These quasi-Fermi levels are measured from their respective band edges. Equation (4.35) clearly shows that $\Sigma_{\vec{k}}^{\mathrm{e/h-out}}/\hbar$ and $\Sigma_{\vec{k}}^{\mathrm{e/h-in}}/\hbar$ represent the electron and hole effective rate of scattering out of and into the $\vec{k}$ state, respectively. The detailed balance equation (4.35) describes a quasi-equilibrium at which the scattering into each state is balanced by the scattering out of that state. Once the time scale of interest is larger than the relaxation time required to reach the balance, i.e., the time for $f_{\vec{k}}^{\mathrm{e/h}} \rightarrow \bar{f}_{\vec{k}}^{\mathrm{e/h}}$ through carrier–carrier interaction, we can always take the carriers at their quasi-equilibrium states which means that although collision happens constantly, there is no change to the carrier number at state $\vec{k}$, since the number of carriers escaping from this state on average is the same as the number of carriers captured by this state. When such a balance is reached for all of the $\vec{k}$ states, carriers must take the quasi-Fermi distribution as given in equation (4.36).

To understand the transient process in which a given initial carrier distribution evolves itself, through electronic injection or optical pumping, into a quasi-Fermi distribution, we can formally integrate the carrier–carrier Boltzmann equation to obtain

$$f_{\vec{k}}^{\mathrm{e/h}}(t) = f_{\vec{k}}^{\mathrm{e/h}}(0)e^{-t/\tau_{\vec{k}}^{\mathrm{r}}} + \frac{\Sigma_{\vec{k}}^{\mathrm{e/h-in}}(\bar{f}_{\vec{k}}^{\mathrm{e/h}})}{\Sigma_{\vec{k}}^{\mathrm{e/h-out}}(\bar{f}_{\vec{k}}^{\mathrm{e/h}}) + \Sigma_{\vec{k}}^{\mathrm{e/h-in}}(\bar{f}_{\vec{k}}^{\mathrm{e/h}})}(1 - \mathrm{e}^{-t/\tau_{\vec{k}}^{\mathrm{r}}}), \tag{4.37}$$

with $f_{\vec{k}}^{\mathrm{e/h}}(0)$ indicating the given initial state. In equation (4.37) we have introduced a relaxation time $\tau_{\vec{k}}^{\mathrm{r}}$ defined as

$$\tau_{\vec{k}}^{\mathrm{r}} \equiv \frac{\hbar}{\Sigma_{\vec{k}}^{\mathrm{e/h-out}}(\bar{f}_{\vec{k}}^{\mathrm{e/h}}) + \Sigma_{\vec{k}}^{\mathrm{e/h-in}}(\bar{f}_{\vec{k}}^{\mathrm{e/h}})}, \tag{4.38}$$

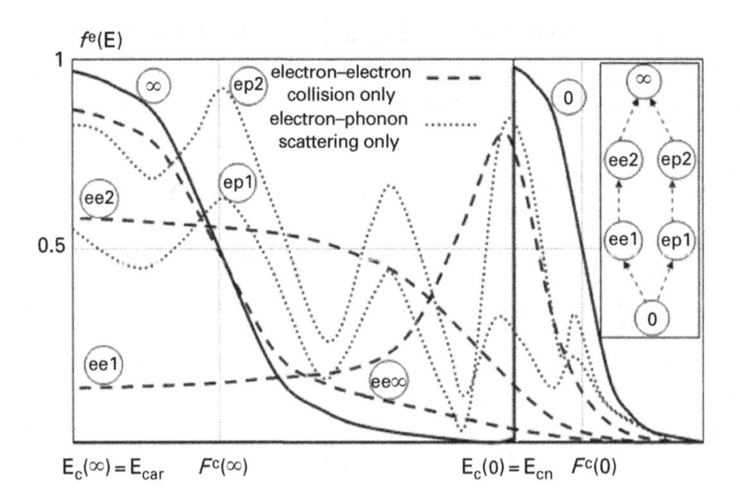

Fig. 4.3. The evolution of the electron distribution. The solid line on the right marked with circled 0 is the initial distribution, which is assumed to be the electron static distribution in an *N* doped region with a broader energy bandgap. The dashed lines marked with circled ee1, ee2, and ee∞ are the electron intermediate and final distributions without the electron–phonon scattering process included, whereas the dotted lines marked with circled ep1 and ep2 are the electron intermediate distributions without the electron–electron collision process included. The solid line on the left marked with circled ∞ is the final distribution as the time goes to infinity (normally only a few picoseconds in a real time scale), which forms the quasi-Fermi distribution in the active region with a narrower energy bandgap. The electron quasi-Fermi levels at the initial and final states are marked as $F^c(0)$ and $F^c(\infty)$, respectively. The conduction band edges of the *N* doped region and the active region are marked as $E_c(0) = E_{cn}$ and $E_c(\infty) = E_{car}$, respectively. The inset sketch on the very right shows the time evolution sequence. Valleys appear in the electron distribution as it evolves through the electron–phonon scattering process with the electron–electron collision process switched off, which indicates the occurrence of LO phonon emission. The spacing of these valleys equals the LO phonon energy. Without electron–phonon scattering, the electrons must be settled at a quasi-Fermi distribution with a higher temperature shown as the curve marked with circled ee∞. With both electron–electron collision and electron–phonon scattering considered, on their way to evolving to the quasi-Fermi distribution, the electrons exchange energies with LO phonons and are cooled down, since the energy exchange leads to a pure energy loss of the electrons. As a result, the electrons will be finally settled at a quasi-Fermi distribution with a lowered temperature that is balanced with the LO phonons, or the lattice. This process also describes electron injection from a heterojunction, as discussed in Chapter 5.

which is the inverse of the summation of the effective electron or hole net decay rates. Solution of equation (4.37) shows that after a few $\tau_{\vec{k}}^{-r}$'s that are normally on a sub-picosecond scale, the carriers reach their steady states described by the quasi-Fermi distributions [3, 6], regardless of their initial distributions, as illustrated by Fig. 4.3.

Finally, the last two terms on the RHS of equations (4.28b&c) describe the carrier–phonon scattering contribution to the carrier number change. We can actually append these two terms to equation (4.34) to obtain the carrier–carrier and carrier–phonon

Boltzmann equation as

$$\frac{\mathrm{d}f_{\vec{k}}^{\mathrm{e/h}}}{\mathrm{d}t} = -\frac{1}{\hbar}(\Sigma_{\vec{k}}^{\mathrm{e/h-out}} + \Sigma_{\vec{k}}^{\mathrm{e/h-in}})f_{\vec{k}}^{\mathrm{e/h}} + \frac{1}{\hbar}\Sigma_{\vec{k}}^{\mathrm{e/h-in}} - \frac{1}{\hbar}\sum_{\vec{q}\neq 0,\pm}\Sigma_{\vec{q},\pm}^{\mathrm{e/h-p}}f_{\vec{q}}^{\mathrm{p}} - \frac{1}{\hbar}\sum_{\vec{q}\neq 0,\pm}\Sigma_{\vec{q},\pm}^{\mathrm{e/h}}. \quad (4.39)$$

According to equations (4.27a&b), and under the assumption that $f_{\vec{q}}^{\mathrm{p}}$ reaches its steady state $\bar{f}_{\vec{q}}^{\mathrm{p}}$ as given in equation (4.25), the added carrier–phonon scattering terms both contribute to the net decay rate and the inhomogeneous driving source, therefore, the scattering process will change the effective carrier relaxation time and the carrier static distribution. Actually, because of the discrete LO phonon frequency, its distribution in the wave vector $\vec{k}$ domain is also discrete. Therefore, the carrier–phonon scattering has peaks in its relaxation process from its initial distribution to a static distribution [3], as illustrated by Fig. 4.3. These peaks appear at the $\vec{k}$ values corresponding to energies at $\varepsilon_{(\mathrm{e/h})\vec{k}} - n\hbar\omega_{\mathrm{LO}}$, with $n = 1, 2, 3, \ldots$ If we consider both carrier–carrier collision and carrier–phonon scattering, the peaks are submerged by the smooth carrier–carrier relaxation [3, 6] as described by equation (4.37), which is also illustrated by Fig. 4.3. It is worth mentioning that carrier–carrier collision is a momentum and kinetic energy conserved process. Therefore, through the relaxation process, although carriers (electron and hole) change profiles from their initial distributions to the quasi-Fermi distributions, their total energy remains the same as the initial excitation energy, which makes the effective plasma temperature well above the semiconductor lattice temperature. Dissipation of the carrier energy, i.e., the plasma cooling, will happen through carrier–photon scattering. Without considering the carrier–phonon scattering process, the carriers would converge to quasi-Fermi distributions with a higher (plasma) temperature, which is certainly not the case in the real world. Only through this scattering process, carriers will lose energy to the LO phonons and will eventually take the lattice temperature and converge to the quasi-Fermi distributions at the cooled lattice temperature. This is the scattering effect on the carrier static distribution, which does not change the (quasi-Fermi) distribution form; rather, it changes the parameter, i.e., the temperature, in the distribution function.

In summary, after a relaxation process, carrier–carrier interaction (collision) makes the carriers take a quasi-Fermi distribution regardless of their initial excitation profiles, whereas carrier–phonon interaction (scattering) cools the carriers' temperature to the lattice temperature so that carriers eventually take a quasi-Fermi distribution with the lattice temperature. This relaxation process is normally on a scale of sub-picoseconds to a few picoseconds.

Once the governing equations (4.28a–c) are solved in a self-consistent manner, we obtain the microscopic polariton numbers and carrier numbers at all $\vec{k}$ states. Through a summation in the wave vector space as shown in equations (4.14) and (4.16), we obtain the macroscopic polarization, electron and hole densities. Consequently, those material optical properties can be extracted from the polarization.

4.2 The free-carrier model as a zeroth order solution

4.2.1 The free-carrier model

The free-carrier model is established under the following assumptions.

(1) Rate equation approximation
Carrier–carrier collision and carrier–phonon scattering are very fast processes, which means that it takes negligible time for the injected electrons and holes to reach their quasi-Fermi distributions at the lattice temperature. The interaction between carriers and polaritons makes carriers only slightly deviate from their steady states. Once such an interaction disappears, carriers should rapidly damp down to their steady states with the quasi-Fermi distributions. Therefore, the carrier–carrier and carrier–phonon interactions in equation (4.28a) are phenomenologically replaced by terms in the form of $-\gamma_a(f_{\vec{k}}^{\mathrm{e/h}} - \bar{f}_{\vec{k}}^{\mathrm{e/h}})$, with γ_a introduced as a phenomenological carrier damping factor.
(2) Full screening approximation
The Coulomb interaction between carriers (electrons and holes) is fully screened. As a consequence, its effects on the transition energy and dipole matrix elements are ignored. However, we phenomenologically introduce a polariton damping factor γ in the time domain to describe its spectral broadening, which is one of the final effects of the Coulomb interaction. With the Coulomb interaction completely ignored, we should have a purely inhomogeneously broadened polariton, hence an optical gain, spectrum. Unfortunately, this is not true in the real world. Actually, the optical gain of semiconductors always shows a short wavelength range homogeneous broadening to some extent [6]. This is the reason we bring in the phenomenological polariton damping factor γ to capture qualitatively the gain broadening behavior without expensive calculations on the many-body Coulomb effect involved.

Under these two assumptions, the governing equations (4.28a–c) reduce to

$$\frac{\mathrm{d}p_{\vec{k}}}{\mathrm{d}t} = -(\mathrm{j}\bar{\omega}_{\vec{k}} + \gamma)p_{\vec{k}} - \frac{\mathrm{j}}{\hbar}(\vec{\mu}_{\vec{k}} \cdot \vec{E})(f_{\vec{k}}^{\mathrm{e}} + f_{\vec{k}}^{\mathrm{h}} - 1), \tag{4.40a}$$

$$\frac{\mathrm{d}f_{\vec{k}}^{\mathrm{e}}}{\mathrm{d}t} = \frac{2}{\hbar}\mathrm{Im}[(\vec{\mu}_{\vec{k}}^{*} \cdot \vec{E})p_{\vec{k}}] - \gamma_a(f_{\vec{k}}^{\mathrm{e}} - \bar{f}_{\vec{k}}^{\mathrm{e}}), \tag{4.40b}$$

$$\frac{\mathrm{d}f_{\vec{k}}^{\mathrm{h}}}{\mathrm{d}t} = \frac{2}{\hbar}\mathrm{Im}[(\vec{\mu}_{\vec{k}}^{*} \cdot \vec{E})p_{\vec{k}}] - \gamma_a(f_{\vec{k}}^{\mathrm{h}} - \bar{f}_{\vec{k}}^{\mathrm{h}}). \tag{4.40c}$$

In the free-carrier model, the phenomenological polariton and carrier damping factor, γ and γ_a in rad/s, can be obtained only through fitting the calculated result to the experimental data.

The solution to these equations can be obtained only in conjunction with the initial conditions. There are two methods of dealing with the initial conditions.

(1) The injection is not directly applied to the active region (i.e., the material that we are studying).
In this case, equation (4.14) has to be used with both $N_{\rm e}(t)$ and $N_{\rm h}(t)$ determined by solving the equations that govern carrier transport in the external region, which will be discussed in Chapter 5. The initial carrier distributions inside the active region must be given as well, which we can assume is at equilibrium, so that electrons and holes follow Fermi–Dirac distributions with the unified Fermi level determined by the charge neutral condition $N_{\rm e}(0) + N_{\rm A}^{-} = N_{\rm h}(0) + N_{\rm D}^{+}$, or

$$\frac{1}{\Omega}\sum_{\vec{k}} f_{\vec{k}}^{\rm e} + N_{\rm A}^{-} = \frac{1}{\Omega}\sum_{\vec{k}} f_{\vec{k}}^{\rm h} + N_{\rm D}^{+}, \tag{4.41}$$

at $t = 0$ with $N_{\rm A}^{-}$ and $N_{\rm D}^{+}$ defined as the ionized acceptor and donor concentrations inside the active region, respectively.
(2) Injection is applied to the active region directly.
In this case, a blocked injection term

$$\Lambda_{\vec{k}}^{\rm e/h} = \frac{\eta_{\rm tr} J}{edN_0}\bar{f}_{\vec{k}0}^{\rm e/h}(1 - f_{\vec{k}}^{\rm e/h}), \tag{4.42}$$

should be introduced as an inhomogeneous driving term in the carrier equations. In equation (4.42), $\eta_{\rm tr}$ indicates the carrier transport efficiency inside the active region, J the injection current density, d the active region thickness, N_0 and $\bar{f}_{\vec{k}0}^{\rm e/h}$ the total carrier density and quasi-Fermi distribution in the absence of the optical field (i.e., $\vec{E} = 0$).

4.2.2 The carrier rate equation

Considering a scheme with direct injection (applied to the active region) and with other carrier consumption mechanisms (such as spontaneous emission and carrier non-radiative decay because of defect capture) included, we can rewrite the carrier equations (4.40b&c) as

$$\frac{{\rm d}f_{\vec{k}}^{\rm e}}{{\rm d}t} = \frac{2}{\hbar}{\rm Im}[(\vec{\mu}_{\vec{k}}^{*}\cdot\vec{E})p_{\vec{k}}] + \Lambda_{\vec{k}}^{\rm e} - {\rm B}_{\vec{k}} f_{\vec{k}}^{\rm e} f_{\vec{k}}^{\rm h} - {\rm A}_e f_{\vec{k}}^{\rm e} - \gamma_a(f_{\vec{k}}^{\rm e} - \bar{f}_{\vec{k}}^{\rm e}), \tag{4.43a}$$

$$\frac{{\rm d}f_{\vec{k}}^{\rm h}}{{\rm d}t} = \frac{2}{\hbar}{\rm Im}[(\vec{\mu}_{\vec{k}}^{*}\cdot\vec{E})p_{\vec{k}}] + \Lambda_{\vec{k}}^{\rm h} - {\rm B}_{\vec{k}} f_{\vec{k}}^{\rm e} f_{\vec{k}}^{\rm h} - {\rm A}_h f_{\vec{k}}^{\rm h} - \gamma_a(f_{\vec{k}}^{\rm h} - \bar{f}_{\vec{k}}^{\rm h}), \tag{4.43b}$$

with ${\rm B}_{\vec{k}}$ and ${\rm A}_{\rm e/h}$ introduced as the spontaneous emission rate constant and electron/hole non-radiative decay constant, respectively. In the comprehensive model described in Section 4.1, carrier radiative recombination through spontaneous emission was not taken into account since we did not quantize the non-coherent spontaneously emitted field. Defects in semiconductors were not considered either, because of their random nature. Therefore, we add on these carrier consumption mechanisms in this chapter in a phenomenological manner.

Making summations over all of the states in the wave vector space and dividing the active region volume on both sides of equations (4.43a&b) gives

$$\frac{\mathrm{d}N_{\mathrm{e}}(t)}{\mathrm{d}t}=\frac{2}{\hbar\Omega}\mathrm{Im}\left[\sum_{\vec{k}}(\vec{\mu}_{\vec{k}}^{*}\cdot\vec{E})p_{\vec{k}}\right]+\frac{1}{\Omega}\sum_{\vec{k}}\Lambda_{\vec{k}}^{\mathrm{e}}-\frac{1}{\Omega}\sum_{\vec{k}}\mathrm{B}_{\vec{k}}f_{\vec{k}}^{\mathrm{e}}f_{\vec{k}}^{\mathrm{h}}-\mathrm{A_e}N_{\mathrm{e}}(t), \tag{4.44a}$$

$$\frac{\mathrm{d}N_{\mathrm{h}}(t)}{\mathrm{d}t}=\frac{2}{\hbar\Omega}\mathrm{Im}\left[\sum_{\vec{k}}(\vec{\mu}_{\vec{k}}^{*}\cdot\vec{E})p_{\vec{k}}\right]+\frac{1}{\Omega}\sum_{\vec{k}}\Lambda_{\vec{k}}^{\mathrm{h}}-\frac{1}{\Omega}\sum_{\vec{k}}\mathrm{B}_{\vec{k}}f_{\vec{k}}^{\mathrm{e}}f_{\vec{k}}^{\mathrm{h}}-\mathrm{A_h}N_{\mathrm{h}}(t). \tag{4.44b}$$

In deriving equations (4.44a&b), we have utilized the fact that the total number of carriers (electrons and holes) must be conserved throughout the carrier–carrier and carrier–phonon interactions (i.e., collision and scattering). Since carriers eventually reach their steady state, $\sum_{\vec{k}} f_{\vec{k}}^{\mathrm{e/h}}=\sum_{\vec{k}}\bar{f}_{\vec{k}}^{\mathrm{e/h}}$.

By assuming

$$\vec{E}(t)=\frac{1}{2}\tilde{\vec{E}}\mathrm{e}^{-\mathrm{j}\omega t}+\mathrm{c.c.}, \tag{4.45}$$

we have

$$\vec{P}(t)=\frac{1}{2}\tilde{\vec{P}}\mathrm{e}^{-\mathrm{j}\omega t}+\mathrm{c.c.}, \tag{4.46}$$

with $\tilde{\vec{E}}$ and $\tilde{\vec{P}}$ introduced as the slow-varying envelope functions.

Comparing equation (4.46) with (4.16) we obtain

$$\tilde{\vec{P}}=\frac{2}{\Omega}\mathrm{e}^{\mathrm{j}\omega t}\sum_{\vec{k}}\vec{\mu}_{\vec{k}}^{*}p_{\vec{k}}. \tag{4.47}$$

We also find

$$\frac{1}{\Omega}\sum_{\vec{k}}\Lambda_{\vec{k}}^{\mathrm{e/h}}=\frac{\eta_{\mathrm{tr}}J}{edN_0\Omega}\sum_{\vec{k}}\bar{f}_{\vec{k}0}^{\mathrm{e/h}}(1-f_{\vec{k}}^{\mathrm{e/h}})\approx\frac{\eta_{\mathrm{tr}}J}{edN_0\Omega}\sum_{\vec{k}}\bar{f}_{\vec{k}0}^{\mathrm{e/h}}(1-\bar{f}_{\vec{k}}^{\mathrm{e/h}})=\frac{\eta_{\mathrm{e/h}}J}{ed}, \tag{4.48}$$

where we have defined the injection efficiency as

$$\eta_{\mathrm{e/h}}\equiv\frac{\eta_{\mathrm{tr}}}{N_0\Omega}\sum_{\vec{k}}\bar{f}_{\vec{k}0}^{\mathrm{e/h}}(1-\bar{f}_{\vec{k}}^{\mathrm{e/h}}). \tag{4.49}$$

Finally, we have

$$\frac{1}{\Omega}\sum_{\vec{k}}\mathrm{B}_{\vec{k}}f_{\vec{k}}^{\mathrm{e}}f_{\vec{k}}^{\mathrm{h}}\approx\frac{1}{\Omega}\sum_{\vec{k}}\mathrm{B}_{\vec{k}}\bar{f}_{\vec{k}}^{\mathrm{e}}\bar{f}_{\vec{k}}^{\mathrm{h}}\approx\frac{\bar{\mathrm{B}}}{\Omega}\sum_{\vec{k}}\bar{f}_{\vec{k}}^{\mathrm{e}}\bar{f}_{\vec{k}}^{\mathrm{h}}\equiv\Gamma_{\mathrm{spon}}. \tag{4.50}$$

Substituting equations (4.47), (4.48) and (4.50) into (4.44) gives the carrier rate equations

$$\frac{\mathrm{d}N_\mathrm{e}(t)}{\mathrm{d}t} = \frac{\vec{E}(t)}{\hbar} \cdot \mathrm{Im}(\vec{\tilde{P}}\mathrm{e}^{-\mathrm{j}\omega t}) + \frac{\eta_\mathrm{e} J}{ed} - \Gamma_\mathrm{spon} - \mathrm{A_e} N_\mathrm{e}(t), \tag{4.51a}$$

$$\frac{\mathrm{d}N_\mathrm{h}(t)}{\mathrm{d}t} = \frac{\vec{E}(t)}{\hbar} \cdot \mathrm{Im}(\vec{\tilde{P}}\mathrm{e}^{-\mathrm{j}\omega t}) + \frac{\eta_\mathrm{h} J}{ed} - \Gamma_\mathrm{spon} - \mathrm{A_h} N_\mathrm{h}(t). \tag{4.51b}$$

For low carrier densities, $\varepsilon_{(\mathrm{e/h})\vec{k}} - F^{\mathrm{c/v}} \gg \mathrm{k_B}T$, equation (4.36) reduces to the Maxwell–Boltzmann distribution

$$\bar{f}_{\vec{k}}^{\mathrm{e/h}} \approx \mathrm{e}^{\frac{F^{\mathrm{c/v}} - \varepsilon_{(\mathrm{e/h})\vec{k}}}{\mathrm{k_B}T}}. \tag{4.52}$$

For bulk semiconductors, substituting equation (4.52) and $\varepsilon_{(\mathrm{e/h})\vec{k}} = \hbar^2 k^2 / 2m_\mathrm{e/h}$ into equation (4.50) yields

$$\Gamma_\mathrm{spon} \equiv \frac{\bar{\mathrm{B}}}{\Omega} \sum_{\vec{k}} \bar{f}_{\vec{k}}^{\mathrm{e}} \bar{f}_{\vec{k}}^{\mathrm{h}} = \frac{\bar{\mathrm{B}}}{\pi^2} \mathrm{e}^{\frac{F^\mathrm{c} + F^\mathrm{v}}{\mathrm{k_B}T}} \int_0^\infty \mathrm{d}k (k^2 \mathrm{e}^{-\frac{\hbar^2 k^2}{2m_\mathrm{r}\mathrm{k_B}T}}) = 2\left[\frac{m_\mathrm{r}\mathrm{k_B}T}{2\pi\hbar^2}\right]^{\frac{3}{2}} \bar{\mathrm{B}} e^{\frac{F^\mathrm{c} + F^\mathrm{v}}{\mathrm{k_B}T}}$$

$$= 4\bar{\mathrm{B}}\left[\frac{\pi\hbar^2}{2(m_\mathrm{e} + m_\mathrm{h})\mathrm{k_B}T}\right]^{\frac{3}{2}} N_\mathrm{e}(t) N_\mathrm{h}(t) = \mathrm{B} N_\mathrm{e}(t) N_\mathrm{h}(t), \tag{4.53}$$

with the spontaneous recombination rate B defined as

$$\mathrm{B} \equiv 4\bar{\mathrm{B}}\left[\frac{\pi\hbar^2}{2(m_\mathrm{e} + m_\mathrm{h})\mathrm{k_B}T}\right]^{\frac{3}{2}} = \frac{\bar{\mathrm{B}}}{2}\left[\frac{1}{(N_\mathrm{c}/2)^{\frac{2}{3}} + (N_\mathrm{v}/2)^{\frac{2}{3}}}\right]^{\frac{3}{2}}. \tag{4.54}$$

In deriving equation (4.53), we have also used

$$N_\mathrm{e/h}(t) = \frac{1}{\Omega} \sum_{\vec{k}} \bar{f}_{\vec{k}}^{\mathrm{e/h}} \approx \frac{1}{\pi^2} \mathrm{e}^{\frac{F^{\mathrm{c/v}}}{\mathrm{k_B}T}} \int_0^\infty \mathrm{d}k \left(k^2 \mathrm{e}^{-\frac{\hbar^2 k^2}{2m_\mathrm{e/h}\mathrm{k_B}T}}\right)$$

$$= 2\left[\frac{m_\mathrm{e/h}\mathrm{k_B}T}{2\pi\hbar^2}\right]^{3/2} \mathrm{e}^{\frac{F^{\mathrm{c/v}}}{\mathrm{k_B}T}} = N_\mathrm{c/v} \mathrm{e}^{\frac{F^{\mathrm{c/v}}}{\mathrm{k_B}T}}, \tag{4.55}$$

with the parabolic conduction and valence band edge densities defined as

$$N_\mathrm{c/v} \equiv 2\left[\frac{m_\mathrm{e/h}\mathrm{k_B}T}{2\pi\hbar^2}\right]^{3/2} \approx 2.51 \times 10^{19} \left(\frac{m_\mathrm{e/h}}{\mathrm{m_0}} \frac{T}{300}\right)^{3/2} \mathrm{cm}^{-3}. \tag{4.56}$$

Plugging equation (4.53) into (4.51a&b) we obtain the carrier rate equations in the most popular form. If the active region is doped, we still have the carrier rate equations in the form of (4.51a&b), but the carrier consumption term because of spontaneous emission on the RHS of these equations must be modified to [6]

$$\Gamma_\mathrm{spon} \equiv \begin{cases} \frac{1}{\Omega} \sum_{\vec{k}} \mathrm{B}_{\vec{k}} (\bar{f}_{\vec{k}}^{\mathrm{e}} - \bar{f}_{\vec{k}}^{\mathrm{D}}) \bar{f}_{\vec{k}}^{\mathrm{h}} & \text{N-doped} \\ \frac{1}{\Omega} \sum_{\vec{k}} \mathrm{B}_{\vec{k}} \bar{f}_{\vec{k}}^{\mathrm{e}} (\bar{f}_{\vec{k}}^{\mathrm{h}} - \bar{f}_{\vec{k}}^{\mathrm{A}}) & \text{P-doped,} \end{cases} \tag{4.57}$$

with $\bar{f}_{\vec{k}}^{D/A}$ denoting the Fermi–Dirac distributions of the ionized donor and acceptor carriers, respectively.

4.2.3 The polariton rate equation

We can formally integrate the polariton equation (4.40a) to obtain [6]

$$p_{\vec{k}} = -\frac{\mathrm{j}}{\hbar}\int_{-\infty}^{t} \mathrm{d}\tau\{[\vec{\mu}_{\vec{k}} \cdot E(\tau)](\bar{f}_{\vec{k}}^{\mathrm{e}} + \bar{f}_{\vec{k}}^{\mathrm{h}} - 1)\mathrm{e}^{(\mathrm{j}\bar{\omega}_{\vec{k}}+\gamma)(\tau-t)}\}, \tag{4.58}$$

where the carrier number expectations are replaced by their steady state quasi-Fermi distributions.

Under the rate equation approximation, the time domain slow-varying factors can be taken out of the integral in equation (4.58). Hence we find

$$\begin{aligned} p_{\vec{k}} &= -\frac{\mathrm{j}}{2\hbar}(\vec{\mu}_{\vec{k}} \cdot \vec{\tilde{E}})(\bar{f}_{\vec{k}}^{\mathrm{e}} + \bar{f}_{\vec{k}}^{\mathrm{h}} - 1)\mathrm{e}^{-(\mathrm{j}\bar{\omega}_{\vec{k}}+\gamma)t}\int_{-\infty}^{t} \mathrm{d}\tau[\mathrm{e}^{(-\mathrm{j}\omega+\mathrm{j}\bar{\omega}_{\vec{k}}+\gamma)\tau} + \mathrm{e}^{(\mathrm{j}\omega+\mathrm{j}\bar{\omega}_{\vec{k}}+\gamma)\tau}] \\ &= -\frac{\mathrm{j}}{2\hbar}(\vec{\mu}_{\vec{k}} \cdot \vec{\tilde{E}})(\bar{f}_{\vec{k}}^{\mathrm{e}} + \bar{f}_{\vec{k}}^{\mathrm{h}} - 1)\left[\frac{\mathrm{e}^{-\mathrm{j}\omega t}}{\mathrm{j}(\bar{\omega}_{\vec{k}} - \omega) + \gamma} + \frac{\mathrm{e}^{\mathrm{j}\omega t}}{\mathrm{j}(\bar{\omega}_{\vec{k}} + \omega) + \gamma}\right]. \end{aligned} \tag{4.59}$$

Under a rotating-wave approximation [8], we obtain

$$p_{\vec{k}} = -\frac{\mathrm{j}}{2\hbar}(\vec{\mu}_{\vec{k}} \cdot \vec{\tilde{E}})(\bar{f}_{\vec{k}}^{\mathrm{e}} + \bar{f}_{\vec{k}}^{\mathrm{h}} - 1)\frac{\mathrm{e}^{-\mathrm{j}\omega t}}{\mathrm{j}(\bar{\omega}_{\vec{k}} - \omega) + \gamma}. \tag{4.60}$$

Therefore, the slow-varying envelope of the macroscopic polarization defined by equation (4.46) is obtained by plugging equation (4.60) into (4.47)

$$\begin{aligned} \tilde{P} &= -\frac{\mathrm{j}}{\hbar\Omega}\tilde{E}_{\mu}\sum_{\vec{k}}|\mu_{\vec{k}}|^2(\bar{f}_{\vec{k}}^{\mathrm{e}} + \bar{f}_{\vec{k}}^{\mathrm{h}} - 1)\frac{1}{\mathrm{j}(\bar{\omega}_{\vec{k}} - \omega) + \gamma} \\ &= -\frac{\tilde{E}_{\mu}}{\hbar\gamma\Omega}\sum_{\vec{k}}|\mu_{\vec{k}}|^2(\bar{f}_{\vec{k}}^{\mathrm{e}} + \bar{f}_{\vec{k}}^{\mathrm{h}} - 1)L(\bar{\omega}_{\vec{k}} - \omega)[(\bar{\omega}_{\vec{k}} - \omega)/\gamma + \mathrm{j}], \end{aligned} \tag{4.61}$$

with $\widetilde{E}_\mu$ defined as the projection of vector $\overset{\rightleftarrows}{E}$ along the direction of vector $\vec{\mu}_{\vec{k}}$. In equation (4.61), we have the Lorentzian line-shape function defined as

$$L(x) = \frac{\gamma^2}{\gamma^2 + x^2}. \tag{4.62}$$

The direction of $\overset{\rightleftarrows}{P}$ is along $\vec{\mu}_{\vec{k}}$ as well.

4.2.4 The susceptibility

By taking the Fourier transform of equation (2.9) we find the frequency domain susceptibility as

$$\vec{P}(\omega) = \varepsilon_0 \widetilde{\chi}(\omega) \vec{E}(\omega). \tag{4.63}$$

Therefore, from

$$\vec{P}(t) = \varepsilon_0 F^{-1}[\widetilde{\chi}(\omega)] \otimes \vec{E}(t) = \frac{1}{2} \overset{\rightleftarrows}{E} \varepsilon_0 F^{-1}[\widetilde{\chi}(\omega)] \otimes \mathrm{e}^{-\mathrm{j}\omega t} + \mathrm{c.c.} = \frac{1}{2} \overset{\rightleftarrows}{P} \mathrm{e}^{-\mathrm{j}\omega t} + \mathrm{c.c.},$$

we obtain $\overset{\rightleftarrows}{P} \mathrm{e}^{-\mathrm{j}\omega t} = \overset{\rightleftarrows}{E} \varepsilon_0 F^{-1}[\widetilde{\chi}(\omega)] \otimes \mathrm{e}^{-\mathrm{j}\omega t}$ or $\overset{\rightleftarrows}{P} \delta(\omega) = \overset{\rightleftarrows}{E} \varepsilon_0 \widetilde{\chi}(\omega) \delta(\omega)$, which yields

$$\widetilde{\chi}(\omega) = \frac{\widetilde{P}}{\varepsilon_0 \widetilde{E}_\mu} = -\frac{1}{\varepsilon_0 \hbar \gamma \Omega} \sum_{\vec{k}} |\mu_{\vec{k}}|^2 (\bar{f}^{\mathrm{e}}_{\vec{k}} + \bar{f}^{\mathrm{h}}_{\vec{k}} - 1)[(\bar{\omega}_{\vec{k}} - \omega)/\gamma + \mathrm{j}] L(\bar{\omega}_{\vec{k}} - \omega), \tag{4.64}$$

in accordance with equation (4.61). In equation (4.64), we have also utilized the fact that $\widetilde{P}$ and $\widetilde{E}_\mu$ are in the same direction as $\overset{\rightleftarrows}{P}$ is along $\vec{\mu}_{\vec{k}}$.

Consequently, we obtain the semiconductor material gain and refractive index change according to equation (2.104)

$$\begin{aligned} g(\omega) &= -\frac{k_0}{n} \mathrm{Im}[\widetilde{\chi}(\omega)] = \frac{\omega}{\varepsilon_0 n c \hbar \gamma \Omega} \sum_{\vec{k}} |\mu_{\vec{k}}|^2 (\bar{f}^{\mathrm{e}}_{\vec{k}} + \bar{f}^{\mathrm{h}}_{\vec{k}} - 1) L(\bar{\omega}_{\vec{k}} - \omega) \\ &= \frac{\omega}{\varepsilon_0 n c \hbar \Omega} \sum_{\vec{k}} |\mu_{\vec{k}}|^2 (\bar{f}^{\mathrm{e}}_{\vec{k}} + \bar{f}^{\mathrm{h}}_{\vec{k}} - 1) \frac{\gamma}{\gamma^2 + (\bar{\omega}_{\vec{k}} - \omega)^2}, \end{aligned} \tag{4.65a}$$

$$\begin{aligned} \Delta n(\omega) &= \frac{1}{2n} \mathrm{Re}[\widetilde{\chi}(\omega)] = -\frac{1}{2\varepsilon_0 n \hbar \gamma \Omega} \sum_{\vec{k}} |\mu_{\vec{k}}|^2 (\bar{f}^{\mathrm{e}}_{\vec{k}} + \bar{f}^{\mathrm{h}}_{\vec{k}} - 1) \frac{\bar{\omega}_{\vec{k}} - \omega}{\gamma} L(\bar{\omega}_{\vec{k}} - \omega) \\ &= -\frac{1}{2\varepsilon_0 n \hbar \Omega} \sum_{\vec{k}} |\mu_{\vec{k}}|^2 (\bar{f}^{\mathrm{e}}_{\vec{k}} + \bar{f}^{\mathrm{h}}_{\vec{k}} - 1) \frac{\bar{\omega}_{\vec{k}} - \omega}{\gamma^2 + (\bar{\omega}_{\vec{k}} - \omega)^2}. \end{aligned} \tag{4.65b}$$

We note that the susceptibility extracted from equation (4.64) is the one induced by carrier injection only, with the background value $n^2 - 1$ excluded. Calculation of the material background refractive index n, or the relative permittivity $\varepsilon_r = \varepsilon/\varepsilon_0 = n^2$, is briefly discussed in Section 4.3.2 in evaluating the screened Coulomb potential. In-depth study of this topic is beyond the scope of this book. One can always refer to [9] or find its value directly in a semiconductor material database [10].

4.3 The screened Coulomb interaction model as a first order solution

4.3.1 The screened Coulomb interaction model

A screened Coulomb interaction model is established under the following assumptions.

(1) Phenomenological collision approximation.
The carrier–carrier collision and carrier–phonon scattering are phenomenologically described by the carriers' rapid relaxation towards their steady states at the lattice temperature, with their effect on the polariton incorporated into a damping factor in the time domain to depict its spectral broadening, as in Section 4.2.

(2) Screened Coulomb potential approximation.
Bare Coulomb interaction between the charged carriers (electrons and holes) is replaced by the plasma screened Coulomb potential applied to every individual charged carrier. As a consequence, effects such as transition energy shift and dipole matrix element renormalization appear because of inclusion of the screened Coulomb force. Actually, the approach of using a screened Coulomb potential to replace the bare one in the last term of a truncated operator–Hamiltonian commutator series is similar to the renormalization method, which is also known as a windowing technique to minimize the truncation error. It is for this reason that we always use the screened Coulomb potential in the highest order interaction energies retained in our governing equations. That is to say, if we retain up to the second order operator–Hamiltonian commutator expansion terms in our governing equations, as we did for equation (4.28a–c), the screened Coulomb potential should be used in the second order interaction energy matrices $\Sigma_{\vec{k}}^{e/h-out}$, $\Sigma_{\vec{k}}^{e/h-in}$, $\Lambda_{\vec{k}\vec{k}'}$, $\Sigma_{\vec{q},\pm}^{e/h-p}$, $\Sigma_{\vec{q},\pm}^{e/h}$, and $\Lambda_{\vec{k}\vec{q}}^{p}$ as shown in equations (4.23a–c) and (4.27a–c), whereas the bare Coulomb potential should still be used in the combined zeroth and first order interaction energy matrix $\Theta_{\vec{k}\vec{k}'}$ and in the Rabi frequency $\Omega_{\vec{k}}$ as shown in equations (4.29) and (4.30), respectively, to avoid double counting the second order effect. However, if we retain only up to the first order operator–Hamiltonian commutator expansion terms in our governing equations, the screened Coulomb potential should be used in $\Theta_{\vec{k}\vec{k}'}$ and $\Omega_{\vec{k}}$ as they are related to the last terms in the truncation.

Under these two assumptions, the governing equations (4.28a–c) reduce to

$$\frac{dp_{\vec{k}}}{dt} = -\frac{j}{\hbar}\sum_{\vec{k}'}(\Theta_{\vec{k}\vec{k}'} - j\hbar\gamma\delta_{\vec{k}\vec{k}'})p_{\vec{k}'} - \frac{j}{\hbar}(\vec{\mu}_{\vec{k}} \cdot \vec{E})(f_{\vec{k}}^{e} + f_{\vec{k}}^{h} - 1), \qquad (4.66a)$$

$$\frac{\mathrm{d}f^{\mathrm{e}}_{\vec{k}}}{\mathrm{d}t} = \frac{2}{\hbar}\mathrm{Im}[\Omega^*_{\vec{k}} p_{\vec{k}}] - \gamma_a(f^{\mathrm{e}}_{\vec{k}} - \bar{f}^{\mathrm{e}}_{\vec{k}}), \tag{4.66b}$$

$$\frac{\mathrm{d}f^{\mathrm{h}}_{\vec{k}}}{\mathrm{d}t} = \frac{2}{\hbar}\mathrm{Im}[\Omega^*_{\vec{k}} p_{\vec{k}}] - \gamma_a(f^{\mathrm{h}}_{\vec{k}} - \bar{f}^{\mathrm{h}}_{\vec{k}}), \tag{4.66c}$$

where $\Theta_{\vec{k}\vec{k}'}$, $\Omega_{\vec{k}}$ are still in the form of equations (4.29) and (4.30), respectively, but with the Coulomb potential given in the screened form rather than in the bare form given as equation (4.9a–d). Thus we have

$$\Theta_{\vec{k}\vec{k}'} \equiv \begin{cases} \hbar\bar{\omega}_{\vec{k}} - \sum_{\vec{l}\neq\vec{k}} \widetilde{V}_{|\vec{l}-\vec{k}|}(f^{\mathrm{e}}_{\vec{l}} + f^{\mathrm{h}}_{\vec{l}} - 1) & \vec{k}' = \vec{k} \\ \widetilde{V}_{|\vec{k}'-\vec{k}|}(f^{\mathrm{e}}_{\vec{k}} + f^{\mathrm{h}}_{\vec{k}} - 1) & \vec{k}' \neq \vec{k} \end{cases}, \tag{4.67}$$

$$\Omega_{\vec{k}} \equiv \frac{1}{\hbar}\left(\vec{\mu}_{\vec{k}} \cdot \vec{E} + \sum_{\vec{l}\neq\vec{k}} \widetilde{V}_{|\vec{l}-\vec{k}|} p_{\vec{l}}\right). \tag{4.68}$$

4.3.2 The screened Coulomb potential

The screened Coulomb potential can be derived from the following self-consistent Hartree–Fock model [6]. In this model, on the one hand, by using the screened Coulomb potential to truncate the operator–Hamiltonian commutator, we can find an expression for the carrier distribution in terms of the screened Coulomb potential. On the other hand, the screened Coulomb potential and the carrier distribution must satisfy Poisson's equation, from which we can find the screened Coulomb potential by eliminating the carrier distribution through substituting it with the previously obtained expression.

Assuming that a test electron at the origin, i.e.,

$$f^{\mathrm{e}}_0(\vec{r}) = \delta(\vec{r}), \tag{4.69}$$

is adiabatically introduced to a background electron plasma distribution $f^{\mathrm{e}}(\vec{r})$. Because of the disturbance, the original background $f^{\mathrm{e}}(\vec{r})$ will be redistributed to $\widetilde{f}^{\mathrm{e}}(\vec{r})$. We can express the screened (by the redistributed background) electrostatic potential $\widetilde{\phi}(\vec{r})$ in Poisson's equation

$$\nabla^2\widetilde{\phi}(\vec{r}) = -\frac{\mathrm{e}}{\varepsilon_0}[f^{\mathrm{e}}_0(\vec{r}) + \widetilde{f}^{\mathrm{e}}(\vec{r})], \tag{4.70}$$

which is derived from the Maxwell equations for an electrostatic field.

Taking the Fourier transform of the above equation yields Poisson's equation in the wave vector domain

$$\widetilde{\phi}_{\vec{k}} = \frac{\mathrm{e}}{\varepsilon_0 k^2}\left(\frac{1}{\Omega} + \widetilde{f}^{\mathrm{e}}_{\vec{k}}\right), \tag{4.71}$$

or

$$\widetilde{V}_{|\vec{k}|} = \mathrm{e}\widetilde{\phi}_{\vec{k}} = \frac{\mathrm{e}^2}{\varepsilon_0 k^2}\left(\frac{1}{\Omega} + \widetilde{f}^{\mathrm{e}}_{\vec{k}}\right), \tag{4.72}$$

where

$$\widetilde{\phi}_{\vec{k}} = \frac{1}{\Omega}\int_{\Omega} \mathrm{d}\vec{r}\,[\mathrm{e}^{-\mathrm{j}\vec{k}\cdot\vec{r}}\widetilde{\phi}(\vec{r})], \quad \widetilde{f}^{\mathrm{e}}_{\vec{k}} = \frac{1}{\Omega}\int_{\Omega} \mathrm{d}\vec{r}\,[\mathrm{e}^{-\mathrm{j}\vec{k}\cdot\vec{r}}\widetilde{f}^{\mathrm{e}}(\vec{r})], \tag{4.73}$$

and according to equation (4.69)

$$\frac{1}{\Omega}\int_{\Omega} d\vec{r}\,[e^{-j\vec{k}\cdot\vec{r}} f_0^e(\vec{r})] = \frac{1}{\Omega}\int_{\Omega} d\vec{r}\,[e^{-j\vec{k}\cdot\vec{r}}\delta(\vec{r})] = \frac{1}{\Omega}.$$

By switching off the background electron plasma, we obtain the unscreened Coulomb potential $V_{|\vec{k}|} = e\phi_{\vec{k}} = e^2/(\varepsilon_0\Omega k^2)$, which is the same as equation (4.9b).

Therefore, the screened Coulomb potential can be rewritten in the form

$$\widetilde{V}_{|\vec{k}|} = \frac{e^2}{\varepsilon_0\Omega k^2}(1 + \widetilde{f}_{\vec{k}}^e\Omega) = V_{|\vec{k}|}(1 + \widetilde{f}_{\vec{k}}^e\Omega). \tag{4.74}$$

On the other hand, the redistribution of the background electron plasma from the original $f^e(\vec{r})$ (or $f_{\vec{k}}^e$ in the wave vector domain) to $\widetilde{f}^e(\vec{r})$ (or $\widetilde{f}_{\vec{k}}^e$ in the wave vector domain) is driven by the screened Coulomb potential

$$\widetilde{V}(\vec{r}) = e\widetilde{\phi}(\vec{r}), \tag{4.75}$$

(or $\widetilde{V}_{|\vec{k}|} = e\widetilde{\phi}_{\vec{k}}$ in the wave vector domain). This process is governed by the time–wave vector domain equation of motion for the electron density operator.

By letting the original background electron plasma number operator be

$$\hat{f}_{\vec{k}}^e \equiv \hat{a}_{\vec{k}}^+\hat{a}_{\vec{k}}, \tag{4.76}$$

and the screened background electron plasma density operator be

$$\vec{F}_{\vec{l}}^e \equiv \sum_{\vec{k}} \hat{a}_{\vec{k}-\vec{l}}^+\hat{a}_{\vec{k}}/\Omega, \tag{4.77}$$

we can write the system Hamiltonian as

$$\vec{H}_{\text{eff}} = \sum_{\vec{k}} \varepsilon_{\vec{k}}^e \hat{a}_{\vec{k}}^+\hat{a}_{\vec{k}} + \Omega\sum_{\vec{l}} \widetilde{V}_{|\vec{l}|}\vec{F}_{-\vec{l}}^e. \tag{4.78}$$

As the expected values of the original background electron plasma number and the screened background electron plasma density operators, we have

$$f_{\vec{k}}^e = \langle\hat{a}_{\vec{k}}^+\hat{a}_{\vec{k}}\rangle, \tag{4.79}$$

and

$$\widetilde{f}_{\vec{k}}^e \equiv \langle\vec{F}_{\vec{k}}^e\rangle, \tag{4.80}$$

where the inverse Fourier transform of $\widetilde{f}_{\vec{k}}^e$ is the background electron plasma redistribution caused by the disturbance of the test electron

$$\widetilde{f}^e(\vec{r}) = \sum_{\vec{l}} e^{j\vec{l}\cdot\vec{r}}\widetilde{f}_{\vec{l}}^e = \sum_{\vec{l}} e^{j\vec{l}\cdot\vec{r}}\langle\vec{F}_{\vec{l}}^e\rangle. \tag{4.81}$$

The equation of motion for the screened background electron plasma number operator $\hat{a}^{+}_{\vec{k}-\vec{k}'}\hat{a}_{\vec{k}}$ in the time–wave vector domain can be expressed in the form of the Heisenberg equation

$$\begin{aligned} \mathrm{j}\hbar\frac{\mathrm{d}}{\mathrm{d}t}\hat{a}^{+}_{\vec{k}-\vec{k}'}\hat{a}_{\vec{k}} &= [\hat{a}^{+}_{\vec{k}-\vec{k}'}\hat{a}_{\vec{k}}, \hat{H}_{\mathrm{eff}}] \\ &= (\varepsilon^{\mathrm{e}}_{\vec{k}} - \varepsilon^{\mathrm{e}}_{\vec{k}-\vec{k}'})\hat{a}^{+}_{\vec{k}-\vec{k}'}\hat{a}_{\vec{k}} + \sum_{\vec{l}} \widetilde{V}_{|\vec{l}|}(\hat{a}^{+}_{\vec{k}-\vec{k}'}\hat{a}_{\vec{k}+\vec{l}} - \hat{a}^{+}_{\vec{k}-\vec{k}'-\vec{l}}\hat{a}_{\vec{k}}). \end{aligned} \tag{4.82}$$

Taking the expected value of equation (4.82) and keeping only those slowly varying terms with $\vec{l} = -\vec{k}'$, we find

$$\begin{aligned} \mathrm{j}\hbar\frac{\mathrm{d}}{\mathrm{d}t}\langle\hat{a}^{+}_{\vec{k}-\vec{k}'}\hat{a}_{\vec{k}}\rangle &= (\varepsilon^{\mathrm{e}}_{\vec{k}} - \varepsilon^{\mathrm{e}}_{\vec{k}-\vec{k}'})\langle\hat{a}^{+}_{\vec{k}-\vec{k}'}\hat{a}_{\vec{k}}\rangle + \widetilde{V}_{|\vec{k}'|}(\langle\hat{a}^{+}_{\vec{k}-\vec{k}'}\hat{a}_{\vec{k}-\vec{k}'}\rangle - \langle\hat{a}^{+}_{\vec{k}}\hat{a}_{\vec{k}}\rangle) \\ &= (\varepsilon^{\mathrm{e}}_{\vec{k}} - \varepsilon^{\mathrm{e}}_{\vec{k}-\vec{k}'})\langle\hat{a}^{+}_{\vec{k}-\vec{k}'}\hat{a}_{\vec{k}}\rangle + \widetilde{V}_{|\vec{k}'|}(f^{\mathrm{e}}_{\vec{k}-\vec{k}'} - f^{\mathrm{e}}_{\vec{k}}). \end{aligned} \tag{4.83}$$

We assume that $\langle\hat{a}^{+}_{\vec{k}-\vec{k}'}\hat{a}_{\vec{k}}\rangle$ has a solution of the form $\mathrm{e}^{(-\mathrm{j}\omega+\delta/\hbar)t}$ (damped harmonic oscillation), where $\delta \to 0$ indicates that the perturbation (test electron) is switched on adiabatically, i.e., we had a homogenous background electron plasma at $t \to -\infty$. We further assume that the original background electron plasma number expectation $f^{\mathrm{e}}_{\vec{k}}$ also follows this response. Under these assumptions, the above equation in the time–wave vector domain can be transformed into the frequency–wave vector domain as

$$\hbar(\omega + \mathrm{j}\delta)\langle\hat{a}^{+}_{\vec{k}-\vec{k}'}\hat{a}_{\vec{k}}\rangle = (\varepsilon^{\mathrm{e}}_{\vec{k}} - \varepsilon^{\mathrm{e}}_{\vec{k}-\vec{k}'})\langle\hat{a}^{+}_{\vec{k}-\vec{k}'}\hat{a}_{\vec{k}}\rangle + \widetilde{V}_{|\vec{k}'|}(f^{\mathrm{e}}_{\vec{k}-\vec{k}'} - f^{\mathrm{e}}_{\vec{k}})$$

or

$$\langle\hat{a}^{+}_{\vec{k}-\vec{k}'}\hat{a}_{\vec{k}}\rangle = \widetilde{V}_{|\vec{k}'|}\frac{f^{\mathrm{e}}_{\vec{k}-\vec{k}'} - f^{\mathrm{e}}_{\vec{k}}}{\hbar\omega + \mathrm{j}\delta + \varepsilon^{\mathrm{e}}_{\vec{k}-\vec{k}'} - \varepsilon^{\mathrm{e}}_{\vec{k}}}. \tag{4.84}$$

Substituting equation (4.77) into (4.80) and replacing the screened background electron plasma number expectation by equation (4.84) we obtain

$$\widetilde{f}^{\mathrm{e}}_{\vec{k}'} = \frac{1}{\Omega}\sum_{\vec{k}}\langle\hat{a}^{+}_{\vec{k}-\vec{k}'}\hat{a}_{\vec{k}}\rangle = \frac{\widetilde{V}|\vec{k}'|}{\Omega}\sum_{\vec{k}}\frac{f^{\mathrm{e}}_{\vec{k}-\vec{k}'} - f^{\mathrm{e}}_{\vec{k}}}{\hbar\omega + \mathrm{j}\delta + \varepsilon^{\mathrm{e}}_{\vec{k}-\vec{k}'} - \varepsilon^{\mathrm{e}}_{\vec{k}}}. \tag{4.85}$$

Plugging equation (4.85) into the screened Coulomb potential formula (4.74) yields

$$\widetilde{V}_{|\vec{k}|} = V_{|\vec{k}|}\left(1 + \widetilde{V}_{|\vec{k}|}\sum_{\vec{l}}\frac{f^{\mathrm{e}}_{\vec{l}-\vec{k}} - f^{\mathrm{e}}_{\vec{l}}}{\hbar\omega + \mathrm{j}\delta + \varepsilon^{\mathrm{e}}_{\vec{l}-\vec{k}} - \varepsilon^{\mathrm{e}}_{\vec{l}}}\right),$$

or

$$\widetilde{V}_{|\vec{k}|} = V_{|\vec{k}|}/\left(1 - V_{|\vec{k}|}\sum_{\vec{l}}\frac{f^{\mathrm{e}}_{\vec{l}-\vec{k}} - f^{\mathrm{e}}_{\vec{l}}}{\hbar\omega + \mathrm{j}\delta + \varepsilon^{\mathrm{e}}_{\vec{l}-\vec{k}} - \varepsilon^{\mathrm{e}}_{\vec{l}}}\right). \tag{4.86}$$

Following a similar approach, we can include the hole contribution to obtain

$$\widetilde{V}_{|\vec{k}|} = \frac{V_{|\vec{k}|}}{\varepsilon_{r,\vec{k}}(\omega)} = V_{|\vec{k}|} \Big/ \left[1 - V_{|\vec{k}|} \sum_{\vec{l}} \sum_{\alpha=\mathrm{e,h}} \frac{f^{\alpha}_{\vec{l}-\vec{k}} - f^{\alpha}_{\vec{l}}}{\hbar\omega + \mathrm{j}\delta + \varepsilon^{\alpha}_{\vec{l}-\vec{k}} - \varepsilon^{\alpha}_{\vec{l}}} \right], \tag{4.87}$$

where

$$\varepsilon_{r,\vec{k}}(\omega) \equiv 1 - V_{|\vec{k}|} \sum_{\vec{l}} \sum_{\alpha=\mathrm{e,h}} \frac{f^{\alpha}_{\vec{l}-\vec{k}} - f^{\alpha}_{\vec{l}}}{\hbar\omega + \mathrm{j}\delta + \varepsilon^{\alpha}_{\vec{l}-\vec{k}} - \varepsilon^{\alpha}_{\vec{l}}}. \tag{4.88}$$

Equation (4.88) is known as the Lindhard formula for the relative permittivity of semiconductors, where the excitonic screening is neglected [11, 12]. Under such an approximation, the screening effect of the electron–hole plasma equals the sum of the effects resulting from the separate electron and hole plasmas.

The Lindhard formula can be further simplified under more specific conditions. For example, under the long wavelength ($\vec{k} \to 0$) limit, the Lindhard formula (4.88) is reduced to the classical Drude formula

$$\varepsilon_{r,\vec{k}=0}(\omega) = 1 - \frac{\omega_{\mathrm{pl}}^2}{\omega^2}, \tag{4.89}$$

where

$$\omega_{\mathrm{pl}}^2 \equiv N\mathrm{e}^2/(\varepsilon_0 m_{\mathrm{r}}) \tag{4.90}$$

with $N = N_{\mathrm{e}} = N_{\mathrm{h}}$ given in equation (4.14) and m_{r} the reduced electron–hole effective mass. Expression (4.89) is the same as the material permittivity dispersion formula derived from a phenomenological dipole oscillator model with the damping ignored.

A better approximation comes from replacing the continuum of electron and hole pair excitations, represented by the continuum of poles in the Lindhard formula (4.88), by a single effective plasmon pole [13]

$$\begin{aligned} \frac{1}{\varepsilon_{r,\vec{k}}(\omega)} &= 1/[1 - V_{|\vec{k}|} \sum_{\vec{l}} \sum_{\alpha=\mathrm{e,h}} \frac{f^{\alpha}_{\vec{l}-\vec{k}} - f^{\alpha}_{\vec{l}}}{\hbar\omega + \mathrm{j}\delta + \varepsilon^{\alpha}_{\vec{l}-\vec{k}} - \varepsilon^{\alpha}_{\vec{l}}}] \\ &\approx 1 + V_{|\vec{k}|} \sum_{\vec{l}} \sum_{\alpha=\mathrm{e,h}} \frac{f^{\alpha}_{\vec{l}-\vec{k}} - f^{\alpha}_{\vec{l}}}{\hbar\omega + \mathrm{j}\delta + \varepsilon^{\alpha}_{\vec{l}-\vec{k}} - \varepsilon^{\alpha}_{\vec{l}}} \\ &\to 1 + \frac{\omega_{\mathrm{pl}}^2}{(\omega + \mathrm{j}\delta/\hbar)^2 - \omega_k^2}, \end{aligned} \tag{4.91}$$

where

$$\omega_k^2 \equiv \omega_{\mathrm{pl}}^2 (1 + \frac{k^2}{\kappa^2}) + \mathrm{C}(\frac{\hbar k^2}{4m_{\mathrm{r}}})^2, \tag{4.92}$$

with

$$\kappa \equiv \sqrt{\left(\frac{\mathrm{e}^2}{\varepsilon_0} \sum_{\alpha=\mathrm{c,v}} \frac{\partial N}{\partial F^{\alpha}} \right)}, \tag{4.93}$$

as the inverse static screening length, and C as a numerical constant between 1 and 4. Under the static plasmon pole approximation, equation (4.91) is further simplified to

$$\frac{1}{\varepsilon_{r,\vec{k}}(\omega)} = 1 - \frac{\omega_{\mathrm{pl}}^2}{\omega_{\vec{k}}^2} = \frac{(\omega_{\mathrm{pl}}^2/\kappa^2) + \mathrm{C}(\hbar/4m_{\mathrm{r}})^2k^2}{\omega_{\mathrm{pl}}^2 + (\omega_{\mathrm{pl}}^2/\kappa^2)k^2 + \mathrm{C}(\hbar/4m_{\mathrm{r}})^2k^4}k^2. \tag{4.94}$$

Hence we find the screened Coulomb potential under the static plasmon pole approximation

$$\widetilde{V}_{|\vec{k}|} = V_{|\vec{k}|}\frac{(\omega_{\mathrm{pl}}^2/\kappa^2) + \mathrm{C}(\hbar/4m_{\mathrm{r}})^2k^2}{\omega_{\mathrm{pl}}^2 + (\omega_{\mathrm{pl}}^2/\kappa^2)k^2 + \mathrm{C}(\hbar/4m_{\mathrm{r}})^2k^4}k^2, \tag{4.95}$$

with ω_{pl}^2 and κ given by equations (4.90) and (4.93), respectively.

4.3.3 Solution under zero injection and the exciton absorption

Under zero injection, equation (4.66a) becomes

$$\frac{\mathrm{d}p_{\vec{k}}}{\mathrm{d}t} = -(\mathrm{j}\bar{\omega}_{\vec{k}} + \frac{\mathrm{j}}{\hbar}\sum_{\vec{l}\neq\vec{k}}\widetilde{V}_{|\vec{l}-\vec{k}|} + \gamma)p_{\vec{k}} + \frac{\mathrm{j}}{\hbar}\vec{\mu}_{\vec{k}}\cdot\vec{E} + \frac{\mathrm{j}}{\hbar}\sum_{\vec{k}'\neq\vec{k}}\widetilde{V}_{|\vec{k}'-\vec{k}|}p_{\vec{k}'}, \tag{4.96}$$

with $\bar{\omega}_{\vec{k}}$ given by equation (4.31).

Under the plane-wave optical field assumption, i.e.,

$$\vec{E}(t) = \frac{1}{2}\vec{E}_0\mathrm{e}^{\mathrm{j}(\vec{k}_0\cdot\vec{r}-\omega t)} + \mathrm{c.c.}, \tag{4.97}$$

with $\vec{k}_0$ indicating the optical wave vector, by taking the Fourier transform of equation (4.96) to convert it from the time–wave vector domain to the frequency–wave vector domain, we find

$$-\mathrm{j}\omega\widetilde{p}_{\vec{k}} = -(\mathrm{j}\bar{\omega}_{\vec{k}} + \frac{\mathrm{j}}{\hbar}\sum_{\vec{l}\neq\vec{k}}\widetilde{V}_{|\vec{l}-\vec{k}|} + \gamma)\widetilde{p}_{\vec{k}} + \frac{\mathrm{j}}{\hbar}(\vec{\mu}_{\vec{k}}\cdot\vec{E}_0)\mathrm{e}^{\mathrm{j}\vec{k}_0\cdot\vec{r}} + \frac{\mathrm{j}}{\hbar}\sum_{\vec{k}'\neq\vec{k}}\widetilde{V}_{|\vec{k}'-\vec{k}|}\widetilde{p}_{\vec{k}'}, \tag{4.98}$$

where

$$p_{\vec{k}} \equiv (1/2)\widetilde{p}_{\vec{k}}\mathrm{e}^{-\mathrm{j}\omega t}, \tag{4.99}$$

is assumed. In (4.98), the space coordinate $\vec{r}$ indicates the distance between the conduction band electron and valence band hole, whereas $\vec{r}$ indicates the space coordinate of the electron–hole pair mass center. Under such an arrangement, the polariton has no dependence on $\vec{r}$. This is the reason a specific slow-varying envelope in the optical field has been assumed to be in the form of equation (4.97). Equation (4.98) can also be rewritten as

$$[\hbar\bar{\omega}_{\vec{k}} + \sum_{\vec{l}\neq\vec{k}}\widetilde{V}_{|\vec{l}-\vec{k}|} - \hbar(\omega + \mathrm{j}\gamma)]\widetilde{p}_{\vec{k}} = (\vec{\mu}_{\vec{k}}\cdot\vec{E}_0)\mathrm{e}^{\mathrm{j}\vec{k}_0\cdot\vec{r}} + \sum_{\vec{k}'\neq\vec{k}}\widetilde{V}_{|\vec{k}'-\vec{k}|}\widetilde{p}_{\vec{k}'}. \tag{4.100}$$

Therefore, by taking the inverse space Fourier transform of equation (4.100) to convert it from the frequency–wave vector domain to the frequency–space domain, we obtain

$$\left[-\frac{\hbar^2\nabla^2}{2m_\mathrm{r}} - \frac{\mathrm{e}^2}{4\pi\varepsilon r} + E_\mathrm{g} - \hbar(\omega + \mathrm{j}\gamma)\right]\widetilde{p}(\vec{r}) = (\vec{\mu}\cdot\vec{E}_0)\mathrm{e}^{\mathrm{j}\vec{k}_0\cdot\vec{r}}\delta(\vec{r})\Omega, \tag{4.101}$$

for bulk semiconductors, where equations (4.87), (4.9b) and (4.32) have been used with $\varepsilon \equiv \varepsilon_0\varepsilon_r(\omega)$. $\varepsilon_r(\omega)$ is the space–frequency domain relative permittivity obtained by taking the inverse space Fourier transform of $\varepsilon_{r,\vec{k}}(\omega)$ given by the Lindhard formula (4.88). The wave vector $\vec{k}$ dependence on the dipole matrix element $\vec{\mu}_{\vec{k}}$ can be ignored provided that only a small $\vec{k}$ value in the neighborhood of the fundamental absorption edge is considered.

To solve the inhomogeneous equation (4.101), the related homogeneous equation in the form

$$\left(-\frac{\hbar^2\nabla^2}{2m_\mathrm{r}} - \frac{\mathrm{e}^2}{4\pi\varepsilon r}\right)\varphi_n(\vec{r}) = \varepsilon_n\varphi_n(\vec{r}), \tag{4.102}$$

must be solved first. Actually, the eigenvalue problem equation (4.102), also known as the Wannier equation [2, 3], takes the form of the Schrödinger equation for the relative motion of an electron and a hole interacting with the attractive Coulomb potential, which is analogous to the electron orbiting problem in a hydrogen atom. The solutions of equation (4.102) are the Wannier excitons [6].

The solution to the inhomogeneous equation (4.101) can therefore be constructed as a linear superposition of the eigenfunctions obtained from equation (4.102)

$$\widetilde{p}(\vec{r}) = \sum_m p_m\varphi_m(\vec{r}). \tag{4.103}$$

Substituting equation (4.103) into (4.101), multiplying by $\varphi_n^*(\vec{r})$ and integrating over the entire space Ω we obtain [2]

$$p_n = \frac{(\vec{\mu}\cdot\vec{E}_0)\Omega\varphi_n^*(0)}{\varepsilon_n + E_\mathrm{g} - \hbar(\omega + \mathrm{j}\gamma)}\mathrm{e}^{\mathrm{j}\vec{k}_0\cdot\vec{r}}, \tag{4.104}$$

where the orthonormal condition for the eigenfunction of equation (4.102)

$$\int_\Omega \mathrm{d}\vec{r}\,[\varphi_m^*(\vec{r})\varphi_n(\vec{r})] = \delta_{mn}, \tag{4.105}$$

has been used.

Replacing the coefficient in equation (4.103) with (4.104), taking the Fourier transform on the obtained expression to find $\widetilde{p}_{\vec{k}}$, and substituting the result into equation (4.99) yields

$$p_{\vec{k}} = \frac{1}{2}\mathrm{e}^{-\mathrm{j}\omega t}F[\widetilde{p}(\vec{r})] = \frac{1}{2}\mathrm{e}^{\mathrm{j}(\vec{k}_0\cdot\vec{r}-\omega t)}\sum_m \frac{(\vec{\mu}\cdot\vec{E}_0)\varphi_m^*(0)}{\varepsilon_m + E_\mathrm{g} - \hbar(\omega + \mathrm{j}\gamma)}\int_\Omega \mathrm{d}\vec{r}\,[\varphi_m(\vec{r})\mathrm{e}^{-\mathrm{j}\vec{k}\cdot\vec{r}}]. \tag{4.106}$$

From equation (4.47), we obtain

$$\widetilde{P} = E_{0\mu} e^{j\vec{k}_0 \cdot \vec{r}} |\mu|^2 \sum_m \frac{\varphi_m^*(0)}{\varepsilon_m + E_g - \hbar(\omega + j\gamma)} \frac{1}{\Omega} \sum_{\vec{k}} \int_\Omega d\vec{r} [\varphi_m(\vec{r}) e^{-j\vec{k}\cdot\vec{r}}]$$

$$= 2\widetilde{E}_\mu |\mu|^2 \sum_m \frac{|\varphi_m^*(0)|^2}{\varepsilon_m + E_g - \hbar(\omega + j\gamma)}, \tag{4.107}$$

where $E_0 e^{j\vec{k}_0 \cdot \vec{r}} = \widetilde{E}$ has been used by comparing equation (4.97) with (4.45). Again, $\widetilde{E}_\mu$ is the projection of vector $\vec{\widetilde{E}}$ along the direction of $\vec{\mu}$. $\widetilde{P}$ and $\widetilde{E}_\mu$ are in the same direction as $\vec{\widetilde{P}}$ is along $\vec{\mu}$.

Therefore, we find the susceptibility given by

$$\widetilde{\chi}(\omega) = \frac{\widetilde{P}}{\varepsilon_0 \widetilde{E}_\mu} = \frac{2|\mu|^2}{\varepsilon_0} \sum_m \frac{|\varphi_m(0)|^2}{\varepsilon_m + E_g - \hbar(\omega + j\gamma)}$$

$$= \frac{2|\mu|^2}{\varepsilon_0 \hbar\gamma} \sum_m |\varphi_m(0)|^2 \left[\left(\frac{\varepsilon_m + E_g}{\hbar} - \omega\right)/\gamma + j\right] L\left(\frac{\varepsilon_m + E_g}{\hbar} - \omega\right), \tag{4.108}$$

with the Lorentzian line-shape function defined by equation (4.62).

The corresponding material absorption and refractive index change are given as

$$\alpha(\omega) = -\frac{k_0}{n} \mathrm{Im}[\widetilde{\chi}(\omega)] = -\frac{2\omega|\mu|^2}{\varepsilon_0 n c \hbar\gamma} \sum_m |\varphi_m(0)|^2 L\left(\frac{\varepsilon_m + E_g}{\hbar} - \omega\right)$$

$$= -\frac{2\omega|\mu|^2}{\varepsilon_0 n c \hbar} \sum_m |\varphi_m(0)|^2 \frac{\gamma}{\gamma^2 + [(\varepsilon_m + E_g)/\hbar - \omega]^2}, \tag{4.109a}$$

$$\Delta n(\omega) = \frac{1}{2n} \mathrm{Re}[\widetilde{\chi}(\omega)] = \frac{|\mu|^2}{\varepsilon_0 n \hbar\gamma} \sum_m |\varphi_m(0)|^2 \frac{(\varepsilon_m + E_g)/\hbar - \omega}{\gamma} L(\frac{\varepsilon_m + E_g}{\hbar} - \omega)$$

$$= \frac{|\mu|^2}{\varepsilon_0 n \hbar} \sum_m |\varphi_m(0)|^2 \frac{(\varepsilon_m + E_g)/\hbar - \omega}{\gamma^2 + [(\varepsilon_m + E_g)/\hbar - \omega]^2}. \tag{4.109b}$$

For bulk semiconductors, the eigenvalue problem equation (4.102) can be solved for both bound (discrete) and continuum states [14–17]. The energies and wave functions at $\vec{r} = 0$ are given as

$$\varepsilon_m = -\frac{\varepsilon_R}{m^2}, \quad |\varphi_m(0)|^2 = \frac{1}{\pi a_0^3 m^3}, \tag{4.110a}$$

with $m = 1, 2, 3, \ldots$ for the bound states with negative eigenvalues ($\varepsilon_m < 0$) and

$$E = \frac{\hbar^2 k^2}{2m_r}, \quad |\varphi_E(0)|^2 = \frac{1}{4\pi\varepsilon_R a_0^3} \frac{e^{\pi/\sqrt{(E/\varepsilon_R)}}}{\sinh(\pi/\sqrt{(E/\varepsilon_R)})}, \tag{4.110b}$$

for the continuum states with positive eigenvalues ($E > 0$), where the Bohr radius and the Rydberg energy are defined as

$$a_0 \equiv \frac{4\pi\varepsilon\hbar^2}{m_r e^2}, \tag{4.111}$$

$$\varepsilon_R \equiv \frac{\hbar^2(1/a_0)^2}{2m_r} = \frac{m_r e^4}{2\hbar^2(4\pi\varepsilon)^2}, \tag{4.112}$$

respectively.

Substituting equations (4.110a&b) into (4.109a), we find

$$\alpha(\omega) = -\alpha_0\{4\sum_m \frac{\bar{\gamma}/m^3}{\bar{\gamma}^2 + [\Delta + 1/m^2]^2} + \int_0^\infty dx[\frac{e^{\pi/\sqrt{x}}}{\sinh(\pi/\sqrt{x})}\frac{\bar{\gamma}}{\bar{\gamma}^2 + [x-\Delta]^2}]\}, \tag{4.113}$$

where the relative transition energy and half-linewidth energy in the Lorentzian line-shape function (both normalized by the Rydberg energy), and the absorption coefficient are defined as

$$\Delta \equiv (\hbar\omega - E_g)/\varepsilon_R, \tag{4.114a}$$

$$\bar{\gamma} \equiv \hbar\gamma/\varepsilon_R, \tag{4.114b}$$

$$\alpha_0 \equiv \frac{\omega|\mu|^2}{2\pi\varepsilon_0 nca_0^3\varepsilon_R}. \tag{4.114c}$$

Equation (4.113) is in the form of the Elliot formula [14], which reveals that the absorption spectrum comprises a set of discrete exciton resonance absorption peaks and a continuous absorption floor due to the ionized states. The exciton absorption peaks are a consequence of the electron–hole Coulomb attraction, which has some unique features. Firstly, the first absorption peak ($m = 1$) appears at $\Delta_1 = -1$, which corresponds to $\hbar\omega_1 = E_g - \varepsilon_R$, or a transition energy below the bandgap energy. The rest of the peaks ($m > 1$) have their transition energies rapidly converging to, but never greater than the bandgap energy in accordance with $\hbar\omega_m = E_g - \varepsilon_R/m^2$. Secondly, as the peak transition energies approach the bandgap, each peak absorption value drops very rapidly in the order of $1/m^3$. For example, the absorption strength of the second peak is only $1/8$ of that of the first peak. As for continuous absorption, equation (4.113) shows a constant value starting from the bandgap ($\Delta = 0$) and near the band edge ($\Delta \to 0$), which indicates that the continuous absorption spectrum is step-function like in the neighborhood of the band edge. This conclusion is justified if we let $\bar{\gamma} \to 0$, hence $\bar{\gamma}/[\bar{\gamma}^2 + [x-\Delta]^2] \to \pi\delta(x-\Delta)$, and the integral in equation (4.113) approaches a constant

$$\int_0^\infty dx[\frac{e^{\pi/\sqrt{x}}}{\sinh(\pi/\sqrt{x})}\frac{\bar{\gamma}}{\bar{\gamma}^2 + [x-\Delta]^2}] \to \int_0^\infty dx\left[\frac{\pi\delta(x-\Delta)e^{\pi/\sqrt{x}}}{\sinh(\pi/\sqrt{x})}\right]$$
$$= \frac{\pi e^{\pi/\sqrt{\Delta}}}{\sinh(\pi/\sqrt{\Delta})} = \frac{2\pi}{1 - e^{-2\pi/\sqrt{\Delta}}} \to 2\pi.$$

In the free-carrier model with the Coulomb interaction ignored, however, an exciton resonance peak does not appear and the continuous absorption spectrum shows a square-root function like shape near the band edge, in accordance with equation (4.65a). From these analyses, we can conclude that it is crucial to include the many-body Coulomb interaction effects under a zero or weak injection where the carrier density is low. Under a strong injection where the carrier density is high, however, the free-carrier model may offer reasonably accurate results. This is because the Coulomb interaction is effectively screened under a high carrier density. This implies that, in real world applications, we should include the many-body Coulomb interaction effects in our material models for the simulation of photodetectors (PDs), electro-absorption modulators (EAMs), and photo-luminescent (PL) assessment of epitaxial wafers where the carrier density remains at a low level. The free-carrier model, on the other hand, can be sufficient for modeling laser diodes, semiconductor optical amplifiers, superluminescent light emitting diodes, and light emitting diodes where the carrier density is at a considerable level because of the injection.

For QW structures, the eigenvalue problem equation (4.102) can be solved for both bound (discrete) and continuum states [18–20]. By following a similar approach, we will be able to find the material absorption and refractive index change including the 2D exciton effect from equations (4.109a&b).

4.3.4 Solution under arbitrary injection

Under an arbitrary injection, the polariton equation (4.66a) can be rewritten as

$$\begin{aligned}\frac{\mathrm{d}p_{\vec{k}}}{\mathrm{d}t} &= -[\mathrm{j}\bar{\omega}_{\vec{k}} - \frac{\mathrm{j}}{\hbar}\sum_{\vec{l}\neq\vec{k}} \widetilde{V}_{|\vec{l}-\vec{k}|}(f^{\mathrm{e}}_{\vec{l}} + f^{\mathrm{h}}_{\vec{l}} - 1) + \gamma]p_{\vec{k}} \\ &\quad - \frac{\mathrm{j}}{\hbar}(\vec{\mu}_{\vec{k}}\cdot\vec{E} + \sum_{\vec{k}'\neq\vec{k}} \widetilde{V}_{|\vec{k}'-\vec{k}|}p_{\vec{k}'})(f^{\mathrm{e}}_{\vec{k}} + f^{\mathrm{h}}_{\vec{k}} - 1). \end{aligned} \tag{4.115}$$

Comparing with the equivalent equation in the free-carrier model (4.40a), we find:

(1) The bandgap has been changed.
The renormalized bandgap becomes

$$\begin{aligned}\bar{\omega}'_{\vec{k}} &\equiv \bar{\omega}_{\vec{k}} - \frac{1}{\hbar}\sum_{\vec{l}\neq\vec{k}} \widetilde{V}_{|\vec{l}-\vec{k}|}(f^{\mathrm{e}}_{\vec{l}} + f^{\mathrm{h}}_{\vec{l}} - 1) \\ &= \frac{1}{\hbar}\left[\frac{\hbar^2k^2}{2m_{\mathrm{r}}} + E_{\mathrm{g}} - \sum_{\vec{l}\neq\vec{k}} \widetilde{V}_{|\vec{l}-\vec{k}|}(f^{\mathrm{e}}_{\vec{l}} + f^{\mathrm{h}}_{\vec{l}} - 1)\right] \\ &= \frac{1}{\hbar}\left[\frac{\hbar^2k^2}{2m_{\mathrm{r}}} + \varepsilon_{\mathrm{g}}\right], \end{aligned} \tag{4.116}$$

where

$$\begin{aligned}\varepsilon_{\mathrm{g}} &\equiv E_{\mathrm{g}} - \sum_{\vec{l}\neq\vec{k}} \widetilde{V}_{|\vec{l}-\vec{k}|}(f_{\vec{l}}^{\mathrm{e}} + f_{\vec{l}}^{\mathrm{h}} - 1) \\ &= E_{\mathrm{g}} + \sum_{\vec{l}\neq\vec{k}} \widetilde{V}_{|\vec{l}-\vec{k}|} - \sum_{\vec{l}\neq\vec{k}} \widetilde{V}_{|\vec{l}-\vec{k}|}(f_{\vec{l}}^{\mathrm{e}} + f_{\vec{l}}^{\mathrm{h}}) \\ &= \varepsilon_{\mathrm{g}}^{\mathrm{V}} + \Delta\varepsilon^{\mathrm{CH}} + \Delta\varepsilon_{\vec{k}}^{\mathrm{SX}},\end{aligned} \tag{4.117}$$

with the Coulomb potential included bandgap energy, the Coulomb hole self-energy (Debye shift), and the screened exchange shift defined as [3, 6]

$$\varepsilon_{\mathrm{g}}^{\mathrm{V}} \equiv E_{\mathrm{g}} + \sum_{\vec{l}\neq\vec{k}} V_{|\vec{l}-\vec{k}|} = E_{\mathrm{g}} + \sum_{\vec{q}\neq 0} V_{|\vec{q}|}, \tag{4.118a}$$

$$\Delta\varepsilon^{\mathrm{CH}} \equiv \sum_{\vec{l}\neq\vec{k}} (\widetilde{V}_{|\vec{l}-\vec{k}|} - V_{|\vec{l}-\vec{k}|}) = \sum_{\vec{q}\neq 0} (\widetilde{V}_{|\vec{q}|} - V_{|\vec{q}|}), \tag{4.118b}$$

$$\Delta\varepsilon_{\vec{k}}^{\mathrm{SX}} \equiv -\sum_{\vec{l}\neq\vec{k}} \widetilde{V}_{|\vec{l}-\vec{k}|}(f_{\vec{l}}^{\mathrm{e}} + f_{\vec{l}}^{\mathrm{h}}), \tag{4.118c}$$

respectively.

If the bare Coulomb potential has been included in the hole kinetic energy (e.g., through the hole effective mass), it should not be double counted and hence we have $\varepsilon_{\mathrm{g}}^{\mathrm{V}} \equiv E_{\mathrm{g}}$ instead of equation (4.118a).

Under the static plasmon pole approximation equation (4.94), the Debye shift can be found as [3]

$$\Delta\varepsilon^{\mathrm{CH}} = -2\varepsilon_{\mathrm{R}} a_0 \kappa / \sqrt{\left(1 + \frac{\sqrt{(\mathrm{C})} a_0^2 \kappa^2 \varepsilon_{\mathrm{R}}}{\hbar\omega_{\mathrm{pl}}}\right)}, \tag{4.119}$$

with κ given by equation (4.93) and C a numerical constant between 1 and 4, as specified in Section 4.3.2.

The screened exchange shift is carrier (electron and hole) dependent, hence it is also $\vec{k}$ dependent.

(2) The inhomogeneous driving source has extra contributions from other polariton components in the wave vector domain, which brings in the coupling between different polaritons.

We still formally integrate equation (4.115) to obtain [3]

$$p_{\vec{k}} = -\frac{\mathrm{j}}{\hbar} \int_{-\infty}^{t} \mathrm{d}\tau \left\{ \left[\vec{\mu}_{\vec{k}} \cdot \vec{E}(\tau) + \sum_{\vec{k}'\neq\vec{k}} \widetilde{V}_{|\vec{k}'-\vec{k}|} p_{\vec{k}'} \right] (f_{\vec{k}}^{\mathrm{e}} + f_{\vec{k}}^{\mathrm{h}} - 1) \mathrm{e}^{(\mathrm{j}\bar{\omega}_{\vec{k}}' + \gamma)(\tau - t)} \right\}, \tag{4.120}$$

where (4.117) is used.

Under the rate equation approximation, those time domain slow-varying factors can be taken out of the integral in equation (4.120). Also, by employing the rotating-wave approximation, equation (4.120) becomes

$$p_{\vec{k}} = -\frac{\mathrm{j}}{\hbar}(\bar{f}^{\mathrm{e}}_{\vec{k}} + \bar{f}^{\mathrm{h}}_{\vec{k}} - 1)\left\{\frac{\vec{\mu}_{\vec{k}} \cdot \vec{\widetilde{E}}}{2} \frac{\mathrm{e}^{-\mathrm{j}\omega t}}{\mathrm{j}(\bar{\omega}_{\vec{k}'} - \omega) + \gamma} + \sum_{\vec{k}' \neq \vec{k}} \widetilde{V}_{|\vec{k}'-\vec{k}|} \int_{-\infty}^{t} \mathrm{d}\tau \left[p_{\vec{k}'}(\tau)\mathrm{e}^{(\mathrm{j}\bar{\omega}_{\vec{k}'}+\gamma)(\tau-t)}\right]\right\}. \quad (4.121)$$

We can now solve equation (4.121) by following a perturbation approach. Actually, as the zeroth order solution with the screened Coulomb potential ignored, we have

$$p^{(0)}_{\vec{k}} = -\frac{\mathrm{j}}{\hbar}(\bar{f}^{\mathrm{e}}_{\vec{k}} + \bar{f}^{\mathrm{h}}_{\vec{k}} - 1)\left[\frac{\vec{\mu}_{\vec{k}} \cdot \vec{\widetilde{E}}}{2} \frac{\mathrm{e}^{-\mathrm{j}\omega t}}{\mathrm{j}(\bar{\omega}_{\vec{k}'} - \omega) + \gamma}\right] = \frac{\widetilde{E}_{\mu}\mathrm{e}^{-\mathrm{j}\omega t}}{2}\chi^{(0)}_{\vec{k}}, \quad (4.122)$$

with the microscopic zeroth order susceptibility defined as

$$\chi^{(0)}_{\vec{k}} \equiv -\frac{\mathrm{j}\mu_{\vec{k}}}{\hbar} \frac{\bar{f}^{\mathrm{e}}_{\vec{k}} + \bar{f}^{\mathrm{h}}_{\vec{k}} - 1}{\mathrm{j}(\bar{\omega}_{\vec{k}'} - \omega) + \gamma}. \quad (4.123)$$

Substituting equation (4.122) into the RHS of (4.121), we find the first order contribution

$$\begin{aligned}
p^{(1)}_{\vec{k}} &= -\frac{\mathrm{j}}{\hbar}(\bar{f}^{\mathrm{e}}_{\vec{k}} + \bar{f}^{\mathrm{h}}_{\vec{k}} - 1) \sum_{\vec{k}' \neq \vec{k}} \widetilde{V}_{|\vec{k}'-\vec{k}|} \int_{-\infty}^{t} \mathrm{d}\tau [p^{(0)}_{\vec{k}'}(\tau)\mathrm{e}^{(\mathrm{j}\bar{\omega}_{\vec{k}'}+\gamma)(\tau-t)}] \\
&= -\frac{\mathrm{j}\widetilde{E}_{\mu}}{2\hbar}\left(\sum_{\vec{k}' \neq \vec{k}} \widetilde{V}_{|\vec{k}'-\vec{k}|}\chi^{(0)}_{\vec{k}'}\right)(\bar{f}^{\mathrm{e}}_{\vec{k}} + \bar{f}^{\mathrm{h}}_{\vec{k}} - 1)\mathrm{e}^{-(\mathrm{j}\bar{\omega}_{\vec{k}'}+\gamma)t} \int_{-\infty}^{t} \mathrm{d}\tau \mathrm{e}^{(-\mathrm{j}\omega+\mathrm{j}\bar{\omega}_{\vec{k}'}+\gamma)\tau} \\
&= -\frac{\mathrm{j}\widetilde{E}_{\mu}}{2\hbar}\left(\sum_{\vec{k}' \neq \vec{k}} \widetilde{V}_{|\vec{k}'-\vec{k}|}\chi^{(0)}_{\vec{k}'}\right)(\bar{f}^{\mathrm{e}}_{\vec{k}} + \bar{f}^{\mathrm{h}}_{\vec{k}} - 1)\frac{\mathrm{e}^{-\mathrm{j}\omega t}}{\mathrm{j}(\bar{\omega}_{\vec{k}'} - \omega) + \gamma} \\
&= \frac{1}{2}\widetilde{E}_{\mu}\mathrm{e}^{-\mathrm{j}\omega t}\chi^{(0)}_{\vec{k}}\left(\frac{1}{\mu_{\vec{k}}} \sum_{\vec{k}' \neq \vec{k}} \widetilde{V}_{|\vec{k}'-\vec{k}|}\chi^{(0)}_{\vec{k}'}\right) = \frac{1}{2}\widetilde{E}_{\mu}\mathrm{e}^{-\mathrm{j}\omega t}\chi^{(0)}_{\vec{k}} q_{\vec{k}},
\end{aligned} \quad (4.124)$$

where

$$q_{\vec{k}} \equiv \frac{1}{\mu_{\vec{k}}} \sum_{\vec{k}' \neq \vec{k}} \widetilde{V}_{|\vec{k}'-\vec{k}|}\chi^{(0)}_{\vec{k}'}. \quad (4.125)$$

Such a process may continue by looking for higher orders so as to obtain a more accurate solution. However, based on the Pade approximation [21], the following summing technique may greatly reduce the computational effort [22]

$$p_{\vec{k}} = p^{(0)}_{\vec{k}} + p^{(1)}_{\vec{k}} + \cdots = \frac{\widetilde{E}_{\mu}\mathrm{e}^{-\mathrm{j}\omega t}}{2}\chi^{(0)}_{\vec{k}}(1 + q_{\vec{k}} + \cdots) \approx \frac{\widetilde{E}_{\mu}\mathrm{e}^{-\mathrm{j}\omega t}}{2}\frac{\chi^{(0)}_{\vec{k}}}{1 - q_{\vec{k}}}. \quad (4.126)$$

Plugging equation (4.126) into (4.47) yields

$$\widetilde{P} = \frac{1}{\Omega}\widetilde{E}_{\mu}\sum_{\vec{k}}\mu_{\vec{k}}^{*}\frac{\chi_{\vec{k}}^{(0)}}{1-q_{\vec{k}}} = -\frac{\mathrm{j}}{\hbar\Omega}\widetilde{E}_{\mu}\sum_{\vec{k}}|\mu_{\vec{k}}|^{2}\frac{\bar{f}_{\vec{k}}^{\mathrm{e}}+\bar{f}_{\vec{k}}^{\mathrm{h}}-1}{\mathrm{j}(\bar{\omega}_{\vec{k}'}-\omega)+\gamma}\frac{1}{1-q_{\vec{k}}}. \quad (4.127)$$

Therefore, we find the susceptibility given by

$$\begin{aligned}\widetilde{\chi}(\omega) &= \frac{\widetilde{P}}{\varepsilon_{0}\widetilde{E}_{\mu}} = -\frac{\mathrm{j}}{\varepsilon_{0}\hbar\Omega}\sum_{\vec{k}}|\mu_{\vec{k}}|^{2}\frac{\bar{f}_{\vec{k}}^{\mathrm{e}}+\bar{f}_{\vec{k}}^{\mathrm{h}}-1}{\mathrm{j}(\bar{\omega}_{\vec{k}'}-\omega)+\gamma}\frac{1}{1-q_{\vec{k}}} \\ &= -\frac{1}{\varepsilon_{0}\hbar\gamma\Omega}\sum_{\vec{k}}|\mu_{\vec{k}}|^{2}(\bar{f}_{\vec{k}}^{\mathrm{e}}+\bar{f}_{\vec{k}}^{\mathrm{h}}-1)[(\bar{\omega}_{\vec{k}'}-\omega)/\gamma+\mathrm{j}]\frac{L(\bar{\omega}_{\vec{k}'}-\omega)}{1-q_{\vec{k}}}. \quad (4.128)\end{aligned}$$

Consequently, we obtain the semiconductor material gain and refractive index change from

$$\begin{aligned}g(\omega) &= -\frac{k_{0}}{n}\mathrm{Im}[\widetilde{\chi}(\omega)] = \frac{\omega}{\varepsilon_{0}nc\hbar\gamma\Omega}\sum_{\vec{k}}|\mu_{\vec{k}}|^{2}(\bar{f}_{\vec{k}}^{\mathrm{e}}+\bar{f}_{\vec{k}}^{\mathrm{h}}-1)\frac{L(\bar{\omega}_{\vec{k}'}-\omega)}{1-q_{\vec{k}}} \\ &= \frac{\omega}{\varepsilon_{0}nc\hbar\Omega}\sum_{\vec{k}}|\mu_{\vec{k}}|^{2}(\bar{f}_{\vec{k}}^{\mathrm{e}}+\bar{f}_{\vec{k}}^{\mathrm{h}}-1)\frac{\gamma}{\gamma^{2}+(\bar{\omega}_{\vec{k}'}-\omega)^{2}}\frac{1}{1-q_{\vec{k}}}, \quad (4.129\mathrm{a})\end{aligned}$$

$$\begin{aligned}\Delta n(\omega) &= \frac{1}{2n}\mathrm{Re}[\widetilde{\chi}(\omega)] = -\frac{1}{2\varepsilon_{0}n\hbar\gamma\Omega}\sum_{\vec{k}}|\mu_{\vec{k}}|^{2}(\bar{f}_{\vec{k}}^{\mathrm{e}}+\bar{f}_{\vec{k}}^{\mathrm{h}}-1)\frac{\bar{\omega}_{\vec{k}'}-\omega}{\gamma}\frac{L(\bar{\omega}_{\vec{k}'}-\omega)}{1-q_{\vec{k}}} \\ &= -\frac{1}{2\varepsilon_{0}n\hbar\Omega}\sum_{\vec{k}}|\mu_{\vec{k}}|^{2}(\bar{f}_{\vec{k}}^{\mathrm{e}}+\bar{f}_{\vec{k}}^{\mathrm{h}}-1)\frac{\bar{\omega}_{\vec{k}'}-\omega}{\gamma^{2}+(\bar{\omega}_{\vec{k}'}-\omega)^{2}}\frac{1}{1-q_{\vec{k}}}. \quad (4.129\mathrm{b})\end{aligned}$$

4.4 The many-body correlation model as a second order solution

4.4.1 The many-body correlation model

In this model, we deal directly with the general governing equations (4.28a–c) derived from the Heisenberg equation with the second order terms in operator–Hamiltonian commutator expansions retained. This is equivalent to removing the first approximation in the screened Coulomb interaction model introduced in Section 4.3 and modifying the second approximation there to gain a higher level of accuracy by adding on the many-body Coulomb correlations.

We rewrite equations (4.28a–c) in the following form to show the governing equations explicitly in the many-body correlation model

$$\frac{\mathrm{d}p_{\vec{k}}}{\mathrm{d}t} = -\frac{1}{\hbar}\sum_{\vec{k}'}(\mathrm{j}\Theta_{\vec{k}\vec{k}'} + \Lambda_{\vec{k}\vec{k}'})p_{\vec{k}'} - \frac{1}{\hbar}\sum_{\vec{q}}\Lambda^{\mathrm{p}}_{\vec{k}\vec{q}}p_{\vec{k}+\vec{q}} - \frac{\mathrm{j}}{\hbar}(\vec{\mu}_{\vec{k}}\cdot\vec{E})(f^{\mathrm{e}}_{\vec{k}} + f^{\mathrm{h}}_{\vec{k}} - 1), \tag{4.130a}$$

$$\begin{aligned}\frac{\mathrm{d}f^{\mathrm{e}}_{\vec{k}}}{\mathrm{d}t} &= -\frac{1}{\hbar}(\Sigma^{\mathrm{e-out}}_{\vec{k}} + \Sigma^{\mathrm{e-in}}_{\vec{k}})f^{\mathrm{e}}_{\vec{k}} + 2\mathrm{Im}(\Omega^*_{\vec{k}}p_{\vec{k}}) \\ &\quad - \frac{1}{\hbar}\sum_{\vec{q}\neq 0,\pm}\Sigma^{\mathrm{e-p}}_{\vec{q},\pm}f^{\mathrm{p}}_{\vec{q}} + \frac{1}{\hbar}\Sigma^{\mathrm{e-in}}_{\vec{k}} - \frac{1}{\hbar}\sum_{\vec{q}\neq 0,\pm}\Sigma^{\mathrm{e}}_{\vec{q},\pm},\end{aligned} \tag{4.130b}$$

$$\begin{aligned}\frac{\mathrm{d}f^{\mathrm{h}}_{\vec{k}}}{\mathrm{d}t} &= -\frac{1}{\hbar}(\Sigma^{\mathrm{h-out}}_{\vec{k}} + \Sigma^{\mathrm{h-in}}_{\vec{k}})f^{\mathrm{h}}_{\vec{k}} + 2\mathrm{Im}[\Omega^*_{\vec{k}}p_{\vec{k}}] \\ &\quad - \frac{1}{\hbar}\sum_{\vec{q}\neq 0,\pm}\Sigma^{\mathrm{h-p}}_{\vec{q},\pm}f^{\mathrm{p}}_{\vec{q}} + \frac{1}{\hbar}\Sigma^{\mathrm{h-in}}_{\vec{k}} - \frac{1}{\hbar}\sum_{\vec{q}\neq 0,\pm}\Sigma^{\mathrm{h}}_{\vec{q},\pm}.\end{aligned} \tag{4.130c}$$

In equations (4.130a–c), the first order interaction energy matrix $\Theta_{\vec{k}\vec{k}'}$ and the Rabi frequency $\Omega_{\vec{k}}$ are given by equations (4.29) and (4.30) with the bare Coulomb potential, whereas the second order interaction energy matrices $\Sigma^{\mathrm{e/h-out}}_{\vec{k}}$, $\Sigma^{\mathrm{e/h-in}}_{\vec{k}}$, $\Lambda_{\vec{k}\vec{k}'}$, $\Sigma^{\mathrm{e/h-p}}_{\vec{q},\pm}$, $\Sigma^{\mathrm{e/h}}_{\vec{q},\pm}$, and $\Lambda^{\mathrm{p}}_{\vec{k}\vec{q}}$ that account for the carrier–carrier collision and carrier–phonon scattering are given by equations (4.23a–c) and (4.27a–c) with the screened Coulomb potential, for the reason we have explained in the beginning of Section 4.3.

4.4.2 A semi-analytical solution

Again, we rewrite equation (4.130a) in the form

$$\begin{aligned}\frac{\mathrm{d}p_{\vec{k}}}{\mathrm{d}t} &= -\frac{1}{\hbar}(\mathrm{j}\Theta_{\vec{k}\vec{k}} + \Lambda_{\vec{k}\vec{k}} + \Lambda^{\mathrm{p}}_{\vec{k}0})p_{\vec{k}} - \frac{\mathrm{j}}{\hbar}(\vec{\mu}_{\vec{k}}\cdot\vec{E})(f^{\mathrm{e}}_{\vec{k}} + f^{\mathrm{h}}_{\vec{k}} - 1) \\ &\quad - \frac{1}{\hbar}\sum_{\vec{k}'\neq\vec{k}}(\mathrm{j}\Theta_{\vec{k}\vec{k}'} + \Lambda_{\vec{k}\vec{k}'})p_{\vec{k}'} - \frac{1}{\hbar}\sum_{\vec{q}\neq 0}\Lambda^{\mathrm{p}}_{\vec{k}\vec{q}}p_{\vec{k}+\vec{q}} \\ &= -[\mathrm{j}\bar{\omega}_{\vec{k}'} + \gamma_{\vec{k}}]p_{\vec{k}} - \frac{\mathrm{j}}{\hbar}(\vec{\mu}_{\vec{k}}\cdot\vec{E} + \sum_{\vec{k}'\neq\vec{k}}V_{|\vec{k}'-\vec{k}|}p_{\vec{k}'})(f^{\mathrm{e}}_{\vec{k}} + f^{\mathrm{h}}_{\vec{k}} - 1) - \sum_{\vec{k}'\neq\vec{k}}\beta_{\vec{k}\vec{k}'}p_{\vec{k}'},\end{aligned} \tag{4.131}$$

where we have defined

$$\bar{\omega}_{\vec{k}'} \equiv \bar{\omega}_{\vec{k}} - \frac{1}{\hbar} \sum_{\vec{l} \neq \vec{k}} V_{|\vec{l}-\vec{k}|}(f^{\rm e}_{\vec{l}} + f^{\rm h}_{\vec{l}} - 1), \tag{4.132a}$$

$$\gamma_{\vec{k}} \equiv \frac{1}{\hbar}(\Lambda_{\vec{k}\vec{k}} + \Lambda^{\rm p}_{\vec{k}0}), \tag{4.132b}$$

$$\beta_{\vec{k}\vec{k}'} \equiv \frac{1}{\hbar}[\Lambda_{\vec{k}\vec{k}'} + \Lambda^{\rm p}_{\vec{k}(\vec{k}'-\vec{k})}]. \tag{4.132c}$$

By formally integrating equation (4.131), we obtain

$$p_{\vec{k}} = -\int_{-\infty}^{t} {\rm d}\tau \Bigg\{ \Bigg[\frac{\rm j}{\hbar}(\vec{\mu}_{\vec{k}} \cdot \vec{E} + \sum_{\vec{k}' \neq \vec{k}} V_{|\vec{k}'-\vec{k}|} p_{\vec{k}'})(f^{\rm e}_{\vec{k}} + f^{\rm h}_{\vec{k}} - 1) + \sum_{\vec{k}' \neq \vec{k}} \beta_{\vec{k}\vec{k}'} p_{\vec{k}'} \Bigg] {\rm e}^{({\rm j}\bar{\omega}_{\vec{k}'} + \gamma_{\vec{k}})(\tau - t)} \Bigg\}. \tag{4.133}$$

If the time scale of interest is significantly longer than a few picoseconds, we can assume that carriers have reached their steady states described by their quasi-Fermi distributions. Therefore, we can take the time domain slow-varying factors out of the integral in equation (4.133). Also under the rotating-wave approximation, equation (4.133) becomes

$$\begin{aligned} p_{\vec{k}} = -\frac{\rm j}{\hbar}(\bar{f}^{\rm e}_{\vec{k}} + \bar{f}^{\rm h}_{\vec{k}} - 1)\{ &\frac{\vec{\mu}_{\vec{k}} \cdot \tilde{\vec{E}}}{2} \frac{{\rm e}^{-{\rm j}\omega t}}{{\rm j}(\bar{\omega}_{\vec{k}'} - \omega) + \gamma_{\vec{k}}} \\ &+ \sum_{\vec{k}' \neq \vec{k}} V_{|\vec{k}'-\vec{k}|} \int_{-\infty}^{t} {\rm d}\tau [p_{\vec{k}'}(\tau) {\rm e}^{({\rm j}\bar{\omega}_{\vec{k}'} + \gamma_{\vec{k}})(\tau - t)}]\} \\ &- \sum_{\vec{k}' \neq \vec{k}} \beta_{\vec{k}\vec{k}'} \int_{-\infty}^{t} {\rm d}\tau [p_{\vec{k}'}(\tau) {\rm e}^{({\rm j}\bar{\omega}_{\vec{k}'} + \gamma_{\vec{k}})(\tau - t)}]. \end{aligned} \tag{4.134}$$

Still following a perturbation approach, we find the zeroth order solution as

$$p^{(0)}_{\vec{k}} = -\frac{j}{\hbar}(\bar{f}^{\rm e}_{\vec{k}} + \bar{f}^{\rm h}_{\vec{k}} - 1) \left[\frac{\vec{\mu}_{\vec{k}} \cdot \tilde{\vec{E}}}{2} \frac{{\rm e}^{-{\rm j}\omega t}}{{\rm j}(\bar{\omega}_{\vec{k}'} - \omega) + \gamma_{\vec{k}}} \right] = \frac{1}{2} \tilde{E}_{\mu} {\rm e}^{-{\rm j}\omega t} \chi^{(0)}_{\vec{k}}, \tag{4.135}$$

with the microscopic zeroth order susceptibility defined as

$$\chi^{(0)}_{\vec{k}} \equiv -\frac{{\rm j}\mu_{\vec{k}}}{\hbar} \frac{\bar{f}^{\rm e}_{\vec{k}} + \bar{f}^{\rm h}_{\vec{k}} - 1}{{\rm j}(\bar{\omega}_{\vec{k}'} - \omega) + \gamma_{\vec{k}}}. \tag{4.136}$$

Noting

$$\int_{-\infty}^{t} \mathrm{d}\tau [p_{\vec{k}'}^{(0)}(\tau) \mathrm{e}^{(\mathrm{j}\bar{\omega}_{\vec{k}'}+\gamma_{\vec{k}})(\tau-t)}]$$

$$= \frac{\widetilde{E}_{\mu}}{2} \chi_{\vec{k}'}^{(0)} \mathrm{e}^{-(\mathrm{j}\bar{\omega}_{\vec{k}'}+\gamma_{\vec{k}})t} \int_{-\infty}^{t} \mathrm{d}\tau \mathrm{e}^{(-\mathrm{j}\omega+\mathrm{j}\bar{\omega}_{\vec{k}'}+\gamma_{\vec{k}})\tau} = \frac{\chi_{\vec{k}'}^{(0)}}{2} \frac{\widetilde{E}_{\mu} \mathrm{e}^{-\mathrm{j}\omega t}}{\mathrm{j}(\bar{\omega}_{\vec{k}'} - \omega) + \gamma_{\vec{k}}},$$

we obtain the first order solution

$$p_{\vec{k}}^{(1)} = -\sum_{\vec{k}' \neq \vec{k}} \left[\frac{\mathrm{j}}{\hbar} (\bar{f}_{\vec{k}}^{\mathrm{e}} + \bar{f}_{\vec{k}}^{\mathrm{h}} - 1) V_{|\vec{k}'-\vec{k}|} + \beta_{\vec{k}\vec{k}'} \right] \int_{-\infty}^{t} \mathrm{d}\tau [p_{\vec{k}'}^{(0)}(\tau) \mathrm{e}^{(\mathrm{j}\bar{\omega}_{\vec{k}'}+\gamma_{\vec{k}})(\tau-t)}]$$

$$= -\frac{1}{2} \frac{\widetilde{E}_{\mu} \mathrm{e}^{-\mathrm{j}\omega t}}{\mathrm{j}(\bar{\omega}_{\vec{k}'} - \omega) + \gamma_{\vec{k}}} \sum_{\vec{k}' \neq \vec{k}} [\frac{\mathrm{j}}{\hbar} (\bar{f}_{\vec{k}}^{\mathrm{e}} + \bar{f}_{\vec{k}}^{\mathrm{h}} - 1) V_{|\vec{k}'-\vec{k}|} + \beta_{\vec{k}\vec{k}'}] \chi_{\vec{k}'}^{(0)}$$

$$= \frac{1}{2} \widetilde{E}_{\mu} \mathrm{e}^{-\mathrm{j}\omega t} \chi_{\vec{k}}^{(0)} [\frac{1}{\mu_{\vec{k}}} \sum_{\vec{k}' \neq \vec{k}} V_{|\vec{k}'-\vec{k}|} \chi_{\vec{k}'}^{(0)} + \frac{\hbar}{\mathrm{j}\mu_{\vec{k}} (\bar{f}_{\vec{k}}^{\mathrm{e}} + \bar{f}_{\vec{k}}^{\mathrm{h}} - 1)} \sum_{\vec{k}' \neq \vec{k}} \beta_{\vec{k}\vec{k}'} \chi_{\vec{k}'}^{(0)}]$$

$$= \frac{1}{2} \widetilde{E}_{\mu} \mathrm{e}^{-\mathrm{j}\omega t} \chi_{\vec{k}}^{(0)} (q_{\vec{k}} + s_{\vec{k}}), \tag{4.137}$$

with $q_{\vec{k}}$ and $s_{\vec{k}}$ defined as

$$q_{\vec{k}} \equiv \frac{1}{\mu_{\vec{k}}} \sum_{\vec{k}' \neq \vec{k}} V_{|\vec{k}'-\vec{k}|} \chi_{\vec{k}'}^{(0)}, \tag{4.138}$$

$$s_{\vec{k}} \equiv \frac{\hbar}{\mathrm{j}\mu_{\vec{k}} (\bar{f}_{\vec{k}}^{\mathrm{e}} + \bar{f}_{\vec{k}}^{\mathrm{h}} - 1)} \sum_{\vec{k}' \neq \vec{k}} \beta_{\vec{k}\vec{k}'} \chi_{\vec{k}'}^{(0)}. \tag{4.139}$$

It is obvious that, as the injection increases, $s_{\vec{k}}$ decreases in accordance with equation (4.139).

Using the same summing technique, we obtain

$$p_{\vec{k}} = p_{\vec{k}}^{(0)} + p_{\vec{k}}^{(1)} + \cdots = \frac{\widetilde{E}_{\mu} \mathrm{e}^{-\mathrm{j}\omega t}}{2} \chi_{\vec{k}}^{(0)} [1 + (q_{\vec{k}} + s_{\vec{k}}) + \cdots] \approx \frac{\widetilde{E}_{\mu} \mathrm{e}^{-\mathrm{j}\omega t}}{2} \frac{\chi_{\vec{k}}^{(0)}}{1 - (q_{\vec{k}} + s_{\vec{k}})}. \tag{4.140}$$

Plugging equation (4.140) into (4.47) yields

$$\widetilde{P} = \frac{1}{\Omega} \widetilde{E}_{\mu} \sum_{\vec{k}} \mu_{\vec{k}}^{*} \frac{\chi_{\vec{k}}^{(0)}}{1 - (q_{\vec{k}} + s_{\vec{k}})}$$

$$= -\frac{\mathrm{j}}{\hbar \Omega} \widetilde{E}_{\mu} \sum_{\vec{k}} |\mu_{\vec{k}}|^2 \frac{\bar{f}_{\vec{k}}^{\mathrm{e}} + \bar{f}_{\vec{k}}^{\mathrm{h}} - 1}{\mathrm{j}(\bar{\omega}_{\vec{k}'} - \omega) + \gamma_{\vec{k}}} \frac{1}{1 - (q_{\vec{k}} + s_{\vec{k}})}. \tag{4.141}$$

Therefore, we find the susceptibility from

$$\widetilde{\chi}(\omega) = \frac{\widetilde{P}}{\varepsilon_0 \widetilde{E}_\mu} = -\frac{\mathrm{j}}{\varepsilon_0 \hbar \Omega} \sum_{\vec{k}} |\mu_{\vec{k}}|^2 \frac{\bar{f}^{\mathrm{e}}_{\vec{k}} + \bar{f}^{\mathrm{h}}_{\vec{k}} - 1}{\mathrm{j}(\bar{\omega}_{\vec{k}'} - \omega) + \gamma_{\vec{k}}} \frac{1}{1 - (q_{\vec{k}} + s_{\vec{k}})}$$
$$= -\frac{1}{\varepsilon_0 \hbar \Omega} \sum_{\vec{k}} |\mu_{\vec{k}}|^2 (\bar{f}^{\mathrm{e}}_{\vec{k}} + \bar{f}^{\mathrm{h}}_{\vec{k}} - 1) \frac{(\bar{\omega}_{\vec{k}'} - \omega)/\gamma_{\vec{k}} + \mathrm{j}}{\gamma_{\vec{k}}} \frac{L_{\gamma_{\vec{k}}}(\bar{\omega}_{\vec{k}'} - \omega)}{[1 - (q_{\vec{k}} + s_{\vec{k}})]}, \tag{4.142}$$

with the $\gamma_{\vec{k}}$ dependent Lorentzian line-shape function defined as

$$L_{\gamma_{\vec{k}}}(x) = \frac{\gamma^2_{\vec{k}}}{\gamma^2_{\vec{k}} + x^2}. \tag{4.143}$$

Consequently, we obtain the semiconductor material gain and refractive index change from

$$g(\omega) = -\frac{k_0}{n} \mathrm{Im}[\widetilde{\chi}(\omega)] = \frac{\omega}{\varepsilon_0 n c \hbar \Omega} \sum_{\vec{k}} |\mu_{\vec{k}}|^2 (\bar{f}^{\mathrm{e}}_{\vec{k}} + \bar{f}^{\mathrm{h}}_{\vec{k}} - 1) \frac{L_{\gamma_{\vec{k}}}(\bar{\omega}_{\vec{k}'} - \omega)}{\gamma_{\vec{k}}[1 - (q_{\vec{k}} + s_{\vec{k}})]}$$
$$= \frac{\omega}{\varepsilon_0 n c \hbar \Omega} \sum_{\vec{k}} |\mu_{\vec{k}}|^2 (\bar{f}^{\mathrm{e}}_{\vec{k}} + \bar{f}^{\mathrm{h}}_{\vec{k}} - 1) \frac{\gamma_{\vec{k}}}{\gamma^2_{\vec{k}} + (\bar{\omega}_{\vec{k}'} - \omega)^2} \frac{1}{[1 - (q_{\vec{k}} + s_{\vec{k}})]}, \tag{4.144a}$$

$$\Delta n(\omega) = \frac{1}{2n} \mathrm{Re}[\widetilde{\chi}(\omega)] = -\frac{1}{2\varepsilon_0 n \hbar \Omega} \sum_{\vec{k}} |\mu_{\vec{k}}|^2 (\bar{f}^{\mathrm{e}}_{\vec{k}} + \bar{f}^{\mathrm{h}}_{\vec{k}} - 1) \frac{\bar{\omega}_{\vec{k}'} - \omega}{\gamma^2_{\vec{k}}} \frac{L_{\gamma_{\vec{k}}}(\bar{\omega}_{\vec{k}'} - \omega)}{[1 - (q_{\vec{k}} + s_{\vec{k}})]}$$
$$= -\frac{1}{2\varepsilon_0 n \hbar \Omega} \sum_{\vec{k}} |\mu_{\vec{k}}|^2 (\bar{f}^{\mathrm{e}}_{\vec{k}} + \bar{f}^{\mathrm{h}}_{\vec{k}} - 1) \frac{\bar{\omega}_{\vec{k}'} - \omega}{\gamma^2_{\vec{k}} + (\bar{\omega}_{\vec{k}'} - \omega)^2} \frac{1}{[1 - (q_{\vec{k}} + s_{\vec{k}})]}. \tag{4.144b}$$

4.4.3 The full numerical solution

Our task is now to solve equations (4.130a–c) in a self-consistent manner for any given total carrier densities $N_\mathrm{e}(t)$ and $N_\mathrm{h}(t)$ defined by equation (4.14), where the optical field in the form of equation (4.45) is also given, which will be cancelled in calculating the susceptibility.

To enlarge the time domain step size in numerical integrations, we will take the fast-varying harmonic wave factor $\mathrm{e}^{-\mathrm{j}\omega t}$ explicitly out of the polaritons and leave the equations for the slow-varying envelope functions. According to equation (4.45), we can assume [6]

$$p_{\vec{k}} = \widetilde{p}_{\vec{k}} \mathrm{e}^{-\mathrm{j}\omega t}, \tag{4.145}$$

where $\widetilde{p}_{\vec{k}}$ is the slow-varying envelope function of the polariton. Under this assumption and by noting that $\mathrm{d}p_{\vec{k}}/\mathrm{d}t = \mathrm{e}^{-\mathrm{j}\omega t}\mathrm{d}\widetilde{p}_{\vec{k}}/\mathrm{d}t - \mathrm{j}\omega \mathrm{e}^{-\mathrm{j}\omega t}\widetilde{p}_{\vec{k}}$, we find

$$\hbar\frac{\mathrm{d}\widetilde{p}_{\vec{k}}}{\mathrm{d}t} = \mathrm{j}\hbar\omega\widetilde{p}_{\vec{k}} - \sum_{\vec{k}'}(\mathrm{j}\Theta_{\vec{k}\vec{k}'} + \Lambda_{\vec{k}\vec{k}'})\widetilde{p}_{\vec{k}'} - \sum_{\vec{q}}\Lambda^{\mathrm{p}}_{\vec{k}\vec{q}}\widetilde{p}_{\vec{k}+\vec{q}} - \frac{\mathrm{j}\mu_{\vec{k}}\widetilde{E}_{\mu}}{2}(f^{\mathrm{e}}_{\vec{k}} + f^{\mathrm{h}}_{\vec{k}} - 1), \tag{4.146a}$$

$$\begin{aligned}\hbar\frac{\mathrm{d}f^{\mathrm{e}}_{\vec{k}}}{\mathrm{d}t} &= -(\Sigma^{\mathrm{e-out}}_{\vec{k}} + \Sigma^{\mathrm{e-in}}_{\vec{k}})f^{\mathrm{e}}_{\vec{k}} + 2\hbar\mathrm{Im}(\Omega^{*}_{\vec{k}}\widetilde{p}_{\vec{k}}\mathrm{e}^{-\mathrm{j}\omega t}) + \Sigma^{\mathrm{e-in}}_{\vec{k}} \\ &\quad - \sum_{\vec{q}\neq 0,\pm}(\Sigma^{\mathrm{e-p}}_{\vec{q},\pm}f^{\mathrm{p}}_{\vec{q}} + \Sigma^{\mathrm{e}}_{\vec{q},\pm}),\end{aligned} \tag{4.146b}$$

$$\begin{aligned}\hbar\frac{\mathrm{d}f^{\mathrm{h}}_{\vec{k}}}{\mathrm{d}t} &= -(\Sigma^{\mathrm{h-out}}_{\vec{k}} + \Sigma^{\mathrm{h-in}}_{\vec{k}})f^{\mathrm{h}}_{\vec{k}} + 2\hbar\mathrm{Im}[\Omega^{*}_{\vec{k}}\widetilde{p}_{\vec{k}}\mathrm{e}^{-\mathrm{j}\omega t}] + \Sigma^{\mathrm{h-in}}_{\vec{k}} \\ &\quad - \sum_{\vec{q}\neq 0,\pm}(\Sigma^{\mathrm{h-p}}_{\vec{q},\pm}f^{\mathrm{p}}_{\vec{q}} + \Sigma^{\mathrm{h}}_{\vec{q},\pm}).\end{aligned} \tag{4.146c}$$

In equation (4.146a), the zeroth and first order contributions are

$$\begin{aligned}\sum_{\vec{k}'}\Theta_{\vec{k}\vec{k}'}\widetilde{p}_{\vec{k}'} &= \Theta_{\vec{k}\vec{k}}\widetilde{p}_{\vec{k}} + \sum_{\vec{k}'\neq\vec{k}}\Theta_{\vec{k}\vec{k}'}\widetilde{p}_{\vec{k}'} \\ &= \hbar\bar{\omega}_{\vec{k}}\widetilde{p}_{\vec{k}} - \sum_{\vec{l}\neq\vec{k}}V_{|\vec{l}-\vec{k}|}(f^{\mathrm{e}}_{\vec{l}} + f^{\mathrm{h}}_{\vec{l}} - 1)\widetilde{p}_{\vec{k}} \\ &\quad + \sum_{\vec{k}'\neq\vec{k}}V_{|\vec{k}'-\vec{k}|}\widetilde{p}_{\vec{k}'}(f^{\mathrm{e}}_{\vec{k}} + f^{\mathrm{h}}_{\vec{k}} - 1) \\ &= (\hbar\bar{\omega}_{\vec{k}} - I^{1}_{\vec{k}})\widetilde{p}_{\vec{k}} + I^{2}_{\vec{k}}(f^{\mathrm{e}}_{\vec{k}} + f^{\mathrm{h}}_{\vec{k}} - 1),\end{aligned} \tag{4.147}$$

where we have defined

$$I^{1}_{\vec{k}} \equiv \sum_{\vec{l}\neq\vec{k}}V_{|\vec{l}-\vec{k}|}(f^{\mathrm{e}}_{\vec{l}} + f^{\mathrm{h}}_{\vec{l}} - 1), \tag{4.148a}$$

$$I^{2}_{\vec{k}} \equiv \sum_{\vec{k}'\neq\vec{k}}V_{|\vec{k}'-\vec{k}|}\widetilde{p}_{\vec{k}'}. \tag{4.148b}$$

The second order carrier–carrier collision contribution is

$$\begin{aligned}
\sum_{\vec{k}'} \Lambda_{\vec{k}\vec{k}'} \widetilde{p}_{\vec{k}'} &= \Lambda_{\vec{k}\vec{k}} \widetilde{p}_{\vec{k}} + \sum_{\vec{k}' \neq \vec{k}} \Lambda_{\vec{k}\vec{k}'} \widetilde{p}_{\vec{k}'} \\
&= \widetilde{p}_{\vec{k}} \sum_{\alpha,\beta=\mathrm{e,h}} \sum_{\vec{k}''} \sum_{\vec{l} \neq 0} (2\widetilde{V}^2_{|\vec{l}|} - \delta_{\alpha\beta} \widetilde{V}_{|\vec{l}|} \widetilde{V}_{|\vec{k}''-\vec{l}-\vec{k}|}) \\
&\quad \times g[\varepsilon_{\alpha\vec{k}} + \varepsilon_{\beta\vec{k}''} - \varepsilon_{\alpha(\vec{k}+\vec{l})} - \varepsilon_{\beta(\vec{k}''-\vec{l})}] \\
&\quad \times [f^{\alpha}_{\vec{k}+\vec{l}}(1 - f^{\beta}_{\vec{k}''}) f^{\beta}_{\vec{k}''-\vec{l}} + (1 - f^{\alpha}_{\vec{k}+\vec{l}}) f^{\beta}_{\vec{k}''}(1 - f^{\beta}_{\vec{k}''-\vec{l}})] \\
&\quad + \sum_{\alpha,\beta=\mathrm{e,h}} \sum_{\vec{k}''} \sum_{\vec{k}' \neq \vec{k}} \widetilde{p}_{\vec{k}'} (2\widetilde{V}^2_{|\vec{k}'-\vec{k}|} - \delta_{\alpha\beta} \widetilde{V}_{|\vec{k}'-\vec{k}|} \widetilde{V}_{|\vec{k}''-\vec{k}'|}) \\
&\quad \times g[-\varepsilon_{\alpha\vec{k}} - \varepsilon_{\beta\vec{k}''} + \varepsilon_{\alpha\vec{k}'} + \varepsilon_{\beta(\vec{k}''-\vec{k}'+\vec{k})}] \\
&\quad \times [(1 - f^{\alpha}_{\vec{k}})(1 - f^{\beta}_{\vec{k}''}) f^{\beta}_{\vec{k}''-\vec{k}'+\vec{k}} + f^{\alpha}_{\vec{k}} f^{\beta}_{\vec{k}''}(1 - f^{\beta}_{\vec{k}''-\vec{k}'+\vec{k}})] \\
&= J^1_{\vec{k}} \widetilde{p}_{\vec{k}} + \sum_{\alpha,\beta=\mathrm{e,h}} [J^2_{\vec{k}\,\alpha\beta}(1 - f^{\alpha}_{\vec{k}}) + J^3_{\vec{k}\,\alpha\beta} f^{\alpha}_{\vec{k}}],
\end{aligned} \tag{4.149}$$

where we have defined

$$\begin{aligned}
J^1_{\vec{k}} \equiv & \sum_{\alpha,\beta=\mathrm{e,h}} \sum_{\vec{k}''} \sum_{\vec{l} \neq 0} (2\widetilde{V}^2_{|\vec{l}|} - \delta_{\alpha\beta} \widetilde{V}_{|\vec{l}|} \widetilde{V}_{|\vec{k}''-\vec{l}-\vec{k}|}) \\
& \times g[\varepsilon_{\alpha\vec{k}} + \varepsilon_{\beta\vec{k}''} - \varepsilon_{\alpha(\vec{k}+\vec{l})} - \varepsilon_{\beta(\vec{k}''-\vec{l})}] \\
& \times [f^{\alpha}_{\vec{k}+\vec{l}}(1 - f^{\beta}_{\vec{k}''}) f^{\beta}_{\vec{k}''-\vec{l}} + (1 - f^{\alpha}_{\vec{k}+\vec{l}}) f^{\beta}_{\vec{k}''}(1 - f^{\beta}_{\vec{k}''-\vec{l}})],
\end{aligned} \tag{4.150a}$$

$$\begin{aligned}
J^2_{\vec{k}\,\alpha\beta} \equiv & \sum_{\vec{k}''} \sum_{\vec{k}' \neq \vec{k}} \widetilde{p}_{\vec{k}'} (2\widetilde{V}^2_{|\vec{k}'-\vec{k}|} - \delta_{\alpha\beta} \widetilde{V}_{|\vec{k}'-\vec{k}|} \widetilde{V}_{|\vec{k}''-\vec{k}'|}) \\
& \times g[-\varepsilon_{\alpha\vec{k}} - \varepsilon_{\beta\vec{k}''} + \varepsilon_{\alpha\vec{k}'} + \varepsilon_{\beta(\vec{k}''-\vec{k}'+\vec{k})}](1 - f^{\beta}_{\vec{k}''}) f^{\beta}_{\vec{k}''-\vec{k}'+\vec{k}},
\end{aligned} \tag{4.150b}$$

$$\begin{aligned}
J^3_{\vec{k}\,\alpha\beta} \equiv & \sum_{\vec{k}''} \sum_{\vec{k}' \neq \vec{k}} \widetilde{p}_{\vec{k}'} (2\widetilde{V}^2_{|\vec{k}'-\vec{k}|} - \delta_{\alpha\beta} \widetilde{V}_{|\vec{k}'-\vec{k}|} \widetilde{V}_{|\vec{k}''-\vec{k}'|}) \\
& \times g[-\varepsilon_{\alpha\vec{k}} - \varepsilon_{\beta\vec{k}''} + \varepsilon_{\alpha\vec{k}'} + \varepsilon_{\beta(\vec{k}''-\vec{k}'+\vec{k})}] f^{\beta}_{\vec{k}''}(1 - f^{\beta}_{\vec{k}''-\vec{k}'+\vec{k}}).
\end{aligned} \tag{4.150c}$$

The second order carrier–phonon scattering contribution is

$$\begin{aligned}
\sum_{\vec{q}} \Lambda^{\mathrm{p}}_{\vec{k}\vec{q}} \widetilde{p}_{\vec{k}+\vec{q}} &= \Lambda^{\mathrm{p}}_{\vec{k}0} \widetilde{p}_{\vec{k}} + \sum_{\vec{q} \neq 0} \Lambda^{\mathrm{p}}_{\vec{k}\vec{q}} \widetilde{p}_{\vec{k}+\vec{q}} \\
&= \widetilde{p}_{\vec{k}} \hbar^2 \sum_{\alpha=\mathrm{e,h}} \sum_{\vec{l} \neq 0} G^2_{|\vec{l}|} \{ g[\varepsilon_{\alpha\vec{k}} - \varepsilon_{\alpha(\vec{k}-\vec{l})} - \hbar\omega_{\mathrm{LO}}] \\
&\qquad \times \left[(1 - f^{\alpha}_{\vec{k}-\vec{l}}) f^{\mathrm{p}}_{\vec{l}} + f^{\alpha}_{\vec{k}-\vec{l}}(1 + f^{\mathrm{p}}_{\vec{l}}) \right]
\end{aligned}$$

$$
\begin{aligned}
&+ g[\varepsilon_{\alpha\vec{k}} - \varepsilon_{\alpha(\vec{k}-\vec{l})} + \hbar\omega_{\rm LO}][(1 - f^{\alpha}_{\vec{k}-\vec{l}})(1 + f^{\rm p}_{\vec{l}}) + f^{\alpha}_{\vec{k}-\vec{l}} f^{\rm p}_{\vec{l}}]\Big\} \\
&+ \hbar^2 \sum_{\alpha={\rm e,h}} \sum_{\vec{q}\neq 0} \widetilde{p}_{\vec{k}+\vec{q}} G^2_{|\vec{q}|} \Big\{ g[\varepsilon_{\alpha\vec{k}} - \varepsilon_{\alpha(\vec{k}-\vec{q})} - \hbar\omega_{\rm LO}] \\
&\times [(1 - f^{\alpha}_{\vec{k}}) f^{\rm p}_{\vec{q}} + f^{\alpha}_{\vec{k}}(1 + f^{\rm p}_{\vec{q}})] \\
&+ g[\varepsilon_{\alpha\vec{k}} - \varepsilon_{\alpha(\vec{k}-\vec{q})} + \hbar\omega_{\rm LO}][(1 - f^{\alpha}_{\vec{k}})(1 + f^{\rm p}_{\vec{q}}) + f^{\alpha}_{\vec{k}} f^{\rm p}_{\vec{q}}]\Big\} \\
&= K^1_{\vec{k}} \widetilde{p}_{\vec{k}} + \sum_{\alpha={\rm e,h}} [K^2_{\vec{k}\alpha}(1 - f^{\alpha}_{\vec{k}}) + K^3_{\vec{k}\alpha} f^{\alpha}_{\vec{k}}],
\end{aligned}
\tag{4.151}
$$

where we have defined

$$
\begin{aligned}
K^1_{\vec{k}} \equiv \hbar^2 \sum_{\alpha={\rm e,h}} \sum_{\vec{l}\neq 0} G^2_{|\vec{l}|} \Big\{ & g[\varepsilon_{\alpha\vec{k}} - \varepsilon_{\alpha(\vec{k}-\vec{l})} - \hbar\omega_{\rm LO}][(1 - f^{\alpha}_{\vec{k}-\vec{l}}) f^{\rm p}_{\vec{l}} + f^{\alpha}_{\vec{k}-\vec{l}}(1 + f^{\rm p}_{\vec{l}})] \\
&+ g[\varepsilon_{\alpha\vec{k}} - \varepsilon_{\alpha(\vec{k}-\vec{l})} + \hbar\omega_{\rm LO}][(1 - f^{\alpha}_{\vec{k}-\vec{l}})(1 + f^{\rm p}_{\vec{l}}) + f^{\alpha}_{\vec{k}-\vec{l}} f^{\rm p}_{\vec{l}}]\Big\},
\end{aligned}
\tag{4.152a}
$$

$$
\begin{aligned}
K^2_{\vec{k}\alpha} \equiv \hbar^2 \sum_{\vec{q}\neq 0} \widetilde{p}_{\vec{k}+\vec{q}} G^2_{|\vec{q}|} \Big\{ & g[\varepsilon_{\alpha\vec{k}} - \varepsilon_{\alpha(\vec{k}-\vec{q})} - \hbar\omega_{\rm LO}] f^{\rm p}_{\vec{q}} \\
&+ g[\varepsilon_{\alpha\vec{k}} - \varepsilon_{\alpha(\vec{k}-\vec{q})} + \hbar\omega_{\rm LO}](1 + f^{\rm p}_{\vec{q}})\Big\},
\end{aligned}
\tag{4.152b}
$$

$$
\begin{aligned}
K^3_{\vec{k}\alpha} \equiv \hbar^2 \sum_{\vec{q}\neq 0} \widetilde{p}_{\vec{k}+\vec{q}} G^2_{|\vec{q}|} \Big\{ & g[\varepsilon_{\alpha\vec{k}} - \varepsilon_{\alpha(\vec{k}-\vec{q})} - \hbar\omega_{\rm LO}](1 + f^{\rm p}_{\vec{q}}) \\
&+ g[\varepsilon_{\alpha\vec{k}} - \varepsilon_{\alpha(\vec{k}-\vec{q})} + \hbar\omega_{\rm LO}] f^{\rm p}_{\vec{q}}\Big\}.
\end{aligned}
\tag{4.152c}
$$

In equations (4.146b&c), the Rabi frequency and the carrier–phonon scattering contribution can be written as

$$
\begin{aligned}
\Omega^*_{\vec{k}} &= \frac{1}{\hbar} \vec{\mu}^*_{\vec{k}} \cdot \vec{E}^*(t) + \frac{1}{\hbar} \sum_{\vec{l}\neq\vec{k}} V^*_{|\vec{l}-\vec{k}|} p^*_{\vec{l}} \\
&= \frac{1}{2\hbar} \vec{\mu}^*_{\vec{k}} \cdot \vec{\widetilde{E}}^* {\rm e}^{{\rm j}\omega t} + \frac{1}{\hbar} \sum_{\vec{l}\neq\vec{k}} V_{|\vec{l}-\vec{k}|} \widetilde{p}^*_{\vec{l}} {\rm e}^{{\rm j}\omega t} \\
&= \frac{{\rm e}^{{\rm j}\omega t}}{\hbar} [\frac{1}{2} (\mu^*_{\vec{k}} \widetilde{E}^*_{\mu}) + I^{2*}_{\vec{k}}],
\end{aligned}
\tag{4.153}
$$

$$
\begin{aligned}
&\sum_{\vec{q}\neq 0,\pm} (\Sigma^{\rm e/h-p}_{\vec{q},\pm} f^{\rm p}_{\vec{q}} + \Sigma^{\rm e/h}_{\vec{q},\pm}) \\
&= 2\pi\hbar^2 \sum_{\vec{q}\neq 0,\pm} G^2_{|\vec{q}|} \delta\left[\varepsilon_{({\rm e/h})\vec{k}} - \varepsilon_{({\rm e/h})(\vec{k}-\vec{q})} \mp \hbar\omega_{\rm LO}\right]
\end{aligned}
$$

$$\times \left\{ \left[f^{e/h}_{\vec{k}}(1 - f^{e/h}_{\vec{k}-\vec{q}}) - (1 - f^{e/h}_{\vec{k}}) f^{e/h}_{\vec{k}-\vec{q}} \right] f^{p}_{\vec{q}} + \left[f^{e/h}_{\vec{k}}(1 - f^{e/h}_{\vec{k}-\vec{q}}) \left(\frac{1}{2} \pm \frac{1}{2} \right) - (1 - f^{e/h}_{\vec{k}}) f^{e/h}_{\vec{k}-\vec{q}} \left(\frac{1}{2} \mp \frac{1}{2} \right) \right] \right\}$$
$$= L^{(e/h)-out,p}_{\vec{k}} f^{e/h}_{\vec{k}} - L^{(e/h)-in,p}_{\vec{k}} (1 - f^{e/h}_{\vec{k}}), \tag{4.154}$$

where we have defined

$$L^{(e/h)-out,p}_{\vec{k}} \equiv 2\pi\hbar^2 \sum_{\vec{q}\neq 0,\pm} G^2_{|\vec{q}|} \delta[\varepsilon_{(e/h)\vec{k}} - \varepsilon_{(e/h)(\vec{k}-\vec{q})} \mp \hbar\omega_{LO}](f^{p}_{\vec{q}} + \frac{1}{2} \pm \frac{1}{2})(1 - f^{e/h}_{\vec{k}-\vec{q}}), \tag{4.155a}$$

$$L^{(e/h)-in,p}_{\vec{k}} \equiv 2\pi\hbar^2 \sum_{\vec{q}\neq 0,\pm} G^2_{|\vec{q}|} \delta[\varepsilon_{(e/h)\vec{k}} - \varepsilon_{(e/h)(\vec{k}-\vec{q})} \mp \hbar\omega_{LO}](f^{p}_{\vec{q}} + \frac{1}{2} \mp \frac{1}{2}) f^{e/h}_{\vec{k}-\vec{q}}. \tag{4.155b}$$

Substituting equations (4.147), (4.149), and (4.151) into (4.146a), and equations (4.153) and (4.154) into (4.146b&c), respectively, we finally obtain

$$\hbar \frac{d\tilde{p}_{\vec{k}}}{dt} = [j(\hbar\omega - \hbar\bar{\omega}_{\vec{k}} + I^1_{\vec{k}}) - J^1_{\vec{k}} - K^1_{\vec{k}}]\tilde{p}_{\vec{k}} - j[\frac{1}{2}(\mu_{\vec{k}} \tilde{E}_{\mu}) + I^2_{\vec{k}}](f^{e}_{\vec{k}} + f^{h}_{\vec{k}} - 1) - \sum_{\alpha,\beta=e,h} [(J^2_{\vec{k}\alpha\beta} + K^2_{\vec{k}\alpha}\delta_{\alpha\beta})(1 - f^{\alpha}_{\vec{k}}) + (J^3_{\vec{k}\alpha\beta} + K^3_{\vec{k}\alpha}\delta_{\alpha\beta}) f^{\alpha}_{\vec{k}}], \tag{4.156a}$$

$$\hbar \frac{d f^{e}_{\vec{k}}}{dt} = -(\Sigma^{e-out}_{\vec{k}} + \Sigma^{e-in}_{\vec{k}} + L^{e-out,p}_{\vec{k}} + L^{e-in,p}_{\vec{k}}) f^{e}_{\vec{k}} + 2\mathrm{Im}[(\frac{1}{2}\mu^*_{\vec{k}} \tilde{E}^*_{\mu} + I^{2*}_{\vec{k}})\tilde{p}_{\vec{k}}] + \Sigma^{e-in}_{\vec{k}} + L^{e-in,p}_{\vec{k}}, \tag{4.156b}$$

$$\hbar \frac{d f^{h}_{\vec{k}}}{dt} = -(\Sigma^{h-out}_{\vec{k}} + \Sigma^{h-in}_{\vec{k}} + L^{h-out,p}_{\vec{k}} + L^{h-in,p}_{\vec{k}}) f^{h}_{\vec{k}} + 2\mathrm{Im}[(\frac{1}{2}\mu^*_{\vec{k}} \tilde{E}^*_{\mu} + I^{2*}_{\vec{k}})\tilde{p}_{\vec{k}}] + \Sigma^{h-in}_{\vec{k}} + L^{h-in,p}_{\vec{k}}. \tag{4.156c}$$

Equations (4.156a–c) form a set of coupled ODEs. In these equations, the coefficients (i.e., the summations in the wave vector domain) may still depend on unknown variables $\tilde{p}_{\vec{k}'\neq\vec{k}}$ and $f^{e/h}_{\vec{k}'\neq\vec{k}}$ at other $\vec{k}$ values in the wave vector domain. Therefore, equations (4.156a–c) are quasi-linear and need to be solved numerically.

Some features of the coefficients (i.e., the wave vector domain summations) are summarized in Table 4.3.

Once equations (4.156a–c) are solved for $\tilde{p}_{\vec{k}}$ and $f^{e/h}_{\vec{k}}$, by following equation (4.47), we will be able to obtain the slow-varying envelope of the macroscopic polarization $\tilde{P}$. Finally, the susceptibility can be extracted by $\tilde{\chi}(\omega) = \tilde{P}/(\varepsilon_0 \tilde{E}_{\mu})$.

Table 4.3. Coulomb interaction energies. The table shows the coefficients in the Heisenberg equations that govern the slow-varying envelope function of the polariton number expectation, and the conduction band electron and valence band hole number expectations

V and $\widetilde{V}$ are the bare and screened Coulomb potential in the wave vector domain.

Integral	Dependence		Coulomb potential	Interaction order	Remarks
	$\widetilde{p}_{\vec{k}'\neq\vec{k}}$	$f^{e/h}_{\vec{k}'\neq\vec{k}}$			
$I^1_{\vec{k}}$	no	yes	V	0th , 1st	energy shift, real
$I^2_{\vec{k}}$	yes	no	V	0th, 1st	dipole renormalization, complex
$J^1_{\vec{k}}$	no	yes	$\widetilde{V}$	2nd	carrier collision, complex
$J^{2/3}_{\vec{k}\alpha\beta}$	yes	yes	$\widetilde{V}$	2nd	carrier collision, complex
$K^1_{\vec{k}}$	no	yes	$\widetilde{V}$	2nd	phonon scattering, complex
$K^{2/3}_{\vec{k}\alpha}$	yes	no	$\widetilde{V}$	2nd	phonon scattering, complex
$\Sigma^{(e/h)-out/in}_{\vec{k}}$	no	yes	$\widetilde{V}$	2nd	carrier collision, real
$L^{(e/h)-out/in,p}_{\vec{k}}$	no	yes	$\widetilde{V}$	2nd	phonon scattering, real

References

[1] L. D. Landau and E. M. Lifshitz, *Quantum Mechanics*, 3rd edn (Oxford, UK: Pergamon Press, 1977).

[2] H. Huag and S. W. Koch, *Quantum Theory of the Optical and Electronic Properties of Semiconductors*, 3rd edn (Singapore: World Scientific, 1994).

[3] W. W. Chow, S. W. Koch, and M. Sargent III, *Semiconductor-Laser Physics*, 1st edn (Berlin: Springer-Verlag, 1994).

[4] C. Y. P. Chao and S. L. Chuang, Momentum-space solution of exciton excited states and heavy-hole-light-hole mixing in quantum wells. *Phys. Rev. B*, **48**:11 (1993), 8210–21.

[5] P. L. Taylor and O. Heinonen, *A Quantum Approach to Condensed Matter Physics*, 1st edn (Cambridge UK: Cambridge University Press, 2002).

[6] W. W. Chow and S. W. Koch, *Semiconductor Laser Fundamentals: Physics of the Gain Materials*, 1st edn (Berlin: Springer-Verlag, 1999).

[7] G D. Mahan, *Many Particle Physics*, 1st edn (New York: Plenum Press, 1981).

[8] M. Sargent III, M. O. Scully, and W. E. Lamb, Jr., *Laser Physics*, 1st edn (Reading, MA: Addison-Wesley, 1974).

[9] S. Adachi, *Physical Properties of III-V Semiconductor Compounds, InP, InAs, GaAs, GaP, InGaAs, and InGaAsP*, 1st edn (New York: John Wiley & Sons, 1992).

[10] O. Madelung, (Ed.) *Semiconductors: Basic Data*, 2nd revised edn (Berlin: Springer-Verlag, 1996).

[11] N. W. Ashcroft and N. D. Mermin, *Solid State Theory*, 1st edn (Philadelphia: Saunders College, 1976).

[12] W. A. Harrison, *Solid State Theory*, 1st edn (New York: Dover Publications, 1980).

[13] R. Zimmermann, *Many-Particle Theory of Highly Excited Semiconductors*, 1st edn (Berlin: Teubner, 1988).

[14] R. J. Elliot, Theory of excitons: I. In *Polarons and Excitons, Scottish Universities' Summer School*, ed. C. G. Kuper and G. D. Whitfield. (New York: Plenum, 1962).

[15] R. S. Knox, Theory of Excitons. In *Solid State Physics, Suppl. 5*. (New York: Academic, 1963).

[16] J. O. Dimmock, Introduction to the theory of exciton states in semiconductors. Chapter 7 in *Semiconductor and Semimetals, Vol. 3*, ed. R. K. Willardson and A. C. Beer. (New York: Academic, 1967).

[17] F. Bassani and G. P. Parravicini, *Electronic States and Optical Transitions in Solids*, 1st edn (Oxford, UK, Pergamon, 1975).

[18] M. Shinada and S. Sugano, Interband optical transitions in extremely anisotropic semiconductors, I: Bound and unbound exciton absorption. *Journal of Phys. Soc. Jpn.*, **21** (1966), 1936–46.

[19] S. L. Chuang, S. Schmitt-Rink, D. A. B. Miller, and D. S. Chemla, Exciton Green's function approach to optical absorption in a quantum well with an applied electric field. *Phys. Rev. B*, **43**:2 (1991), 1500–9.

[20] C. Y. P. Chao and S. L. Chuang, Analytical and numerical solutions for a two-dimensional exciton in momentum space. *Phys. Rev. B*, **43**:8 (1991), 6530–43.

[21] P. R. Gaves-Morris, (Ed.) *Pade Approximants and Their Application*, 1st edn (New York: Academic Press, 1973).

[22] H. Huag and S. W. Koch, Semiconductor laser theory with many-body effects. *Phys. Rev. A*, **39** (1989), 1887–98.

5 Carrier transport and thermal diffusion models

5.1 The carrier transport model

5.1.1 Poisson and carrier continuity equations

The electrical signals involved in optoelectronic devices are either DC (for biasing) or at a relatively low frequency (for modulation) compared with the optical frequency. As such, the time-dependent change in the magnetic field is negligible. The Maxwell equations are simplified to

$$\nabla \times \vec{E}(\vec{r}, t) = 0, \tag{5.1}$$

$$\nabla \times \vec{H}(\vec{r}, t) = \frac{\partial \vec{D}(\vec{r}, t)}{\partial t} + \vec{J}(\vec{r}, t), \tag{5.2}$$

$$\nabla \cdot \vec{D}(\vec{r}, t) = \rho(\vec{r}, t), \tag{5.3}$$

Equation (5.1) suggests that we can introduce an electrostatic potential $\Phi(\vec{r}, t)$ in V by defining

$$\vec{E}(\vec{r}, t) = -\nabla \Phi(\vec{r}, t), \tag{5.4}$$

as it is usually easier to deal with a scalar variable.

Knowing that the material dispersion is negligible within the electric signal bandwidth, or

$$\vec{D}(\vec{r}, t) = \int_{-\infty}^{t} \varepsilon(\vec{r}, t - \tau)\vec{E}(\vec{r}, \tau)\mathrm{d}\tau \approx \tilde{\varepsilon}(\vec{r})\vec{E}(\vec{r}, t), \tag{5.5}$$

with $\tilde{\varepsilon}$ indicating the frequency domain permittivity of the host medium measured at a low (electrical signal) frequency, we obtain from equations (5.2)–(5.5)

$$\begin{aligned} \frac{\partial \nabla \cdot \vec{D}(\vec{r}, t)}{\partial t} + \nabla \cdot \vec{J}(\vec{r}, t) &= \frac{\partial \rho(\vec{r}, t)}{\partial t} + \nabla \cdot \vec{J}(\vec{r}, t) \\ &= \nabla \cdot [\nabla \times \vec{H}(\vec{r}, t)] = 0, \end{aligned} \tag{5.6}$$

$$\nabla \cdot \tilde{\varepsilon}(\vec{r})\vec{E}(\vec{r}, t) = -\nabla \cdot \tilde{\varepsilon}(\vec{r})\nabla\Phi(\vec{r}, t) = \rho(\vec{r}, t). \tag{5.7}$$

In semiconductor optoelectronic devices, the total current density consists of the electron and hole current densities

$$\vec{J}(\vec{r}, t) = \vec{J}_{\mathrm{e}}(\vec{r}, t) + \vec{J}_{\mathrm{h}}(\vec{r}, t), \tag{5.8}$$

while the total charge density consists of the motionless space charge densities and the mobile electron and hole densities

$$\rho(\vec{r}, t) = \mathrm{e}[N_\mathrm{h}(\vec{r}, t) - N_\mathrm{e}(\vec{r}, t) + N_\mathrm{D}^+(\vec{r}) - N_\mathrm{A}^-(\vec{r})]. \tag{5.9}$$

Note that $\vec{J}_\mathrm{e}$ and $\vec{J}_\mathrm{h}$ in equation (5.8) indicate the electron and hole current densities in $\mathrm{A/m^2}$, N_e and N_h in equation (5.9) the electron and hole densities in $1/\mathrm{m}^3$, and N_D^+ and N_A^- the ionized donor and acceptor concentrations in $1/\mathrm{m}^3$, respectively.

Plugging equations (5.8) and (5.9) into (5.6) and (5.7) we obtain

$$\mathrm{e}\frac{\partial}{\partial t}[N_\mathrm{h}(\vec{r}, t) - N_\mathrm{e}(\vec{r}, t)] + \nabla \cdot [\vec{J}_\mathrm{e}(\vec{r}, t) + \vec{J}_\mathrm{h}(\vec{r}, t)] = 0, \tag{5.10}$$

$$\nabla \cdot \tilde{\varepsilon}(\vec{r})\nabla\Phi(\vec{r}, t) = -\mathrm{e}[N_\mathrm{h}(\vec{r}, t) - N_\mathrm{e}(\vec{r}, t) + N_\mathrm{D}^+(\vec{r}) - N_\mathrm{A}^-(\vec{r})]. \tag{5.11}$$

By further introducing an electron–hole pair recombination rate $R(\vec{r}, t)$, we can split equation (5.10) into two equations to govern electrons and holes, respectively

$$\nabla \cdot \vec{J}_\mathrm{e}(\vec{r}, t) - \mathrm{e}\frac{\partial}{\partial t}N_\mathrm{e}(\vec{r}, t) = \mathrm{e}R(\vec{r}, t), \tag{5.12a}$$

$$\nabla \cdot \vec{J}_\mathrm{h}(\vec{r}, t) + \mathrm{e}\frac{\partial}{\partial t}N_\mathrm{h}(\vec{r}, t) = -\mathrm{e}R(\vec{r}, t). \tag{5.12b}$$

Equation (5.11), known as Poisson's equation, and (5.12a&b), known as the carrier (electron and hole) continuity equations, form the classical carrier transport model. They are directly derived from the Maxwell equations for quasi-electrostatic fields. However, this classical model is not sufficient for modeling carrier transport as there are two vectorial and three scalar unknown variables but only three scalar equations. We still need two vectorial equations that link the current density, the carrier density and the potential distributions together. Such a relation, known as the carrier transport equation, can be established at either microscopic or macroscopic level, depending on the area we are dealing with inside the device.

5.1.2 The drift and diffusion model for a non-active region

Although in principle the carrier transport equation should be established at the microscopic level so as to take into account all the interactions in the transport, such as carrier–carrier collision and carrier–phonon scattering, we can find the corresponding macroscopic quantity through statistics. We have discussed this model in Chapter 4 where the Boltzmann equation resulted from the Heisenberg equation. Despite its accuracy, since it is derived from first principles, a major drawback of this model is the huge burden involved in numerical computation. In the non-active region of a bulk semiconductor in an optoelectronic device, this model seems to be an overkill as the only effect in such a region on the carrier and on the optical wave is a combined sum loss, i.e., carrier consumption in the transport process (through various non-radiative and spontaneous emission recombinations) and optical wave attenuation (through free-carrier absorption).

There is no direct interaction between the carrier and the optical wave. That is to say, in the non-active region, the optoelectronic device is no more than an electronic device. Therefore, a phenomenological drift and diffusion model that has been extremely successful in modeling bulk semiconductor based electronic devices can be employed here justifiably.

In bulk semiconductors, carriers can be viewed as classical particles with negligible matter wave behavior. Phenomenologically, the driving force of the current flow comprises the potential gradient (i.e., charged carrier drift under the electric field), the carrier diffusion (because of its non-uniform distribution), and the thermal gradient (because of the Seebeck effect), hence we can write

$$\vec{J} = \mathrm{e}DN\left(-\frac{\nabla\Phi}{V_T} + \frac{\nabla N}{N} + \frac{\nabla T}{2T}\right),$$

where it is not difficult to deduce that coefficient D should be the carrier diffusivity in m^2/s by dimension matching, since the driving forces are all normalized in 1/m. By utilizing the well-known Einstein relation $D/\mu = k_\mathrm{B}T/\mathrm{e} = V_\mathrm{T}$ and $D^\mathrm{T} = D/2T$ [1] with μ and D^T denoted as the carrier mobility in $\mathrm{m}^2/(\mathrm{V\,s})$ and thermal diffusivity in $\mathrm{m}^2/(\mathrm{K\,s})$, we can write the carrier transport equations for electron and hole, respectively

$$\vec{J}_\mathrm{e} = -\mathrm{e}\mu_\mathrm{e}N_\mathrm{e}(\vec{r},t)\nabla[\Phi(\vec{r},t) + \Phi_\mathrm{C}(\vec{r})] + \mathrm{e}D_\mathrm{e}\nabla N_\mathrm{e}(\vec{r},t) + \mathrm{e}D_\mathrm{e}^\mathrm{T}N_\mathrm{e}(\vec{r},t)\nabla T(\vec{r},t), \tag{5.13a}$$

$$\vec{J}_\mathrm{h} = -\mathrm{e}\mu_\mathrm{h}N_\mathrm{h}(\vec{r},t)\nabla[\Phi(\vec{r},t) - \Phi_\mathrm{V}(\vec{r})] - \mathrm{e}D_\mathrm{h}\nabla N_\mathrm{h}(\vec{r},t) - \mathrm{e}D_\mathrm{h}^\mathrm{T}N_\mathrm{h}(\vec{r},t)\nabla T(\vec{r},t). \tag{5.13b}$$

In equations (5.13a&b), the conduction and valence band edge variations at heterojunctions are given by [2–4]

$$\mathrm{e}\Phi_\mathrm{C}(\vec{r}) = \chi - \chi_\mathrm{r} + k_\mathrm{B}T\ln\left(\frac{N_\mathrm{c}}{N_\mathrm{cr}}\right), \tag{5.14a}$$

$$\mathrm{e}\Phi_\mathrm{V}(\vec{r}) = -(E_\mathrm{g} - E_\mathrm{gr}) - (\chi - \chi_\mathrm{r}) + k_\mathrm{B}T\ln\left(\frac{N_\mathrm{v}}{N_\mathrm{vr}}\right), \tag{5.14b}$$

with χ indicating the electron affinity in eV, E_g the bandgap energy in eV, and $N_{\mathrm{c/v}}$ the parabolic conduction and valence band edge densities defined as in equation (6.56), respectively. Subscript r on χ, E_g and $N_{\mathrm{c/v}}$ indicates the corresponding quantities in the reference material, which is normally selected as the substrate.

Therefore, in the non-active region, Poisson's equation (5.11), the carrier continuity equations (5.12a&b), which are universally valid, plus the phenomenological drift and diffusion equations (5.13a&b) form a complete model that governs carrier transport in the non-active region. We usually call this the classical carrier transport model.

By substituting the current densities in equations (5.13a&b) into the carrier continuity equations (5.12a&b), we can write the classical carrier transport equations in a rather compact form with the current densities eliminated

$$\nabla \cdot \tilde{\varepsilon}(\vec{r})\nabla\Phi(\vec{r},t) = -\mathrm{e}[N_{\mathrm{h}}(\vec{r},t) - N_{\mathrm{e}}(\vec{r},t) + N_{\mathrm{D}}^{+}(\vec{r}) - N_{\mathrm{A}}^{-}(\vec{r})], \tag{5.15a}$$

$$\frac{\partial N_{\mathrm{e}}(\vec{r},t)}{\partial t} = -R(\vec{r},t) - \nabla \cdot \mu_{\mathrm{e}}(\vec{r})N_{\mathrm{e}}(\vec{r},t)\nabla[\Phi(\vec{r},t) + \Phi_{\mathrm{C}}(\vec{r})] + \nabla \cdot D_{\mathrm{e}}(\vec{r})\nabla N_{\mathrm{e}}(\vec{r},t) + \nabla \cdot D_{\mathrm{e}}^{\mathrm{T}}(\vec{r})N_{\mathrm{e}}(\vec{r},t)\nabla T(\vec{r},t), \tag{5.15b}$$

$$\frac{\partial N_{\mathrm{h}}(\vec{r},t)}{\partial t} = -R(\vec{r},t) + \nabla \cdot \mu_{\mathrm{h}}(\vec{r})N_{\mathrm{h}}(\vec{r},t)\nabla[\Phi(\vec{r},t) - \Phi_{\mathrm{V}}(\vec{r})] + \nabla \cdot D_{\mathrm{h}}(\vec{r})\nabla N_{\mathrm{h}}(\vec{r},t) + \nabla \cdot D_{\mathrm{h}}^{\mathrm{T}}(\vec{r})N_{\mathrm{h}}(\vec{r},t)\nabla T(\vec{r},t), \tag{5.15c}$$

where we have assigned the spatial dependence to every material parameter as, depending on the device structure design, these parameters can be very different for layers with different material composition or different doping concentration.

5.1.3 The carrier transport model for the active region

As discussed in Chapter 4, it is critical to include the many-body Coulomb effect inside the active region where the interband carrier–photon interaction, i.e., the stimulated emission, takes place. Macroscopic models unfortunately cannot handle such an effect, hence we have to use the set of equations derived in Chapter 4 to find the microscopic carrier (electron and hole) and polariton number expectations in a self-consistent manner, and we will be able to find the macroscopic carrier densities and polarization through summations in the wave vector domain.

Unless the carriers are directly injected into the active region, which is not possible in reality, the injection condition for the active region is given by the carrier transport result in the non-active region which bridges the electrodes and the active region. Therefore, we have to solve the problem of how to link the macroscopic carrier densities that we find through the drift and diffusion carrier transport model in the non-active region to the microscopic carrier number expectations in the active region.

As illustrated by Fig. 5.1, to solve this problem, we consider a four-band model inside the active region. The top conduction band electrons and bottom valence band holes have their energies matching with the conduction band electrons and valence band holes in the non-active region, respectively. For these electrons and holes, known as the hot carriers because of their higher kinetic energies, we still have the classical carrier transport equations introduced in Section 5.1.2. The lower conduction band (or energy level in QW) electrons and upper valence band (or energy level in QW) holes, known as the cold carriers, interact with the optical wave through stimulated emission and absorption, hence the microscopic model introduced in Chapter 4 has to be adopted with the many-body Coulomb effect included in case it is necessary. Since the origin of a cold carrier is a hot carrier when it loses part of its energy through the carrier–phonon scattering process, we can establish a coupling between the hot and cold carriers phenomenologically by modifying their governing equations.

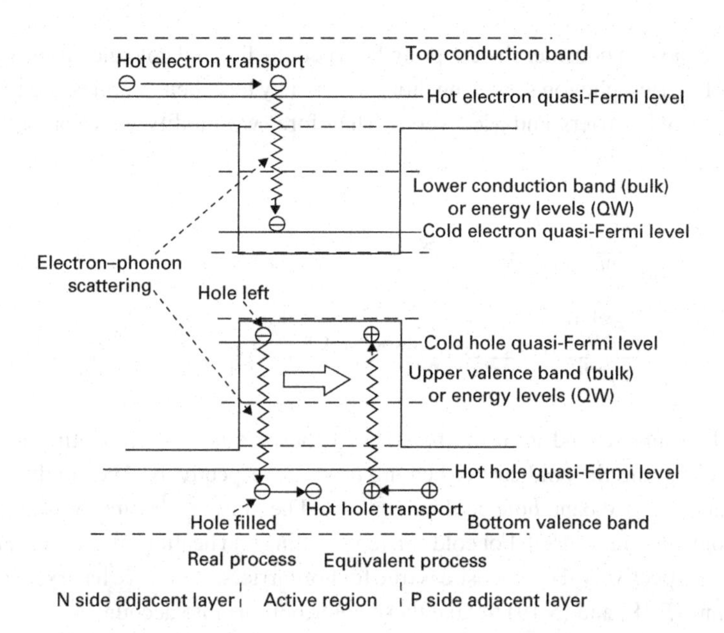

Fig. 5.1. The four-band model for the active region. The top conduction band and hot electron quasi-Fermi level are aligned with those in the N side adjacent layer, while the bottom valence band and hot hole quasi-Fermi level are aligned with those in the P side adjacent layer. Through the carrier–phonon scattering process, the hot electrons drop to the lower conduction band (bulk) or energy levels (QW) and the hot holes jump to the upper valence band (bulk) or energy levels (QW), between which the stimulated emission takes place.

Actually, without the coupling introduced, we can rewrite the general governing equations (4.28a–c) for cold carriers in a more compact form as

$$\frac{\mathrm{d}f_{\vec{k}}^{\mathrm{e/h,C}}}{\mathrm{d}t} = F^{\mathrm{e/h}}(f_{\vec{k}}^{\mathrm{e/k,C}}), \tag{5.16}$$

with $F^{\mathrm{e/h}}(\ldots)$ representing the functions on the RHS of equations (4.28b&c), respectively. Similarly, the microscopic governing equations for hot carriers can be written as

$$\frac{\mathrm{d}f_{\vec{k}}^{\mathrm{e/h,H}}}{\mathrm{d}t} = G^{\mathrm{e/h}}(f_{\vec{k}}^{\mathrm{e/k,H}}) - \gamma_{\mathrm{H}}^{\mathrm{e/h}}(f_{\vec{k}}^{\mathrm{e/k,H}} - \overline{f}_{\vec{k}}^{\mathrm{e/k,H}}), \tag{5.17}$$

with $G^{\mathrm{e/h}}(\ldots)$ representing a certain function which converges to the RHS of equations (5.15b&c) after a summation in the wave vector domain, $\gamma_{\mathrm{H}}^{\mathrm{e/h}}$ a phenomenologically introduced damping factor, and $\overline{f}_{\vec{k}}^{\mathrm{e/k,H}}$ the hot electron and hole steady state quasi-Fermi distributions, respectively. If we make a summation over all of the states in the wave vector domain, equation (5.17) reduces to (5.15b&c) for the hot electrons and holes, respectively.

Now, we have to consider the coupling between the hot and cold electrons and holes. Because of carrier–phonon scattering, hot carriers may lose their energies and turn themselves into cold carriers and vice versa. Therefore, we modify equations (5.16) and (5.17) to

$$\frac{\mathrm{d}f_{\vec{k}}^{\mathrm{e/h,C}}}{\mathrm{d}t} = \frac{f_{\vec{k}}^{\mathrm{e/h,H}}}{\tau_{\mathrm{cl}}^{\mathrm{e/h}}} - \frac{f_{\vec{k}}^{\mathrm{e/h,C}}}{\tau_{\mathrm{ht}}^{\mathrm{e/h}}} + F^{\mathrm{e/h}}(f_{\vec{k}}^{\mathrm{e/k,C}}), \tag{5.18}$$

$$\frac{\mathrm{d}f_{\vec{k}}^{\mathrm{e/h,H}}}{\mathrm{d}t} = -\frac{f_{\vec{k}}^{\mathrm{e/h,H}}}{\tau_{\mathrm{cl}}^{\mathrm{e/h}}} + \frac{f_{\vec{k}}^{\mathrm{e/h,C}}}{\tau_{\mathrm{ht}}^{\mathrm{e/h}}} + G^{\mathrm{e/h}}(f_{\vec{k}}^{\mathrm{e/k,H}}) - \gamma_{\mathrm{H}}^{\mathrm{e/h}}(f_{\vec{k}}^{\mathrm{e/k,H}} - \overline{f}_{\vec{k}}^{\mathrm{e/k,H}}), \tag{5.19}$$

with the first and second term on the RHS of these equations accounting for the hot carrier cooling and the cold carrier heating process, respectively. The conduction band electron and valence band hole cooling ($\tau_{\mathrm{cl}}^{\mathrm{e/h}}$) and heating ($\tau_{\mathrm{ht}}^{\mathrm{e/h}}$) time constant are phenomenologically introduced. For cold carriers, cooling and heating are gaining and losing processes, respectively. The opposite is true for hot carriers. Hence we have selected signs in equations (5.18) and (5.19) to take these contributions into account.

We have no intention of solving the extra equation (5.19), hence we make a summation over the wave vector domain and divide the active region volume on both sides to obtain

$$\begin{aligned}\frac{\partial N_{\mathrm{e}}^{\mathrm{H}}(\vec{r},t)}{\partial t} &= -\frac{N_{\mathrm{e}}^{\mathrm{H}}(\vec{r},t)}{\tau_{\mathrm{cl}}^{\mathrm{e}}} + \frac{N_{\mathrm{e}}^{\mathrm{C}}(\vec{r},t)}{\tau_{\mathrm{ht}}^{\mathrm{e}}} \\ &\quad - R(\vec{r},t) - \nabla\cdot\mu_{\mathrm{e}}(\vec{r})N_{\mathrm{e}}^{\mathrm{H}}(\vec{r},t)\nabla[\Phi(\vec{r},t)+\Phi_{\mathrm{C}}(\vec{r})] \\ &\quad + \nabla\cdot D_{\mathrm{e}}(\vec{r})\nabla N_{\mathrm{e}}^{\mathrm{H}}(\vec{r},t) + \nabla\cdot D_{\mathrm{e}}^{\mathrm{T}}(\vec{r})N_{\mathrm{e}}^{\mathrm{H}}(\vec{r},t)\nabla T(\vec{r},t),\end{aligned} \tag{5.20a}$$

$$\begin{aligned}\frac{\partial N_{\mathrm{h}}^{\mathrm{H}}(\vec{r},t)}{\partial t} &= -\frac{N_{\mathrm{h}}^{\mathrm{H}}(\vec{r},t)}{\tau_{\mathrm{cl}}^{\mathrm{h}}} + \frac{N_{\mathrm{h}}^{\mathrm{C}}(\vec{r},t)}{\tau_{\mathrm{ht}}^{\mathrm{h}}} \\ &\quad - R(\vec{r},t) + \nabla\cdot\mu_{\mathrm{h}}(\vec{r})N_{\mathrm{h}}^{\mathrm{H}}(\vec{r},t)\nabla[\Phi(\vec{r},t)-\Phi_{\mathrm{V}}(\vec{r})] \\ &\quad + \nabla\cdot D_{\mathrm{h}}(\vec{r})\nabla N_{\mathrm{h}}^{\mathrm{H}}(\vec{r},t) + \nabla\cdot D_{\mathrm{h}}^{\mathrm{T}}(\vec{r})N_{\mathrm{h}}^{\mathrm{H}}(\vec{r},t)\nabla T(\vec{r},t),\end{aligned} \tag{5.20b}$$

where we have assumed that, within the time scale of interest, the hot carriers have reached their steady states with quasi-Fermi distributions. Hence equation (5.18) can be further written as

$$\frac{\mathrm{d}f_{\vec{k},\vec{r}}^{\mathrm{e/h,C}}}{\mathrm{d}t} = \frac{\overline{f}_{\vec{k},\vec{r}}^{\mathrm{e/h,H}}}{\tau_{\mathrm{cl}}^{\mathrm{e/h}}} - \frac{f_{\vec{k},\vec{r}}^{\mathrm{e/h,C}}}{\tau_{\mathrm{ht}}^{\mathrm{e/h}}} + F^{\mathrm{e/h}}(f_{\vec{k},\vec{r}}^{\mathrm{e/k,C}}), \tag{5.21}$$

where

$$\overline{f}_{\vec{k},\vec{r}}^{\mathrm{e/h,H}}(t) = \frac{1}{\mathrm{e}^{\frac{\varepsilon_{(\mathrm{e/h})\vec{k}}^{\mathrm{H}} - F_{\mathrm{H}}^{\mathrm{c/v}}(\vec{r},t)}{\mathrm{k_B}T}} + 1}, \tag{5.22}$$

with $\varepsilon_{(\mathrm{e/h})\vec{k}}^{\mathrm{H}}$ and $F_{\mathrm{H}}^{\mathrm{c/v}}$ denoted as the hot carrier conduction and valence band energies and quasi-Fermi levels measured from their band edges, respectively.

In equations (5.21) and (5.22), the space vector $\vec{r}$ can be viewed as a parameter assigned to the hot carrier quasi-Fermi distribution since at different locations inside the active region, the macroscopic hot carrier densities are generally not uniform, hence their quasi-Fermi levels are $\vec{r}$ dependent. Although the microscopic governing equations for the cold carrier number expectations $f^{e/h}_{\vec{k},\vec{r}}$ (and the polariton number expectation $p_{\vec{k},\vec{r}}$) are ODEs with no space dependence, for every space location $\vec{r}$ inside the active region, we still have to solve an individual set of ODEs in the wave vector domain, which has $\vec{r}$ dependent parameters. These $\vec{r}$ dependent parameters come from the non-uniform hot carrier coupling contribution as shown in equations (5.21) and (5.22), and from the non-uniform optical field distribution according to the general governing equations (4.28a–c).

The hot carrier conduction and valence band energies can be found by solving the single electron eigenstates in the N and P side materials adjacent to the active region, respectively. If the bands that host the hot carriers (i.e., the top conduction band and bottom valence band) are parabolic, we further have $\varepsilon^{\mathrm{H}}_{(\mathrm{e/h})\vec{k}} = \hbar^2k^2/(2m^{\mathrm{H}}_{\mathrm{e/h}})$ with $m^{\mathrm{H}}_{\mathrm{e/h}}$ indicating the hot carrier conduction and valence band effective mass, respectively.

The hot carrier quasi-Fermi levels are determined by

$$N^{\mathrm{H}}_{\mathrm{e/h}}(\vec{r},t) = \frac{1}{\Omega}\sum_{\vec{k}} \frac{1}{\mathrm{e}^{\frac{\varepsilon^{\mathrm{H}}_{(\mathrm{e/h})\vec{k}} - F^{\mathrm{c/v}}_{\mathrm{H}}(\vec{r},t)}{\mathrm{k_B}T}} + 1}, \tag{5.23}$$

with the macroscopic hot carrier densities on the LHS given as the solution of the modified classical carrier transport equations (5.20a&b). Note that in equation (5.23), the time scale must be longer than a few picoseconds, to ensure that the hot carriers are settled in their steady states before $N^{\mathrm{H}}_{\mathrm{e/h}}$ changes.

The macroscopic cold carrier densities in equation (5.20a&b) are linked to the microscopic cold carrier number expectations through

$$N^{\mathrm{C}}_{\mathrm{e/h}}(\vec{r},t) = \frac{1}{\Omega}\sum_{\vec{k}} f^{\mathrm{e/h,C}}_{\vec{k},\vec{r}}. \tag{5.24}$$

Finally, we have to modify Poisson's equation (5.15a) to take account of the contribution from both carriers

$$\begin{aligned}\nabla\cdot\tilde{\varepsilon}(\vec{r})\nabla\Phi(\vec{r},t) = -\mathrm{e}[&N^{\mathrm{C}}_{\mathrm{h}}(\vec{r},t) - N^{\mathrm{C}}_{\mathrm{e}}(\vec{r},t)\\ &+ N^{\mathrm{H}}_{\mathrm{h}}(\vec{r},t) - N^{\mathrm{H}}_{\mathrm{e}}(\vec{r},t) + N^{+}_{\mathrm{D}}(\vec{r}) - N^{-}_{\mathrm{A}}(\vec{r})].\end{aligned} \tag{5.25}$$

Now, the modified classical carrier transport equations (5.25) and (5.20a&b) for the hot carriers, and the modified microscopic many-body model equations (4.28a) and (5.21) for the cold carriers form a complete set that models the carrier behavior inside the active region. The hot and cold carriers are described in the macroscopic and microscopic domains respectively, which couple to each other through phenomenologically introduced cooling and heating damping terms. The macroscopic to microscopic domain conversion of the hot carrier is given by equations (5.22) and (5.23), whereas the microscopic to macroscopic domain conversion of the cold carrier is given by equation (5.24).

5.1.4 Simplifications of the carrier transport model

If the cooling time constant is much shorter, the first terms on the RHS of equations (5.20a&b) are dominant, hence we find

$$\frac{\partial N_{\mathrm{e/h}}^{\mathrm{H}}(\vec{r},t)}{\partial t} \approx -\frac{N_{\mathrm{e/h}}^{\mathrm{H}}(\vec{r},t)}{\tau_{\mathrm{cl}}^{\mathrm{e/h}}},$$

which means

$$N_{\mathrm{e/h}}^{\mathrm{H}}(\vec{r},t) \approx N_{\mathrm{e/h}}^{\mathrm{H}}(\vec{r},0)\mathrm{e}^{-\frac{t}{\tau_{\mathrm{cl}}^{\mathrm{e/h}}}} \to 0 \text{ as } \tau_{\mathrm{cl}}^{\mathrm{e/h}} \to 0. \tag{5.26}$$

Since the total carrier number must be conserved in the coupling process between the hot and cold carriers, we must conclude that all the hot carriers have turned themselves into cold carriers within a negligible time period scaled by $\tau_{\mathrm{cl}}^{\mathrm{e/h}}$.

Next we will introduce two different time scales, one for the classical hot carrier transport process denoted as t and the other for the microscopic cold carrier equations (4.28b&c) denoted as t'. The short time scale t' varies from 0 to T' with T' denoted as the time period required for the microscopic cold carriers to reach their static quasi-Fermi distribution. The long time scale t is conventional but any macroscopic variables including those involved in the classical carrier transport equations can be evaluated only at a discrete set $t = 0, T', 2T', 3T', \ldots$

Now instead of equation (5.21), we still have the original governing equations (4.28b&c) for the cold carrier number expectations in time scale t', but with their initial conditions given by the static hot carrier distributions with their quasi-Fermi levels determined by the total hot carrier density. As a result, equations (4.28b&c) can now be solved through

$$f_{\vec{k},\vec{r}}^{\mathrm{e/h,C}}(t'=0) = \frac{1}{\mathrm{e}^{\frac{\varepsilon_{(\mathrm{e/h})\vec{k}}^{\mathrm{H}} - F_{\mathrm{H}}^{\mathrm{c/v}}(\vec{r},t=mT')}{\mathrm{k_B}T}}+1}. \tag{5.27}$$

The quasi-Fermi level in equation (5.27) can be determined by equation (5.23) at $t = mT'$ with $m = 0, 1, 2, 3, \ldots$ Once the cold carriers reach their steady state after the time period T' in accordance with the solution of equations (4.28b&c), we can solve the classical carrier transport equation again for the next time instant in the t scale, i.e., at $t = (m+1)T'$. In repeating this process, we must retain condition (5.26), i.e., there are no hot carriers left inside the active region as they are all turned into cold carriers instantaneously. And as such, we do not need to modify Poisson's equation (5.25). The only carriers inside the active region are now the cold carriers given by equation (5.24). At the active region boundary, we must match the carrier densities in the non-active region with the cold carrier densities inside the active region

$$N_{\mathrm{e}}(\vec{r}=\vec{r}_{\mathrm{n}},t=mT')=N_{\mathrm{e}}^{\mathrm{C}}(\vec{r}=\vec{r}_{\mathrm{n}},t=mT'), \tag{5.28a}$$

$$N_{\mathrm{h}}(\vec{r}=\vec{r}_{\mathrm{p}},t=mT')=N_{\mathrm{h}}^{\mathrm{C}}(\vec{r}=\vec{r}_{\mathrm{p}},t=mT'), \tag{5.28b}$$

with $\vec{r}_{\mathrm{n/p}}$ denoting the space points at the active region boundaries adjacent to the N and P side materials, respectively, and $m = 0, 1, 2, 3, \ldots$

We must note that the cold carrier densities are still $\vec{r}$ dependent because of the non-uniform optical field distribution inside the active region according to the microscopic model equations (4.28a–c). Therefore, as a direct consequence of zero hot carriers inside the active region, we can conclude that the non-uniform (cold) carrier distribution inside the active region is dictated by the optical field only, and has nothing to do with carrier transport inside the active region.

One advantage of using this model instead of the more comprehensive model introduced in Section 5.1.3 is that the microscopic governing equations introduced in Chapter 4, i.e., equations (4.28a–c), are directly applicable. Hence the associated solution techniques are still valid. Besides, there is no need to assume any cooling and heating time constants, which are difficult to obtain anyway.

A major drawback of this model is that it completely ignores the competing cold carrier heating process, which may not be true for some device structures, especially when operated under a high lattice temperature because of the raised ambient temperature or strong injection.

In edge emitting devices, although the current and potential distributions are generally non-uniform along the device longitudinal (wave propagation) direction, which leads to non-zero $\partial\Phi/\partial z$, $\partial N_{\mathrm{e,h}}/\partial z$ and $\partial T/\partial z$ in the classical carrier transport model, in the active region where there is stimulated emission, these terms are negligible. This is because the carrier redistribution due to such a direct transport process is much slower than the redistribution brought about through the longitudinal spatial hole burning (LSHB) effect, which can be understood as photon-assisted carrier transport. Because of LSHB, the carrier density must be low in the place where the optical field is strong, since more carriers are consumed through the stimulated emission process, and vice versa. As such, the carrier density distribution is dictated by the optical field distribution along the wave propagation direction, since the optical wave propagates much faster than the direct carrier transport. That is to say, the longitudinal carrier density distribution is determined by the carrier–photon–carrier (i.e., stimulated emission/absorption–wave propagation–stimulated emission/absorption) process rather than by the direct carrier transport process. As such, the carrier transport process can be considered only in the cross-sectional area, hence equations (5.25) and (5.20a&b) all reduce to 2D in the xy plane perpendicular to the optical wave propagation direction z inside the active region

$$\begin{aligned}\nabla_t\cdot\tilde{\varepsilon}(\vec{r})\nabla_t\Phi(\vec{r},t)=-\mathrm{e}[&N_{\mathrm{h}}^{\mathrm{C}}(\vec{r},t)-N_{\mathrm{e}}^{\mathrm{C}}(\vec{r},t)+N_{\mathrm{h}}^{\mathrm{H}}(\vec{r},t)\\&-N_{\mathrm{e}}^{\mathrm{H}}(\vec{r},t)+N_{\mathrm{D}}^{+}(\vec{r})-N_{\mathrm{A}}^{-}(\vec{r})],\end{aligned} \tag{5.29a}$$

$$\frac{\partial N_{\rm e}^{\rm H}(\vec{r},t)}{\partial t} = -\frac{N_{\rm e}^{\rm H}(\vec{r},t)}{\tau_{\rm cl}^{\rm e}} + \frac{N_{\rm e}^{\rm C}(\vec{r},t)}{\tau_{\rm ht}^{\rm e}} - R(\vec{r},t)$$
$$- \nabla_t \cdot \mu_{\rm e}(\vec{r}) N_{\rm e}^{\rm H}(\vec{r},t)$$
$$\times \nabla_t [\Phi(\vec{r},t) + \Phi_{\rm C}(\vec{r})] + \nabla_t \cdot D_{\rm e}(\vec{r}) \nabla_t N_{\rm e}^{\rm H}(\vec{r},t)$$
$$+ \nabla_t \cdot D_{\rm e}^{\rm T}(\vec{r}) N_{\rm e}^{\rm H}(\vec{r},t) \nabla_t T(\vec{r},t), \tag{5.29b}$$

$$\frac{\partial N_{\rm h}^{\rm H}(\vec{r},t)}{\partial t} = -\frac{N_{\rm h}^{\rm H}(\vec{r},t)}{\tau_{\rm cl}^{\rm h}} + \frac{N_{\rm h}^{\rm C}(\vec{r},t)}{\tau_{\rm ht}^{\rm h}} - R(\vec{r},t)$$
$$+ \nabla_t \cdot \mu_{\rm h}(\vec{r}) N_{\rm h}^{\rm H}(\vec{r},t)$$
$$\times \nabla_t [\Phi(\vec{r},t) - \Phi_{\rm V}(\vec{r})] + \nabla_t \cdot D_{\rm h}(\vec{r}) \nabla_t N_{\rm h}^{\rm H}(\vec{r},t)$$
$$+ \nabla_t \cdot D_{\rm h}^{\rm T}(\vec{r}) N_{\rm h}^{\rm H}(\vec{r},t) \nabla_t T(\vec{r},t), \tag{5.29c}$$

with $\nabla_t \equiv \vec{x}\partial/\partial x + \vec{y}\partial/\partial y$. Because of the non-uniformity of the optical wave intensity along the wave propagation direction z, the cold carrier densities $N_{\rm e/h}^{\rm C}$ will generally be z dependent. Therefore, every variable in equations (5.29a–c) still has z dependence, although the carrier transport effect along z is ignored. However, for uniform device structures along the waveguide direction z, the material parameters such as the mobilities, the carrier and thermal diffusivities, and the heterojunction band edge variations are z independent. In this sense, the 2D carrier transport equations (5.29a–c) are valid individually in each cross-sectional sheet (i.e., the xy plane) along the optical wave propagation direction (z) and are coupled to each other only through the optical wave from sheet to sheet.

5.1.5 The free-carrier transport model

Under the free-carrier assumption, we can rewrite the microscopic governing equations (4.43a&b) for the cold carriers by replacing the direct injection term $\Lambda_{\vec{k}}^{\rm e/h}$ with the hot–cold carrier coupling term as shown in equation (5.21)

$$\frac{{\rm d} f_{\vec{k},\vec{r}}^{\rm e/h,C}}{{\rm d}t} = \frac{\overline{f}_{\vec{k},\vec{r}}^{\rm e/h,H}}{\tau_{\rm cl}^{\rm e/h}} - \frac{f_{\vec{k},\vec{r}}^{\rm e/h,C}}{\tau_{\rm ht}^{\rm e/h}} + \frac{2}{\hbar}{\rm Im}[(\vec{\mu}_{\vec{k}}^{*} \cdot \vec{E}) p_{\vec{k}}]$$
$$- B_{\vec{k}} f_{\vec{k}}^{\rm e,C} f_{\vec{k}}^{\rm h,C} - A_{\rm e} f_{\vec{k}}^{\rm e/h,C} - \gamma_a (f_{\vec{k}}^{\rm e/h,C} - \overline{f}_{\vec{k}}^{\rm e/h,C}). \tag{5.30}$$

Making a summation over the wave vector domain and dividing the active region volume on both sides of equation (5.30) yields

$$\frac{\partial N_{\rm e/h}^{\rm C}(\vec{r},t)}{\partial t} = \frac{N_{\rm e/h}^{\rm H}(\vec{r},t)}{\tau_{\rm cl}^{\rm e/h}} - \frac{N_{\rm e/h}^{\rm C}(\vec{r},t)}{\tau_{\rm ht}^{\rm e/h}} + \frac{\vec{E}(\vec{r},t)}{\hbar} \cdot {\rm Im}[\vec{\tilde{P}}(\vec{r},t){\rm e}^{-{\rm j}\omega t}]$$
$$- B N_{\rm e}^{\rm C}(\vec{r},t) N_{\rm h}^{\rm C}(\vec{r},t) - A_{\rm e/h} N_{\rm e/h}^{\rm C}(\vec{r},t), \tag{5.31}$$

where equations (5.22)–(5.24), (4.47) and (4.53) have been used.

If we define the stimulated emission and non-radiative plus spontaneous emission recombination rates as [5–10]

$$R_{st}(\vec{r},t) \equiv -\frac{\vec{E}(\vec{r},t)}{\hbar}\cdot \mathrm{Im}[\tilde{\vec{P}}(\vec{r},t)\mathrm{e}^{-\mathrm{j}\omega t}], \tag{5.32a}$$

$$R^{\mathrm{e/h}}_{\mathrm{nr+sp}}(\vec{r},t) \equiv A_{\mathrm{e/h}}N^{\mathrm{C}}_{\mathrm{e/h}}(\vec{r},t) + BN^{\mathrm{C}}_{\mathrm{e}}(\vec{r},t)N^{\mathrm{C}}_{\mathrm{h}}(\vec{r},t), \tag{5.32b}$$

and equation (5.31) becomes

$$\frac{\partial N^{\mathrm{C}}_{\mathrm{e}}(\vec{r},t)}{\partial t} = \frac{N^{\mathrm{H}}_{\mathrm{e}}(\vec{r},t)}{\tau^{\mathrm{e}}_{\mathrm{cl}}} - \frac{N^{\mathrm{C}}_{\mathrm{e}}(\vec{r},t)}{\tau^{\mathrm{e}}_{\mathrm{ht}}} - R_{st}(\vec{r},t) - R^{\mathrm{e}}_{\mathrm{nr+sp}}(\vec{r},t), \tag{5.33a}$$

$$\frac{\partial N^{\mathrm{C}}_{\mathrm{h}}(\vec{r},t)}{\partial t} = \frac{N^{\mathrm{H}}_{\mathrm{h}}(\vec{r},t)}{\tau^{\mathrm{h}}_{\mathrm{cl}}} - \frac{N^{\mathrm{C}}_{\mathrm{h}}(\vec{r},t)}{\tau^{\mathrm{h}}_{\mathrm{ht}}} - R_{st}(\vec{r},t) - R^{\mathrm{h}}_{\mathrm{nr+sp}}(\vec{r},t). \tag{5.33b}$$

Finally, either equations (5.20a&b) and (5.25), or equations (5.29a–c) together with equations (5.33a&b), form a complete carrier transport model inside the active region under the free-carrier assumption.

For the active region of bulk semiconductors, the hot and cold electrons and holes stay in the same conduction and valence band but with different wave vectors and kinetic energies, respectively. As such, the transition between hot and cold carriers takes negligible time compared with any other processes. Therefore, it becomes unnecessary to separate the hot and cold carriers. Actually, by adding equations (5.20a) and (5.33a), (5.20b) and (5.33b), respectively, we obtain

$$\begin{aligned}\frac{\partial N_{\mathrm{e}}(\vec{r},t)}{\partial t} &= -R_{st}(\vec{r},t) - R^{\mathrm{e}}_{\mathrm{nr+sp}}(\vec{r},t) - \nabla\cdot\mu_{\mathrm{e}}(\vec{r})N_{\mathrm{e}}(\vec{r},t)\nabla[\Phi(\vec{r},t)+\Phi_{\mathrm{C}}(\vec{r})] \\ &\quad + \nabla\cdot D_{\mathrm{e}}(\vec{r})\nabla N_{\mathrm{e}}(\vec{r},t) + \nabla\cdot D^{\mathrm{T}}_{\mathrm{e}}(\vec{r})N_{\mathrm{e}}(\vec{r},t)\nabla T(\vec{r},t),\end{aligned} \tag{5.34a}$$

$$\begin{aligned}\frac{\partial N_{\mathrm{h}}(\vec{r},t)}{\partial t} &= -R_{st}(\vec{r},t) - R^{\mathrm{h}}_{\mathrm{nr+sp}}(\vec{r},t) + \nabla\cdot\mu_{\mathrm{h}}(\vec{r})N_{\mathrm{h}}(\vec{r},t)\nabla[\Phi(\vec{r},t)-\Phi_{\mathrm{V}}(\vec{r})] \\ &\quad + \nabla\cdot D_{\mathrm{h}}(\vec{r})\nabla N_{\mathrm{h}}(\vec{r},t) + \nabla\cdot D^{\mathrm{T}}_{\mathrm{h}}(\vec{r})N_{\mathrm{h}}(\vec{r},t)\nabla T(\vec{r},t),\end{aligned} \tag{5.34b}$$

where we have defined the total electron and hole densities inside the bulk active region as

$$N_{\mathrm{e/h}}(\vec{r},t) \equiv N^{\mathrm{C}}_{\mathrm{e/h}}(\vec{r},t) + N^{\mathrm{H}}_{\mathrm{e/h}}(\vec{r},t). \tag{5.35}$$

In deriving equations (5.34a&b), we have merged the non-radiative and spontaneous emission recombination rates from both hot and cold carriers and have utilized the fact that the hot carriers make no contribution to the stimulated emission rate, whereas the cold carriers make no contribution to the current density divergence, i.e.,

$$\nabla\cdot\vec{J}^{\mathrm{C}}_{\mathrm{e/h}} = -\mathrm{e}\mu_{\mathrm{e/h}}N^{\mathrm{C}}_{\mathrm{e/h}}\nabla[\Phi\pm\Phi_{\mathrm{C/V}}] \pm \mathrm{e}D_{\mathrm{e/h}}\nabla N^{\mathrm{C}}_{\mathrm{e/h}} \pm \mathrm{e}D^{\mathrm{T}}_{\mathrm{e/h}}N^{\mathrm{C}}_{\mathrm{e/h}}\nabla T = 0,$$

according to equations (5.33a&b).

Accordingly, Poisson's equation (5.25) becomes

$$\nabla \cdot \tilde{\varepsilon}(\vec{r})\nabla\Phi(\vec{r},t) = -\mathrm{e}[N_\mathrm{h}(\vec{r},t) - N_\mathrm{e}(\vec{r},t) + N_\mathrm{D}^+(\vec{r}) - N_\mathrm{A}^-(\vec{r})]. \tag{5.36}$$

Equations (5.36) and (5.34a&b) form the set of governing equations in the free-carrier model for the bulk active region [11, 12]. They take the same form as the governing equations for the non-active region as shown in equations (5.15a–c), with the only difference being an extra term added to account for the contribution of the stimulated emission to the carrier consumption, which exists only inside the active region.

For the active region of a semiconductor QW structure, the hot carriers stay in the continuum bands, whereas the cold carriers are at the discrete energy levels bound by the QW. As a result, there might be noticeable electron and hole cooling and heat time constants, known as the capture (by the QW) and escape (from the QW) times [13–22]. Therefore, we have to keep the hot and cold carriers separated and solve the Poisson equation (5.25) or (5.29a), the hot carrier transport equations (5.20a&b) or (5.29b&c), and the cold carrier rate equations (5.33a&b) simultaneously under the 3D or 2D carrier transport case, respectively.

5.1.6 Recombination rates

As mentioned previously in this chapter, carriers are consumed through radiative and non-radiative processes: the radiative process comprises stimulated and spontaneous emissions, only the former interacting with the optical wave guided by the device.

Carrier stimulated emission recombination happens only inside the active region and is described by equation (5.32a). Utilizing the relation between the slow-varying envelopes of the polarization and the optical field, equation (4.64), and noting that it holds at every space point, i.e., $\tilde{P}(\leftharpoondown r,t) = \varepsilon_0\tilde{\chi}(\vec{r},\omega)\tilde{E}_\mu(\vec{r},t)$, we obtain

$$\begin{aligned} R_\mathrm{st}(\vec{r},t) &= -\frac{|\tilde{E}_\mu(\vec{r},t)|\cos(\omega t - \varphi_{\tilde{E}})}{\hbar}\mathrm{Im}[\varepsilon_0\tilde{\chi}(\vec{r},\omega)\tilde{E}_\mu(\vec{r},t)\mathrm{e}^{-\mathrm{j}\omega t}] \\ &= -\frac{\varepsilon_0|\tilde{E}_\mu(\vec{r},t)|^2}{\hbar}\mathrm{Im}[\tilde{\chi}(\vec{r},\omega)\mathrm{e}^{-\mathrm{j}(\omega t-\varphi_{\tilde{E}})}\cos(\omega t - \varphi_{\tilde{E}})] \\ &= -\frac{\varepsilon_0|\tilde{E}_\mu(\vec{r},t)|^2}{2\hbar}\{\mathrm{Im}[\tilde{\chi}(\vec{r},\omega)] + \mathrm{Im}[\tilde{\chi}(\vec{r},\omega)\mathrm{e}^{-\mathrm{j}2(\omega t-\varphi_{\tilde{E}})}]\}, \end{aligned} \tag{5.37}$$

with $\varphi_{\tilde{E}}$ denoted as the phase of $\tilde{E}_\mu$. In deriving equation (5.37), we have explicitly shown the space and slow-varying time dependences of $\tilde{P}$ and $\tilde{E}_\mu$.

Noting that carriers cannot follow the fast change in the second term on the RHS of equation (5.37) (in order to be consistent with the rotating-wave approximation) and following equation (2.104), we find

$$R_\mathrm{st}(\vec{r},t) \approx -\frac{\varepsilon_0|\tilde{E}_\mu(\vec{r},t)|^2}{2\hbar}\mathrm{Im}[\tilde{\chi}(\vec{r},\omega)] = \frac{n}{2\hbar\omega_0}\sqrt{\left(\frac{\varepsilon_0}{\mu_0}\right)}g(\vec{r},\omega)|\tilde{E}_\mu(\vec{r},t)|^2. \tag{5.38}$$

The spontaneous emission and non-radiative recombinations happen in both the active and non-active regions. Several mechanisms contribute to the non-radiative recombinations. Other than phenomenologically introduced carrier decay, known as the Shockley–Read–Hall (SRH) recombination, as shown in equation (5.32b) there are still Auger recombinations, including band to bound state transitions and band to band Auger impact ionizations [5, 7, 23, 24]. The contribution of the former processes to the recombination rate has its highest order proportional to the carrier density squared, whereas the contribution of the latter processes to the recombination rate has its highest order proportional to the carrier density cubed. Therefore, we merge the contribution of the former processes with that of the SRH and spontaneous emission recombinations, and add the latter contribution to the non-radiative plus spontaneous emission recombination rate. Then equation (5.32b) is modified to

$$R^{e/h}_{nr+sp}(\vec{r},t) = A_{e/h}N_{e/h}(\vec{r},t) + BN_e(\vec{r},t)N_h(\vec{r},t) + C_{e/h}N_e(\vec{r},t)N_h(\vec{r},t)N_{e/h}(\vec{r},t), \quad (5.39)$$

with $C_{e/h}$ introduced as the electron and hole Auger recombination constants, respectively.

However, equation (5.39) gives generally different non-radiative plus spontaneous emission recombination rates for electrons and holes, which is in conflict with the condition which applied when we split equation (5.10) into (5.12a&b). This problem comes from the fact that equation (5.39) only considers electron and hole capture, which is a consumption of the electrons and holes of interest. The recombination rate in equations (5.12a&b), however, is introduced as a net contribution that accounts for the rate of loss minus gain of electrons and holes through these non-radiative processes. Only the spontaneous emission recombination is automatically balanced as electrons and holes are consumed in pairs through this process. To solve this problem, we must include the electron and hole emission in the SRH and Auger recombinations as well, which gives the gain in the electrons and holes of interest. Therefore, by considering the net carrier consumption, we obtain the non-radiative plus spontaneous emission recombination rate as

$$R_{nr+sp}(\vec{r},t) = [N_e(\vec{r},t)N_h(\vec{r},t) - N_{e0}(\vec{r})N_{h0}(\vec{r})] \times \left[\frac{1}{\tau_e N_h(\vec{r},t) + \tau_h N_e(\vec{r},t) + N_{e0}(\vec{r})\tau_e A_e/A_h + N_{h0}(\vec{r})\tau_h A_h/A_e} + B + C_e N_e(\vec{r},t) + C_h N_h(\vec{r},t)\right], \quad (5.40)$$

with $A_{e/h}$ in $1/s$, $\tau_{e/h}$ in s, B in m^3/s and $C_{e/h}$ in m^6/s defined as the electron and hole SRH decay rate, SRH time constant, spontaneous emission and Auger recombination coefficients, and $N_{e0/h0}$ the electron and hole densities at their thermal equilibrium states, respectively. Actually, the product of the electron and hole densities at thermal equilibrium is usually negligibly small. For properly designed optoelectronic devices, the non-radiative plus spontaneous emission recombination is approximately zero in the non-active region, since in the N side material the hole density N_h is low whereas in the P side material the electron density N_e is low, which leads to a small product of the electron and hole densities $N_e N_h$ everywhere outside the active region. Inside the active region, however, both the electrons and holes have high densities, hence the non-radiative plus spontaneous emission recombination cannot be significant according to equation (5.40).

All of the recombination rates (R_{st} and R_{nr+sp}) have dimensions $1/m^3s$.

5.2 The carrier rate equation model

As mentioned in Section 5.1.4, the carrier distribution inside the active region is dictated by the distribution of optical wave intensity due to the LSHB. If the carrier transport effect in the cross-sectional region is negligible, we just need a 1D model along the longitudinal (i.e., the wave propagation) direction. Actually, a carrier rate equation can be extracted phenomenologically through balancing the carrier generation and recombination rates of time-dependent processes, as energy must be conserved for any process.

It is obvious that the carrier change rate must be equal to the total carrier generation rates minus the total carrier consumption rates. The carrier change rate is its time derivative. Carriers can be generated only through current injection, whereas both stimulated emission and non-radiative plus spontaneous emission recombinations consume carriers. Therefore, we obtain the following carrier rate equation

$$\frac{\partial N(z,t)}{\partial t} = \frac{\eta J(z,t)}{ed} - [AN(z,t) + BN^2(z,t) + CN^3(z,t)] - \frac{n_{eff}}{2\hbar\omega_0}\sqrt{\left(\frac{\varepsilon_0}{\mu_0}\right)}\frac{\Gamma}{\Sigma_{ar}} g(z,\omega_0)|e^f(z,t)e^{j\beta_0 z} + e^b(z,t)e^{-j\beta_0 z}|^2, \quad (5.41)$$

where N is the minority carrier density in $1/m^3$ inside the active region, J the injection current density in A/m^2, η the injection efficiency, A $1/s$, B m^3/s and C m^6/s the minority carrier SRH, spontaneous emission, and Auger recombination coefficients, respectively. In addition, n_{eff} denotes the effective index of the guided optical mode, Γ the confinement factor, d the active region thickness in $1/m$, and Σ_{ar} the cross-sectional area of the active region in $1/m^2$, respectively.

In equation (5.41) we have assumed a charge neutrality condition inside the active region so that the electron and hole densities are equalized [25]. We have also modified the non-radiative plus spontaneous emission recombination rate accordingly by merging the electron and hole contributions and dropping the product of electron and hole density at the thermal equilibrium. We have used equation (5.38) for the stimulated emission recombination rate with $\widetilde{E}_\mu$ substituted by the slow-varying envelope function of the optical field in Chapter 2. Since the reference frequency of the optical field is now given at ω_0, we have to take the material optical gain at the same frequency. Our intention in using an arbitrary reference frequency ω in Chapter 4 and in Section 5.1 is to find the whole material gain profile and refractive index change dispersion curve. In deriving the optical wave and carrier governing equations, however, we need to deal with just one (or a few) specific reference frequency at which the device operates.

In looking at the device performance and the device cavity (i.e., the structure along the longitudinal direction) design problems, equation (5.41) is often used in conjunction with the optical governing equations given in Chapter 2. In this approach the material optical gain is supplied by semi-analytical expressions obtained in Chapter 4 from the free-carrier gain model or many-body gain models under Pade approximations (with only $\vec{k}$ domain summations or converted integrals involved), or even by analytical formulas that are further extracted or simplified from these expressions, as will be seen in Chapter 7. Similarly to the design of the cross-sectional structure including the active region, we can work on the material gain model (for the active region design), the 2D carrier transport model in the cross-section (for current injection design through the PN junction), and the 2D optical field eigenvalue problem in the cross-section (for the optical wave guide design) separately. In device modeling, these calculations then provide the effective parameters for the longitudinal 1D equations. As such, we just need to handle a few individual 2D problems (for gain, carrier and optical field) in the cross-section plus a 1D self-consistent problem with every aspect (i.e., gain, carrier, and optical field in cross-section) incorporated along the longitudinal (wave propagation) direction, instead of solving the full 3D problem with every aspect coupled together. This method seems to offer us an excellent balance between accuracy and efficiency in practice. As will be seen in examples given in Chapters 10, 11, and 12, we will mainly rely on this approach to model our optoelectronic device.

Finally, it is worth mentioning that, if there exists a gain grating inside the active region as given by equation (2.74), the carrier density will exhibit a grating pattern as well because of the LSHB effect according to equation (5.41) [26, 27]. Although the carrier transport process tends to smear out such rapid local change, it cannot catch up with the burning speed. Hence a static carrier grating pattern is sustained inside the cavity. Since the carriers contribute to the gain, the gain coupling strength becomes field dependent. As such, the gain-coupled grating becomes non-linear as its coupling coefficient depends on the optical field intensity. This effect, on the one hand, complicates device operation, on the other hand it leaves us increased flexibility to tailor the structure and operating conditions for better performance [28–33].

5.3 The thermal diffusion model

5.3.1 The classical thermal diffusion model

In the material and carrier transport models we have discussed so far, the temperature has been assumed as a constant everywhere, which is unfortunately not true. Because of various non-uniformly distributed heating sources, the temperature inside the optoelectronic device is not uniform and is unbalanced with respect to the ambient temperature, especially under a high current injection. Since the material gain and refractive index change are both sensitive functions of temperature, it is crucial to find the accurate temperature distribution inside the device.

The temperature distribution is governed by the classical thermal diffusion equation [34]

$$\rho_T(\vec{r})C_T(\vec{r})\frac{\partial T(\vec{r},t)}{\partial t} = \nabla \cdot \kappa_T(\vec{r})\nabla T(\vec{r},t) + H(\vec{r},t), \tag{5.42}$$

where ρ_T is the density of the semiconductor in $\mathrm{kg/m^3}$, C_T the specific heat capacity in J/K kg and κ_T the thermal conductivity in W/Km.

Both the lattice and carriers contribute to the specific heat capacity and thermal conductivity, but in semiconductors, the carrier contributions are negligible [35], which greatly simplifies the solution technique required to solve equation (5.42). With the carrier contributions ignored, the total specific heat capacity and thermal conductivity are replaced by the lattice specific heat capacity and thermal conductivity constants which can be found from the semiconductor material databases [36, 37] directly.

In equation (5.42), $H(\vec{r},t)$ is the heat generation rate that serves as the driving force and is measured by the power density in $\mathrm{W/m^3}$. Equation (5.42) shows a power density balance as required by the energy conservation condition. The heating sources inside the device include Joule heating, recombination heating, Thomson heating, and optical absorption heating. The Thomson heating is in a higher order, thus it is negligible. With the rest of the heating sources taken into consideration, we can express the heat generation rate as

$$\begin{aligned} H(\vec{r},t) = {} & \frac{|J_e(\vec{r},t)|^2}{e\mu_e N_e(\vec{r},t)} + \frac{|J_h(\vec{r},t)|^2}{e\mu_h N_h(\vec{r},t)} \\ & + R_{nr+sp}(\vec{r},t)\{F^c(\vec{r},t) - F^v(\vec{r},t) + T(\vec{r},t)[P_e(\vec{r}) + P_h(\vec{r})]\} \\ & + (1-\eta_{ext})R_{st}(\vec{r},t)\{F^c(\vec{r},t) - F^v(\vec{r},t) + T(\vec{r},t)[P_e(\vec{r}) + P_h(\vec{r})]\}, \end{aligned} \tag{5.43}$$

with $J_{e/h}$ denoting the electron and hole current density, η_{ext} the total external quantum efficiency, $F^{c/v}$ the conduction band electron and valence band hole quasi-Fermi level, and $P_{e/h}$ the electron and hole thermoelectric power in J/K, respectively.

Once the electron and hole carrier densities are obtained, the electron and hole current densities can readily be obtained from the drift and diffusion model, equations (5.13a&b). The electron and hole thermoelectric powers can be linked to the electron and hole thermal diffusivities through

$$P_{e/h} = \frac{eD^T_{e/h}}{\mu_{e/h}}. \tag{5.44}$$

In equation (5.43), the first two terms on the RHS account for Joule heating due to current flow through the non-active region. The third term on the RHS describes recombination heating through the non-radiative and spontaneous emission process.

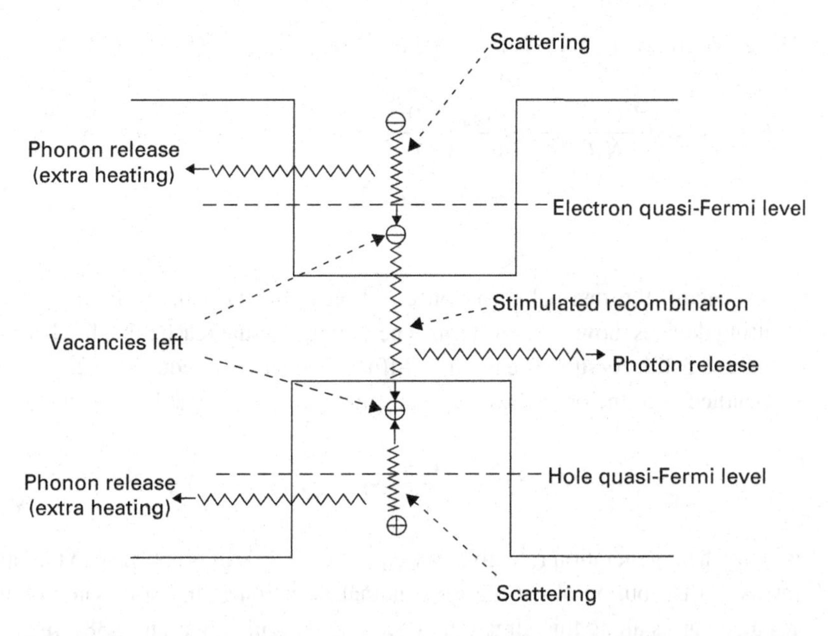

Fig. 5.2. Extra heating through carrier cooling. The non-radiative and radiative recombination processes leave vacancies below the electron quasi-Fermi level and above the hole quasi-Fermi level. The above quasi-Fermi level hot electrons and below quasi-Fermi level hot holes fill the vacancies by giving up their energies to the lattice (i.e., through phonon release), which results in extra heating.

The last term on the RHS denotes optical absorption heating. Note also that it is not only the conduction band electrons below the quasi-Fermi level F^c and the valence band holes above the quasi-Fermi level F^v that contribute to heat generation through non-radiative recombinations and part of the radiative recombinations (through spontaneous and stimulated emissions) because of optical absorption. The conduction band electrons above the quasi-Fermi level F^c and valence band holes below the quasi-Fermi level F^v also contribute. This is because the non-radiative, spontaneous emission and stimulated emission leave vacancies for electrons below F^c and for holes above F^v, therefore, electrons above F^c and holes below F^v have the chance to occupy these vacancies by releasing their energies to the lattice, as illustrated by Fig. 5.2. Hence extra heat is generated through such a carrier cooling process. This extra contribution is accounted for by the terms related to thermoelectric power on the RHS of equation (5.43).

Actually, by utilizing the relations between thermal diffusivity, carrier diffusivity and carrier mobility revealed in Section 5.1.2, we find

$$\begin{aligned} T(\vec{r},t)[P_e(\vec{r})+P_h(\vec{r})] &= eT(\vec{r},t)\left[\frac{D_e^T(\vec{r})}{\mu_e(\vec{r})}+\frac{D_h^T(\vec{r})}{\mu_h(\vec{r})}\right] \\ &= \frac{e}{2}\left[\frac{D_e(\vec{r})}{\mu_e(\vec{r})}+\frac{D_h(\vec{r})}{\mu_h(\vec{r})}\right] = k_B T(\vec{r},t). \end{aligned} \tag{5.45}$$

Therefore equation (5.43) is finally reduced to

$$H(\vec{r},t) = \frac{|J_{\rm e}(\vec{r},t)|^2}{e\mu_{\rm e}N_{\rm e}(\vec{r},t)} + \frac{|J_{\rm h}(\vec{r},t)|^2}{e\mu_{\rm h}N_{\rm h}(\vec{r},t)} + [R_{\rm nr+sp}(\vec{r},t) + (1-\eta_{\rm ext})R_{\rm st}(\vec{r},t)][F^{\rm c}(\vec{r},t) - F^{\rm v}(\vec{r},t) + k_{\rm B}T(\vec{r},t)]. \quad (5.46)$$

In general, the thermal diffusion equation is in 3D with $\partial T/\partial z$ included. In edge emitting devices, however, this term is negligible, for the reason that has been explained in Section 5.1.4. As such, the thermal diffusion equation is reduced to 2D in the xy plane perpendicular to the optical wave propagation direction z inside the active region

$$\rho_{\rm T}(\vec{r})C_{\rm T}(\vec{r})\frac{\partial T(\vec{r},t)}{\partial t} = \nabla_t \cdot \kappa_{\rm T}(\vec{r})\nabla_t T(\vec{r},t) + H(\vec{r},t), \quad (5.47)$$

with the heat generation rate given by equation (5.46). In proportion to the optical field intensity distribution along the longitudinal direction z, the stimulated emission rate $R_{\rm st}$ also varies along this direction in accordance with equation (5.38). In addition, the electron and hole current and carrier densities are all z dependent. The heating source H, therefore, becomes z dependent. As such, every variable in equation (5.47) still has z dependence, although the direct thermal diffusion effect along z is ignored. However, for uniform device structures along the waveguide direction where the material parameters are z independent, the 2D thermal equation (5.47) holds locally in each individual cross-sectional sheet (i.e., the xy plane) along the optical wave propagation direction (z). They are coupled to each other only through the inhomogeneous heating source.

5.3.2 A one-dimensional thermal diffusion model

As a further simplification of the 2D thermal model, we can ignore the lateral (x) heat conduction by introducing adiabatic boundaries and assuming a uniform temperature distribution in this direction (x). Further, if we ignore the difference in thermal conductivity for layers with different material compositions inside the device, equation (5.47) is reduced to [38]

$$\frac{\partial T(y,z,t)}{\partial t} = \frac{\kappa_{\rm T}}{\rho_{\rm T}C_{\rm T}}\frac{\partial^2 T(y,z,t)}{\partial y^2} + \frac{1}{\rho_{\rm T}C_{\rm T}}H(y,z,t). \quad (5.48)$$

In equation (5.48), the second order derivative along the vertical direction y (i.e., along the thermal dissipation direction) can be integrated through a space Fourier transform (by assuming a constant $\rho_{\rm T}C_{\rm T}$ for all the layers) or through a transfer matrix method (if $\rho_{\rm T}C_{\rm T}$ is piecewisely uniform within each layer). Therefore, equation (5.48) becomes an ODE and can readily be used as the governing equation for a thermal model. To be able to use equation (5.48) in conjunction with the 1D optical and 1D carrier rate equation model to complete our device simulation loop, we further write the heat generation rate in a form consistent with those models

$$H(y, z, t) = \frac{|J(z, t)|^2}{\mathrm{e}\mu_r N(z, t)} + \Big[AN(z, t) + BN^2(z, t) + CN^3(z, t)$$
$$+ (1 - \eta_{\mathrm{ext}})\frac{n_{\mathrm{eff}}}{2\hbar\omega_0}\sqrt{\left(\frac{\varepsilon_0}{\mu_0}\right)}\frac{\Gamma}{\Sigma_{\mathrm{ar}}} g(z, \omega_0)|\mathrm{e}^f(z, t)\mathrm{e}^{\mathrm{j}\beta_0 z} + \mathrm{e}^b(z, t)\mathrm{e}^{-\mathrm{j}\beta_0 z}|^2\Big]$$
$$\times [E_g + k_B T(y, z, t)], \tag{5.49}$$

with the effective mobility defined as

$$\frac{1}{\mu_r} \equiv \frac{1}{\mu_e} + \frac{1}{\mu_h}. \tag{5.50}$$

Once we substitute equation (5.49) into (5.48) and integrate the resulting equation in y, we can find the temperature inside the active region for a given boundary condition (e.g., the heat sink temperature). The active region temperature then becomes z and t dependent only, hence can be solved with the carrier density, the material gain and the optical field in a self-consistent manner.

In many real applications, the device series resistance is easily measurable, therefore, we can present the heat generation rate in terms of all measurable parameters

$$H(z, t) = [I^2(z, t)R_s + \frac{E_g}{\mathrm{e}}I(z, t) - P_{\mathrm{op}}]/\Omega_{\mathrm{ar}}, \tag{5.51}$$

with R_s denoting the device series resistance in Ω, P_{op} the total output optical power in W, Ω_{ar} the active region volume in m^3, and I the bias current in A. Equation (5.51) can easily be obtained through the energy conservation condition.

References

[1] R. Stratton, Semiconductor current-flow equations (diffusion and degeneracy). *IEEE Tran. on Electron. Dev.*, **ED-19**:12 (1972), 1288–92.

[2] A. Champagne, R. Maciejko, and J. Glinski, The performance of double active region InGaAsP lasers. *IEEE Journal of Quantum Electron.*, **QE-27**:10 (1991), 2238–47.

[3] M. Gault, P. Mawby, A. R. Adams, and M. Towers, Two-dimensional simulation of constricted-mesa InGaAsP/InP buried-heterostructure lasers. *IEEE Journal of Quantum Electron.*, **QE-30**:8 (1994), 1691–700.

[4] A. R. Zakharian and J. V. Moloney, Comparison of contact design for improved stability of broad area semiconductor lasers. *IEEE Photon. Tech. Lett.*, **13**:12 (2001), 1280–82.

[5] T. Kamiya, Radiative and non-radiative recombination in semiconductors. Chapter 4 in *Handbook of Semiconductor Lasers and Photonic Integrated Circuits*, ed. Y. Suematsu and A. R. Adams. (London: Chapman & Hall, 1994).

[6] S. Wang, *Fundamentals of Semiconductor Theory and Device Physics*, 1st edn (Englewood Cliffs, NJ: Prentice, 1989).

[7] C. T. Sah, *Fundamentals of Solid-State Electronics*, 1st edn (Singapore: World Scientific, 1991).

[8] W. W. Chow and S. W. Koch, *Semiconductor Laser Fundamentals: Physics of the Gain Materials*, 1st edn (Berlin: Springer-Verlag, 1999).

[9] O. Hess and T. Kuhn, Maxwell-Bloch equations for spatially inhomogeneous semiconductor lasers. I. Theoretical formulation. *Phys. Rev. A*, **54**:4 (1996), 3347–59.

[10] E. Gehrig, O. Hess, and R. Wallenstein, Modeling of the performance of high-power diode amplifier systems with an optothermal microscopic spatio-temporal theory. *IEEE Journal of Quantum Electron.*, **QE-35**:3 (1999), 320–31.

[11] Z.-M. Li, K. M. Dzurko, A. Delage, and S. P. McAlister, A self-consistent two-dimensional model of quantum-well semiconductor lasers: optimization of a GRIN-SCH SQW laser structure. *IEEE Journal of Quantum Electron.*, **QE-28**:4 (1992), 792–803.

[12] Z.-M. Li, Two-dimensional numerical simulation of semiconductor lasers. Chapter 7 in *PIER 11 Methods for Modeling and Simulation of Guided-Wave Optoelectronic Devices: Part II Waves and Interactions*, ed. W.-P. Huang. (Cambridge, MA: EMW Publishing, 1995).

[13] N. Tessler and G. Eisenstein, On carrier injection and gain dynamics in quantum well lasers. *IEEE Journal of Quantum Electron.*, **QE-29**:6 (1993), 1586–94.

[14] N. Tessler and G. Eisenstein, Modeling carrier dynamics and small-signal modulation response in quantum-well lasers. *Optical and Quantum Electron.*, **26** (1994), S767–87.

[15] R. Nagarajan, T. Fukushima, S. W. Corzine, and J. E. Bowers, Effects of carrier transport on high-speed quantum well lasers. *Appl. Phys. Lett.*, **59**:15 (1991), 1835–7.

[16] R. Nagarajan, Carrier transport effects in quantum well lasers: an overview. *Optical and Quantum Electron.*, **26** (1994), S647–66.

[17] W. Rideout, W. F. Sharfin, E. S. Koteles, M. O. Vassell, and B Elman, Well-barrier hole burning in quantum well lasers. *IEEE Photon. Tech. Lett.*, **3**:9 (1991), 784–6.

[18] H. Hirayama, J. Yoshida, Y. Miyake, and M. Asada, Carrier capture time and its effect on the efficiency of quantum-well lasers. *IEEE Journal of Quantum Electron.*, **QE-30**:1 (1994), 54–62.

[19] M. Abou-Khalil, M. Goano, A. Champagne, and Roman Maciejko, Capture and escape in quantum wells as scattering events in Monte Carlo simulation. *IEEE Photon. Tech. Lett.*, **8**:1 (1996), 19–21.

[20] B. Deveaud, D. Morris, A. Regreny, *et al.*, Quantum-mechanical versus semiclassical capture and transport properties in quantum well laser structures. *Optical and Quantum Electron.*, **26** (1994), S679–89.

[21] C.-Y. Tsai, C.-Y. Tsai, Y.-H. Lo, and L. F. Eastman, Carrier DC and AC capture and escape times in quantum-well lasers. *IEEE Photon. Tech. Lett.*, **7**:6 (1995), 599–601.

[22] A. M. Fox, D. A. B. Miller, G. Livescu, J. E. Cunningham, and W. Y. Jin, Quantum well carrier sweep out: relation to electroabsorption and exciton saturation. *IEEE Journal of Quantum Electron.*, **QE-27**:10 (1991), 2281–95.

[23] W. L. Engl, H. K. Dirks, and B. Meinerzhagen, Device modeling. *Proc. IEEE*, **71** (1983), 10–33.

[24] S. L. Chuang, *Physics of Optoelectronic Devices*, 1st edn (New York: John Wiley & Sons, 1995).

[25] D. Sadovnikov, X. Li, and W.-P. Huang, A two-dimensional DFB laser model accounting for carrier transport effects. *IEEE Journal of Quantum Electron.*, **QE-31**:10 (1995), 1856–62.

[26] T. W. Johannes and W. Harth, Spontaneous emission rate of gain-coupled DFB lasers: loss grating against gain grating. *Electron. Lett.*, **28** (1992), 1347–9.

[27] X. Pan, B. Tromborg, H. Olesen, and H. E. Lassen, Effective linewidth enhancement factor and spontaneous emission rate of DFB lasers with gain coupling. *IEEE Photon. Tech. Lett.*, **4**:11 (1992), 1213–5.

[28] K. David, J. Buus, and R. G. Baets, Basic analysis of AR-coated, partly gain-coupled DFB lasers: the standing wave effect. *IEEE Journal of Quantum Electron.*, **QE-28**:2 (1992), 427–33.

[29] K. David, G. Morthier, P. Vankwikelberge, *et al.*, Gain-coupled DFB lasers versus index-coupled and phase-shifted DFB lasers: a comparison based on spatial hole burning corrected yield. *IEEE Journal of Quantum Electron.*, **QE-27**: 6 (1991), 1714–23.

[30] J. Hong, K. W. Leong, T. Makino, X. Li, and W.-P. Huang, Impact of random facet phases on modal properties of partly gain-coupled distributed-feedback lasers. *IEEE Journal of Select. Topics in Quantum Electron.*, **3**:2 (1997), 555–68.

[31] J. Zoz, T. W. Johannes, A. Rast, *et al.*, Dynamics and stability of complex-coupled DFB lasers with absorptive gratings. *IEEE Journal of Quantum Electron.*, **QE-31**:8 (1995), 1432–42.

[32] A. J. Lowery and D. Novak, Enhanced maximum intrinsic modulation bandwidth of complex-coupled DFB semiconductor lasers. *Electron. Lett.*, **29** (1993), 461–2.

[33] K. Kudo, J. I. Shim, K. Komori, and S. Arai, Reduction of effective linewidth enhancement factor α_{eff} of DFB lasers with complex coupling coefficients. *IEEE Photon. Tech. Lett.*, **4**:6 (1992), 531–4.

[34] H. S. Carslaw and J. C. Jaeger, *Conduction of Heat in Solids*, 1st edn (Oxford, UK: Oxford University Press, 1959).

[35] G. Wachutka, Rigorous thermodynamics treatment of heat generation and conduction in semiconductor modeling. *IEEE Trans. on Comp. Aided Design*, **9** (1990), 1141–9.

[36] Landolt-Bornstein, Vol. 17 Semiconductors, ed. O. Madelung, M. Schulz, and H. Weiss, in *Numerical Data and Fundamental Relationships in Science and Technology*, ed. K. H. Hellwege. (Berlin: Springer-Verlag, 1982).

[37] S. Adachi, *Physical Properties of III-V Semiconductor Compounds, InP, InAs, GaAs, GaP, InGaAs, and InGaAsP*, 1st edn (New York: John Wiley & Sons, 1992).

[38] X. Li and W.-P. Huang, Simulation of DFB lasers incorporating thermal effects. *IEEE Journal of Quantum Electron.*, **QE-31**:10 (1995), 1848–55.

6 Solution techniques for optical equations

6.1 The optical mode in the cross-sectional area

There are numerous optical mode solvers that deal with optical eigenvalue problems in a scalar form as shown in equation (2.34), or in a more comprehensive vectorial version [1–7], in order to obtain the optical field distribution (i.e., the optical mode) in the cross-sectional area. In contrast to dealing with a 1D slab waveguide problem where an analytical approach (such as a transfer matrix method) exists, a general 2D eigenvalue problem has to be treated numerically. Among many different numerical approaches, the finite difference method (FD) seems to be a popular one for its balance between implementation complexity, computational efficiency and accuracy.

In the 2D domain, equation (2.34) is posed as a boundary value problem of a PDE of elliptical type. Hence, discretization of equation (2.34) under the FD scheme is fairly straightforward since stability is not a concern. For example, we can always stay with the center discretization scheme to gain a second order accuracy. The boundary treatment, however, is crucial especially for those 2D structures with piecewise uniformity, which is most commonly seen in semiconductor optoelectronic devices.

At physical boundaries inside the computation domain, the refractive index is discontinuous, whereas as the solution to equation (2.34), the scalar optical mode must be continuous. Therefore, special treatments are necessary at such boundary points [8–11]. For example, in many numerical solvers we do not select any mesh grid point at the boundary points to avoid refractive index assignment at these boundaries. A different treatment is to assign the arithmetically or harmonically averaged refractive index to the boundary points, depending on the discretization scheme.

Although the optical field decays exponentially for guided modes outside the waveguide core, it extends to infinity in the cross-sectional area in dielectric waveguides. Therefore, we have to truncate the computation domain by introducing an artificial window. As such, artificial boundary conditions must be introduced to prevent any unphysical reflections of the optical wave at the computation window edge. Among the available selections, such as the transparent boundary condition (TBC) [12], the absorbing boundary condition (ABC) [13] and the perfectly matched layer (PML) boundary condition [14], the latter seems to be the best choice for its great success in solving many waveguide problems.

Numerical techniques for solving the scalar mode problem are fairly mature and so we do not intend to repeat them in this book. The solution technique for the semi-vectorial modes (e.g., the modes in a ridge waveguide structure) is virtually the same, in that we need to solve just two scalar mode equations [15, 16]. Open problems still exist for full vectorial modes in strongly guided or asymmetric waveguides where different discretization and boundary treatment skills are necessary. However, this topic is beyond the scope of this book.

6.2 Traveling wave equations

6.2.1 The finite difference method

Different traveling wave equations have been extracted in Chapter 2 to treat different device longitudinal structures. However, they all fit into the same format as shown here

$$\left(\frac{1}{v_g}\frac{\partial}{\partial t} \pm \frac{\partial}{\partial z}\right)\begin{bmatrix} F(z,t) \\ R(z,t) \end{bmatrix} = \begin{bmatrix} A^{11}(z,t) & A^{12}(z,t) \\ A^{21}(z,t) & A^{22}(z,t) \end{bmatrix}\begin{bmatrix} F(z,t) \\ R(z,t) \end{bmatrix} + \begin{bmatrix} C^{\mathrm{f}}(z,t) \\ C^{\mathrm{r}}(z,t) \end{bmatrix}, \tag{6.1}$$

where f means forward propagating and r means reverse propagating.

We will still follow a FD approach to seek the numerical solution of equation (6.1). As shown in Fig. 6.1, by setting up a mesh defined as

$$t = k\Delta t, \quad k = 0, 1, 2, \ldots, \quad z = n\Delta z, \quad n = 1, 2, 3, \ldots, N+1, \tag{6.2}$$

and by following an upwind scheme [17, 18], i.e.,

$$\begin{aligned} \frac{\partial F}{\partial t} &= \frac{F_{n+1,k+1} - F_{n+1,k}}{\Delta t} & \frac{\partial F}{\partial z} &= \frac{F_{n+1,k} - F_{n,k}}{\Delta z} \\ \frac{\partial R}{\partial t} &= \frac{R_{n-1,k+1} - R_{n-1,k}}{\Delta t} & \frac{\partial R}{\partial z} &= \frac{R_{n,k} - R_{n-1,k}}{\Delta z}, \end{aligned} \tag{6.3}$$

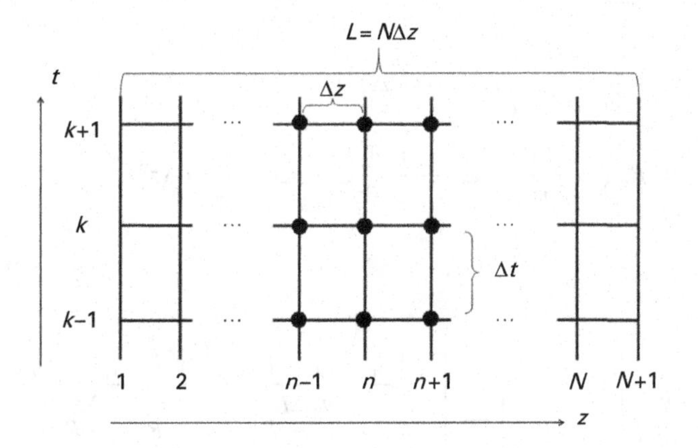

Fig. 6.1. The mesh set-up for finite difference discretization.

equation (6.1) can be discretized as

$$\begin{bmatrix} F_{n+1,k+1} \\ R_{n-1,k+1} \end{bmatrix} = (1-\eta)\begin{bmatrix} F_{n+1,k} \\ R_{n-1,k} \end{bmatrix} + \eta\begin{bmatrix} 1+A^{11}_{n,k}\Delta z & A^{12}_{n,k}\Delta z \\ A^{21}_{n,k}\Delta z & 1+A^{22}_{n,k}\Delta z \end{bmatrix}\begin{bmatrix} F_{n,k} \\ R_{n,k} \end{bmatrix} + \eta\Delta z\begin{bmatrix} C^{\mathrm{f}}_{n,k} \\ C^{\mathrm{r}}_{n,k} \end{bmatrix}, \tag{6.4}$$

where $n = 1, 2, 3, \ldots, N$ for F_n with $F_1 = R_l R_1$, and $n = 2, 3, \ldots, N+1$ for R_n with $R_{N+1} = R_r F_{N+1}$. We have also defined

$$\eta = v_{\mathrm{g}}\Delta t/\Delta z, \tag{6.5}$$

and have used subscripts n and k to represent the corresponding value sampled at the mesh point $z = n\Delta z$ and $t = n\Delta t$.

Since the upwind scheme is forward-time, backward-space (FT–BS) for both the right-going wave F and the left-going wave R, it is straightforward to prove its consistency. Actually, through the following Taylor expansions

$$F_{n+1,k+1} = F_{n,k} + \left.\frac{\partial F}{\partial t}\right|_{n,k}\Delta t + \frac{1}{2}\left.\frac{\partial^2 F}{\partial t^2}\right|_{n,k}\Delta t^2 + O(\Delta t^3) + \left.\frac{\partial F}{\partial z}\right|_{n,k}\Delta z + \frac{1}{2}\left.\frac{\partial^2 F}{\partial z^2}\right|_{n,k}\Delta z^2 + O(\Delta z^3)$$

$$F_{n+1,k} = F_{n,k} + \left.\frac{\partial F}{\partial z}\right|_{n,k}\Delta z + \frac{1}{2}\left.\frac{\partial^2 F}{\partial z^2}\right|_{n,k}\Delta z^2 + O(\Delta z^3),$$

$$R_{n-1,k+1} = R_{n,k} + \left.\frac{\partial R}{\partial t}\right|_{n,k}\Delta t + \frac{1}{2}\left.\frac{\partial^2 R}{\partial t^2}\right|_{n,k}\Delta t^2 + O(\Delta t^3) - \left.\frac{\partial R}{\partial z}\right|_{n,k}\Delta z + \frac{1}{2}\left.\frac{\partial^2 R}{\partial z^2}\right|_{n,k}\Delta z^2 + O(\Delta z^3)$$

$$R_{n-1,k} = R_{n,k} - \left.\frac{\partial R}{\partial z}\right|_{n,k}\Delta z + \frac{1}{2}\left.\frac{\partial^2 R}{\partial z^2}\right|_{n,k}\Delta z^2 + O(\Delta z^3),$$

we find

$$\left(\frac{1}{v_{\mathrm{g}}}\frac{\partial}{\partial t} \pm \frac{\partial}{\partial z}\right)\begin{bmatrix} F(z,t) \\ R(z,t) \end{bmatrix} \to \begin{bmatrix} \dfrac{F_{n+1,k+1} - F_{n+1,k}}{v_{\mathrm{g}}\Delta t} + \dfrac{F_{n+1,k} - F_{n,k}}{\Delta z} \\ \dfrac{R_{n-1,k+1} - R_{n-1,k}}{v_{\mathrm{g}}\Delta t} - \dfrac{R_{n,k} - R_{n-1,k}}{\Delta z} \end{bmatrix}$$

$$= \begin{bmatrix} \frac{1}{v_g} \frac{\partial F}{\partial t}\Big|_{n,k} + \frac{1}{2v_g} \frac{\partial^2 F}{\partial t^2}\Big|_{n,k} \Delta t + O(\Delta t^2) \\ +O\left(\frac{\Delta z^3}{\Delta t}\right) + \frac{\partial F}{\partial z}\Big|_{n,k} + \frac{1}{2} \frac{\partial^2 F}{\partial z^2}\Big|_{n,k} \Delta z + O(\Delta z^2) \\ \frac{1}{v_g} \frac{\partial R}{\partial t}\Big|_{n,k} + \frac{1}{2v_g} \frac{\partial^2 R}{\partial t^2}\Big|_{n,k} \Delta t + O(\Delta t^2) \\ +O\left(\frac{\Delta z^3}{\Delta t}\right) - \frac{\partial R}{\partial z}\Big|_{n,k} + \frac{1}{2} \frac{\partial^2 R}{\partial z^2}\Big|_{n,k} \Delta z + O(\Delta z^2) \end{bmatrix}$$

$$\xrightarrow{\Delta t \to 0, \Delta z \to 0} \begin{bmatrix} \frac{1}{v_g} \frac{\partial F}{\partial t}\Big|_{n,k} + \frac{\partial F}{\partial z}\Big|_{n,k} \\ \frac{1}{v_g} \frac{\partial R}{\partial t}\Big|_{n,k} - \frac{\partial R}{\partial z}\Big|_{n,k} \end{bmatrix},$$

and

$$\begin{bmatrix} A^{11}(z,t) & A^{12}(z,t) \\ A^{21}(z,t) & A^{22}(z,t) \end{bmatrix} \begin{bmatrix} F(z,t) \\ R(z,t) \end{bmatrix} \to \begin{bmatrix} A^{11}_{n,k} & A^{12}_{n,k} \\ A^{21}_{n,k} & A^{22}_{n,k} \end{bmatrix} \begin{bmatrix} F_{n,k} \\ R_{n,k} \end{bmatrix}.$$

Therefore, the discretization scheme, equation (6.4), is consistent with the original equation (6.1) [18]. This analysis also gives us a hint that the upwind scheme is of first order accuracy in terms of the mesh size Δt and Δz.

The stability condition of this scheme can be found through the well-known von Neumann analysis. Actually, we know the conclusion that the necessary and sufficient condition for the homogeneous version of equation (6.1)

$$\left(\frac{1}{v_g}\frac{\partial}{\partial t} \pm \frac{\partial}{\partial z}\right) \begin{bmatrix} F(z,t) \\ R(z,t) \end{bmatrix} = \begin{bmatrix} A^{11}(z,t) & A^{12}(z,t) \\ A^{21}(z,t) & A^{22}(z,t) \end{bmatrix} \begin{bmatrix} F(z,t) \\ R(z,t) \end{bmatrix}, \tag{6.6}$$

to be stable under a consistent one-step scheme is that the equation

$$\left(\frac{1}{v_g}\frac{\partial}{\partial t} \pm \frac{\partial}{\partial z}\right) \begin{bmatrix} F(z,t) \\ R(z,t) \end{bmatrix} = 0, \tag{6.7}$$

is stable under the same scheme [18]. On the other hand, by letting

$$\begin{bmatrix} F_{n,k} \\ R_{n,k} \end{bmatrix} \to \begin{bmatrix} f_0 \xi^k (e^{j\beta\Delta z})^n \\ r_0 \xi^k (e^{-j\beta\Delta z})^n \end{bmatrix}, \tag{6.8}$$

and substituting equation (6.8) into (6.6), we find

$$\xi = 1 - \eta + \frac{\Delta}{2} \pm \sqrt{\left(\left(\frac{\Delta}{2}\right)^2 - \eta^2[1 + (A^{11}_{n,k} + A^{22}_{n,k})\Delta z + (A^{11}_{n,k}A^{22}_{n,k} - A^{12}_{n,k}A^{21}_{n,k})\Delta z^2]\right)}. \tag{6.9}$$

where

$$\frac{\Delta}{2} = \eta[\cos(\beta\Delta z) + \frac{A^{11}_{n,k} + A^{22}_{n,k}}{2}\Delta z\cos(\beta\Delta z) - \mathrm{j}\frac{A^{11}_{n,k} - A^{22}_{n,k}}{2}\Delta z\sin(\beta\Delta z)]. \quad (6.10)$$

By setting $A^{11}_{n,k}$, $A^{12}_{n,k}$, $A^{21}_{n,k}$, and $A^{22}_{n,k}$ to zero, we find $\xi = 1 - \eta + \eta e^{\pm \mathrm{j}\beta\Delta z}$, or

$$|\xi|^2 = 1 - 4\eta(1-\eta)\sin^2\left(\frac{\beta\Delta z}{2}\right). \quad (6.11)$$

Therefore, by letting $\eta \le 1$, known as the Courant–Friedrichs–Lewy stability criterion, we have $|\xi|^2 \le 1$ from equation (6.11), which means that the upwind scheme is stable for equation (6.7) and hence for (6.6).

By retaining up to the first order of Δz terms in equations (6.9) and (6.10), we find

$$\xi \approx 1 + \eta\Delta z\left[\frac{A^{11}_{n,k} + A^{22}_{n,k}}{2} \pm \sqrt{\left(\left(\frac{A^{11}_{n,k} + A^{22}_{n,k}}{2}\right)^2 + A^{12}_{n,k}A^{21}_{n,k}\right)}\right]. \quad (6.12)$$

Since the above scheme is stable, we have to drop the solution corresponding to the plus sign in the bracket on the RHS of equation (6.12), which will lead to $|\xi|^2 > 1$, hence the result is in conflict. As such, if $A^{12}_{n,k}A^{21}_{n,k} = 0$, our selection of η has nothing to do with ξ, so that we can always set $\eta = 1$ to gain maximum efficiency in our simulation. However, if $A^{12}_{n,k}A^{21}_{n,k} \neq 0$, we might want to select $\eta < 1$ to reinforce the stability. The reason is that the optical equation may not be the only equation that we are dealing with. If it needs to be solved simultaneously with other equations, implicit dependence of the parameters $A^{11}_{n,k}$, $A^{12}_{n,k}$, $A^{21}_{n,k}$, and $A^{22}_{n,k}$ on the field components F and R exists, which implies a weak non-linearity of the optical equation. Therefore, running at the edge, i.e., setting $\eta = 1$, could bring instability.

Although we have discussed the stability of the homogeneous equation (6.6), whereas the optical equation is inhomogeneous as shown in equation (6.1), the conclusion of the von Neumann analysis is not affected. Hence the upwind scheme is also stable for equation (6.1). Finally, the convergence of the upwind scheme is guaranteed by its consistency and stability.

The upwind (FT–BS) scheme can also be implemented as

$$\begin{aligned}\frac{\partial F}{\partial t} &= \frac{F_{n,k+1} - F_{n,k}}{\Delta t} \quad \frac{\partial F}{\partial z} = \frac{F_{n,k} - F_{n-1,k}}{\Delta z}\\ \frac{\partial R}{\partial t} &= \frac{R_{n,k+1} - R_{n,k}}{\Delta t} \quad \frac{\partial R}{\partial z} = \frac{R_{n+1,k} - R_{n,k}}{\Delta z}.\end{aligned} \quad (6.13)$$

In accordance with equation (6.13), (6.1) becomes

$$\begin{bmatrix} F_{n,k+1} \\ R_{n,k+1} \end{bmatrix} = \eta \begin{bmatrix} F_{n-1,k} \\ R_{n+1,k} \end{bmatrix} + \eta \begin{bmatrix} \frac{1-\eta}{\eta} + A^{11}_{n,k}\Delta z & A^{12}_{n,k}\Delta z \\ A^{21}_{n,k}\Delta z & \frac{1-\eta}{\eta} + A^{22}_{n,k}\Delta z \end{bmatrix} \begin{bmatrix} F_{n,k} \\ R_{n,k} \end{bmatrix}$$
$$+ \eta\Delta z \begin{bmatrix} C^{\mathrm{f}}_{n,k} \\ C^{\mathrm{r}}_{n,k} \end{bmatrix}. \tag{6.14}$$

Similarly to equation (6.4), this scheme is consistent and stable once $\eta \leq 1$, hence equation (6.14) gives the convergence solution to (6.1).

A combination of equations (6.4) and (6.14) can be used to smooth the field distribution if the parameters A^{11}, A^{12}, A^{21}, and A^{22} change rapidly as functions of z. Actually, by moving one spatial step forward (i.e., $n \to n + 1$), and one spatial step backward (i.e., $n \to n - 1$) in the first and second equation of (6.14), respectively, we obtain

$$\begin{bmatrix} F_{n+1,k+1} \\ R_{n-1,k+1} \end{bmatrix} = \eta \begin{bmatrix} F_{n,k} \\ R_{n,k} \end{bmatrix} + (1-\eta) \begin{bmatrix} F_{n+1,k} \\ R_{n-1,k} \end{bmatrix}$$
$$+ \eta\Delta z \begin{bmatrix} A^{11}_{n+1,k} & A^{12}_{n+1,k} & 0 & 0 \\ 0 & 0 & A^{21}_{n-1,k} & A^{22}_{n-1,k} \end{bmatrix} \begin{bmatrix} F_{n+1,k} \\ R_{n+1,k} \\ F_{n-1,k} \\ R_{n-1,k} \end{bmatrix} + \eta\Delta z \begin{bmatrix} C^{\mathrm{f}}_{n+1,k} \\ C^{\mathrm{r}}_{n-1,k} \end{bmatrix}. \tag{6.15}$$

Taking the arithmetic average of equations (6.4) and (6.15) we obtain

$$\begin{bmatrix} F_{n+1,k+1} \\ R_{n-1,k+1} \end{bmatrix} = \eta \begin{bmatrix} F_{n,k} \\ R_{n,k} \end{bmatrix} + (1-\eta) \begin{bmatrix} F_{n+1,k} \\ R_{n-1,k} \end{bmatrix} + \frac{\eta\Delta z}{2} \begin{bmatrix} A^{11}_{n,k} & A^{12}_{n,k} \\ A^{21}_{n,k} & A^{22}_{n,k} \end{bmatrix} \begin{bmatrix} F_{n,k} \\ R_{n,k} \end{bmatrix}$$
$$+ \frac{\eta\Delta z}{2} \begin{bmatrix} A^{11}_{n+1,k} & A^{12}_{n+1,k} & 0 & 0 \\ 0 & 0 & A^{21}_{n-1,k} & A^{22}_{n-1,k} \end{bmatrix} \begin{bmatrix} F_{n+1,k} \\ R_{n+1,k} \\ F_{n-1,k} \\ R_{n-1,k} \end{bmatrix} + \frac{\eta\Delta z}{2} \begin{bmatrix} C^{\mathrm{f}}_{n+1,k} + C^{\mathrm{f}}_{n,k} \\ C^{\mathrm{r}}_{n-1,k} + C^{\mathrm{r}}_{n,k} \end{bmatrix}. \tag{6.16}$$

It should be noted that equation (6.16) is still a first order accuracy scheme as there is no new mesh point as defined in equation (6.2) involved. An obvious difference between equations (6.16) and both (6.4) and (6.14) is that not only are the parameter values at the space point n, but also their values at a forward space step $n + 1$ for F and a backward space step $n - 1$ for R are involved. Hence equation (6.16) gives a better average once the parameters A^{11}, A^{12}, A^{21}, and A^{22} are not uniform along z. This scheme is particularly useful once we need to select $\eta \ll 1$ to ensure stability, while there is no need to select a very small Δt, as in this case we will have to deal with the large space step $\Delta z \gg v_{\mathrm{g}}\Delta t$.

It is possible to use a different combination of equations (6.4) and (6.14) to gain a second order accuracy in space as there are three different points in space involved in evaluating these equations. However, the increased complication in implementing such a scheme does not seem to be justified by the accuracy gained.

The reason that we have considered only the upwind (i.e., FT–BS) scheme is because the forward-time and forward-space (FT–FS) scheme is unstable although it is consistent, since, following the von Neumann analysis, we have

$$|\xi|^2 = 1 + 4\eta(1+\eta)\sin^2\left(\frac{\beta\Delta z}{2}\right), \tag{6.17}$$

which leads to $|\xi|^2 > 1$ for any $\eta > 0$.

With a modification known as the Lax–Friedrichs method [19], the FT–FS scheme becomes stable, hence it converges. However, it provides poor accuracy since the Lax–Friedrichs scheme brings in the numerical dissipation through an equivalent diffusion term which decreases the optical field intensity spuriously.

There are also backward-time schemes such as the backward-time, backward-space (BT–BS) discretization

$$\begin{aligned}
\frac{\partial F}{\partial t} &= \frac{F_{n,k+1} - F_{n,k}}{\Delta t} & \frac{\partial F}{\partial z} &= \frac{F_{n,k+1} - F_{n-1,k+1}}{\Delta z} \\
\frac{\partial R}{\partial t} &= \frac{R_{n,k+1} - R_{n,k}}{\Delta t} & \frac{\partial R}{\partial z} &= \frac{R_{n+1,k+1} - R_{n,k+1}}{\Delta z},
\end{aligned} \tag{6.18}$$

and the backward-time, forward-space (BT–FS) discretization

$$\begin{aligned}
\frac{\partial F}{\partial t} &= \frac{F_{n,k+1} - F_{n,k}}{\Delta t} & \frac{\partial F}{\partial z} &= \frac{F_{n+1,k+1} - F_{n,k+1}}{\Delta z} \\
\frac{\partial R}{\partial t} &= \frac{R_{n,k+1} - R_{n,k}}{\Delta t} & \frac{\partial R}{\partial z} &= \frac{R_{n,k+1} - R_{n-1,k+1}}{\Delta z}.
\end{aligned} \tag{6.19}$$

In accordance with equations (6.18) and (6.19), (6.1) becomes

$$\begin{aligned}
&\begin{bmatrix} (1+\eta)F_{n,k+1} - \eta F_{n-1,k+1} \\ (1+\eta)R_{n,k+1} - \eta R_{n+1,k+1} \end{bmatrix} \\
&= \begin{bmatrix} 1 + \eta A^{11}_{n,k}\Delta z & \eta A^{12}_{n,k}\Delta z \\ \eta A^{21}_{n,k}\Delta z & 1 + \eta A^{22}_{n,k}\Delta z \end{bmatrix} \begin{bmatrix} F_{n,k} \\ R_{n,k} \end{bmatrix} + \eta\Delta z \begin{bmatrix} C^{\mathrm{f}}_{n,k} \\ C^{\mathrm{r}}_{n,k} \end{bmatrix},
\end{aligned} \tag{6.20}$$

and

$$\begin{aligned}
&\begin{bmatrix} \eta F_{n+1,k+1} + (1-\eta)F_{n,k+1} \\ \eta R_{n-1,k+1} + (1-\eta)R_{n,k+1} \end{bmatrix} \\
&= \begin{bmatrix} 1 + \eta A^{11}_{n,k}\Delta z & \eta A^{12}_{n,k}\Delta z \\ \eta A^{21}_{n,k}\Delta z & 1 + \eta A^{22}_{n,k}\Delta z \end{bmatrix} \begin{bmatrix} F_{n,k} \\ R_{n,k} \end{bmatrix} + \eta\Delta z \begin{bmatrix} C^{\mathrm{f}}_{n,k} \\ C^{\mathrm{r}}_{n,k} \end{bmatrix},
\end{aligned} \tag{6.21}$$

respectively.

Table 6.1. A summary of the features of first order finite difference discretization schemes

	FS	BS
FT (explicit)	consistent but unstable, becomes stable under L–F modification, but still has power dissipation	consistent, stable, hence converges if $v_g \Delta t \leq \Delta z$
BT (implicit)	consistent, stable, hence converges if $v_g \Delta t > \Delta z$	consistent, unconditionally stable, hence converges

It is straightforward to prove that both the BT–BS and BT–FS schemes are consistent. By following the von Neumann analysis, we also have

$$|\xi|^2 = \frac{1}{1 + 4\eta(1+\eta)\sin^2(\frac{\beta \Delta z}{2})}, \tag{6.22}$$

and

$$|\xi|^2 = \frac{1}{1 - 4\eta(1-\eta)\sin^2(\frac{\beta \Delta z}{2})}, \tag{6.23}$$

for the BT–BS and BT–FS schemes, respectively.

Therefore, we come to the conclusion that the BT–BS scheme is unconditionally stable, hence equation (6.20) always converges to (6.1). The BT–FS scheme, however, is stable under the condition $\eta > 1$, hence equation (6.21) converges to (6.1) if $v_g \Delta t > \Delta z$.

In contrast to the schemes of equations (6.4), (6.14), or (6.16) where the field at the next time step is explicitly expressed by the field at the current time step, the schemes of equations (6.20) and (6.21) are implicit, since there are only relations connecting the fields at the next and current time steps. In this case, we have to solve a system of linear equations, or invert a matrix, so as to obtain the field at the next time step for a field given at the current time step. This approach increases the complexity of implementation and computational burden significantly. Despite this drawback, we may still find some applications of these implicit schemes where the given problem requires $v_g \Delta t > \Delta z$. For example, in DC biased or low speed modulated optoelectronic devices with a strong LSHB effect, we have to use a small space step to capture the highly non-uniform field distribution along the cavity, whereas there is no need to take a small time step, and either BT–FS or BT–BS can be found to be useful. Another potential application of these backward-time implicit schemes is in dealing with the inverse propagation problem, where we may need to find out which initial function will lead to a given final state, for tasks like signal waveform optimization.

The features of the different schemes are summarized in Table 6.1.

In order to gain a higher (second) order of accuracy, one may think to use the center difference scheme. However, it is well known that the forward-time, explicit center-space (FT–ECS) scheme with the center-space difference evaluated at the previous time step, is unstable. The forward-time, implicit center-space (FT–ICS) scheme represented by

$$\begin{aligned}\frac{\partial F}{\partial t} &= \frac{F_{n,k+1} - F_{n,k}}{\Delta t} \quad \frac{\partial F}{\partial z} = \frac{F_{n+1,k+1} - F_{n-1,k+1}}{2\Delta z} \\ \frac{\partial R}{\partial t} &= \frac{R_{n,k+1} - R_{n,k}}{\Delta t} \quad \frac{\partial R}{\partial z} = \frac{R_{n+1,k+1} - R_{n-1,k+1}}{2\Delta z},\end{aligned} \tag{6.24}$$

will lead to

$$\begin{aligned}&\begin{bmatrix} \frac{\eta}{2}F_{n+1,k+1} + F_{n,k+1} - \frac{\eta}{2}F_{n-1,k+1} \\ -\frac{\eta}{2}R_{n+1,k+1} + R_{n,k+1} + \frac{\eta}{2}R_{n-1,k+1} \end{bmatrix} \\ &= \begin{bmatrix} 1+\eta A^{11}_{n,k}\Delta z & \eta A^{12}_{n,k}\Delta z \\ \eta A^{21}_{n,k}\Delta z & 1+\eta A^{22}_{n,k}\Delta z \end{bmatrix}\begin{bmatrix} F_{n,k} \\ R_{n,k} \end{bmatrix} + \eta \begin{bmatrix} C^{\mathrm{f}}_{n,k}\Delta z \\ C^{\mathrm{r}}_{n,k}\Delta z \end{bmatrix}.\end{aligned} \tag{6.25}$$

Since

$$\begin{aligned}\left(\frac{1}{v_g}\frac{\partial}{\partial t} \pm \frac{\partial}{\partial z}\right)\begin{bmatrix} F(z,t) \\ R(z,t) \end{bmatrix} &\to \begin{bmatrix} \dfrac{F_{n,k+1} - F_{n,k}}{v_g \Delta t} + \dfrac{F_{n+1,k+1} - F_{n-1,k+1}}{2\Delta z} \\ \dfrac{R_{n,k+1} - R_{n,k}}{v_g \Delta t} - \dfrac{R_{n+1,k+1} - R_{n-1,k+1}}{2\Delta z} \end{bmatrix} \\ &= \begin{bmatrix} \dfrac{1}{v_g}\dfrac{\partial F}{\partial t}\Big|_{n,k} + \dfrac{\partial F}{\partial z}\Big|_{n,k} + \dfrac{1}{2v_g}\dfrac{\partial^2 F}{\partial t^2}\Big|_{n,k}\Delta t + O(\Delta t^2) + O(\Delta z^2) \\ \dfrac{1}{v_g}\dfrac{\partial R}{\partial t}\Big|_{n,k} - \dfrac{\partial R}{\partial z}\Big|_{n,k} + \dfrac{1}{2v_g}\dfrac{\partial^2 R}{\partial t^2}\Big|_{n,k}\Delta t + O(\Delta t^2) + O(\Delta z^2) \end{bmatrix} \\ &\xrightarrow{\Delta t \to 0, \Delta z \to 0} \begin{bmatrix} \dfrac{1}{v_g}\dfrac{\partial F}{\partial t}\Big|_{n,k} + \dfrac{\partial F}{\partial z}\Big|_{n,k} \\ \dfrac{1}{v_g}\dfrac{\partial R}{\partial t}\Big|_{n,k} - \dfrac{\partial R}{\partial z}\Big|_{n,k} \end{bmatrix},\end{aligned}$$

and

$$\begin{bmatrix} A^{11}(z,t) & A^{12}(z,t) \\ A^{21}(z,t) & A^{22}(z,t) \end{bmatrix}\begin{bmatrix} F(z,t) \\ R(z,t) \end{bmatrix} \to \begin{bmatrix} A^{11}_{n,k} & A^{12}_{n,k} \\ A^{21}_{n,k} & A^{22}_{n,k} \end{bmatrix}\begin{bmatrix} F_{n,k} \\ R_{n,k} \end{bmatrix},$$

the discretization scheme, equation (6.25), is consistent with the original equation (6.1) [18]. The von Neumann analysis also shows that this scheme is unconditionally stable since

$$|\xi|^2 = \frac{1}{1+\eta^2 \sin^2(\beta \Delta z)} \leq 1, \tag{6.26}$$

hence equation (6.25) gives the convergent solution to (6.1).

Unfortunately, these backward-time schemes are all implicit so that we have to solve an extra system of linear equations, or to invert a matrix at each time step, which increases the implementation complexity and brings in a considerable extra computational burden. The accuracy gained usually does not offset the extra cost.

The staggered leapfrog method provides a second order accuracy in both time and space, and has been proven to be a consistent and stable scheme. Unfortunately, it often fails to converge when equation (6.1) has to be solved with the carrier equations in a self-consistent manner because of decoupling between the adjacent mesh points. Although by introducing an artificial diffusion term, or through using a two-step Lax–Wendroff scheme [19], we can have the decoupling problem solved, the numerical dissipation appears again.

The Crank–Nicholson scheme has second order accuracy in both time and space, and is consistent and unconditionally stable. Actually, under discretization [18]

$$\frac{\partial F}{\partial t} = \frac{F_{n,k+1} - F_{n,k}}{\Delta t} \quad \frac{\partial F}{\partial z} = \frac{1}{2}\left(\frac{F_{n+1,k+1} - F_{n-1,k+1}}{2\Delta z} + \frac{F_{n+1,k} - F_{n-1,k}}{2\Delta z}\right)$$
$$\frac{\partial R}{\partial t} = \frac{R_{n,k+1} - R_{n,k}}{\Delta t} \quad \frac{\partial R}{\partial z} = \frac{1}{2}\left(\frac{R_{n+1,k+1} - R_{n-1,k+1}}{2\Delta z} + \frac{R_{n+1,k} - R_{n-1,k}}{2\Delta z}\right), \tag{6.27}$$

at mesh point n and $k + 1/2$, equation (6.1) becomes

$$\eta\begin{bmatrix} F_{n+1,k+1} - F_{n-1,k+1} \\ -R_{n+1,k+1} + R_{n-1,k+1} \end{bmatrix} + 2\begin{bmatrix} 2 - \eta A^{11}_{n,k+1}\Delta z & -\eta A^{12}_{n,k+1}\Delta z \\ -\eta A^{21}_{n,k+1}\Delta z & 2 - \eta A^{22}_{n,k+1}\Delta z \end{bmatrix}\begin{bmatrix} F_{n,k+1} \\ R_{n,k+1} \end{bmatrix}$$
$$= \eta\begin{bmatrix} -F_{n+1,k} + F_{n-1,k} \\ R_{n+1,k} - R_{n-1,k} \end{bmatrix} + 2\begin{bmatrix} 2 + \eta A^{11}_{n,k}\Delta z & \eta A^{12}_{n,k}\Delta z \\ \eta A^{21}_{n,k}\Delta z & 2 + \eta A^{22}_{n,k}\Delta z \end{bmatrix}\begin{bmatrix} F_{n,k} \\ R_{n,k} \end{bmatrix}$$
$$+ 2\eta\Delta z\begin{bmatrix} C^{\mathrm{f}}_{n,k+1} + C^{\mathrm{f}}_{n,k} \\ C^{\mathrm{r}}_{n,k+1} + C^{\mathrm{r}}_{n,k} \end{bmatrix}. \tag{6.28}$$

It is straightforward to prove that equation (6.28) is consistent with (6.1). The von Neumann analysis shows that

$$|\xi|^2 = \left|\frac{1 \mp \mathrm{j}\frac{\eta}{2}\sin(\beta\Delta z)}{1 \pm \mathrm{j}\frac{\eta}{2}\sin(\beta\Delta z)}\right|^2 \equiv 1. \tag{6.29}$$

Therefore, the Crank–Nicholson scheme, equation (6.28), converges to (6.1). The only drawback of this scheme is its implicit nature.

An approach similar to the Crank–Nicholson scheme is the box scheme with discretization [18]

$$\frac{\partial F}{\partial t} = \frac{1}{2}\left(\frac{F_{n+1,k+1} - F_{n+1,k}}{\Delta t} + \frac{F_{n,k+1} - F_{n,k}}{\Delta t}\right)$$
$$\frac{\partial F}{\partial z} = \frac{1}{2}\left(\frac{F_{n+1,k+1} - F_{n,k+1}}{\Delta z} + \frac{F_{n+1,k} - F_{n,k}}{\Delta z}\right)$$
$$\frac{\partial R}{\partial t} = \frac{1}{2}\left(\frac{R_{n+1,k+1} - R_{n+1,k}}{\Delta t} + \frac{R_{n,k+1} - R_{n,k}}{\Delta t}\right)$$
$$\frac{\partial R}{\partial z} = \frac{1}{2}\left(\frac{R_{n+1,k+1} - R_{n,k+1}}{\Delta z} + \frac{R_{n+1,k} - R_{n,k}}{\Delta z}\right), \tag{6.30}$$

at mesh point $n + 1/2$ and $k + 1/2$, or

$$\begin{bmatrix} A^{11}(z,t) & A^{12}(z,t) \\ A^{21}(z,t) & A^{22}(z,t) \end{bmatrix} \begin{bmatrix} F(z,t) \\ R(z,t) \end{bmatrix} + \begin{bmatrix} C^{\mathrm{f}}(z,t) \\ C^{\mathrm{r}}(z,t) \end{bmatrix}$$
$$\rightarrow \frac{1}{4} \begin{bmatrix} A^{11}_{n+1,k+1} & A^{12}_{n+1,k+1} \\ A^{21}_{n+1,k+1} & A^{22}_{n+1,k+1} \end{bmatrix} \begin{bmatrix} F_{n+1,k+1} \\ R_{n+1,k+1} \end{bmatrix}$$
$$+ \frac{1}{4} \begin{bmatrix} A^{11}_{n+1,k} & A^{12}_{n+1,k} \\ A^{21}_{n+1,k} & A^{22}_{n+1,k} \end{bmatrix} \begin{bmatrix} F_{n+1,k} \\ R_{n+1,k} \end{bmatrix}$$
$$+ \frac{1}{4} \begin{bmatrix} A^{11}_{n,k+1} & A^{12}_{n,k+1} \\ A^{21}_{n,k+1} & A^{22}_{n,k+1} \end{bmatrix} \begin{bmatrix} F_{n,k+1} \\ R_{n,k+1} \end{bmatrix} + \frac{1}{4} \begin{bmatrix} A^{11}_{n,k} & A^{12}_{n,k} \\ A^{21}_{n,k} & A^{22}_{n,k} \end{bmatrix} \begin{bmatrix} F_{n,k} \\ R_{n,k} \end{bmatrix}$$
$$+ \frac{1}{4} \begin{bmatrix} C^{\mathrm{f}}_{n+1,k+1} \\ C^{\mathrm{r}}_{n+1,k+1} \end{bmatrix} + \frac{1}{4} \begin{bmatrix} C^{\mathrm{f}}_{n+1,k} \\ C^{\mathrm{r}}_{n+1,k} \end{bmatrix} + \frac{1}{4} \begin{bmatrix} C^{\mathrm{f}}_{n,k+1} \\ C^{\mathrm{r}}_{n,k+1} \end{bmatrix} + \frac{1}{4} \begin{bmatrix} C^{\mathrm{f}}_{n,k} \\ C^{\mathrm{r}}_{n,k} \end{bmatrix}. \quad (6.31)$$

The expression for this scheme is therefore obtained by plugging equations (6.30) and (6.31) into (6.1).

Through similar consistency and von Neumann stability analyses as shown above, we find that the box scheme also has second order accuracy in both time and space, and is consistent and unconditionally stable, hence giving a convergent solution to equation (6.1). The drawback of the box scheme, again, is its implicit nature.

Although the Crank–Nicholson and box schemes offer higher accuracy, larger discretization steps should be allowed. However, the non-uniformity of the parameters A^{11}, A^{12}, A^{21}, and A^{22} affects the accuracy as well, since we have used a linear interpolation of these parameters at the half mesh points, $(n, k + 1/2)$ for the Crank–Nicholson and $(n + 1/2, k + 1/2)$ for the box schemes, respectively. In this sense, if the parameters are highly non-uniform in space (z) and time (t), we still have to choose small Δz and Δt in these schemes to ensure accuracy. However, if the parameters are only highly non-uniform in space (z), but change slowly in time (t), we can use the Crank–Nicholson scheme where the parameters are not interpolated in space, and hence less error is involved.

Other than the direct discretization methods that have been successfully implemented in a variety of different versions in laser modeling [20–23], the indirect discretization methods based on the operator split [24], e.g., split-step, alternating-direction implicit (ADI), and integral transformation or eigenfunction expansion, are also gaining popularity because of their balance between computational accuracy and efficiency [25]. The integral transformation or eigenfunction expansion method is actually a standing wave approach which will be discussed in Section 6.3. The ADI method based on rational (Pade) factorizations is more like a pure numerical algorithm with little physics, and hence it will not be discussed in this book. We next focus on the split-step method.

6.2.2 The split-step method

In dealing with the optical traveling wave equation (6.1), we may rewrite it in the form

$$\pm\frac{\partial}{\partial z}\begin{bmatrix} F(z,t) \\ R(z,t) \end{bmatrix} = \begin{bmatrix} A^{11}(z,t) - \frac{1}{v_g}\frac{\partial}{\partial t} & A^{12}(z,t) \\ A^{21}(z,t) & A^{22}(z,t) - \frac{1}{v_g}\frac{\partial}{\partial t} \end{bmatrix}\begin{bmatrix} F(z,t) \\ R(z,t) \end{bmatrix} + \begin{bmatrix} C^{\mathrm{f}}(z,t) \\ C^{\mathrm{r}}(z,t) \end{bmatrix}. \tag{6.32}$$

If we consider the nth subsection in a waveguide with length Δz as shown in Fig. 6.2, temporarily ignoring the variable and parameter dependence on time at a frozen time instant $t = k\Delta t$, ignoring the z dependence of the parameters, and ignoring the inhomogeneous spontaneous emission contributions, we find that the fields at the entrance and exit of this subsection, i.e., $F_{n,k}$, $R_{n,k}$ and $F_{n+1,k}$, $R_{n+1,k}$, are governed by

$$\pm\frac{\mathrm{d}}{\mathrm{d}z}\begin{bmatrix} F(z) \\ R(z) \end{bmatrix} = \begin{bmatrix} a_{11} & a_{12} \\ a_{21} & a_{22} \end{bmatrix}\begin{bmatrix} F(z) \\ R(z) \end{bmatrix}, \tag{6.33}$$

with

$$a_{11} \equiv A^{11}_{n,k} - \frac{1}{v_g}\frac{\partial}{\partial t}, \quad a_{12} \equiv A^{12}_{n,k}, \quad a_{21} \equiv A^{21}_{n,k}, \quad a_{22} \equiv A^{22}_{n,k} - \frac{1}{v_g}\frac{\partial}{\partial t}, \tag{6.34}$$

introduced as constants. The right- and left-going waves F and R are assumed to be z dependent only, evaluated at time $t = k\Delta t$.

Equation (6.33) can readily be solved subject to the given boundary conditions at the entrance of this subsection, i.e.,

$$\begin{bmatrix} F(0) \\ R(0) \end{bmatrix} = \begin{bmatrix} F_{n,k} \\ R_{n,k} \end{bmatrix},$$
$$\frac{\mathrm{d}}{\mathrm{d}z}\begin{bmatrix} F(z) \\ R(z) \end{bmatrix}_{z=0} = \begin{bmatrix} a_{11} & a_{12} \\ -a_{21} & -a_{22} \end{bmatrix}\begin{bmatrix} F(0) \\ R(0) \end{bmatrix} = \begin{bmatrix} a_{11} & a_{12} \\ -a_{21} & -a_{22} \end{bmatrix}\begin{bmatrix} F_{n,k} \\ R_{n,k} \end{bmatrix}, \tag{6.35}$$

with $F_{n,k}$ and $R_{n,k}$ taken as given variables.

Actually, equation (6.33) can be made decoupled as

$$\frac{\mathrm{d}^2 F(z)}{\mathrm{d}z^2} - (a_{11} - a_{22})\frac{\mathrm{d}F(z)}{\mathrm{d}z} + (a_{12}a_{21} - a_{11}a_{22})F(z) = 0$$
$$\frac{\mathrm{d}^2 R(z)}{\mathrm{d}z^2} - (a_{11} - a_{22})\frac{\mathrm{d}R(z)}{\mathrm{d}z} + (a_{12}a_{21} - a_{11}a_{22})R(z) = 0. \tag{6.36}$$

The general solution to equation (6.36) can be found as

$$F(z) = \mathrm{e}^{\frac{a_{11}-a_{22}}{2}z}[C_1\cosh(\gamma z) + C_2\sinh(\gamma z)]$$
$$R(z) = \mathrm{e}^{\frac{a_{11}-a_{22}}{2}z}[C_3\cosh(\gamma z) + C_4\sinh(\gamma z)], \tag{6.37}$$

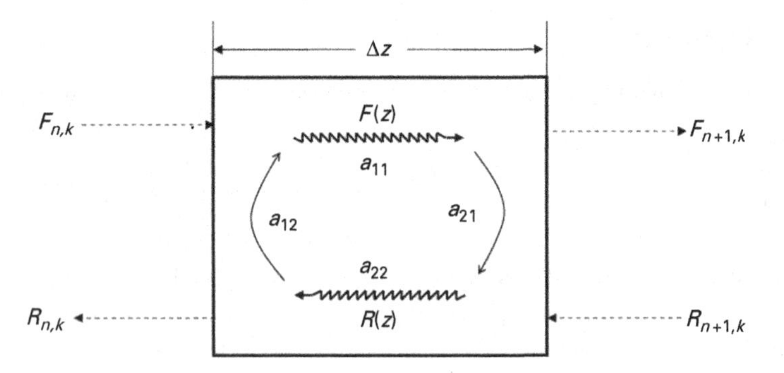

Fig. 6.2. Coupled wave integration in the nth subsection at a frozen time instant $t = k\Delta t$.

where

$$\gamma = \pm\sqrt{\left(\left(\frac{a_{11}+a_{22}}{2}\right)^2 - a_{12}a_{21}\right)}. \tag{6.38}$$

Plugging equation (6.37) into (6.35) yields

$$F(0) = C_1 = F_{n,k}, \quad \left.\frac{\mathrm{d}F}{\mathrm{d}z}\right|_{z=0} = \frac{a_{11}-a_{22}}{2}C_1 + \gamma C_2 = a_{11}F_{n,k} + a_{12}R_{n,k},$$

$$R(0) = C_3 = R_{n,k}, \quad \left.\frac{\mathrm{d}R}{\mathrm{d}z}\right|_{z=0} = \frac{a_{11}-a_{22}}{2}C_3 + \gamma C_4 = -a_{21}F_{n,k} - a_{22}R_{n,k},$$

or

$$C_1 = F_{n,k}, \quad C_2 = \frac{a_{11}+a_{22}}{2\gamma}F_{n,k} + \frac{a_{12}}{\gamma}R_{n,k},$$
$$C_3 = R_{n,k}, \quad C_4 = -\frac{a_{21}}{\gamma}F_{n,k} - \frac{a_{11}+a_{22}}{2\gamma}R_{n,k}. \tag{6.39}$$

Substituting equation (6.39) into (6.37) and setting $z = \Delta z$, we obtain

$$\begin{bmatrix} F_{n+1,k} \\ R_{n+1,k} \end{bmatrix} = \mathrm{e}^{\frac{a_{11}-a_{22}}{2}\Delta z} \begin{bmatrix} a'_{11} & a'_{12} \\ a'_{21} & a'_{22} \end{bmatrix} \begin{bmatrix} F_{n,k} \\ R_{n,k} \end{bmatrix}, \tag{6.40}$$

where

$$a'_{11} \equiv \cosh(\gamma\Delta z) + \frac{a_{11}+a_{22}}{2\gamma}\sinh(\gamma\Delta z), \quad a'_{12} \equiv \frac{a_{12}}{\gamma}\sinh(\gamma\Delta z),$$
$$a'_{21} \equiv -\frac{a_{21}}{\gamma}\sinh(\gamma\Delta z), \quad a'_{22} \equiv \cosh(\gamma\Delta z) - \frac{a_{11}+a_{22}}{2\gamma}\sinh(\gamma\Delta z). \tag{6.41}$$

In considering the time dependence of these traveling waves, we need to make explicit the time derivative operator hidden in those matrix elements. Noting that, with respect to the right- and left-going waves, F travels from n to $n+1$ but R travels from $n+1$ to n, we need to express $F_{n+1,k}$ and $R_{n,k}$ in terms of $F_{n,k}$ and $R_{n+1,k}$, since both the former and the latter group of variables should be aligned in the time domain. This can readily be realized by converting equation (6.40) to

$$\begin{bmatrix} F_{n+1,k} \\ R_{n,k} \end{bmatrix} = \frac{1}{a'_{22}} \begin{bmatrix} e^{\frac{a_{11}-a_{22}}{2}\Delta z} & a'_{12} \\ -a'_{21} & e^{-\frac{a_{11}-a_{22}}{2}\Delta z} \end{bmatrix} \begin{bmatrix} F_{n,k} \\ R_{n+1,k} \end{bmatrix}$$
$$= \frac{1}{\cosh(\gamma\Delta z) - \frac{a_{11}+a_{22}}{2\gamma}\sinh(\gamma\Delta z)} \begin{bmatrix} e^{\frac{a_{11}-a_{22}}{2}\Delta z} & \frac{a_{12}}{\gamma}\sinh(\gamma\Delta z) \\ \frac{a_{21}}{\gamma}\sinh(\gamma\Delta z) & e^{-\frac{a_{11}-a_{22}}{2}\Delta z} \end{bmatrix} \begin{bmatrix} F_{n,k} \\ R_{n+1,k} \end{bmatrix}. \tag{6.42}$$

From equations (6.34) and (6.32), we know that a_{12} and a_{21}, i.e., $A^{12}_{n,k}$ and $A^{21}_{n,k}$, are the cross-coupling (through, e.g., reflection from the grating) coefficients between the right- and left-going waves. They are normally smaller than the self-coupling (i.e., amplitude gain or attenuation and phase shift through propagation) coefficients a_{11} and a_{22}. Therefore, we can ignore the cross-coupling term in equation (6.42) in restoring the time dependence. As a result, equation (6.42) becomes

$$\begin{bmatrix} F_{n+1,k} \\ R_{n,k} \end{bmatrix} \approx \frac{1}{\cosh(\gamma\Delta z) - \sinh(\gamma\Delta z)} \begin{bmatrix} e^{\frac{a_{11}-a_{22}}{2}\Delta z} & 0 \\ 0 & e^{-\frac{a_{11}-a_{22}}{2}\Delta z} \end{bmatrix} \begin{bmatrix} F_{n,k} \\ R_{n+1,k} \end{bmatrix}$$
$$= e^{\gamma\Delta z} \begin{bmatrix} e^{\frac{a_{11}-a_{22}}{2}\Delta z} & 0 \\ 0 & e^{-\frac{a_{11}-a_{22}}{2}\Delta z} \end{bmatrix} \begin{bmatrix} F_{n,k} \\ R_{n+1,k} \end{bmatrix}$$
$$\approx e^{\frac{a_{11}+a_{22}}{2}\Delta z} \begin{bmatrix} e^{\frac{a_{11}-a_{22}}{2}\Delta z} & 0 \\ 0 & e^{-\frac{a_{11}-a_{22}}{2}\Delta z} \end{bmatrix} \begin{bmatrix} F_{n,k} \\ R_{n+1,k} \end{bmatrix}$$
$$= \begin{bmatrix} e^{a_{11}\Delta z} & 0 \\ 0 & e^{a_{22}\Delta z} \end{bmatrix} \begin{bmatrix} F_{n,k} \\ R_{n+1,k} \end{bmatrix} = e^{-\frac{\Delta z}{v_g}\frac{\partial}{\partial t}} \begin{bmatrix} e^{A^{11}_{n,k}\Delta z} & 0 \\ 0 & e^{A^{22}_{n,k}\Delta z} \end{bmatrix} \begin{bmatrix} F_{n,k} \\ R_{n+1,k} \end{bmatrix}, \tag{6.43}$$

where we have utilized $\gamma \approx (a_{11} + a_{22})/2$ according to equations (6.38) and (6.34). Equation (6.43) can then be written as

$$e^{\frac{\Delta z}{v_g}\frac{\partial}{\partial t}} \begin{bmatrix} F_{n+1,k} \\ R_{n,k} \end{bmatrix} = \begin{bmatrix} e^{A^{11}_{n,k}\Delta z} & 0 \\ 0 & e^{A^{22}_{n,k}\Delta z} \end{bmatrix} \begin{bmatrix} F_{n,k} \\ R_{n+1,k} \end{bmatrix},$$

or

$$\begin{bmatrix} F_{n+1,k+1} \\ R_{n,k+1} \end{bmatrix} = \begin{bmatrix} e^{A^{11}_{n,k}\Delta z} & 0 \\ 0 & e^{A^{22}_{n,k}\Delta z} \end{bmatrix} \begin{bmatrix} F_{n,k} \\ R_{n+1,k} \end{bmatrix}, \tag{6.44}$$

where the relation

$$e^{t_0\frac{\partial}{\partial t}} f(t) = f(t + t_0), \tag{6.45}$$

has been used. In equation (6.44), the time step Δt must be selected such that

$$\Delta t = \Delta z / v_g. \tag{6.46}$$

Equation (6.44) clearly shows that, to reflect the time dependence in wave propagation through subsection Δz, we just need to restore a proper time delay $\Delta t = \Delta z / v_g$ between the waves at the exit and at the entrance. Since F and R travel in opposite directions, their entrances and exits are swapped. This is the reason we have used equation (6.42) instead of (6.40) to sort the waves according to their time sequence.

Substituting equation (6.34) into (6.42) with the time derivative operator ignored, we find the space-independent part of equation (6.42) as

$$\begin{bmatrix} F_{n+1,k} \\ R_{n,k} \end{bmatrix} = \frac{1}{\cosh(\gamma_{n,k}\Delta z) - \frac{A^{11}_{n,k}+A^{22}_{n,k}}{2\gamma_{n,k}}\sinh(\gamma_{n,k}\Delta z)} \begin{bmatrix} e^{\frac{A^{11}_{n,k}-A^{22}_{n,k}}{2}\Delta z} & \frac{A^{12}_{n,k}}{\gamma_{n,k}}\sinh(\gamma_{n,k}\Delta z) \\ \frac{A^{21}_{n,k}}{\gamma_{n,k}}\sinh(\gamma_{n,k}\Delta z) & e^{-\frac{A^{11}_{n,k}-A^{22}_{n,k}}{2}\Delta z} \end{bmatrix} \begin{bmatrix} F_{n,k} \\ R_{n+1,k} \end{bmatrix}, \tag{6.47}$$

with $\gamma_{n,k}$ redefined as

$$\gamma_{n,k} \equiv \pm\sqrt{\left(\frac{A^{11}_{n,k}+A^{22}_{n,k}}{2}\right)^2 - A^{12}_{n,k}A^{21}_{n,k}}. \tag{6.48}$$

Finally, combining the space-dependent part (6.47) with the time-dependent part (6.44) and counting in the inhomogeneous spontaneous emission contributions, we obtain

$$\begin{bmatrix} F_{n+1,k+1} \\ R_{n,k+1} \end{bmatrix} = \frac{1}{\cosh(\gamma_{n,k}\Delta z) - \frac{A^{11}_{n,k}+A^{22}_{n,k}}{2\gamma_{n,k}}\sinh(\gamma_{n,k}\Delta z)} \times \begin{bmatrix} e^{\frac{A^{11}_{n,k}-A^{22}_{n,k}}{2}\Delta z} & \frac{A^{12}_{n,k}}{\gamma_{n,k}}\sinh(\gamma_{n,k}\Delta z) \\ \frac{A^{21}_{n,k}}{\gamma_{n,k}}\sinh(\gamma_{n,k}\Delta z) & e^{-\frac{A^{11}_{n,k}-A^{22}_{n,k}}{2}\Delta z} \end{bmatrix} \begin{bmatrix} F_{n,k} \\ R_{n+1,k} \end{bmatrix} + \Delta z \begin{bmatrix} C^{\mathrm{f}}_{n,k} \\ C^{\mathrm{r}}_{n,k} \end{bmatrix}. \tag{6.49}$$

Equation (6.49) forms a general split-step algorithm where the wave dependences on time and on space are treated sequentially rather than simultaneously within every subsection. Without the distributed coupling between the contra-propagated waves through reflection, and under the assumption that the parameters are constants within a subsection, such a treatment is accurate because the time and space operators are commutable. With distributed coupling, however, equation (6.49) is obtained with the distributed delay along the subsection counted as a maximum delay. Obviously, as the coupling becomes stronger, we have to reduce the subsection length Δz to ensure accuracy.

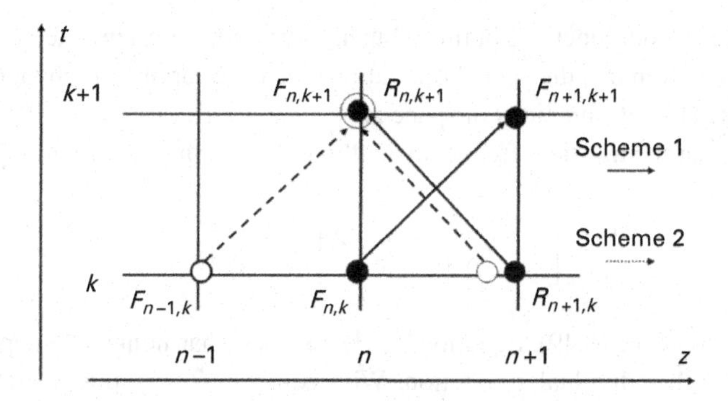

Fig. 6.3. Comparison between two split-step schemes. Parameters in the first scheme, equation (6.49), are consistent in space (z), but inconsistent in time, whereas parameters in the second scheme, equation (6.51), are consistent in time, but inconsistent in space (z).

Because of the symmetry between the time and space operators in the traveling wave equation, we can also write equation (6.1) as

$$\frac{1}{v_g}\frac{\partial}{\partial t}\begin{bmatrix} F(z,t) \\ R(z,t) \end{bmatrix} = \begin{bmatrix} A^{11}(z,t) - \frac{\partial}{\partial z} & A^{12}(z,t) \\ A^{21}(z,t) & A^{22}(z,t) + \frac{\partial}{\partial z} \end{bmatrix} \begin{bmatrix} F(z,t) \\ R(z,t) \end{bmatrix} + \begin{bmatrix} C^{\mathrm{f}}(z,t) \\ C^{\mathrm{r}}(z,t) \end{bmatrix}. \tag{6.50}$$

By following a similar approach, we find

$$\begin{bmatrix} F_{n,k+1} \\ R_{n,k+1} \end{bmatrix} = \mathrm{e}^{\frac{A^{11}_{n,k}+A^{22}_{n,k}}{2}\Delta z} \sinh(\gamma_{n,k}\Delta z)$$
$$\times \begin{bmatrix} \coth(\gamma_{n,k}\Delta z) + \frac{A^{11}_{n,k}-A^{22}_{n,k}}{2\gamma_{n,k}} & \frac{A^{12}_{n,k}}{\gamma_{n,k}} \\ \frac{A^{21}_{n,k}}{\gamma_{n,k}} & \coth(\gamma_{n,k}\Delta z) - \frac{A^{11}_{n,k}-A^{22}_{n,k}}{2\gamma_{n,k}} \end{bmatrix} \begin{bmatrix} F_{n-1,k} \\ R_{n+1,k} \end{bmatrix} + \Delta z \begin{bmatrix} C^{\mathrm{f}}_{n,k} \\ C^{\mathrm{r}}_{n,k} \end{bmatrix}, \tag{6.51}$$

where equation (6.46) still holds and

$$\gamma_{n,k} \equiv \pm\sqrt{\left(\left(\frac{A^{11}_{n,k} - A^{22}_{n,k}}{2}\right)^2 + A^{12}_{n,k}A^{21}_{n,k}\right)}. \tag{6.52}$$

As split-step algorithms, schemes (6.49) and (6.51) are illustrated in Fig. 6.3.

Through these derivations, we find that, in equation (6.49), the space dependence of the parameters A^{11}, A^{12}, A^{21}, A^{22} and the inhomogeneous spontaneous emission contributions is consistent, whereas their time dependence is not. On the contrary, in equation (6.51), the time dependence of these parameters is consistent, whereas their space dependence is not. One way to improve accuracy is to use their arithmetically averaged values at mesh points (n, k) and $(n, k+1)$, and (n, k) and $(n+1, k)$ to replace the original values at (n, k) in equations (6.49) and (6.51), respectively. However, parameter values at mesh point $(n, k+1)$ are usually unknown, which means such an improvement is only viable for equation (6.51). We may also conclude that equation (6.49) is preferred if the parameters are smooth functions of time t but highly non-uniform in space z, whereas equation (6.51) is preferred in the opposite situation where the parameters change rapidly with time t but are smooth functions of space z.

It is also clear that the transient frequency method [26, 27] through assuming

$$\frac{1}{v_g}\frac{\partial}{\partial t}\begin{bmatrix} F(z,t) \\ R(z,t) \end{bmatrix} \approx \beta(z,t)\begin{bmatrix} F(z,t) \\ R(z,t) \end{bmatrix}, \tag{6.53}$$

is a special case of scheme (6.49). Equation (6.53) assumes a harmonic time dependence of the waves in each individual subsection. With equation (6.45), this assumption is not necessary, which makes the split-step schemes, equations (6.49) and (6.51), more comprehensive. For example, equations (6.49) and (6.51) can be employed either to model wave propagation in the laser diode, where the wave is indeed quasi-harmonic, or to model pulse propagation in the SOA, where the wave is far from harmonic.

This method also bridges the traveling wave and standing wave approaches. If we solve equation (6.33) as an eigenvalue problem for the entire device subject to the boundary condition defined at the two facets, to obtain the eigenvalues and eigenfunctions known as the net complex gain (i.e., amplitude gain and frequency detuning) and the longitudinal mode, we can construct our solution to equation (6.32) as a superposition of these eigenfunctions, simply because the eigenfunction set is complete. In this sense we can view the standing wave approach as a one-step split-step method at steady state with $\partial/\partial t \rightarrow 0$ to obtain all possible spatial distributions that are allowed by the longitudinal structure and boundary condition, plus a rate equation analysis to find out how these distributions are combined at any time instant.

There is also a high order split-step method that uses the Lie–Trotter–Suzuki product formula to approximate the time evolution operator, where the stability of the algorithm is guaranteed [28]. Since we use this method just as a mathematical tool, we will skip any detailed discussion on this topic. The application of this method can be found from, e.g., [29].

6.2.3 Time domain convolution through the digital filter

In solving the broadband optical wave equation (2.97), we have to find the polarization given in equation (2.98), which reflects the delayed response of the material. Equation (2.98) can conveniently be implemented in the frequency domain. However, we need to perform the fast Fourier transform (FFT) algorithm back and forth at every progressing time step, which not only brings a considerable computational burden, but also causes a possible aliasing problem in the time domain because of the frequency domain discretization. Therefore, the following time domain convolution is realized through a digital filtering approach [30, 31].

Actually, by truncating the frequency domain susceptibility (as can be seen in Chapter 4, the susceptibility is usually obtained from the material model in the frequency domain directly) beyond a bandwidth of our interest, we construct two periodic functions in the frequency domain which match the real and imaginary parts of the

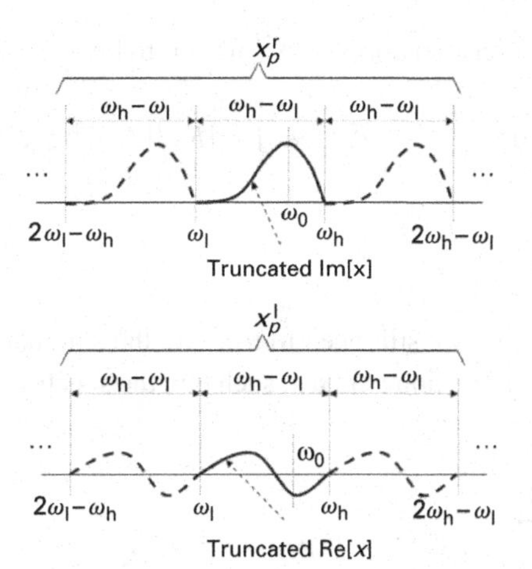

Fig. 6.4. Construction of the periodic susceptibility for digital filter extraction.

retained susceptibility within the band of interest to us, as shown in Fig. 6.4. Namely, we construct

$$\Delta\widetilde{\chi}_P(z,\omega) = \Delta\widetilde{\chi}^{\mathrm{r}}_P(z,\omega) + j\Delta\widetilde{\chi}^{\mathrm{i}}_P(z,\omega), \tag{6.54}$$

with a period of

$$\omega_{\mathrm{B}} \equiv (\omega_0 + \omega_{\mathrm{h}}) - (\omega_0 + \omega_{\mathrm{l}}) = \omega_{\mathrm{h}} - \omega_{\mathrm{l}}, \tag{6.55}$$

where ω_{h} and ω_{l} are the upper and lower cut-off frequencies at which we truncate the frequency domain susceptibility. Within the frequency band of interest, i.e., for $\omega_{\mathrm{l}} < \omega < \omega_{\mathrm{h}}$, we let

$$\Delta\widetilde{\chi}_P(z,\omega) = \Delta\widetilde{\chi}(z,\omega), \tag{6.56}$$

where $\Delta\widetilde{\chi}(z,\omega)$ can be obtained from equation (2.90) once $\widetilde{\chi}(z,\omega)$ is obtained from the material solver.

Since $\Delta\widetilde{\chi}(z,\omega)$ is periodic in terms of ω, through the Fourier expansion, we obtain

$$\Delta\widetilde{\chi}^{\mathrm{r}}_P(z,\omega) = \sum_{n=-\infty}^{\infty} c^{\mathrm{r}}_n(z)\mathrm{e}^{\mathrm{j}nT\omega}, \tag{6.57a}$$

$$\Delta\widetilde{\chi}^{\mathrm{i}}_P(z,\omega) = \sum_{n=-\infty}^{\infty} c^{\mathrm{i}}_n(z)\mathrm{e}^{\mathrm{j}nT\omega}, \tag{6.57b}$$

where

$$c^{\mathrm{i}}_n(z) = \frac{T}{2\pi}\int_{\omega_{\mathrm{l}}}^{\omega_{\mathrm{h}}} \Delta\widetilde{\chi}^{\mathrm{r}}_P(z,\omega)\mathrm{e}^{-\mathrm{j}nTt}\mathrm{d}t, \tag{6.58a}$$

$$c^{\mathrm{i}}_n(z) = \frac{T}{2\pi}\int_{\omega_{\mathrm{l}}}^{\omega_{\mathrm{h}}} \Delta\widetilde{\chi}^{\mathrm{i}}_P(z,\omega)\mathrm{e}^{-\mathrm{j}nTt}\mathrm{d}t, \tag{6.58b}$$

and

$$T \equiv 2\pi/\omega_{\mathrm{B}}. \tag{6.59}$$

Therefore, we find the inverse Fourier transform of $\Delta\tilde{\chi}_P(z,\omega)$ to be

$$\Delta\chi_P(z,t) = F^{-1}[\Delta\tilde{\chi}_P(z,t)] = F^{-1}[\Delta\tilde{\chi}_P^{\mathrm{r}}(z,\omega)] + \mathrm{j}F^{-1}[\Delta\tilde{\chi}_P^{-\mathrm{i}}(z,\omega)]$$
$$= \sum_{n=-\infty}^{\infty} [c_n^{\mathrm{r}}(z) + \mathrm{j}c_n^{\mathrm{i}}(z)]\delta(t-nT). \tag{6.60}$$

To make equation (6.60) computable, we still need to truncate the summation on its RHS. Actually, by introducing a set of real numbers w_n, such that the spectrum (Fourier transform) of the truncated series

$$\Delta\chi_P(z,t) \approx \sum_{n=-N}^{N} w_n[c_n^{\mathrm{r}}(z) + \mathrm{j}c_n^{\mathrm{i}}(z)]\delta(t-nT), \tag{6.61}$$

will have minimum error in comparing with $\Delta\tilde{\chi}(z,\omega)$ within the frequency band of interest, i.e., $\omega_{\mathrm{l}} < \omega < \omega_{\mathrm{h}}$, we find w_n by minimizing

$$\varepsilon_{\mathrm{error}} = \sum_{m=1}^{M}\left\{\sum_{n=-N}^{N} w_n[c_n^{\mathrm{r}}(z) + \mathrm{j}c_n^{\mathrm{i}}(z)]\mathrm{e}^{\mathrm{j}nT\omega_m} - \Delta\tilde{\chi}(z,\omega_m)\right\}^2, \tag{6.62}$$

with $\omega_{\mathrm{l}} < \omega_m < \omega_{\mathrm{h}}$, $m = 1, 2, 3, \ldots, M$ denoting a set of frequency samples within the band of interest. Hence equation (6.61) is fully determined with a minimum error for an integer N selected in truncating equation (6.60).

In computing convolution equation (2.98), we can therefore use $\Delta\chi_P(z,t)$ in equation (6.61) to replace $\Delta\chi(z,t)$. The only difference is that $\Delta\chi(z,t)$ is continuous, whereas $\Delta\chi_P(z,t)$ is discrete. The latter is a sampling version of the former with the sampling rate guaranteed dense enough to capture any fast change in time domain as we need, since we can always set ω_{B} sufficiently large. As a result, equation (2.98) becomes

$$\Delta p^{\mathrm{f,b}}(z,t) \approx \varepsilon_0 \int_{-\infty}^{t} \Delta\chi_P(z,t-\tau)e^{\mathrm{f,b}}(z,\tau)\mathrm{e}^{\mathrm{j}\omega_0(t-\tau)}\mathrm{d}\tau$$
$$= \varepsilon_0 \int_{-\infty}^{t} \sum_{n=-M}^{M} w_n[c_n^{\mathrm{r}}(z) + \mathrm{j}c_n^{\mathrm{i}}(z)]\delta(t-nT-\tau)e^{\mathrm{f,b}}(z,\tau)\mathrm{e}^{\mathrm{j}\omega_0(t-\tau)}\mathrm{d}\tau$$
$$= \varepsilon_0 \sum_{n=-M}^{M} w_n \mathrm{e}^{\mathrm{j}nT\omega_0}[c_n^{\mathrm{r}}(z) + \mathrm{j}c_n^{\mathrm{i}}(z)]e^{\mathrm{f,b}}(z,t-nT). \tag{6.63}$$

Note that, for a given material system, $\Delta\tilde{\chi}(z,\omega)$ is determined, hence extraction of the coefficients in equation (6.63), i.e., c_n^{r}, c_n^{i}, w_n, $n = 0, \pm1, \pm2, \ldots, \pm N$, needs to be executed only once. As opposed to integration (2.98) in its original form, equation (6.63) can be computed in a much more efficient way, as we need simply to perform a series of shift and add operations.

6.3 Standing wave equations

In dealing with the standing wave model, we have to solve equations (2.135), (2.139) and (2.126). Equation (2.139) involves only numerical quadrature that has been extensively discussed in, e.g., [19]. Equation (2.126) is a set of ODEs, the solution technique for which will be discussed in Chapter 7. As for equation (2.135), it fits into a form similar to equation (6.32), hence the related solution technique can be built in a similar way, known as the transfer matrix method. However, there are two unique problems that we face in solving equation (2.135) and we will focus on these two points in this section.

The first problem is how to split the terms in equation (2.122). Depending on how we select the reference for the external bias, we have three different strategies.

The first strategy is to use the zero bias as the reference, known as the "cold cavity" model [32]. The advantage of this model is that the parameters in equation (2.135) are not bias dependent, hence they are not time dependent. Therefore, we just need to solve equation (2.135) once. The drawback, however, is that usually a large number of the eigenfunctions (i.e., the longitudinal optical modes) obtained as the solution to equation (2.135) have to be retained in the mode expansion (2.125a) to ensure accuracy. This increases the computational burden of equations (2.139) and (2.126) as the computational complexity is proportional to the square of (not linearly proportional to) the number of the modes retained, because of the cross-coupling terms in equations (2.139) and (2.126). For this reason, as the number of modes retained increases, the required computation time grows fairly rapidly. It is clear that once the square root of the number of modes retained becomes comparable to the total time steps, the advantage gained by not solving the eigenvalue problem, equation (2.135), repeatedly at every time step will be cancelled out.

The second strategy is to use the adiabatic bias as the reference, known as the "floating bias" model [33]. As opposed to the "cold cavity" model, the advantage of the "floating bias" is that only a minimum number of modes needs to be retained in equation (2.125a). For single mode operated devices such as DFB lasers, we appreciate this feature as only one mode needs to be retained, which makes solving equations (2.139) and (2.126) extremely efficient. The major drawback, however, is that we have to solve equation (2.135) repeatedly at every time step. In reality, we may find a situation in which the bias is adiabatic for the whole time range (e.g., in DC analysis) or within a time period (e.g., in the top or bottom flat part of a square pulsed bias). Since there is no need to solve equation (2.135) repeatedly for adiabatic biases without an appreciable difference, by storing the modes for consecutive steps where the adiabatic bias is invariant, we may reduce the computation time significantly. As a last point to make this model clear, the adiabatic bias can be obtained through the step-function decomposition for an arbitrary given time-dependent bias function

$$V(t) = V(0)U(t) + \int_0^t \frac{\mathrm{d}V(\tau)}{\mathrm{d}\tau} U(t-\tau)\mathrm{d}\tau, \tag{6.64}$$

where $U(t)$ is the step function and the integral should be replaced by a corresponding summation in numerical implementations.

The third strategy is somewhere between the other two; i.e., we use a fixed, but not necessarily zero adiabatic bias as the reference, known as the "hot cavity" model. Using such a reference, equation (2.135) is still solved only once, which is similar to the "cold cavity" model. However, since we choose a reference that is in the neighborhood of an area in which the optical field undergoes little change, there is no need to retain a large number of modes in equation (2.125a) either, as the chance that we may miss the operating mode is slight. An obvious selection in laser modeling is to use the threshold bias as the reference. It is well known that for a bias above the threshold, the carrier density and hence the gain is somehow clamped, especially when the LSHB effect is not severe. As a result, the longitudinal optical field distribution, i.e., the longitudinal optical mode, undergoes little change from its shape at the threshold bias. Once the modes are solved at the threshold bias, we are certain that the optical field under above-threshold biases will not skip the combination of the few modes that have been found at the threshold. To determine the threshold bias, one can always use the uniform field assumption where the solution can be obtained analytically. The exact equation (2.135) will then be solved under such an obtained threshold for the "hot cavity" modes. The drawback of this model lies in the fact that it is sometimes difficult to find such a reference bias point, especially for device structures with a weak resonance or low quality factor. The standing wave model is suitable only for resonant structures, as otherwise we would not even be able to find any modes, but for the device structures to which this model is applicable, usually we can find a non-zero reference bias at which the required number of modes in the expansion is greatly suppressed.

These three strategies are illustrated in Fig. 6.5. We can actually establish a mapping between these models and well-known methods in modeling passive optical waveguides. The "cold cavity" and "hot cavity" models correspond to "global mode" expansion, whereas the "floating bias" model is similar to "local mode" expansion. For example, in an analysis of the horn waveguide with the mode expansion method, we could use a set of modes obtained from (1) one end of the structure, (2) the structure with an averaged width, or (3) the structure with the exact width at each cross-section, where (1) and (2) are known as the "global mode" (though with different reference), and (3) is known as the "local mode" expansion method, respectively. With our bias mapped to the width of the horn waveguide, we find that the "cold cavity" and "hot cavity" models correspond to methods (1) and (2) or "global mode" expansion, while the "floating bias" model corresponds to method (3) or "local mode" expansion.

The second problem is related to solving equation (2.135) itself. Although the formulations related to equation (2.135) are similar to equations (6.40) and (6.41) from subsection to subsection, the solution techniques are completely different as equation (2.135) is posed as an eigenvalue problem subject to the boundary condition at the two ends of the device, whereas problem equation (6.32), as the origin of equations (6.40) and (6.41), is not.

Generally, with the coupling between the right- and left-going traveling waves considered to be caused by the distributed reflection, equation (2.135) can be posed in a

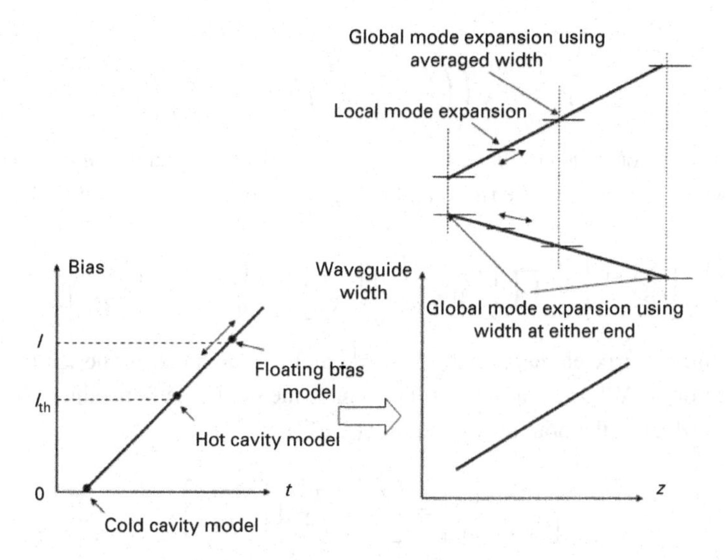

Fig. 6.5. Modal expansions taking different references.

similar form to equation (6.32) for any given adiabatic bias, i.e.,

$$\pm\frac{\mathrm{d}}{\mathrm{d}z}\begin{bmatrix} F(z) \\ R(z) \end{bmatrix} = \begin{bmatrix} A^{11}(z) & A^{12}(z) \\ A^{21}(z) & A^{22}(z) \end{bmatrix}\begin{bmatrix} F(z) \\ R(z) \end{bmatrix}, \tag{6.65}$$

with the unknown eigenvalue embedded in A^{11} and A^{22}. Because of the generally non-uniform dependence of the parameters on space z, we do not expect any analytical solutions to equation (6.65). Following the transfer matrix approach [34], we subdivide the device into subsections along z, with each subsection short enough such that the parameters can be approximated as constants. As such, the solution to equation (6.65) in each subsection takes a form similar to equations (6.40) and (6.41)

$$\begin{bmatrix} F_{n+1} \\ R_{n+1} \end{bmatrix} = \begin{bmatrix} a_n^{11} & a_n^{12} \\ a_n^{21} & a_n^{22} \end{bmatrix}\begin{bmatrix} F_n \\ R_n \end{bmatrix}, \tag{6.66}$$

where

$$a_n^{11} \equiv [\cosh(\gamma_n\Delta z) + \frac{A_n^{11}+A_n^{22}}{2\gamma_n}\sinh(\gamma_n\Delta z)]\mathrm{e}^{\frac{A_n^{11}-A_n^{22}}{2}\Delta z}$$

$$a_n^{12} \equiv \frac{A_n^{12}}{\gamma_n}\sinh(\gamma_n\Delta z)\mathrm{e}^{\frac{A_n^{11}-A_n^{22}}{2}\Delta z}$$

$$a_n^{21} \equiv -\frac{A_n^{21}}{\gamma_n}\sinh(\gamma_n\Delta z)\mathrm{e}^{\frac{A_n^{11}-A_n^{22}}{2}\Delta z}$$

$$a_n^{22} \equiv [\cosh(\gamma_n\Delta z) - \frac{A_n^{11}+A_n^{22}}{2\gamma_n}\sinh(\gamma_n\Delta z)]\mathrm{e}^{\frac{A_n^{11}-A_n^{22}}{2}\Delta z}, \tag{6.67}$$

and

$$\gamma_n \equiv \pm\sqrt{\left(\left(\frac{A_n^{11}+A_n^{22}}{2}\right)^2 - A_n^{12}A_n^{21}\right)}. \tag{6.68}$$

In equation (6.66), the coefficient matrix on the RHS is named as the transfer matrix, through which we can link the right- and left-going waves at the two ends of the device through

$$\begin{bmatrix} F_{N+1} \\ R_{N+1} \end{bmatrix} = \prod_{n=N}^{1} \begin{bmatrix} a_n^{11} & a_n^{12} \\ a_n^{21} & a_n^{22} \end{bmatrix} \begin{bmatrix} F_1 \\ R_1 \end{bmatrix} = \begin{bmatrix} a^{11} & a^{12} \\ a^{21} & a^{22} \end{bmatrix} \begin{bmatrix} F_1 \\ R_1 \end{bmatrix}, \tag{6.69}$$

with the final matrix elements a^{11}, a^{12}, a^{21}, and a^{22} obtained numerically from the multiplication of N 2×2 matrices. At resonance, the field is self-sustained without any input. According to the boundary condition (2.137), we have

$$\begin{bmatrix} F_{N+1} \\ R_r F_{N+1} \end{bmatrix} = \begin{bmatrix} a^{11} & a^{12} \\ a^{21} & a^{22} \end{bmatrix} \begin{bmatrix} R_l R_1 \\ R_1 \end{bmatrix}. \tag{6.70}$$

Therefore, the resonance condition is obtained by

$$R_r R_l a^{11} + R_r a^{12} - R_l a^{21} - a^{22} = 0. \tag{6.71}$$

As the unknown, the eigenvalue is embedded in equation (6.71) through the matrix elements. Hence equation (6.71) is the eigenequation that determines the eigenvalues.

Once the eigenvalues are found from equation (6.71) through a root-searching algorithm, the corresponding field distributions, or the optical modes, can be found through

$$\begin{bmatrix} F'_{m+1} \\ R'_{m+1} \end{bmatrix} = \prod_{n=m}^{1} \begin{bmatrix} a_n^{11} & a_n^{12} \\ a_n^{21} & a_n^{22} \end{bmatrix} \begin{bmatrix} R_l \\ 1 \end{bmatrix}, \tag{6.72}$$

for $m = 1, 2, 3, \ldots, N$. For the kth mode, we should use the kth eigenvalue obtained as the solution to equation (6.71) in those elements in the transfer matrices from subsection to subsection.

Once the field prior to normalization is found through equation (6.72), the normalization factor can be computed through

$$C = \sqrt{\left(1 + |R_l|^2 + \sum_{m=1}^{N} (|F'_{m+1}|^2 + |R'_{m+1}|^2)\right)}. \tag{6.73}$$

Finally, the normalized optical mode is found through

$$\begin{bmatrix} F_1 \\ R_1 \end{bmatrix} = \frac{1}{C} \begin{bmatrix} R_l \\ 1 \end{bmatrix}, \quad \begin{bmatrix} F_{m+1} \\ R_{m+1} \end{bmatrix} = \frac{1}{C} \begin{bmatrix} F'_{m+1} \\ R'_{m+1} \end{bmatrix}, \tag{6.74}$$

for $m = 1, 2, 3, \ldots, N$.

In solving equation (6.71), a complex variable numerical root-searching routine has to be employed. Usually, Muller's method [35, 36] is preferred in developing such a routine because it is possible to obtain complex roots even with a real number initial guess, whereas in the more popular Newton's method or the secant method, one can never find a complex root starting from a real number initial guess. One can also employ the approach in [37] for complex root-searching.

Depending on the complexity of the structure, in case the searching fails or hits the wrong root, we can always rely on a successive scanning algorithm that guarantees the search's success. Actually, we can always start searching from a simple grating structure (e.g., with zero facet reflection) for which we know the solution. Taking this solution as the initial guess and altering the parameters one by one, gradually towards the given structure, we can always find the correct root as long as we keep updating the initial guess with the previous searching outcome.

To find the exact solution as the initial guess, we can always look at the simplest case where the parameters in equation (6.65) are constants with zero reflections at the device facets.

Noting that

$$\begin{bmatrix} a^{11}(x) & a^{12}(x) \\ a^{21}(x) & a^{22}(x) \end{bmatrix} \begin{bmatrix} a^{11}(y) & a^{12}(y) \\ a^{21}(y) & a^{22}(y) \end{bmatrix} = \begin{bmatrix} a^{11}(x+y) & a^{12}(x+y) \\ a^{21}(x+y) & a^{22}(x+y) \end{bmatrix}, \tag{6.75}$$

where

$$\begin{aligned} a^{11}(x) &\equiv \left[\cosh(\gamma x) + \frac{A^{11}+A^{22}}{2\gamma}\sinh(\gamma x)\right] e^{\frac{A^{11}-A^{22}}{2}x} \\ a^{22}(x) &\equiv \frac{A^{12}}{\gamma}\sinh(\gamma x) e^{\frac{A^{11}-A^{22}}{2}x} \\ a^{21}(x) &\equiv -\frac{A^{21}}{\gamma}\sinh(\gamma x) e^{\frac{A^{11}-A^{22}}{2}x} \\ a^{22}(x) &\equiv \left[\cosh(\gamma x) - \frac{A^{11}+A^{22}}{2\gamma}\sinh(\gamma x)\right] e^{\frac{A^{11}-A^{22}}{2}x}, \end{aligned} \tag{6.76}$$

with γ still defined by equation (6.68) with constant parameters A^{11}, A^{12}, A^{21}, and A^{22}, we find

$$\prod_{n=N}^{1} \begin{bmatrix} a^{11}(\Delta z) & a^{12}(\Delta z) \\ a^{21}(\Delta z) & a^{22}(\Delta z) \end{bmatrix} = \begin{bmatrix} a^{11}(L) & a^{12}(L) \\ a^{21}(L) & a^{22}(L) \end{bmatrix}, \tag{6.77}$$

with L introduced as the total device length in the z direction.

Since the eigenequation (6.71) reduces to $a^{22} = 0$ under the zero reflection assumption, we find from equations (6.76) and (6.77)

$$\cosh(\gamma L) - \frac{A^{11}+A^{22}}{2\gamma}\sinh(\gamma L) = 0, \text{ or } \gamma\coth(\gamma L) = \frac{A^{11}+A^{22}}{2}, \tag{6.78}$$

which is the same as [38] when we apply equation (6.65) to DFB laser diodes.

Equation (6.78) is given in a transcendental form and we are able to find some properties possessed by the eigenvalues. For example, in DFB laser diodes with a first order grating, from equations (2.84) and (2.135), we find

$$\frac{\mathrm{d}e^{\mathrm{f}}(z)}{\mathrm{d}z} = \left[\mathrm{j}\delta + \mathrm{j}k_0\Gamma\Delta n_V(z,\omega_0) + \frac{1}{2}\Gamma g_V(z,\omega_0) - \frac{1}{2}\bar{\alpha}(z) - \mathrm{j}\frac{\beta_V^2}{2\beta_0} \right] e^{\mathrm{f}}(z)$$
$$+ \mathrm{j}\kappa_1 e^{\mathrm{b}}(z,t) - \frac{\mathrm{d}e^{\mathrm{b}}(z)}{\mathrm{d}z}$$
$$= \left[\mathrm{j}\delta + \mathrm{j}k_0\Gamma\Delta n_V(z,\omega_0) + \frac{1}{2}\Gamma g_V(z,\omega_0) - \frac{1}{2}\bar{\alpha}(z) - \mathrm{j}\frac{\beta_V^2}{2\beta_0} \right] e^{\mathrm{b}}(z)$$
$$+ \mathrm{j}\kappa_{-1} e^{\mathrm{f}}(z,t). \tag{6.79}$$

By comparing equation (6.79) with (6.65) and ignoring the space dependence of the parameters, we find

$$A^{11} = \mathrm{j}\delta + \mathrm{j}k_0\Gamma\Delta n_V + \frac{1}{2}\Gamma g_V - \frac{1}{2}\bar{\alpha} - \mathrm{j}\frac{\beta_V^2}{2\beta_0},\ A^{12} = \mathrm{j}\kappa_1$$
$$A^{21} = \mathrm{j}\kappa_{-1},\ A^{22} = \mathrm{j}\delta + \mathrm{j}k_0\Gamma\Delta n_V + \frac{1}{2}\Gamma g_V - \frac{1}{2}\bar{\alpha} - \mathrm{j}\frac{\beta_V^2}{2\beta_0}, \tag{6.80}$$

once we map $e^{\mathrm{f}}(z)$ and $e^{\mathrm{b}}(z)$ to $F(z)$ and $R(z)$, respectively. Hence equation (6.78) becomes

$$\gamma\coth(\gamma L) = \alpha + \mathrm{j}\beta,\ \text{ or } \gamma^2\coth^2(\gamma L) = \gamma^2 - \kappa_1\kappa_{-1}, \tag{6.81}$$

with

$$\gamma = \pm\sqrt{\left((\alpha + \mathrm{j}\beta)^2 + \kappa_1\kappa_{-1}\right)},\ \text{ or } \alpha + \mathrm{j}\beta = \pm\sqrt{\left(\gamma^2 - \kappa_1\kappa_{-1}\right)}, \tag{6.82}$$

and the virtual gain and detuning defined as

$$\alpha \equiv \frac{1}{2}\Gamma g_V - \frac{1}{2}\bar{\alpha} + \frac{\mathrm{Im}(\beta_V^2)}{2\beta_0}$$
$$\beta \equiv \delta + k_0\Gamma\Delta n_V - \frac{\mathrm{Re}(\beta_V^2)}{2\beta_0}. \tag{6.83}$$

Once the virtual gain and detuning are found as the root of equation (6.81) for any given coupling coefficients κ_1 and κ_{-1}, we obtain the imaginary and real part of the eigenvalue.

In the purely index coupled grating, once the grating is symmetric or anti-symmetric, we have $\kappa_1 = \kappa_{-1} = |\kappa|$ or $\kappa_1 = -\kappa_{-1} = \mathrm{j}|\kappa|$. In either case, equations (6.81) and (6.82) become

$$\gamma^2\coth^2(\gamma L) = \gamma^2 - |\kappa|^2, \tag{6.84}$$

and

$$\alpha + \mathrm{j}\beta = \pm\sqrt{(\gamma^2 - |\kappa|^2)}. \tag{6.85}$$

It is clear that, according to equation (6.84), if γ is a solution, $-\gamma$ and $\pm\gamma^*$ are all solutions, since $(\gamma^2)^* = (\gamma^*)^2$, $[\coth^2(\gamma)]^* = \coth^2(\gamma^*)$. As a result, we find from equation (6.85) that if (α, β) is a solution, $(\pm\alpha, \pm\beta)$ will be solutions as well, since $\sqrt{\left(\gamma^2 - |\kappa|^2\right)^*} = \sqrt{\left((\gamma^*)^2 - |\kappa|^2\right)}$.

Similarly, in purely gain or loss coupled gratings, once the grating is symmetric or anti-symmetric, we have $\kappa_1 = \kappa_{-1} = \mathrm{j}|\kappa|$ or $\kappa_1 = -\kappa_{-1} = |\kappa|$. In either case, equations (6.81) and (6.82) become

$$\gamma^2 \coth^2(\gamma L) = \gamma^2 + |\kappa|^2, \tag{6.86}$$

and

$$\alpha + \mathrm{j}\beta = \pm\sqrt{\left(\gamma^2 + |\kappa|^2\right)}. \tag{6.87}$$

Thus we still have a similar conclusion. An apparent difference between these two particular grating structures is that in the purely index coupled grating, $\beta = 0$ is not a solution as, otherwise, γ will be real according to equation (6.85), which makes the LHS of equation (6.84) $> \gamma^2$ but the RHS of (6.84) $< \gamma^2$. However, for purely gain or loss coupled gratings, real γ with $\beta = 0$ can be a solution according to equations (6.87) and (6.86).

We may conclude that, if $\kappa_1\kappa_{-1}$ is real, as the root of equation (6.81), γ and hence (α, β) appear in quadriads. That is to say, the solution to the eigenequation (6.78) has quadruple-degeneracy once the product of A^{12} and A^{21} is real. If $A^{12}A^{21}$ is complex or the parameters in equation (6.65) have space dependence, although such degeneracy breaks, we can still find three other accompanying roots in the neighborhood of $(\alpha, -\beta)$ and $(-\alpha, \pm\beta)$ for every root at (α, β). Following this guidance, it is unlikely that we will miss any root if Muller's algorithm is adopted for root-searching.

A final problem is to determine which root should be qualified as the eigenvalue with its corresponding eigenfunction retained as an optical mode in the expansion (2.125a). This is not a mathematical problem and so we need the physics on the mode ranking. Actually, for roots γ with similar β, we should retain those with smaller α, while for roots γ with similar α, we should retain those with smaller $|\beta|$, since the modes with smaller α and $|\beta|$ need smaller gain and are less detuned, which qualifies them as the lower order modes to be retained. Particularly in lasers, the eigenequation is the oscillation condition. Modes with smaller α and $|\beta|$ have lower thresholds and are closer to phase matching. As the injection increases from zero, these modes will reach their amplitude and phase oscillation conditions first and hence become the lasing modes. For roots with large α but small $|\beta|$, or small α but large $|\beta|$, we should rank the lower order modes as those with smaller $|\gamma|$. Lastly, for modes with similar $|\gamma|$ but different α and $|\beta|$, we should leave all of them as the modes in the expansion (2.125a).

References

[1] Working group I, COST-216, Comparison of different modeling techniques for longitudinally invariant integrated optical waveguides. *IEE Proc. J*, **136** (1989), 273–80.

[2] W.-P. Huang, (Ed.) *PIER 10 Method for Modeling and Simulation of Guided-Wave Optoelectronic Devices: Part I Modes and Couplings*, 1st edn (Cambridge, MA: EMW Publishing, 1995).

[3] C. Vassallo, 1993-1995 optical mode solvers. *Optical and Quantum Electron.*, **29** (1997), 95–114.

[4] G. Zhou and X. Li, Wave equation based semi-vectorial compact 2D-FDTD method for optical waveguide modal analysis. *IEEE/OSA Journal of Lightwave Tech.*, **LT-22**:2 (2004), 677–83.

[5] M. Wik, D. Dumas, and D. Yevick, Comparison of vector finite-difference techniques for modal analysis. *Journal of Opt. Soc. Am. A*, **22**:7 (2005), 1341–7.

[6] N. Thomas, P. Sewell, and T. Benson, A new full-vectorial high order finite-difference scheme for the modal analysis of rectangular dielectric waveguides. *IEEE/OSA Journal of Lightwave Tech.*, **LT-25**:9 (2007), 2563–70.

[7] P.-J. Chiang, C.-L. Wu, C.-H. Teng, C.-S. Yang, and H.-C. Zhang, Full-vectorial optical waveguide mode solvers using multidomain pseudospectral frequency-domain (PSFD) formulations. *IEEE Journal of Quantum Electron.*, **QE-44**:1 (2008), 56–66.

[8] K. Bierwirth, N. Schulz, and F. Arndt, Finite-difference analysis of rectangular dielectric waveguide structures. *IEEE Trans. on Microwave Theory and Tech.*, **MTT-34**:11 (1986), 1104–13.

[9] P. Lusse, P. Stuwe, J. Schule, and H.-G. Unger, Analysis of vectorial mode fields in optical waveguides by a new finite difference method. *IEEE/OSA Journal of Lightwave Tech.*, **LT-12**:3 (1994), 487–94.

[10] G. R. Hadley, High-accuracy finite-difference equations for dielectric waveguide analysis II: dielectric corners. *IEEE/OSA Journal of Lightwave Tech.*, **LT-20**:7 (2002), 1219–31.

[11] T. Lu and D. Yevick, Comparative evaluation of a novel series approximation for electromagnetic fields at dielectric corners with boundary element method applications. *IEEE/OSA Journal of Lightwave Tech.*, **LT-22**:5 (2004), 1426–32.

[12] C. Vassallo and J. M. van der Keur, Comparison of a few transparent boundary conditions for finite-difference optical mode-solvers. *IEEE/OSA Journal of Lightwave Tech.*, **LT-15**:2 (1997), 397–402.

[13] D. Jimenez and F. Perez-Murano, Comparison of highly efficient absorbing boundary conditions for the beam propagation method. *Journal of Opt. Soc. Am. A*, **18**:8 (2001), 2015–25.

[14] J.-P. Berenger, A perfect matched layer for the absorption of electromagnetic waves. *Journal of Comp. Phys.*, **114** (1994), 185–200.

[15] M. S. Stern, Semivectorial polarized finite difference method for optical waveguides with arbitrary index profiles. *IEE Proc. J*, **135** (1988), 56–63.

[16] C. Vassallo, *Optical Waveguide Concepts*, 1st edn (Amsterdam: Elsevier, 1991).

[17] J. W. Thomas, *Numerical Partial Differential Equations: Finite Difference Methods*, 1st edn (New York: Springer-Verlag, 1995).

[18] J. C. Strikwerda, *Finite Difference Schemes and Partial Differential Equations*, 1st edn (New York: Chapman & Hall, 1989).

[19] W. H. Press, S. A. Teukolsky, W. T. Vetterling, and B. P. Flannery, *Numerical Recipes in Fortran: The Art of Scientific Computing*, 2nd edn (Cambridge, UK: Cambridge University Press, 1992).

[20] L. M. Zhang, S. F. Yu, M. C. Nowell, *et al.*, Dynamic analysis of radiation and side-mode suppression in a second-order DFB laser using time-domain large-signal traveling wave model. *IEEE Journal of Quantum Electron.*, **QE-30**:6 (1994), 1389–95.

[21] W. Li, W.-P. Huang, X. Li, and J. Hong, Multiwavelength gain-coupled DFB laser cascade: design modeling and simulation. *IEEE Journal of Quantum Electron.*, **QE-36**:10 (2000), 1110–6.

[22] C. F. Tsang, D. D. Marcenac, J. E. Carroll, and L. M. Zhang, Comparison between "power matrix model" and "time domain model" in modeling large signal responses of DFB lasers. *IEE Proc. Optoelectron.*, **141**:2 (1994), 89–96.

[23] G. C. Dente and M. L. Tilton, Modeling multiple-longitudinal-mode dynamics in semiconductor lasers. *IEEE Journal of Quantum Electron.*, **QE-34**:2 (1998), 325–35.

[24] H. De Raedt, J. S. Kole, K. F. L. Michielsen, and M. T. Figge, Unified framework for numerical methods to solve the time-dependent Maxwell equations. *Comp. Phys. Comm.*, **156** (2003) 43–61.

[25] B.-S. Kim, Y. Chung, and J.-S. Lee, An efficient split-step time-domain dynamic modeling of DFB/DBR laser diodes. *IEEE Journal of Quantum Electron.*, **QE-36**:7 (2000), 787–94.

[26] D. J. Jones, L. M. Zhang, J. E. Carroll, and D. D. Marcenac, Dynamic of monolithic passively mode-locked semiconductor lasers. *IEEE Journal of Quantum Electron.*, **QE-31**:6 (1995), 1051–8.

[27] S. F. Yu, Time-domain traveling-wave algorithms on the analysis of distributed feedback lasers. *IEE Proc. Optoelectron.*, **150**:3 (2003), 266–72.

[28] M. Suzuki, Decomposition formulas of exponential operators and Lie exponentials with some applications to quantum mechanics and statistical physics. *Journal of Math. Phys.*, **26**:4 (1985), 601–12.

[29] Y.-P. Xi, W.-P. Huang, and X. Li, High-order split-step schemes for time-dependent coupled-wave equations. *IEEE Journal of Quantum Electron.*, **QE-43**:5 (2007), 419–25.

[30] M. Kolesik and J. V. Moloney, Spatial digital filter method for broad-band simulation of semiconductor lasers. *IEEE Journal of Quantum Electron.*, **QE-37**:7 (2001), 936–44.

[31] W. Li, W.-P. Huang, and X. Li, Digital filter approach for simulation of a complex integrated laser diode based on the traveling-wave model. *IEEE Journal of Quantum Electron.*, **QE-40**:5 (2004), 473–80.

[32] Y.-P. Xi, X. Li, and W.-P. Huang, Time-domain standing-wave approach based on cold cavity modes for simulation of DFB lasers. *IEEE Journal of Quantum Electron.*, **QE-44**: 10 (2008), 931–7.

[33] X. Li, A. D. Sadovnikov, W.-P. Huang, and T. Makino, A physics-based three-dimensional model for distributed feedback (DFB) laser diodes. *IEEE Journal of Quantum Electron.*, **QE-34**:9 (1998), 1545–53.

[34] M. Yamada and K. Sakuda, Analysis of almost-periodic distributed feedback slab waveguides via a fundamental matrix approach. *Appl. Optics*, **26**:16 (1987), 3474–8.

[35] D. E. Muller, A method for solving algebraic equations using an automatic computer. *MTAC*, **10** (1956), 208–15.

[36] S. D. Conte and C. De Boor, *Elementary Numerical Analysis – An Algorithm Approach*, 1st edn (New York: McGraw-Hill, 1980).

[37] L. C. Botten, M. S. Craig, and R. C. McPhedran, Complex zeros of analytic functions. *Comp. Phys. Comm.*, **29** (1983), 245–59.

[38] H. Kogelnik and C. V. Shank, Coupled-wave theory of distributed feedback lasers. *Journal of Appl. Phys.*, **43**:5 (1972), 2327–35.

7 Solution techniques for material gain equations

7.1 Single electron band structures

For bulk semiconductors, a band structure calculation is made to diagonalize the Luttinger–Kohn Hamiltonian matrix to find the energies as the eigenvalues and the electron (and hole) wave functions as the eigenfunctions. For low dimension matrices, we can pose the eigenvalue problem as $\det|\overline{H}^{\mathrm{LK}} - E\overline{I}| = 0$ and use a root-searching routine to find the eigenvalues as the roots of this equation. For high dimension matrices, however, the root-searching scheme is less efficient and hence we should switch to the matrix computation methods. Since the matrices involved are Hermitian, a number of algorithms can be used, such as the Jacobian transformation, or the Householder reduction (to tridiagonal matrix) plus orthogonal-lower-triangular (QL) iteration [1, 2].

For QW structures, we have to solve the 1D coupled ODEs posed as the boundary value problem. Either the FD method with the center discretization scheme or the transfer matrix method [3–5] can be employed. In some device applications (e.g., in EAMs and PDs), the potential applied to the QW changes with the external bias, which deforms the QW and unbinds the conduction band electron and valence band hole. As a result, the conduction band electron and valence band hole may stay at the partially bound state, a status similar to the leaky mode in a slanted or curved dielectric waveguide. In looking for the energy bands and wave functions through the FD method for the conduction band electrons and valence band holes as the eigensolutions to the effective mass equations, i.e., equations (3.129) for electrons and (3.134) for holes, we have to incorporate the absorbing boundary condition or the more efficient PML boundary condition in a similar way to [6, 7] to truncate the computation window. We will skip the details of the single electron band structure calculation since there is a vast amount of related work in open literature [8].

7.2 Material gain calculations

7.2.1 The free-carrier gain model

The material susceptibility is generally computed in the wave vector domain by solving a group of simultaneous ODEs. In the free-carrier model with the many-body effect ignored, the material gain and refractive index change are given as the wave vector domain summations shown in equations (4.65a&b). Since the band structure of the active region material must be solved anyway, we usually perform the summation (or integration) with respect to the transition energy ($E = \hbar\bar{\omega}_{\vec{k}}$) rather than to the wave vector ($\vec{k}$) by utilizing the pre-solved band structure, or the dispersion relation between E and $\vec{k}$. The reason is that the summation (or integration) over E has a limited lower bound (i.e., the bandgap energy) while the integrand drops to zero rapidly near the upper bound.

Actually, for bulk semiconductors, we have

$$\sum_{\vec{k}} = 2\int_{-\infty}^{\infty}\int_{-\infty}^{\infty}\int_{-\infty}^{\infty} \mathrm{d}n_x\, \mathrm{d}n_y\, \mathrm{d}n_z = 2\int_{-\infty}^{\infty}\int_{-\infty}^{\infty}\int_{-\infty}^{\infty} \mathrm{d}k_x\, \mathrm{d}k_y\, \mathrm{d}k_z\left(\frac{\mathrm{d}n_x}{\mathrm{d}k_x}\frac{\mathrm{d}n_y}{\mathrm{d}k_y}\frac{\mathrm{d}n_z}{\mathrm{d}k_z}\right)$$

$$= 2\int_{-\infty}^{\infty}\int_{-\infty}^{\infty}\int_{-\infty}^{\infty} \mathrm{d}k_x\, \mathrm{d}k_y\, \mathrm{d}k_z\left(\frac{L_x}{2\pi}\frac{L_y}{2\pi}\frac{L_z}{2\pi}\right) = \frac{2\Omega}{(2\pi)^3}\int_{-\infty}^{\infty}\int_{-\infty}^{\infty}\int_{-\infty}^{\infty} \mathrm{d}k_x\, \mathrm{d}k_y\, \mathrm{d}k_z$$

$$= \frac{2\Omega}{(2\pi)^3}\int_0^{\infty}\int_0^{\pi}\int_0^{2\pi} k_r^2\, \mathrm{d}k_r \sin\theta_k\, \mathrm{d}\theta_k\, \mathrm{d}\varphi_k, \tag{7.1a}$$

with n_x, n_y, and n_z indicating the number of states in the wave vector domain along the x, y, and z directions, respectively, L_x, L_y, and L_z the dimensions of the bulk in a real 3D space domain where $L_xL_yL_z = \Omega$.

With spherical symmetry, equation (7.1a) reduces to

$$\sum_{\vec{k}} = \frac{\Omega}{\pi^2}\int_0^{\infty} k_r^2\, \mathrm{d}k_r \equiv \frac{\Omega}{\pi^2}\int_0^{\infty} k^2\, \mathrm{d}k. \tag{7.1b}$$

For QW structures, we have

$$\sum_{\vec{k}} = \sum_{n_z} 2\int_{-\infty}^{\infty}\int_{-\infty}^{\infty} \mathrm{d}n_x\, \mathrm{d}n_y = 2\sum_{n_z}\int_{-\infty}^{\infty}\int_{-\infty}^{\infty} \mathrm{d}k_x\, \mathrm{d}k_y\left(\frac{\mathrm{d}n_x}{\mathrm{d}k_x}\frac{\mathrm{d}n_y}{\mathrm{d}k_y}\right)$$

$$= 2\sum_{n_z}\int_{-\infty}^{\infty}\int_{-\infty}^{\infty} \mathrm{d}k_x\, \mathrm{d}k_y\left(\frac{L_x}{2\pi}\frac{L_y}{2\pi}\right) = \frac{2\Sigma}{(2\pi)^2}\sum_{n_z}\int_{-\infty}^{\infty}\int_{-\infty}^{\infty} \mathrm{d}k_x\, \mathrm{d}k_y$$

$$= \frac{2\Sigma}{(2\pi)^2}\sum_{n_z}\int_0^{\infty}\int_0^{2\pi} k_t \mathrm{d}k_t\, \mathrm{d}\varphi_k. \tag{7.2a}$$

With cylindrical symmetry, equation (7.2a) reduces to

$$\sum_{\vec{k}} = \frac{\Sigma}{\pi}\sum_{n_z}\int_0^{\infty} k_t\, \mathrm{d}k_t. \tag{7.2b}$$

In accordance with equation (4.32), for parabolic bands, the dispersion relation for bulk semiconductors can be written as

$$E \equiv \hbar\bar{\omega}_{\vec{k}} = E_{\mathrm{g}} + \varepsilon_{\mathrm{e}\vec{k}} + \varepsilon_{\mathrm{h}\vec{k}} = \left(E_{\mathrm{g}} + \frac{\hbar^2k^2}{2m_{\mathrm{e}}} + \frac{\hbar^2k^2}{2m_{\mathrm{h}}}\right) = E_{\mathrm{g}} + \frac{\hbar^2k^2}{2m_{\mathrm{r}}}, \tag{7.3}$$

with the wave vector k defined in a 3D domain.

For QW structures, we have

$$E \equiv \hbar\overline{\omega}_{\vec{k}} = E_{\mathrm{g}} + \varepsilon_{\mathrm{e}\vec{k}} + \varepsilon_{\mathrm{h}\vec{k}} = \left(E_{\mathrm{g}} + \varepsilon_m^{\mathrm{c}} + \frac{\hbar^2 k_t^2}{2m_{\mathrm{e}}} + \varepsilon_n^{\mathrm{v}} + \frac{\hbar^2 k_t^2}{2m_{\mathrm{h}}}\right) = E_{\mathrm{g}}^{mn} + \frac{\hbar^2 k_t^2}{2m_{\mathrm{r}}}, \tag{7.4}$$

with $\varepsilon_m^{\mathrm{c}}$ and $\varepsilon_n^{\mathrm{v}}$ indicating the energy of the mth and nth discrete states bound by the QW on the conduction and valence band side, respectively, k_t the wave vector defined in a 2D domain (i.e., in the $k_x k_y$ plane), and

$$E_{\mathrm{g}}^{mn} \equiv E_{\mathrm{g}} + \varepsilon_m^{\mathrm{c}} + \varepsilon_n^{\mathrm{v}}. \tag{7.5}$$

Therefore, for bulk semiconductors with spherical symmetry, equation (7.1b) becomes

$$\sum_{\vec{k}} = \frac{\Omega}{\pi^2}\int_0^\infty k^2\,\mathrm{d}k = \frac{\Omega}{2\pi^2}\left(\frac{2m_{\mathrm{r}}}{\hbar^2}\right)^{\frac{3}{2}}\int_{E_{\mathrm{g}}}^\infty \sqrt{(E-E_{\mathrm{g}})}\,\mathrm{d}E = \Omega\int_{E_{\mathrm{g}}}^\infty \rho_{\mathrm{3D}}(E)\,\mathrm{d}E, \tag{7.6}$$

with the 3D density of states in terms of the transition energy E defined as

$$\rho_{\mathrm{3D}}(E) \equiv \frac{1}{2\pi^2}\left(\frac{2m_{\mathrm{r}}}{\hbar^2}\right)^{\frac{3}{2}}\sqrt{(E-E_{\mathrm{g}})}. \tag{7.7}$$

For QW structures with cylindrical symmetry, equation (7.2b) becomes

$$\sum_{\vec{k}} = \frac{\Sigma}{\pi}\sum_{n_z}\int_0^\infty k_t\,\mathrm{d}k_t = \frac{m_{\mathrm{r}}\Sigma}{\pi\hbar^2}\sum_{m,n}\int_{E_{\mathrm{g}}^{mn}}^\infty \mathrm{d}E = \Omega\rho_{\mathrm{2D}}\sum_{m,n}\int_{E_{\mathrm{g}}^{mn}}^\infty \mathrm{d}E, \tag{7.8}$$

with the 2D density of states in terms of the transition energy E defined as

$$\rho_{\mathrm{2D}} \equiv \frac{m_{\mathrm{r}}}{\pi\hbar^2 L_z}, \tag{7.9}$$

where L_z is the thickness of the QW. We find from equation (7.9) that the 2D density of states is a constant without dependence on the transition energy.

For bulk semiconductors, we can also rewrite the quasi-Fermi distributions for the conduction band electrons and valence band holes given in equation (4.36) as functions of the transition energy

$$f_{\mathrm{c}}(E) \equiv \overline{f}_{\vec{k}}^{\mathrm{e}} = \frac{1}{\mathrm{e}^{\frac{\varepsilon_{\mathrm{e}\vec{k}}-F^{\mathrm{c}}}{\mathrm{k_B}T}}+1} = \frac{1}{\mathrm{e}^{\frac{\left(\frac{\hbar^2k^2}{2m_{\mathrm{e}}}+E_{\mathrm{g}}\right)-(F^{\mathrm{c}}+E_{\mathrm{g}})}{\mathrm{k_B}T}}+1}$$

$$= \frac{1}{\mathrm{e}^{\frac{\frac{m_{\mathrm{r}}}{m_{\mathrm{e}}}(E-E_{\mathrm{g}})+E_{\mathrm{g}}-F_{\mathrm{c}}}{\mathrm{k_B}T}}+1} = \frac{1}{\mathrm{e}^{\frac{\frac{m_{\mathrm{r}}}{m_{\mathrm{e}}}E+\frac{m_{\mathrm{r}}}{m_{\mathrm{h}}}E_{\mathrm{g}}-F_{\mathrm{c}}}{\mathrm{k_B}T}}+1}, \tag{7.10a}$$

and

$$f_v(E) \equiv \overline{f}_{\vec{k}}^{h} = \frac{1}{e^{\frac{\varepsilon_{h\vec{k}} - F^v}{k_B T}} + 1} = \frac{1}{e^{\frac{\frac{\hbar^2 k^2}{2m_h} + F_v}{k_B T}} + 1} = \frac{1}{e^{\frac{\frac{m_r}{m_h} E - \frac{m_r}{m_h} E_g + F_v}{k_B T}} + 1}, \tag{7.10b}$$

with $F_c \equiv F^c + E_g$ and $F_v \equiv -F^v$ introduced as the conduction and valence band quasi-Fermi levels measured universally from the valence band edge at $\vec{k} = 0$.

The dipole matrix element defined in equation (4.7) can also be written in terms of the transition energy E once the single electron band structure is solved. Hence we record $\mu_{\vec{k}}$ as $\mu(E)$.

According to equations (4.65a&b), the bulk semiconductor material gain and refractive index change are computed by

$$g(\omega) = \frac{\omega}{\varepsilon_0 n c} \int_{E_g}^{\infty} dE \left\{ \rho_{3D}(E)|\mu(E)|^2 [f_c(E) + f_v(E) - 1] \frac{\gamma\hbar}{(\hbar\gamma)^2 + (E - \hbar\omega)^2} \right\}, \tag{7.11a}$$

$$\Delta n(\omega) = -\frac{1}{2\varepsilon_0 n} \int_{E_g}^{\infty} dE \left\{ \rho_{3D}(E)|\mu(E)|^2 [f_c(E) + f_v(E) - 1] \frac{E - \hbar\omega}{(\hbar\gamma)^2 + (E - \hbar\omega)^2} \right\}. \tag{7.11b}$$

Noting that the downward transition is scaled by the product of the conduction band electron and valence band hole populations, i.e., $f_c(E) f_v(E)$, while the upward transition is scaled by the product of the valence band electron and conduction band hole populations, i.e., $[1 - f_c(E)][1 - f_v(E)]$, we find that equation (7.11a) clearly shows that the stimulated emission gain is proportional to the difference between the downward transition and the upward transition, i.e.,

$$f_c(E) f_v(E) - [1 - f_c(E)][1 - f_v(E)] = f_c(E) + f_v(E) - 1.$$

Since the spontaneous emission gain is proportional to the downward transition only, which is scaled by $f_c(E) f_v(E)$, and the spontaneous emission gain must have the same coefficient as the stimulated emission gain according to the Einstein relation, we can conclude that, for bulk semiconductors, the spontaneous emission gain is given as

$$g_{sp}(\omega) = \frac{\omega}{\varepsilon_0 n c} \int_{E_g}^{\infty} dE \left\{ \rho_{3D}(E)|\mu(E)|^2 f_c(E) f_v(E) \frac{\gamma\hbar}{(\hbar\gamma)^2 + (E - \hbar\omega)^2} \right\}. \tag{7.12}$$

According to equation (4.14), the conduction and valence band quasi-Fermi levels are linked to the conduction band electron and valence band hole densities through

$$N_{e/h} = \frac{1}{\Omega} \sum_{\vec{k}} \overline{f}_{\vec{k}}^{e/h} = \frac{1}{\pi^2} \int_0^{\infty} dk \frac{k^2}{e^{\frac{\varepsilon_{(e/h)\vec{k}} - F^c}{k_B T}} + 1} = \frac{1}{\pi^2} \int_0^{\infty} dk \frac{k^2}{e^{\frac{\hbar^2 k^2/(2m_{e/h}) - F^{c/v}}{k_B T}} + 1}$$

$$= \frac{(2m_{e/h})^{3/2}}{2\pi^2 \hbar^3} \int_0^{\infty} dE \frac{\sqrt{E}}{e^{\frac{E - F^{c/v}}{k_B T}} + 1} = N_{c/v} F_{1/2}\left(\frac{F^{c/v}}{k_B T}\right), \tag{7.13}$$

with the Fermi–Dirac integral given as

$$F_p(x) \equiv \frac{1}{\Gamma(p+1)} \int_0^\infty \mathrm{d}u \frac{u^{\mathrm{p}}}{\mathrm{e}^{u-x}+1}. \tag{7.14}$$

Also, in equation (7.13), the parabolic conduction and valence band edge densities $N_{\mathrm{c/v}}$ have been defined by equation (4.56).

For QW structures, the quasi-Fermi distributions in equation (4.36) become

$$f_{\mathrm{c}}^{m}(E) \equiv \overline{f}_{\vec{k}}^{\mathrm{e},m} = \frac{1}{\mathrm{e}^{\frac{\varepsilon_{\mathrm{e}\vec{k}} - F^{\mathrm{c}}}{\mathrm{k_B}T}} + 1} = \frac{1}{\mathrm{e}^{\frac{(\varepsilon_m^{\mathrm{c}} + \frac{\hbar^2 k_t^2}{2m_{\mathrm{e}}} + E_{\mathrm{g}}) - (F^{\mathrm{c}} + E_{\mathrm{g}})}{\mathrm{k_B}T}} + 1}$$
$$= \frac{1}{\mathrm{e}^{\frac{\varepsilon_m^{\mathrm{c}} + \frac{m_{\mathrm{r}}}{m_{\mathrm{e}}}(E - E_{\mathrm{g}}^{mn}) + E_{\mathrm{g}} - F_{\mathrm{c}}}{\mathrm{k_B}T}} + 1} \tag{7.15a}$$

and

$$f_{\mathrm{v}}^{n}(E) \equiv \overline{f}_{\vec{k}}^{\mathrm{h},n} = \frac{1}{\mathrm{e}^{\frac{\varepsilon_{\mathrm{h}\vec{k}} - F^{\mathrm{v}}}{\mathrm{k_B}T}} + 1} = \frac{1}{\mathrm{e}^{\frac{\varepsilon_n^{\mathrm{v}} + \frac{\hbar^2 k_t^2}{2m_{\mathrm{h}}} + F_{\mathrm{v}}}{\mathrm{k_B}T}} + 1}$$
$$= \frac{1}{\mathrm{e}^{\frac{\varepsilon_n^{\mathrm{v}} + \frac{m_{\mathrm{r}}}{m_{\mathrm{h}}}(E - E_{\mathrm{g}}^{mn}) + F_{\mathrm{v}}}{\mathrm{k_B}T}} + 1}. \tag{7.15b}$$

Therefore, from equation (4.65a&b) we find that, for QW structures, the material gain and refractive index change are computed by

$$g(\omega) = \frac{\omega\rho_{2\mathrm{D}}}{\varepsilon_0 nc} \sum_{m,n} \int_{E_{\mathrm{g}}^{mn}}^{\infty} \mathrm{d}E \left\{ |\mu(E)|^2 [f_{\mathrm{c}}^{m}(E) + f_{\mathrm{v}}^{n}(E) - 1] \frac{\gamma\hbar}{(\hbar\gamma)^2 + (E - \hbar\omega)^2} \right\}, \tag{7.16a}$$

$$\Delta n(\omega) = -\frac{\rho_{2\mathrm{D}}}{2\varepsilon_0 n} \sum_{m,n} \int_{E_{\mathrm{g}}^{mn}}^{\infty} \mathrm{d}E \left\{ |\mu(E)|^2 [f_{\mathrm{c}}^{m}(E) + f_{\mathrm{v}}^{n}(E) - 1] \frac{E - \hbar\omega}{\hbar\gamma^2 + (E - \hbar\omega)^2} \right\}. \tag{7.16b}$$

Consequently, the QW structure spontaneous emission gain is given as

$$g_{\mathrm{sp}}(\omega) = \frac{\omega\rho_{2\mathrm{D}}}{\varepsilon_0 nc} \sum_{m,n} \int_{E_{\mathrm{g}}^{mn}}^{\infty} \mathrm{d}E \left\{ |\mu(E)|^2 f_{\mathrm{c}}^{m}(E) f_{\mathrm{v}}^{n}(E) \frac{\gamma\hbar}{(\hbar\gamma)^2 + (E - \hbar\omega)^2} \right\}. \tag{7.17}$$

According to equation (4.14), the conduction and valence band quasi-Fermi levels are linked to the conduction band electron and valence band hole densities through

$$N_{\mathrm{e/h}} = \frac{1}{\Sigma}\sum_{\vec{k}} \overline{f}_{\vec{k}}^{\mathrm{e/h}} = \frac{1}{\Sigma}\sum_{\vec{k}} \frac{1}{\mathrm{e}^{\frac{\varepsilon_{(\mathrm{e/h})\vec{k}} - F^{\mathrm{c/v}}}{\mathrm{k_B}T}} + 1}$$

$$= \frac{1}{\pi}\sum_{m,n}\int_0^{\infty} \mathrm{d}k_t \left[\frac{k_t}{\mathrm{e}^{\frac{\varepsilon_{m/n}^{\mathrm{c/v}} + \hbar^2 k_t^2/(2m_{\mathrm{e/h}}) - F^{\mathrm{c/v}}}{\mathrm{k_B}T}} + 1}\right]$$

$$= \frac{m_{\mathrm{e/h}}}{\pi\hbar^2}\sum_{m,n}\int_0^{\infty} \mathrm{d}\varepsilon \frac{1}{\mathrm{e}^{\frac{\varepsilon + \varepsilon_{m/n}^{\mathrm{c/v}} - F^{\mathrm{c/v}}}{\mathrm{k_B}T}} + 1} = 2\left(\frac{N_{\mathrm{c/v}}}{2}\right)^{2/3}\sum_{m,n}\ln\left(1 + \mathrm{e}^{\frac{F^{\mathrm{c/v}} - \varepsilon_{m/n}^{\mathrm{c/v}}}{\mathrm{k_B}T}}\right), \tag{7.18}$$

with the parabolic conduction and valence band edge densities $N_{\mathrm{c/v}}$ defined in equation (4.56). Note that, similarly to the 2D carrier densities introduced for the QW, the dimension of the carrier densities defined in the first equation on the RHS of (7.18) is $1/\mathrm{cm}^2$.

7.2.2 The screened Coulomb interaction gain model

In the screened Coulomb interaction model, the material gain and refractive index change are given by equations (4.129a&b), respectively. In computing $\overline{\omega}_{\vec{k}'}$ and $q_{\vec{k}}$, we must note that the screened Coulomb potential $\tilde{V}_{|\vec{k}'-\vec{k}|}$ is a function of the difference between two wave vectors $\vec{k}' - \vec{k}$. As a result, in using equations (7.1a) and (7.2a) to convert the summations into integrals in the wave vector domain, we have neither spherical symmetry in bulk nor cylindrical symmetry in the QW. For this reason, we only replace the summations in equations (4.116), (4.117), (4.125), and (4.129a&b) with the corresponding integrals as shown in equations (7.1a) and (7.2a) for bulk semiconductors and QW structures, respectively, without further transforming the integrals from the wave vector domain into the energy domain. For details on handling these wave vector domain integrals, one can refer to, e.g., [9, 10].

7.2.3 The many-body gain model

In the semi-analytical many-body gain model, the material gain and refractive index change are calculated using equations (4.144a&b), respectively. Again, in computing $\overline{\omega}_{\vec{k}'}$, $\gamma_{\vec{k}}$, $q_{\vec{k}}$, and $\overline{s}_{\vec{k}}$ using equations (4.132a&b), (4.138), (4.139), and (4.132c), since both the screened and bare Coulomb potentials depend on the differences between wave vectors, there is no symmetry that can be utilized to simplify the integration in the wave vector domain. Therefore, after converting the summations into integrals through equations (7.1a) and (7.2a) for bulk semiconductors and QW structures, respectively, we have to compute those integrals in the wave vector domain directly. These computations only involve numerical quadratures, hence we will not discuss the details. One can always refer to [11, 12] for various related techniques.

In the full numerical approach, a necessary step is to evaluate the parameters that appeared in the system of ODEs given as equations (4.146a)–(4.156c) or their modified versions as described in Section 5.1.3. These parameters are all evaluated by the wave vector domain integrals converted from the corresponding summations by equations (7.1a) and (7.2a) for bulk semiconductors and QW structures, respectively. These converted integrals are all given in Appendix B.

Since these parameters have to be evaluated for every $\vec{k}$ value and updated at every time step along with the integration of the ODEs, efficient computation of these parameters through the wave vector domain integrals is crucial. Actually, in computing the parameters that account for the second order Coulomb interactions listed in Table 4.3, we can always isolate the integral in the neighborhood of the singular point from the rest. For the integration in the neighborhood of the singular point, we can convert it back into a summation of only a few terms, hence the contribution of the singular point is obtained virtually analytically. For the integration in areas where the integrand has no singular point, we can always implement the Monte-Carlo integration algorithm [8] since the integrand in the left domain is not strongly peaked locally. In applying the Monte-Carlo method, a uniform sampling scheme can be applied directly to the whole integration domain with the isolated singular point neighborhood included as well. The integrand value will then be set to zero if the sample indeed falls inside the neighborhood, to avoid double counting. This strategy simplifies the implementation without much cost in this particular case. We do not lose accuracy by doing this since the isolated singular point neighborhood takes a very small fraction of the entire integration domain, which makes the probability that the sampling point will fall into this neighborhood negligible, hence the sampling is still efficient.

A step-by-step procedure for the full numerical implementation of this model for QW structures is shown below.

(1) Initialization.

Input the external optical field: E (from the optical solver).

Input the electron and hole densities: N_{e} and N_{h} (from the carrier transport solver).

Input the conduction band electron and valence band hole energies: $\varepsilon_{(\mathrm{e/h})\vec{k}}$ (from the band structure solver).

Input the conduction band electron and valence band hole wave functions: $\langle \vec{r} \mid \phi_n^{q(\mathrm{c/v})}(k_x, k_y, z)\rangle$ (from the band structure solver).

Input the dipole matrix element: $\mu_{\vec{k}}$ (from the band structure solver).

Under the parabolic band assumption, according to equation (7.4), we have

$$\varepsilon_{(\mathrm{e/h})\vec{k}} = \varepsilon_{m/n}^{\mathrm{c/v}} + \hbar^2 k_t^2/(2m_{\mathrm{e/h}}).$$

At the steady state, the carrier densities are linked to their quasi-Fermi levels through equation (7.18), or

$$N_{e/h}(t' \to T') = 2\left(\frac{N_{c/v}}{2}\right)^{2/3} \sum_{m,n} \ln\left(1 + e^{\frac{F_{T'}^{c/v} - \varepsilon_{m/n}^{c/v}}{k_B T}}\right). \tag{7.19}$$

In order to be consistent with the macroscopic carrier transport model, the microscopic processes are described in a short time scale $t' \subset [0, T']$ as introduced in Section 5.1.4.

If there is only one pair of bound electron and hole states inside the QW, we can readily obtain the static quasi-Fermi levels through

$$F_{T'}^{c/v} = \varepsilon_{1/1}^{c/v} + k_B T \ln\left[e^{\frac{2N_{e/h}}{(N_{c/v}/2)^{2/3}}} - 1\right]. \tag{7.20}$$

Otherwise, we have to solve the transcendental equation (7.19) to find the static quasi-Fermi levels corresponding to the given steady state electron and hole densities. These static quasi-Fermi levels will be used later for setting up the integration bounds in the wave vector domain.

The initial carrier (electron and hole) distributions will be assumed to be at the quasi-equilibrium states with their initial quasi-Fermi levels aligned with those in the adjacent layers (bulk semiconductors). For the given carrier densities $N_{(e/h)0}$ in the adjacent N and P type layers, we can find their quasi-Fermi levels by reversing equation (7.13)

$$F_0^{c/v} = k_B T F_{1/2}^{-1}\left[\frac{N_{(e/h)0}}{N_{c/v}}\right], \tag{7.21}$$

with $F_{1/2}^{-1}[\ldots]$ indicating the inverse Fermi–Dirac integral in the order of $1/2$. Since $F_0^{c/v}$ are the quasi-Fermi levels measured from the adjacent layer conduction and valence band edges, respectively, when measured by the QW conduction and valence band edges, the quasi-Fermi levels should be given as

$$F^{c/v} = F_0^{c/v} + \Delta E^{c/v} = k_B T F_{1/2}^{-1}\left[\frac{N_{(e/h)0}}{N_{c/v}}\right] + \Delta E^{c/v}, \tag{7.22}$$

where $\Delta E^{c/v}$ indicates the conduction and valence band edge offset between the N and P side adjacent layers and the QW, respectively, as illustrated by Fig. 7.1.

Therefore, we find the initial conditions for the conduction band electron and valence band hole number expectations

$$f_{\vec{k}}^{e/h}(0) = \frac{1}{e^{\frac{\hbar^2 k_t^2/(2m_{e/h}) + \varepsilon_{m/n}^{c/v} - \Delta E^{c/v}}{k_B T} - F_{1/2}^{-1}\left(\frac{N_{(e/h)0}}{N_{c/v}}\right)} + 1}. \tag{7.23}$$

We also set the initial condition for the polariton number expectation as

$$p_{\vec{k}}(0) = 0. \tag{7.24}$$

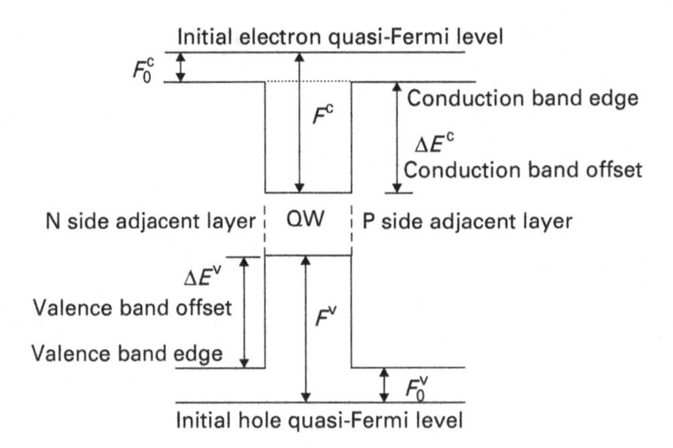

Fig. 7.1. Initial electron and hole quasi-Fermi levels with respect to different references.

(2) Wave vector domain discretization.

For QW structures, the wave vector domain is in the 2D $k_x k_y$ plane. Under the polar coordinate system, we set

$$\varphi_k \text{ from } 0 \text{ to } 2\pi, \; \mathrm{d}\varphi_k = 2\pi/N_\varphi. \tag{7.25}$$

$$k_t \text{ from } 0 \text{ to } \pi/d, \; \mathrm{d}k_t = \max[0.2\pi/d, \max(\sqrt{\left(2m_\mathrm{e}F_{T'}^\mathrm{c}\right)}/\hbar, \sqrt{\left(2m_\mathrm{h}F_{T'}^\mathrm{v}\right)}/\hbar)]/N_k. \tag{7.26}$$

In equation (7.26), the static quasi-Fermi levels $F_{T'}^{\mathrm{c/v}}$ are computed in the initialization step.

(3) Integration.

As an efficient and accurate numerical quadrature approach, the Gaussian–Legendre method [1] can be implemented to evaluate the integrals reflecting the zeroth and first order Coulomb interaction contributions listed in Table 4.3. Similarly to the second order Coulomb interaction terms, we can use the singular point isolation plus the Monte-Carlo integration method [8] as previously introduced.

(4) Solving system of ODEs.

For a system of ODEs given in the form

$$\frac{\mathrm{d}}{\mathrm{d}t} y_k(t') = F_k[y_k(t'), t'], \tag{7.27}$$

with $k = 1, 2, 3, \ldots, K$, and known initial values $y_k(t'_j)$, there are several different schemes to obtain $y_k(t'_{j+1})$. By continuously marching in the time domain following such a scheme, we will be able to find all the unknowns $y_k(t')$ with $k = 1, 2, 3, \ldots, K$ in the time domain.

The major difference, however, between these schemes lies in the different ways of compromising between the number of evaluations on the derivative functions (the right hand side of the ODEs) at each marching step and the marching step size. The latter method (e.g., Bulirsch–Stoer's implementation of the Richardson extrapolation [1]) uses fewer evaluations of the derivative functions but the marching step size cannot be large. On the contrary, the former method (e.g., Cash–Karp's implementation of the fifth order Runge–Kutta method [1]) takes the maximum possible marching step size by an adaptive approach which guarantees a preset accuracy. Its drawback, however, is in the repeated evaluation of the derivative functions.

In dealing with equations (4.156a–c) or their modified versions as described in Section 5.1.3, those ODEs take simple derivative function forms with their parameters being very difficult to evaluate, since time-consuming integrations in the wave vector domain are involved in computing these parameters. Because of the dependence of the parameters on the unknown variables themselves, i.e., the implicit non-linearity of the ODEs, we have to update these parameters at every time step. Moreover, the accuracy of the unknown functions computed at each converged time step is crucial, as otherwise the parameters evaluated based on these functions would be spoiled as well, which amplifies the error at each following time step and rapidly makes the result diverge. On the contrary, evaluation of the derivative functions on the RHS of ODEs, equation (7.27), can be executed efficiently. For this reason, the adaptive Runge–Kutta algorithm is the preferred approach for saving computation time. With those algorithms based on the Richardson extrapolation, however, it is most likely that we will have to update the parameters more often in order to secure accuracy, which will result in extremely low efficiency, since the time required for evaluating those wave vector domain integrals is overwhelming, despite all the measures we can take as described previously.

At time step t'_j with the converged result $[y_k(t'_j), t'_j]$, the fifth order Runge–Kutta algorithm is implemented giving

$$y_k(t'_{j+1}) = y_k(t'_j) + \sum_{i=1}^{6} c_i d_{ki} + O(h^6), \tag{7.28a}$$

with the embedded fourth order formula in the form

$$y^*_k(t'_{j+1}) = y_k(t'_j) + \sum_{i=1}^{6} c^*_i d_{ki} + O(h^5). \tag{7.28b}$$

Hence the estimated error is

$$\Delta_k \equiv y_k(t'_{j+1}) - y^*_k(t'_{j+1}) = \sum_{i=1}^{6} (c_i - c^*_i) d_{ki}. \tag{7.29}$$

In equations (7.28a&b) and (7.29), as given in Appendix C, parameters d_{ki}, $i = 1, 2, 3, 4, 5, 6$, are obtained through evaluations of the derivative functions on the RHS of ODEs, equation (7.27). Constants c_i and c^*_i are also listed in Appendix C.

Since such an error is in the order of h^5, we obtain

$$h_0 = h \min_k |\frac{\Delta_{0k}}{\Delta_k}|^{0.2}. \tag{7.30}$$

For an initial forward time marching step h, equation (7.28a) provides us the solution of ODEs at $t'_{j+1} = t'_j + h$, while equation (7.30) supplies us the step h_0 at which the error is bound by a given tolerance Δ_{0k}.

Finally, a flow chart is given in Fig. 7.2, which summarizes numerical computation of the material susceptibility (i.e., the gain and refractive index change) using the many-body gain model.

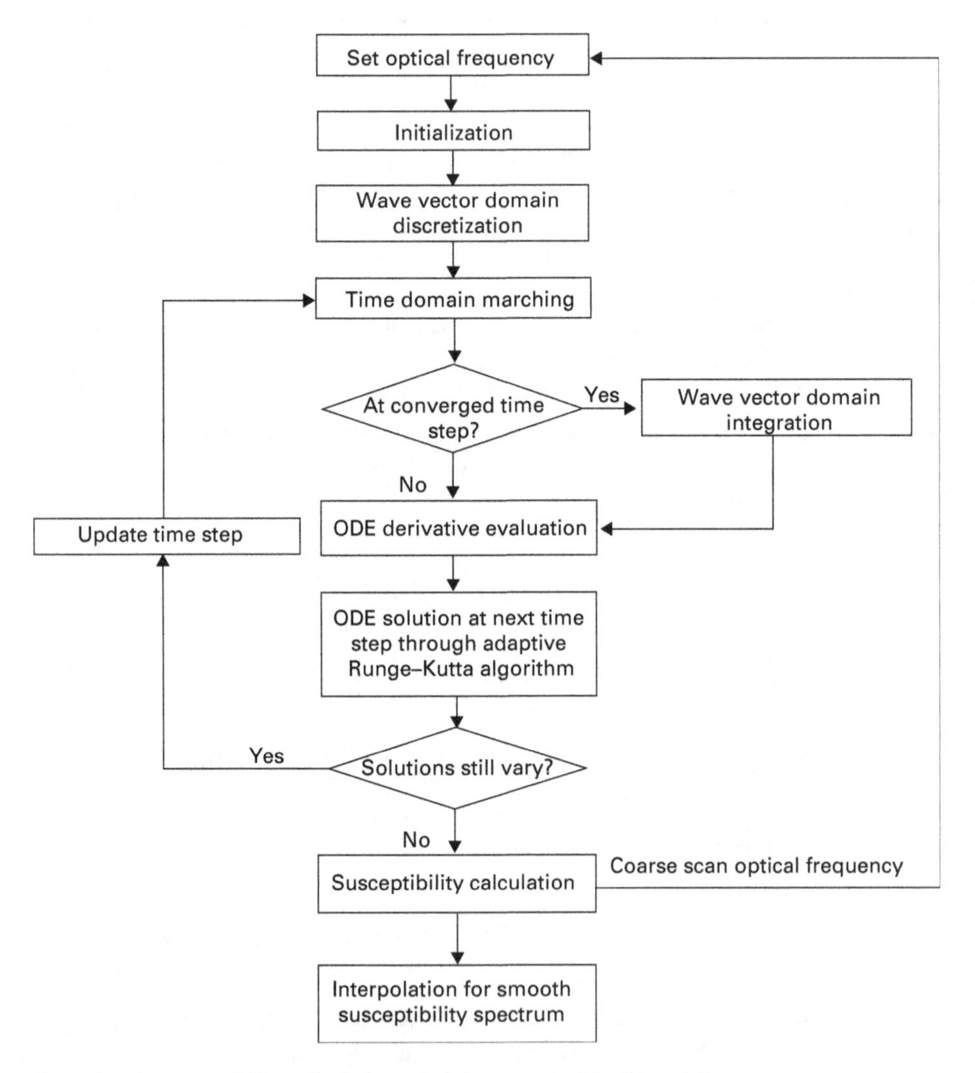

Fig. 7.2. Procedure for susceptibility calculation using the many-body gain model.

7.3 Parameterization of material properties

In modeling of the device performance, the material optical properties, including the stimulated emission gain, spontaneous emission gain, and refractive index change, will be most frequently invoked. Therefore, these calculations have to be done analytically. Otherwise, we cannot afford the time spent on the computation. For example, if we take 30 points along the horizontal direction, 100 points along the vertical direction and 20 sections along the longitudinal direction, we get a total of 60 000 points. Suppose we need ten iterations on average to make all the equations converge at a single operating point; in total we would need to calculate the material gains and refractive index change 600 000 times. Assuming that the material model in its original form is computed numerically, and assuming that the numerical model takes only 1 minute to complete all the executions, we still need at least 600 000 minutes or 10 000 hours for calculation of the device performance at just a single operating point. This is obviously not viable.

The reason that we suffer an extremely low efficiency is because there is a huge redundancy if we use the above strategy. Actually, we do not expect that the material properties are all different at those points. There must be many different points with the same material properties. However, in the above strategy, the rigorous material model is called for anyway for every point, regardless of whether this call exactly repeats a previous calculation or not. As a result, most of the time is wasted on repeated calls.

To solve this problem, we should follow a different strategy.

(1) Identify all the possible variables that can cause material property change, e.g., the carrier densities, the temperature, and the operating wavelength.
(2) Estimate the maximum varying range of each variable.
(3) Identify a mesh grid in the variable space, confined by the estimated range in each variable dimension.
(4) Compute the material properties on this grid by invoking the numerical model; it is guaranteed that there is no redundancy in this case, as the material properties must be different at different grid points once the grid is constructed stop by step as per (1)–(3).
(5) Extract analytical formulas by searching for the best fit between such formulas and the material properties found at the mesh grid.
(6) Use such extracted analytical formulas for device performance computation.

The above steps (1)–(5) need to be executed just once for a given material and active region structure. Compared with the old strategy, with the help of the sophisticated interpolation routine, the mesh grid in step (3) comprises many fewer points at which the rigorous material model needs to be invoked. For example, ten different carrier density levels, five different temperature levels, and five different operating wavelengths are normally enough for the extraction of sufficiently accurate analytical formulas (with error $<5\%$) covering a feasible device operating range (e.g., carrier density from 0.5×10^{18} cm^3 to 5×10^{18} cm^3, temperature from 273–353 K, and a wavelength spectral range around 50 nm). The total number of times the rigorous material model is invoked is only 250, which makes this approach far more efficient.

We normally assume that the analytical formulas for the stimulated emission and spontaneous emission gains take a polynomial form as functions of the carrier densities, with their orders and coefficients all extracted through fitting to the gains calculated by the numerical model (i.e., the rigorous material model) at a set of equally spaced carrier densities. By assuming an exponential dependence of every coefficient on the temperature, we can have all the characteristic temperature parameters extracted through fitting to the gains calculated by the numerical model under a set of equally spaced temperatures again. Finally, we use rational forms to express the frequency dependence of the gains. Again, the orders and coefficients can be extracted by fitting the analytical form to the gain profiles obtained numerically.

Following a similar approach, the analytical formula for the material refractive index change can be obtained.

The extracted analytical formulas for the material gains and refractive index change generally take a universal form as shown [13]

$$\left[\sum_{i=0}^{I} a_i e^{\frac{T-T_{\text{ref}}}{T_i^{\text{a}}}} (N_{\text{e}})^i + \sum_{i=0}^{I} b_i e^{\frac{T-T_{\text{ref}}}{T_i^{\text{b}}}} (N_{\text{h}})^i\right] \frac{\sum_{l=0}^{L} c_l e^{\frac{T-T_{\text{ref}}}{T_l^{\text{c}}}} \omega^l}{\sum_{m=0}^{M} d_m e^{\frac{T-T_{\text{ref}}}{T_m^{\text{d}}}} \omega^m}. \tag{7.31}$$

The orders (I, L, M) and coefficients ($a_i, b_i, c_l, d_m, T_i^{\text{a}}, T_i^{\text{b}}, T_l^{\text{c}}, T_m^{\text{d}}$) are all extracted through fittings as mentioned above. Note that T_{ref} is a reference temperature and is normally chosen to be room temperature (e.g., 300 K).

Finally, a phenomenological non-linear gain saturation factor in the form

$$\frac{1}{1 + \varepsilon_{\text{sat}} |\phi(x, y)|^2 [|e^{\text{f}}(z, t)|^2 + |e^{\text{b}}(z, t)|^2]}, \tag{7.32}$$

is multiplied into the analytical gain formula to count in the strong field-induced spectral hole burning (SHB) effect, which has been neglected in the gain models introduced in Chapter 4. In equation (7.32), ε_{sat} is introduced as the non-linear gain saturation coefficient.

References

[1] W. H. Press, S. A. Teukolsky, W. T. Vetterling, and B. P. Flannery, *Numerical Recipes in Fortran: The Art of Scientific Computing*, 2nd edn (Cambridge, UK: Cambridge University Press, 1992).

[2] G. H. Golub and C. F. Van Loan, *Matrix Computations*, 2nd edn (Baltimore: The John Hopkins University Press, 1989).

[3] S. L. Chuang, *Physics of Optoelectronic Devices*, 1st edn (New York: John Wiley & Sons, 1995).

[4] A. K. Ghatak, K. Thyagarajan, and M. R. Shenoy, A novel numerical technique for solving the one-dimensional Schrödinger equation using matrix approach - application to quantum well structures. *IEEE Journal of Quantum Electron.*, **QE-24**:8 (1988), 1524–31.
[5] I. A. Shelykh and V. K. Ivanov, To the development of the transfer matrix formalism. *Phys. Stat. Sol.*, (c) **1**:6 (2004), 1549–53.
[6] D. Neuhauser and M. Baer, The time-dependent Schrödinger equation: application of absorbing boundary conditions. *Journal of Chem. Phys.*, **90**:8 (1989), 4351–5.
[7] C. Farrell and U. Leonhardt, The perfectly matched layer in numerical simulation of nonlinear and matter waves. *Journal of Optics B*, **7** (2005), 1–4.
[8] J. M. Thijssen, *Computational Physics*, 1st edn (Cambridge, UK: Cambridge University Press, 1999).
[9] W. W. Chow and S. W. Koch, *Semiconductor Laser Fundamentals: Physics of the Gain Materials*, 1st edn (Berlin: Springer-Verlag, 1999).
[10] W. W. Chow, S. W. Koch, and M. Sargent III, *Semiconductor-Laser Physics*, 1st edn (Berlin: Springer-Verlag, 1994).
[11] G. Monegato and L. Scuderi, Numerical integration of functions with endpoint singularities and/or complex poles in 3D Galerkin boundary element methods. *Publ. RIMS Kyoto Univ.*, **41** (2005), 869–95.
[12] G. Evans, *Practical Numerical Integration*, 1st edn (New York: John Wiley & Sons, 1993).
[13] X. Li, A. D. Sadovnikov, W.-P. Huang, and T. Makino, A physics-based three-dimensional model for distributed feedback (DFB) laser diodes. *IEEE Journal of Quantum Electron.*, **QE-34**:9 (1998), 1545–53.

8 Solution techniques for carrier transport and thermal diffusion equations

8.1 The static carrier transport equation

In describing the solution technique for the classical carrier transport equation in a steady state, we follow a step-by-step FD procedure that is commonly used in many numerical approaches for solving PDEs. Without losing generality, we will use the 2D equations (5.29a–c) and (5.33a&b) as our example, which is the model most commonly used for solving the cross-sectional carrier transport problem in semiconductor lasers [1–11]. At the steady state, these equations can be written as

$$\nabla_t \cdot \tilde{\varepsilon}(\vec{r})\nabla_t\Phi(\vec{r}) = -\mathrm{e}[N_\mathrm{h}^\mathrm{C}(\vec{r}) - N_\mathrm{e}^\mathrm{C}(\vec{r}) + N_\mathrm{h}^\mathrm{H}(\vec{r}) - N_\mathrm{e}^\mathrm{H}(\vec{r}) + N_\mathrm{D}^+(\vec{r}) - N_\mathrm{A}^-(\vec{r})], \tag{8.1a}$$

$$0 = -\frac{N_\mathrm{e}^\mathrm{H}(\vec{r})}{\tau_\mathrm{cl}^\mathrm{e}} + \frac{N_\mathrm{e}^\mathrm{C}(\vec{r})}{\tau_\mathrm{ht}^\mathrm{e}} - R(\vec{r}) - \nabla_t \cdot \mu_\mathrm{e}(\vec{r})N_\mathrm{e}^\mathrm{H}(\vec{r})\nabla_t[\Phi(\vec{r}) + \Phi_\mathrm{C}(\vec{r})] + \nabla_t \cdot D_\mathrm{e}(\vec{r})\nabla_t N_\mathrm{e}^\mathrm{H}(\vec{r}) + \nabla_t \cdot D_\mathrm{e}^\mathrm{T}(\vec{r})N_\mathrm{e}^\mathrm{H}(\vec{r})\nabla_t T(\vec{r}), \tag{8.1b}$$

$$0 = -\frac{N_\mathrm{h}^\mathrm{H}(\vec{r})}{\tau_\mathrm{cl}^\mathrm{h}} + \frac{N_\mathrm{h}^\mathrm{C}(\vec{r})}{\tau_\mathrm{ht}^\mathrm{h}} - R(\vec{r}) + \nabla_t \cdot \mu_\mathrm{h}(\vec{r})N_\mathrm{h}^\mathrm{H}(\vec{r})\nabla_t[\Phi(\vec{r}) - \Phi_V(\vec{r})] + \nabla_t \cdot D_\mathrm{h}(\vec{r})\nabla_t N_\mathrm{h}^\mathrm{H}(\vec{r}) + \nabla_t \cdot D_\mathrm{h}^\mathrm{T}(\vec{r})N_\mathrm{h}^\mathrm{H}(\vec{r})\nabla_t T(\vec{r}), \tag{8.1c}$$

$$0 = \frac{N_\mathrm{e}^\mathrm{H}(\vec{r})}{\tau_\mathrm{cl}^\mathrm{e}} - \frac{N_\mathrm{e}^\mathrm{C}(\vec{r})}{\tau_\mathrm{ht}^\mathrm{e}} - R_\mathrm{st}(\vec{r}) - R_\mathrm{nr+sp}^\mathrm{e}(\vec{r}), \tag{8.1d}$$

$$0 = \frac{N_\mathrm{h}^\mathrm{H}(\vec{r})}{\tau_\mathrm{cl}^\mathrm{h}} - \frac{N_\mathrm{h}^\mathrm{C}(\vec{r})}{\tau_\mathrm{ht}^\mathrm{h}} - R_\mathrm{st}(\vec{r}) - R_\mathrm{nr+sp}^\mathrm{h}(\vec{r}). \tag{8.1e}$$

Adding equations (8.1b) to (8.1d), and (8.1c) to (8.1e), respectively, ignoring the hot carrier non-radiative plus spontaneous emission recombination contribution R in (8.1b) and (8.1c) and supplying these contributions for the cold carriers in (8.1d) and (8.1e) with equation (5.40), ignoring the cold carrier contributions to the electrostatic potential, since they only exist in the active region and cancel out each other under the quasi-charge

Table 8.1. Variable scaling chart

Variable	Scaling factor	Value
x, y	x_0	$\max \lvert x - y \rvert, x, y \in \Sigma$
Φ, Φ_C, Φ_V	Φ_0	kT/e
$N_e^H, N_h^H, N_D^+, N_A^-$	C_0	$\max \lvert N_D^+(x, y) - N_A^-(x, y) \rvert, x, y \in \Sigma$
D_e, D_h	D_0	$\max[D_e(x, y), D_h(x, y)], x, y \in \Sigma$
μ_e, μ_h	μ_0	D_0/Φ_0
R_{st}, R_{nr+sp}	R_0	$D_0 C_0/x_0^2$
t	T_0	x_0^2/D_0

neutrality condition $N_e^C \approx N_h^C$, and ignoring the Seebeck effect yields

$$\nabla_t \cdot \tilde{\varepsilon}(\vec{r})\nabla_t\Phi(\vec{r}) = -e[N_h^H(\vec{r}) - N_e^H(\vec{r}) + N_D^+(\vec{r}) - N_A^-(\vec{r})], \tag{8.2a}$$

$$\nabla_t \cdot D_e(\vec{r})\nabla_t N_e^H(\vec{r}) - \nabla_t \cdot \mu_e(\vec{r})N_e^H(\vec{r})\nabla_t[\Phi(\vec{r}) + \Phi_C(\vec{r})] = R_{st}(\vec{r}) + R_{nr+sp}(\vec{r}), \tag{8.2b}$$

$$\nabla_t \cdot D_h(\vec{r})\nabla_t N_h^H(\vec{r}) + \nabla_t \cdot \mu_h(\vec{r})N_h^H(\vec{r})\nabla_t[\Phi(\vec{r}) - \Phi_V(\vec{r})] = R_{st}(\vec{r}) + R_{nr+sp}(\vec{r}), \tag{8.2c}$$

Further noting that these equations will be solved in a series of cross-sectional sheets (i.e., in 2D xy planes) perpendicular to the wave propagation direction z, we can take the space argument z as a parameter hence $\vec{r}$ only indicates $x\vec{x} + y\vec{y}$ in dealing with equation (8.2a–c).

8.1.1 Scaling

As the unknown variables, the electrostatic potential Φ, the electron and hole densities N_e^H and N_h^H in equations (8.2a–c) are of different orders of magnitude when measured by their conventional units and show different behaviors in regions with different space charge densities and different recombination rates. The first step is to scale these equations appropriately to avoid any overflow or underflow problems or loss of significant digits. A typical scaling method is given in Table 8.1 [12].

Equations (8.2a–c) will be transformed into the following form after such scaling

$$\nabla_t \cdot [\lambda^2(x, y)\nabla_t\phi(x, y)] - [n(x, y) - p(x, y) - C(x, y)] = 0, \tag{8.3a}$$

$$\nabla_t \cdot d_n(x, y)\nabla_t n(x, y) - \nabla_t \cdot \mu_n(x, y)n(x, y)\nabla_t[\phi(x, y) + \phi_C(x, y)] - R(x, y) = 0, \tag{8.3b}$$

$$\nabla_t \cdot d_p(x, y)\nabla_t p(x, y) + \nabla_t \cdot \mu_p(x, y)p(x, y)\nabla_t[\phi(x, y) - \phi_V(x, y)] - R(x, y) = 0, \tag{8.3c}$$

with the scaled variables and parameters defined as

$$\begin{aligned}
x &\equiv x/x_0 \\
y &\equiv y/x_0 \\
t &\equiv t/T_0 \\
\phi(x, y) &\equiv \Phi(x, y)/\Phi_0 \\
n(x, y) &\equiv N_{\mathrm{e}}^{\mathrm{H}}(x, y)/C_0 \\
p(x, y) &\equiv H_{\mathrm{h}}^{\mathrm{H}}(x, y)/C_0 \\
\lambda^2(x, y) &\equiv \Phi_0\widetilde{\varepsilon}(x, y)/(\mathrm{e}x_0^2 C_0) \\
C(x, y) &\equiv [N_{\mathrm{D}}^{+}(x, y) - N_{\mathrm{A}}^{-}(x, y)]/C_0 \\
d_{n,p}(x, y) &\equiv D_{\mathrm{e,h}}(x, y)/D_0 \\
\mu_{n,p}(x, y) &\equiv \mu_{\mathrm{e,h}}(x, y)/\mu_0 \\
\phi_{\mathrm{C,V}}(x, y) &\equiv \Phi_{\mathrm{C,V}}(x, y)/\Phi_0 \\
R(x, y) &\equiv [R_{\mathrm{st}}(x, y) + R_{\mathrm{nr+sp}}(x, y)]/R_0.
\end{aligned} \tag{8.4}$$

8.1.2 Boundary conditions

Equations (8.3a–c) are posed as a boundary value problem in a 2D domain representing the device geometry. In principle, the boundaries can be split into two parts, as set out below. One represents the real physical boundaries, such as contacts and interfaces to insulating material and between semiconductors with different material compositions or doping concentrations, whereas the other represents the artificial boundaries that have to be introduced to truncate the computation window.

(1) Artificial boundary.

At the artificial boundary, we assume either natural boundary conditions that guarantee that the domain under consideration is self-contained

$$\frac{\partial \phi}{\partial \vec{n}} = 0, \quad \frac{\partial n}{\partial \vec{n}} = 0, \text{ and } \frac{\partial p}{\partial \vec{n}} = 0, \tag{8.5a}$$

with $\vec{n}$ indicating the boundary surface normal direction, or specify the pre-estimated Dirichlet values for the electrostatic potential and carrier densities

$$\phi = \phi_0, \quad n = n_0, \text{ and } p = p_0. \tag{8.5b}$$

The applicability of these boundary conditions has to be justified by physical reasoning.

(2) Physical boundary 1 – Ohmic contact.

Note that

$$\phi(t) - \phi_b = \phi_{bias}(t),$$

$$np - n_i^2 = 0 \text{ and } n - p - C = 0, \tag{8.6a}$$

or

$$n = 0.5[\surd\left(C^2 + 4n_i^2\right) + C], \tag{8.6b}$$

$$p = 0.5[\surd\left(C^2 + 4n_i^2\right) - C], \tag{8.6c}$$

where

$$n_i^2 = N_c N_v e^{-E_g/kT}/C_o^2, \tag{8.7}$$

with ϕ_b, ϕ_{bias} introduced as the built-in potential and the external bias voltage, and $N_{c,v}$ the parabolic conduction and valence band edge densities given in equation (4.56), respectively.

(3) Physical boundary 2 – Schottky contact.

Note that

$$\phi(t) - \phi_b + \phi_s = \phi_{bias}(t), \tag{8.8a}$$

with ϕ_s defined as the Schottky barrier height. Under the reverse bias

$$\vec{J}_n \cdot \vec{n} = -ev_n[n - 0.5(\surd\left(C^2 + 4n_i^2\right) + C)], \tag{8.8b}$$

$$\vec{J}_p \cdot \vec{n} = ev_p[p - 0.5(\surd\left(C^2 + 4n_i^2\right) - C)], \tag{8.8c}$$

where $\vec{J}_{n,p}$ are the electron and hole current densities, $v_{n,p}$ the electron and hole thermionic recombination velocities at the contact, respectively.

(4) Physical boundary 3 – the insulator contact.

Note that

$$\varepsilon_{sem}\frac{\partial\phi}{\partial\vec{n}} - \varepsilon_{ins}\frac{\phi_{bias}(t) - \phi}{d_{ins}} = Q_{int}, \tag{8.9a}$$

where ε_{sem} and ε_{ins} represent the dielectric constant of the semiconductor and the insulator, respectively, d_{ins} the thickness of the insulator, ϕ_{bias} the electrostatic potential applied to the insulator, Q_{int} the charges at the surface. Quite often the existence of the surface charge can be neglected and the insulator is very thick; under these assumptions, equation (8.9a) reduces to

$$\frac{\partial\phi}{\partial\vec{n}} = 0. \tag{8.9b}$$

We still have

$$\vec{J}_n \cdot \vec{n} = -eR^{\text{surf}}, \tag{8.9c}$$

$$\vec{J}_p \cdot \vec{n} = eR^{\text{surf}}, \tag{8.9d}$$

with R^{surf} defined as the surface recombination rate. It is also quite often the case that surface recombination can be ignored. This will lead to a zero current flow condition in the boundary surface normal direction.

(5) Physical internal interface – the heterojunction.

In compositionally non-uniform semiconductors, forces in addition to the electrostatic field apply to electrons and holes at the heterojunction boundaries, and the non-uniform densities-of-states modify electron and hole diffusion. Both effects, known as the rigid band and density-of-states effects, have been included by the additional terms $\phi_{C,V}$ in equations (8.3b&c) [13]. These additional terms $\phi_{C,V}$ have been given in equations (5.14a&b).

8.1.3 The initial solution

The scaled equations (8.3a–c) constitute a singularly perturbed boundary value problem with λ as the perturbation parameter. We can therefore solve equation (8.3a) for $\lambda = 0$ (charge neutral solution) first and use this solution as our initial guess to start the iterations in the numerical approach.

8.1.4 The finite difference discretization

We will use a rectangular-shaped computation window with N_x mesh lines parallel to the x axis, N_y mesh lines parallel to y axis, and with the first and the last line coinciding with the computation window boundaries. Therefore, we have $N_x N_y$ mesh grid points on which an approximate solution for the PDEs (8.3a–c) needs to be found.

To deal with such boundary value problems, we will use the most popular center discretization scheme, with non-uniform mesh allowed. In the 2D version of this scheme, for any inner point, we replace the differential equations by difference equations where only the nearest four neighboring points are invoked. Hence this scheme is also known as the classical five-point discretization.

By introducing [12]

$$h_i = x_{i+1} - x_i, i = 1, \ldots, N_x - 1, \tag{8.10}$$

$$k_j = y_{j+1} - y_j, j = 1, \ldots, N_y - 1, \tag{8.11}$$

and

$$u_{i,j} = u(x_i, y_j), i = 1, \ldots, N_x, \; j = 1, \ldots, N_y, \tag{8.12a}$$

$$u_{i+1/2,j} = u[(x_i + x_{i+1})/2, y_j], \; i = 1, \ldots, N_x - 1, \; j = 1, \ldots, N_y, \tag{8.12b}$$

$$u_{i,j+1/2} = u[x_i, (y_j + y_{j+1})/2], \; i = 1, \ldots, N_x, \; j = 1, \ldots, N_y - 1, \tag{8.12c}$$

with u representing any of the unknown variables ϕ, n, or p in equations (8.3a–c), we replace all the first order partial derivatives through the Taylor expansion

$$\frac{\partial u}{\partial x}|_{i,j} = \frac{u_{i+1/2,j} - u_{i-1/2,j}}{(h_i + h_{i-1})/2} + \frac{h_{i-1} - h_i}{4}\frac{\partial^2 u}{\partial x^2}|_{i,j} + O(h^2), \tag{8.13a}$$

$$\frac{\partial u}{\partial y}|_{i,j} = \frac{u_{i,j+1/2} - u_{i,j-1/2}}{(k_j + k_{j-1})/2} + \frac{k_{j-1} - k_j}{4}\frac{\partial^2 u}{\partial y^2}|_{i,j} + O(k^2). \tag{8.13b}$$

In this scheme, the local truncation error for a uniform or quasi-uniform mesh, e.g., a mesh defined as $h_{i+1}, k_{j+1} = h_i, k_j[1 + O(h_i, k_j)]$, is of the second order in the mesh spacing. In dealing with the carrier transport equations, however, a strongly non-uniform mesh is often mandatory since the solution may exhibit a smooth change in some regions of the device, whereas in others it may vary rapidly. We can therefore only expect a truncation error of the first order in terms of the mesh spacing.

According to equations (8.13a&b), Poisson's equation (8.3a) becomes

$$\frac{\lambda^2_{i+1/2,j}\frac{\partial\phi}{\partial x}|_{i+1/2,j} - \lambda^2_{i-1/2,j}\frac{\partial\phi}{\partial x}|_{i-1/2,j}}{(h_i + h_{i-1})/2} + O(h)$$
$$+ \frac{\lambda^2_{i,j+1/2}\frac{\partial\phi}{\partial y}|_{i,j+1/2} - \lambda^2_{i,j-1/2}\frac{\partial\phi}{\partial y}|_{i,j-1/2}}{(k_j + k_{j-1})/2} + O(k) - n_{i,j} + p_{i,j} + C_{i,j} = 0. \tag{8.14}$$

The mid-interval potential derivative values can be further estimated through

$$\frac{\partial\phi}{\partial x}|_{i+1/2,j} = \frac{\phi_{i+1,j} - \phi_{i,j}}{h_i} + O(h^2), \quad \frac{\partial\phi}{\partial x}|_{i-1/2,j} = \frac{\phi_{i,j} - \phi_{i-1,j}}{h_{i-1}} + O(h^2),$$
$$\frac{\partial\phi}{\partial y}|_{i,j+1/2} = \frac{\phi_{i,j+1} - \phi_{i,j}}{k_j} + O(k^2), \quad \frac{\partial\phi}{\partial y}|_{i,j-1/2} = \frac{\phi_{i,j} - \phi_{i,j-1}}{k_{j-1}} + O(k^2). \tag{8.15}$$

Substituting equation (8.15) into (8.14) we obtain

$$\frac{\lambda^2_{i+1/2,j}\frac{\phi_{i+1,j}-\phi_{i,j}}{h_i} - \lambda^2_{i-1/2,j}\frac{\phi_{i,j}-\phi_{i-1,j}}{h_{i-1}}}{(h_i + h_{i-1})/2}$$
$$+ \frac{\lambda^2_{i,j+1/2}\frac{\phi_{i,j+1}-\phi_{i,j}}{k_j} - \lambda^2_{i,j-1/2}\frac{\phi_{i,j}-\phi_{i,j-1}}{k_{j-1}}}{(k_j + k_{j-1})/2} - n_{i,j} + p_{i,j} + C_{i,j} = 0. \tag{8.16}$$

This is the final discretized form of Poisson's equation (8.3a) with the local truncation error linearly proportional to the mesh spacing.

Taking $J_0 = \mathrm{e}D_0C_0/x_0$ as the normalization constant, we reintroduce the normalized electron and hole current densities as the intermediate variables according to the drift and diffusion model

$$J_{nx} = -\left[d_n\frac{\partial n}{\partial x} - \mu_n n\frac{\partial(\phi+\phi_\mathrm{C})}{\partial x}\right], \quad J_{ny} = -\left[d_n\frac{\partial n}{\partial y} - \mu_n n\frac{\partial(\phi+\phi_\mathrm{C})}{\partial y}\right],$$
$$J_{px} = d_p\frac{\partial p}{\partial x} + \mu_p p\frac{\partial(\phi-\phi_\mathrm{V})}{\partial x}, \quad J_{py} = d_p\frac{\partial p}{\partial y} + \mu_p p\frac{\partial(\phi-\phi_\mathrm{V})}{\partial y}. \tag{8.17}$$

According to equations (8.13a&b) and (8.17), carrier continuity equations (8.3b&c) become

$$\frac{(-J_{nx})|_{i+1/2,\,j} - (-J_{nx})|_{i-1/2,\,j}}{(h_i+h_{i-1})/2} + O(h)$$
$$+ \frac{(-J_{ny})|_{i,\,j+1/2} - (-J_{ny})|_{i,\,j-1/2}}{(k_j+k_{j-1})/2} + O(k) - R|_{i,\,j} = 0 \tag{8.18a}$$

$$\frac{J_{px}|_{i+1/2,\,j} - J_{px}|_{i-1/2,\,j}}{(h_i+h_{i-1})/2} + O(h)$$
$$+ \frac{J_{py}|_{i,\,j+1/2} - J_{py}|_{i,\,j-1/2}}{(k_j+k_{j-1})/2} + O(k) - R|_{i,\,j} = 0. \tag{8.18b}$$

The discretization of the carrier continuity equations (8.18a&b) is more crucial because of their embedded non-linearities when solved in a self-consistent manner with Poisson's equation (8.16) through the drift and diffusion model, equation (8.17).

In conventional FD schemes, differential equations obtained at an intermediate step are discretized again through an FD scheme, which, however, is only suitable in dealing with linear PDEs as the discretization order in terms of the mesh spacing is preserved. In non-linear equations, however, such a repeating FD discretization strategy could bring in huge errors unless higher order FD discretization is introduced, depending on the non-linear function dependence, since the discretization order in terms of the mesh spacing could be reduced through the non-linear operation. Using the electron current density in the x direction as an example, according to equations (8.17), (8.13a) and (8.12b), we find

$$(-J_{nx})|_{i+1/2,\,j} = d_n|_{i+1/2,\,j}\left[\frac{n_{i+1,\,j}-n_{i,\,j}}{h_i} + O(h^2)\right]$$
$$- \mu_n|_{i+1/2,\,j}\left[\frac{n_{i+1,\,j}+n_{i,\,j}}{2} + O(h^2)\right]\left[\frac{\phi_{i+1,\,j}+\phi_{\mathrm{C}i+1,\,j}-\phi_{i,\,j}-\phi_{\mathrm{C}i,\,j}}{h_i} + O(h^2)\right]$$
$$= d_n|_{i+1/2,\,j}\left[\frac{n_{i+1,\,j}-n_{i,\,j}}{h_i} + O(h^2)\right]$$
$$- \mu_n|_{i+1/2,\,j}\left[\frac{n_{i+1,\,j}+n_{i,\,j}}{2}\,\frac{\phi_{i+1,\,j}+\phi_{\mathrm{C}i+1,\,j}-\phi_{i,\,j}-\phi_{\mathrm{C}i,\,j}}{h_i} + O(h)\right], \tag{8.19a}$$

$$(-J_{nx})|_{i-1/2,\,j} = d_n|_{i-1/2,\,j}[\frac{n_{i,\,j} - n_{i-1,\,j}}{h_{i-1}} + O(h^2)]$$

$$- \mu_n|_{i-1/2,\,j}[\frac{n_{i,\,j} + n_{i-1,\,j}}{2} + O(h^2)][\frac{\phi_{i,\,j} + \phi_{Ci,\,j} - \phi_{i-1,\,j} - \phi_{Ci-1,\,j}}{h_{i-1}} + O(h^2)]$$

$$= d_n|_{i-1/2,\,j}[\frac{n_{i,\,j} - n_{i-1,\,j}}{h_{i-1}} + O(h^2)]$$

$$- \mu_n|_{i-1/2,\,j}[\frac{n_{i,\,j} + n_{i-1,\,j}}{2} \frac{\phi_{i,\,j} + \phi_{Ci,\,j} - \phi_{i-1,\,j} - \phi_{Ci-1,\,j}}{h_{i-1}} + O(h)]. \quad (8.19b)$$

Substituting equations (8.19a&b) into the first term on the RHS of (8.18a) yields

$$\frac{(-J_{nx})|_{i+1/2,\,j} - (-J_{nx})|_{i-1/2,\,j}}{(h_i + h_{i-1})/2} = d_n|_{i+1/2,\,j}\left[\frac{n_{i+1,\,j} - n_{i,\,j}}{h_i(h_i + h_{i-1})/2} + O(h)\right]$$

$$- \mu_n|_{i+1/2,\,j}\left[\frac{n_{i+1,\,j} + n_{i,\,j}}{2} \frac{\phi_{i+1,\,j} + \phi_{Ci+1,\,j} - \phi_{i,\,j} - \phi_{Ci,\,j}}{h_i(h_i + h_{i-1})/2} + O(1)\right]$$

$$- d_n|_{i-1/2,\,j}\left[\frac{n_{i,\,j} - n_{i-1,\,j}}{h_{i-1}(h_i + h_{i-1})/2} + O(h)\right]$$

$$+ \mu_n|_{i-1/2,\,j}\left[\frac{n_{i,\,j} + n_{i-1,\,j}}{2} \frac{\phi_{i,\,j} + \phi_{Ci,\,j} - \phi_{i-1,\,j} - \phi_{Ci-1,\,j}}{h_{i-1}(h_i + h_{i-1})/2} + O(1)\right]. \quad (8.20)$$

It is obvious that, as $h \to 0$, the discretization error on the RHS of equation (8.20) generally does not approach zero. Therefore, we conclude that equation (8.18a) does not converge under the conventional FD scheme.

Although higher order FD schemes can solve this problem, it needs more mesh grid points to be invoked, which will greatly complicate the implementation and make the matrix involved no longer sparse. And the latter in turn reduces the computational efficiency.

Therefore, in order to make equations (8.18a&b) converge, we have to find a solution to the normalized current densities at the intermediate step in the order of $O(h^2)$ and $O(k^2)$. Actually, by following the well-known Scharfetter and Gummel approach [14], we can find the normalized current densities by integrating equation (8.17) along each mesh interval. Such current densities obtained have a quadratic dependence of the local truncation error on the mesh spacing, which ensures the convergence of equations (8.18a&b). This algorithm is implemented through the following steps.

As the first step, we expand the current densities

$$J_{nx}(x, y_j) = J_{nx}|_{i+1/2,\,j} + [x - (x_i + h_i/2)]\frac{\partial J_{nx}}{\partial x}|_{i+1/2,\,j} + O(h^2)$$

$$J_{ny}(x_i, y) = J_{ny}|_{i,\,j+1/2} + [y - (y_j + k_j/2)]\frac{\partial J_{ny}}{\partial y}|_{i,\,j+1/2} + O(k^2), \quad (8.21a)$$

$$J_{px}(x, y_j) = J_{px}|_{i+1/2,\,j} + [x - (x_i + h_i/2)]\frac{\partial J_{px}}{\partial x}|_{i+1/2,\,j} + O(h^2)$$

$$J_{py}(x_i, y) = J_{py}|_{i,\,j+1/2} + [y - (y_j + k_j/2)]\frac{\partial J_{py}}{\partial y}|_{i,\,j+1/2} + O(k^2), \tag{8.21b}$$

within the mesh square $x \in [x_i, x_{i+1}]$, $y \in [y_j, y_{j+1}]$ centralized at $[x_i+h_i/2, y_j+k_j/2]$.

In the next step, by plugging equation (8.17) into (8.21a&b), we obtain a set of first order differential equations for the carrier densities defined inside the mesh square $x \in [x_i, x_{i+1}]$, $y \in [y_j, y_{j+1}]$, if we ignore the higher order terms in mesh spacing, i.e., $O(h^2)$ and $O(k^2)$. Taking the electron current density in the x direction as an example again, within the mesh interval $x \in [x_i, x_{i+1}]$, $y = y_j$, we have

$$\mu_n n\frac{\partial(\phi + \phi_C)}{\partial x} - d_n\frac{\partial n}{\partial x} = J_{nx}|_{i+1/2,\,j} + [x - (x_i + h_i/2)]\frac{\partial J_{nx}}{\partial x}|_{i+1/2,\,j}, \tag{8.22}$$

subject to the boundary conditions

$$n(x_i, y_j) = n_{i,\,j}, \text{ and } n(x_{i+1}, y_j) = n_{i+1,\,j}. \tag{8.23}$$

Actually, by assuming that $J_{nx}|_{i+1/2,\,j}$, $\partial J_{nx}/\partial x|_{i+1/2,\,j}$, and the partial derivative of the electrostatic potential in equation (8.22) are all constants within $[x_i, x_{i+1}]$, which is the assumption that we have already invoked in equation (8.21a) for obtaining (8.22) itself and in equation (8.15) for obtaining the discretized Poisson's equation (8.16), respectively, we can solve equation (8.22) as a first order linear ODE to determine the variation of the electron density along the path $x \in [x_i, x_{i+1}]$, $y = y_j$. In doing this, a scaled Einstein relation is assumed for the scaled carrier diffusivities and mobilities and both quantities are assumed to be constant in the mesh interval $[x_i, x_{i+1}]$. The general solution to equation (8.22) is

$$n(x, y_j) = Ce^{\phi(x,y_j)+\phi_C(x,y_j)} + h_i\frac{J_{nx}|_{i+1/2,\,j}}{\mu_n|_{i+1/2,\,j}}\frac{1 - e^{\phi(x,y_j)+\phi_C(x,y_j)}}{\phi_{i+1,\,j} + \phi_{Ci+1,\,j} - \phi_{i,\,j} - \phi_{Ci,\,j}} + O(h^3), \tag{8.24}$$

with $x \in [x_i, x_{i+1}]$. By matching the solution in equation (8.24) to the boundary conditions (8.23), we obtain

$$n_{i,\,j} = Ce^{\phi_{i,\,j}+\phi_{Ci,\,j}} + h_i\frac{J_{nx}|_{i+1/2,\,j}}{\mu_n|_{i+1/2,\,j}}\frac{1 - e^{\phi_{i,\,j}+\phi_{Ci,\,j}}}{\phi_{i+1,\,j} + \phi_{Ci+1,\,j} - \phi_{i,\,j} - \phi_{Ci,\,j}}, \tag{8.25a}$$

$$n_{i+1,\,j} = Ce^{\phi_{i+1,\,j}+\phi_{Ci+1,\,j}} + h_i\frac{J_{nx}|_{i+1/2,\,j}}{\mu_n|_{i+1/2,\,j}}\frac{1 - e^{\phi_{i+1,\,j}+\phi_{Ci+1,\,j}}}{\phi_{i+1,\,j} + \phi_{Ci+1,\,j} - \phi_{i,\,j} - \phi_{Ci,\,j}}. \tag{8.25b}$$

By eliminating the integration constant C from equations (8.25a&b), we obtain

$$J_{nx}|_{i+1/2,\,j} = \frac{\mu_n|_{i+1/2,\,j}}{h_i}[B(\phi_{i,\,j} + \phi_{Ci,\,j} - \phi_{i+1,\,j} - \phi_{Ci+1,\,j})n_{i,\,j}$$
$$- B(\phi_{i+1,\,j} + \phi_{Ci+1,\,j} - \phi_{i,\,j} - \phi_{Ci,\,j})n_{i+1,\,j}], \tag{8.26}$$

with the Bernoulli function defined as

$$B(x) \equiv \frac{x}{e^x - 1}. \tag{8.27}$$

From equations (8.25a) and (8.26), we find

$$C = n_{i,j} e^{-(\phi_{i,j}+\phi_{Ci,j})} - h_i \frac{J_{nx}|_{i+1/2,j}}{\mu_n|_{i+1/2,j}} \frac{e^{-(\phi_{i,j}+\phi_{Ci,j})} - 1}{\phi_{i+1,j} + \phi_{Ci+1,j} - \phi_{i,j} - \phi_{Ci,j}}. \tag{8.28}$$

Substituting equation (8.28) into (8.24) yields

$$\begin{aligned}
n(x, y_j) &= e^{\phi(x,y_j)+\phi_C(x,y_j)-(\phi_{i,j}+\phi_{Ci,j})} n_{i,j} - h_i \frac{J_{nx}|_{i+1/2,j}}{\mu_n|_{i+1/2,j}} \frac{e^{\phi(x,y_j)+\phi_C(x,y_j)-(\phi_{i,j}+\phi_{Ci,j})} - 1}{\phi_{i+1,j} + \phi_{Ci+1,j} - \phi_{i,j} - \phi_{Ci,j}} \\
&= \frac{e^{\phi(x,y_j)+\phi_C(x,y_j)-(\phi_{i+1,j}+\phi_{Ci+1,j})} - 1}{e^{\phi_{i,j}+\phi_{Ci,j}-\phi_{i+1,j}-\phi_{Ci+1,j}} - 1} n_{i,j} + \frac{e^{\phi(x,y_j)+\phi_C(x,y_j)-(\phi_{i,j}+\phi_{Ci,j})} - 1}{e^{\phi_{i+1,j}+\phi_{Ci+1,j}-\phi_{i,j}-\phi_{Ci,j}} - 1} n_{i+1,j} \\
&= \frac{1 - e^{(\phi_{i+1,j}+\phi_{Ci+1,j}-\phi_{i,j}-\phi_{Ci,j})\frac{x-x_i-h_i}{h_i}}}{1 - e^{\phi_{i,j}+\phi_{Ci,j}-\phi_{i+1,j}-\phi_{Ci+1,j}}} n_{i,j} + \frac{1 - e^{(\phi_{i+1,j}+\phi_{Ci+1,j}-\phi_{i,j}-\phi_{Ci,j})\frac{x-x_i}{h_i}}}{1 - e^{\phi_{i+1,j}+\phi_{Ci+1,j}-\phi_{i,j}-\phi_{Ci,j}}} n_{i+1,j} \\
&= [1 - g_{i,j}(x,\phi)] n_{i,j} + g_{i,j}(x,\phi) n_{i+1,j},
\end{aligned} \tag{8.29}$$

where equation (8.26) and following expansions

$$\begin{aligned}
\phi(x, y_j) + \phi_C(x, y_j) &= \phi_{i,j} + \phi_{Ci,j} + \frac{\partial(\phi + \phi_C)}{\partial x}\Big|_{i,j}(x - x_i) + \cdots \\
&= \phi_{i,j} + \phi_{Ci,j} + \frac{\phi_{i+1,j} + \phi_{Ci+1,j} - (\phi_{i,j} + \phi_{Ci,j})}{h_i}(x - x_i) + O(h^2) \\
\phi(x, y_j) + \phi_C(x, y_j) &= \phi_{i+1,j} + \phi_{Ci+1,j} + \frac{\partial(\phi + \phi_C)}{\partial x}\Big|_{i+1,j}(x - x_{i+1}) + \cdots \\
&= \phi_{i+1,j} + \phi_{Ci+1,j} + \frac{\phi_{i+1,j} + \phi_{Ci+1,j} - (\phi_{i,j} + \phi_{Ci,j})}{h_i} \\
&\quad \times (x - x_i - h_i) + O(h^2)
\end{aligned}$$

have been used with the growth function in equation (8.29) introduced as

$$g_{i,j}(x,\phi) \equiv \frac{1 - e^{(\phi_{i+1,j}+\phi_{Ci+1,j}-\phi_{i,j}-\phi_{Ci,j})\frac{x-x_i}{h_i}}}{1 - e^{(\phi_{i+1,j}+\phi_{Ci+1,j}-\phi_{i,j}-\phi_{Ci,j})}}. \tag{8.30}$$

According to equation (8.30), the growth function reduces to a linear function

$$g_{i,j}(x,\phi) = \frac{x - x_i}{h_i}, \tag{8.31}$$

once $\phi_{i+1,j} + \phi_{Ci+1,j} = \phi_{i,j} + \phi_{Ci,j}$. This indicates a linear interpolation in equation (8.29) or a consistent FD scheme, which is expected since the carrier density equations (8.3b&c) all degenerate to linear equations under zero electrostatic potential change (i.e., zero field and zero drift).

By following fully analogous approaches we obtain expressions for $n(x_i, y)$, $p(x, y_j)$, and $p(x_i, y)$, where $x \in [x_i, x_{i+1}]$ and $y \in [y_j, y_{j+1}]$.

In the third step, by following equation (8.17), we express all of the current densities required in equations (8.18a&b) in terms of the electrostatic potential and carrier densities obtained in the form of equation (8.26) and its analogs.

In the final step, by substituting the current densities obtained above into equations (8.18a&b), we obtain the discretized carrier continuity equations

$$
\begin{aligned}
&\mu_n|_{i+1/2,j}\frac{B(\phi_{i+1,j}+\phi_{Ci+1,j}-\phi_{i,j}-\phi_{Ci,j})n_{i+1,j}-B(\phi_{i,j}+\phi_{Ci,j}-\phi_{i+1,j}-\phi_{Ci+1,j})n_{i,j}}{h_i(h_i+h_{i-1})/2}\\
&-\mu_n|_{i-1/2,j}\frac{B(\phi_{i,j}+\phi_{Ci,j}-\phi_{i-1,j}-\phi_{Ci-1,j})n_{i,j}-B(\phi_{i-1,j}+\phi_{Ci-1,j}-\phi_{i,j}-\phi_{Ci,j})n_{i-1,j}}{h_{i-1}(h_i+h_{i-1})/2}\\
&+\mu_n|_{i,j+1/2}\frac{B(\phi_{i,j+1}+\phi_{Ci,j+1}-\phi_{i,j}-\phi_{Ci,j})n_{i,j+1}-B(\phi_{i,j}+\phi_{Ci,j}-\phi_{i,j+1}-\phi_{Ci,j+1})n_{i,j}}{k_j(k_j+k_{j-1})/2}\\
&-\mu_n|_{i,j-1/2}\frac{B(\phi_{i,j}+\phi_{Ci,j}-\phi_{i,j-1}-\phi_{Ci,j-1})n_{i,j}-B(\phi_{i,j-1}+\phi_{Ci,j-1}-\phi_{i,j}-\phi_{Ci,j})n_{i,j-1}}{k_{j-1}(k_j+k_{j-1})/2}\\
&-R|_{i,j}=0,
\end{aligned}
\tag{8.32a}
$$

$$
\begin{aligned}
&\mu_p|_{i+1/2,j}\frac{B(\phi_{i,j}-\phi_{Vi,j}-\phi_{i+1,j}+\phi_{Vi+1,j})p_{i+1,j}-B(\phi_{i+1,j}-\phi_{Vi+1,j}-\phi_{i,j}+\phi_{Vi,j})p_{i,j}}{h_i(h_i+h_{i-1})/2}\\
&-\mu_p|_{i-1/2,j}\frac{B(\phi_{i-1,j}-\phi_{Vi-1,j}-\phi_{i,j}+\phi_{Vi,j})p_{i,j}-B(\phi_{i,j}-\phi_{Vi,j}-\phi_{i-1,j}+\phi_{Vi-1,j})p_{i-1,j}}{h_{i-1}(h_i+h_{i-1})/2}\\
&+\mu_p|_{i,j+1/2}\frac{B(\phi_{i,j}-\phi_{Vi,j}-\phi_{i,j+1}+\phi_{Vi,j+1})p_{i,j+1}-B(\phi_{i,j+1}-\phi_{Vi,j+1}-\phi_{i,j}+\phi_{Vi,j})p_{i,j}}{k_j(k_j+k_{j-1})/2}\\
&-\mu_p|_{i,j-1/2}\frac{B(\phi_{i,j-1}-\phi_{Vi,j-1}-\phi_{i,j}+\phi_{Vi,j})p_{i,j}-B(\phi_{i,j}-\phi_{Vi,j}-\phi_{i,j-1}+\phi_{Vi,j-1})p_{i,j-1}}{k_{j-1}(k_j+k_{j-1})/2}\\
&-R|_{i,j}=0.
\end{aligned}
\tag{8.32b}
$$

Once the electrostatic potential is solved using equation (8.16) on the mesh grid at the first order of accuracy in terms of the mesh spacing, the discretized carrier continuity equations (8.32a&b) have a local truncation error linearly proportional to the mesh spacing. If the mesh is uniform or quasi-uniform, equations (8.32a&b) will be at the first order of accuracy in terms of the mesh spacing anyway, regardless of the accuracy achieved by the electrostatic potential.

The boundary conditions must be incorporated in this discretization scheme as well. For edge emitting optoelectronic devices, we usually impose the Ohmic, infinity insulator, and artificial boundaries, as illustrated in Fig. 8.1.

For the top Ohmic contact along the horizontal direction (x), we have the Dirichlet boundary condition

$$\phi_{i,j}=\phi_{\text{bias}}|_{i,j}+\phi_b|_{i,j}, \tag{8.33a}$$

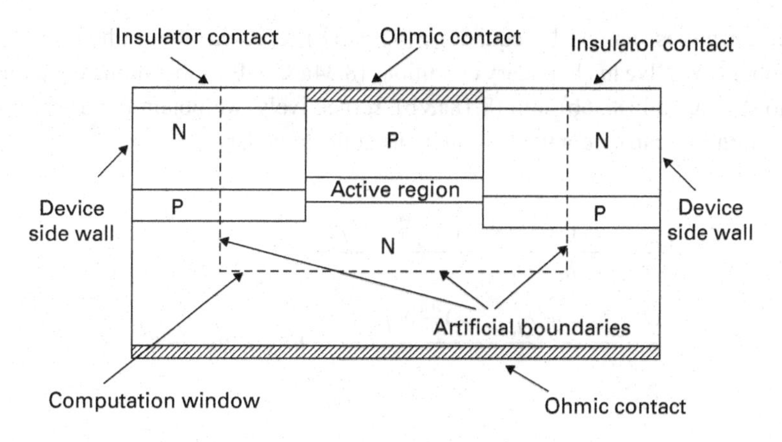

Fig. 8.1. Boundary conditions imposed on the carrier transport solver.

and

$$n_{i,j} = C_{i,j}, \text{ and } p_{i,j} = n_i^2|_{i,j}/C_{i,j}, \tag{8.33b}$$

if $C_{i,j} > 0$, or

$$p_{i,j} = |C_{i,j}|, \text{ and } n_{i,j} = n_i^2|_{i,j}/|C_{i,j}|, \tag{8.33c}$$

if $C_{i,j} < 0$.

Equations (8.33b&c) are obtained from (8.6b&c) based on the assumption that $|C| \gg 2n_i$.

For the top infinity insulator contacts along the horizontal direction (x) without surface charge and recombination, we have the Neumann boundary conditions

$$\frac{\partial \phi}{\partial y}|_{i,j} = 0, \tag{8.34a}$$

$$J_{ny}|_{i,j} = 0, \text{ and } J_{py}|_{i,j} = 0. \tag{8.34b}$$

A mirror image approach can be taken to discretize these quantities at the boundary point. For the equidistance grid, by applying the center linear interpolation

$$u_{i,j} = 0.5[u_{i,j+1/2} + u_{i,j-1/2}] + O(k_j^2), \tag{8.35}$$

to equations (8.34a&b), we find

$$\frac{\partial \phi}{\partial y}|_{i,j+1/2} = -\frac{\partial \phi}{\partial y}|_{i,j-1/2} + O(k_j^2), \tag{8.36a}$$

and

$$J_{ny}|_{i,j+1/2} = -J_{ny}|_{i,j-1/2} + O(k_j^2)$$
$$J_{py}|_{i,j+1/2} = -J_{py}|_{i,j-1/2} + O(k_j^2). \tag{8.36b}$$

The quantities defined by equations (8.36a&b) represent the artificial images which implicitly resolve the boundary conditions (8.34a&b). By substituting equations (8.36a) into (8.14), and (8.36b) into (8.18a&b), respectively, we obtain the discrete Poisson's equation and carrier continuity equations at the boundary

$$\frac{\lambda^2_{i+1/2,\,j}\frac{\partial\phi}{\partial x}|_{i+1/2,\,j} - \lambda^2_{i-1/2,\,j}\frac{\partial\phi}{\partial x}|_{i-1/2,\,j}}{(h_i + h_{i-1})/2} - \frac{(\lambda^2_{i,\,j+1/2} + \lambda^2_{i,\,j-1/2})\frac{\partial\phi}{\partial y}|_{i,\,j-1/2}}{k_{j-1}} - n_{i,\,j} + p_{i,\,j} + C_{i,\,j} = 0, \tag{8.37a}$$

and

$$\frac{(-J_{nx})|_{i+1/2,\,j} - (-J_{nx})|_{i-1/2,\,j}}{(h_i + h_{i-1})/2} - \frac{2(-J_{ny})|_{i,\,j-1/2}}{k_{j-1}} - R|_{i,\,j} = 0, \tag{8.37b}$$

$$\frac{J_{px}|_{i+1/2,\,j} - J_{px}|_{i-1/2,\,j}}{(h_i + h_{i-1})/2} - \frac{2J_{py}|_{i,\,j-1/2}}{k_{j-1}} - R|_{i,\,j} = 0. \tag{8.37c}$$

Following a similar approach to that used in obtaining equation (8.16) and (8.32a&b), we finally have

$$\frac{\lambda^2_{i+1/2,\,j}\frac{\phi_{i+1,\,j}-\phi_{i,\,j}}{h_i} - \lambda^2_{i-1/2,\,j}\frac{\phi_{i,\,j}-\phi_{i-1,\,j}}{h_{i-1}}}{(h_i + h_{i-1})/2} - \frac{(\lambda^2_{i,\,j+1/2} + \lambda^2_{i,\,j-1/2})(\phi_{i,\,j} - \phi_{i,\,j-1})}{k^2_{j-1}} - n_{i,\,j} + p_{i,\,j} + C_{i,\,j} = 0, \tag{8.38a}$$

$$\begin{aligned}
&\mu_n|_{i+1/2,\,j}\frac{B(\phi_{i+1,\,j} + \phi_{\mathrm{C}i+1,\,j} - \phi_{i,\,j} - \phi_{\mathrm{C}i,\,j})n_{i+1,\,j} - B(\phi_{i,\,j} + \phi_{\mathrm{C}i,\,j} - \phi_{i+1,\,j} - \phi_{\mathrm{C}i+1,\,j})n_{i,\,j}}{h_i(h_i + h_{i-1})/2} \\
&- \mu_n|_{i-1/2,\,j}\frac{B(\phi_{i,\,j} + \phi_{\mathrm{C}i,\,j} - \phi_{i-1,\,j} - \phi_{\mathrm{C}i-1,\,j})n_{i,\,j} - B(\phi_{i-1,\,j} + \phi_{\mathrm{C}i-1,\,j} - \phi_{i,\,j} - \phi_{\mathrm{C}i,\,j})n_{i-1,\,j}}{h_{i-1}(h_i + h_{i-1})/2} \\
&- \mu_n|_{i,\,j-1/2}\frac{B(\phi_{i,\,j} + \phi_{\mathrm{C}i,\,j} - \phi_{i,\,j-1} - \phi_{\mathrm{C}i,\,j-1})n_{i,\,j} - B(\phi_{i,\,j-1} + \phi_{\mathrm{C}i,\,j-1} - \phi_{i,\,j} - \phi_{\mathrm{C}i,\,j})n_{i,\,j-1}}{k^2_{j-1}/2} \\
&- R|_{i,\,j} = 0,
\end{aligned} \tag{8.38b}$$

$$\begin{aligned}
&\mu_p|_{i+1/2,\,j}\frac{B(\phi_{i,\,j} - \phi_{\mathrm{V}i,\,j} - \phi_{i+1,\,j} + \phi_{\mathrm{V}i+1,\,j})p_{i+1,\,j} - B(\phi_{i+1,\,j} - \phi_{\mathrm{V}i+1,\,j} - \phi_{i,\,j} + \phi_{\mathrm{V}i,\,j})p_{i,\,j}}{h_i(h_i + h_{i-1})/2} \\
&- \mu_p|_{i-1/2,\,j}\frac{B(\phi_{i-1,\,j} - \phi_{\mathrm{V}i-1,\,j} - \phi_{i,\,j} + \phi_{\mathrm{V}i,\,j})p_{i,\,j} - B(\phi_{i,\,j} - \phi_{\mathrm{V}i,\,j} - \phi_{i-1,\,j} + \phi_{\mathrm{V}i-1,\,j})p_{i-1,\,j}}{h_{i-1}(h_i + h_{i-1})/2} \\
&- \mu_p|_{i,\,j-1/2}\frac{B(\phi_{i,\,j-1} - \phi_{\mathrm{V}i,\,j-1} - \phi_{i,\,j} + \phi_{\mathrm{V}i,\,j})p_{i,\,j} - B(\phi_{i,\,j} - \phi_{\mathrm{V}i,\,j} - \phi_{i,\,j-1} + \phi_{\mathrm{V}i,\,j-1})p_{i,\,j-1}}{k^2_{j-1}/2} \\
&- R|_{i,\,j} = 0.
\end{aligned} \tag{8.38c}$$

For the bottom artificial boundary along the horizontal direction (x), we usually apply the Neumann boundary conditions specified by equation (8.5a), or

$$\frac{\partial \phi}{\partial y}|_{i,\,j} = 0, \quad \frac{\partial n}{\partial y}|_{i,\,j} = 0, \text{ and } \frac{\partial p}{\partial y}|_{i,\,j} = 0. \tag{8.39}$$

According to equation (8.17), equation (8.39) is equivalent to (8.34a&b). Hence equations (8.38a–c) apply to the bottom artificial boundary as well.

For the artificial boundaries along the vertical direction (y), we still have the Neumann boundary conditions specified by equation (8.5a), or

$$\frac{\partial \phi}{\partial x}|_{i,\,j} = 0, \quad \frac{\partial n}{\partial x}|_{i,\,j} = 0, \text{ and } \frac{\partial p}{\partial x}|_{i,\,j} = 0. \tag{8.40}$$

According to equation (8.17), (8.40) is equivalent to

$$\frac{\partial \phi}{\partial x}|_{i,\,j} = 0, \tag{8.41a}$$

$$J_{nx}|_{i,\,j} = 0, \text{ and } J_{px}|_{i,\,j} = 0. \tag{8.41b}$$

Therefore, by following an approach similar to that used in obtaining equation (8.35) through (8.38c), we obtain

$$\begin{aligned}
&-\frac{(\lambda^2_{i+1/2,\,j} + \lambda^2_{i-1/2,\,j})(\phi_{i,\,j} - \phi_{i-1,\,j})}{h^2_{i-1}} \\
&+ \frac{\lambda^2_{i,\,j+1/2}\frac{\phi_{i,\,j+1}-\phi_{i,\,j}}{k_j} - \lambda^2_{i,\,j-1/2}\frac{\phi_{i,\,j}-\phi_{i,\,j-1}}{k_{j-1}}}{(k_j + k_{j-1})/2} - n_{i,\,j} + p_{i,\,j} + C_{i,\,j} = 0,
\end{aligned} \tag{8.42a}$$

$$\begin{aligned}
&-\mu_n|_{i-1/2,\,j}\frac{B(\phi_{i,\,j} + \phi_{Ci,\,j} - \phi_{i-1,\,j} - \phi_{Ci-1,\,j})n_{i,\,j} - B(\phi_{i-1,\,j} + \phi_{Ci-1,\,j} - \phi_{i,\,j} - \phi_{Ci,\,j})n_{i-1,\,j}}{h^2_{i-1}/2} \\
&+ \mu_n|_{i,\,j+1/2}\frac{B(\phi_{i,\,j+1} + \phi_{Ci,\,j+1} - \phi_{i,\,j} - \phi_{Ci,\,j})n_{i,\,j+1} - B(\phi_{i,\,j} + \phi_{Ci,\,j} - \phi_{i,\,j+1} - \phi_{Ci,\,j+1})n_{i,\,j}}{k_j(k_j + k_{j-1})/2} \\
&- \mu_n|_{i,\,j-1/2}\frac{B(\phi_{i,\,j} + \phi_{Ci,\,j} - \phi_{i,\,j-1} - \phi_{Ci,\,j-1})n_{i,\,j} - B(\phi_{i,\,j-1} + \phi_{Ci,\,j-1} - \phi_{i,\,j} - \phi_{Ci,\,j})n_{i,\,j-1}}{k_{j-1}(k_j + k_{j-1})/2} \\
&- R|_{i,\,j} = 0,
\end{aligned} \tag{8.42b}$$

$$\begin{aligned}
&-\mu_p|_{i-1/2,\,j}\frac{B(\phi_{i-1,\,j} - \phi_{Vi-1,\,j} - \phi_{i,\,j} + \phi_{Vi,\,j})p_{i,\,j} - B(\phi_{i,\,j} - \phi_{Vi,\,j} - \phi_{i-1,\,j} + \phi_{Vi-1,\,j})p_{i-1,\,j}}{h^2_{i-1}/2} \\
&+ \mu_p|_{i,\,j+1/2}\frac{B(\phi_{i,\,j} - \phi_{Vi,\,j} - \phi_{i,\,j+1} + \phi_{Vi,\,j+1})p_{i,\,j+1} - B(\phi_{i,\,j+1} - \phi_{Vi,\,j+1} - \phi_{i,\,j} + \phi_{Vi,\,j})p_{i,\,j}}{k_j(k_j + k_{j-1})/2} \\
&- \mu_p|_{i,\,j-1/2}\frac{B(\phi_{i,\,j-1} - \phi_{Vi,\,j-1} - \phi_{i,\,j} + \phi_{Vi,\,j})p_{i,\,j} - B(\phi_{i,\,j} - \phi_{Vi,\,j} - \phi_{i,\,j-1} + \phi_{Vi,\,j-1})p_{i,\,j-1}}{k_{j-1}(k_j + k_{j-1})/2} \\
&- R|_{i,\,j} = 0.
\end{aligned} \tag{8.42c}$$

Under such imposed boundary conditions and the associated discretization schemes, there is no inconsistency problem at the four corner points. That is to say, either from equations (8.38a–c) or (8.42a–c), we have

$$-\frac{(\lambda_{i+1/2,\,j}^2+\lambda_{i-1/2,\,j}^2)(\phi_{i,\,j}-\phi_{i-1,\,j})}{h_{i-1}^2}$$

$$-\frac{(\lambda_{i,\,j+1/2}^2+\lambda_{i,\,j-1/2}^2)(\phi_{i,\,j}-\phi_{i,\,j-1})}{k_{j-1}^2}-n_{i,\,j}+p_{i,\,j}+C_{i,\,j}=0, \tag{8.43a}$$

$$-\mu_n|_{i-1/2,\,j}\frac{B(\phi_{i,\,j}+\phi_{Ci,\,j}-\phi_{i-1,\,j}-\phi_{Ci-1,\,j})n_{i,\,j}-B(\phi_{i-1,\,j}+\phi_{Ci-1,\,j}-\phi_{i,\,j}-\phi_{Ci,\,j})n_{i-1,\,j}}{h_{i-1}^2/2}$$

$$-\mu_n|_{i,\,j-1/2}\frac{B(\phi_{i,\,j}+\phi_{Ci,\,j}-\phi_{i,\,j-1}-\phi_{Ci,\,j-1})n_{i,\,j}-B(\phi_{i,\,j-1}+\phi_{Ci,\,j-1}-\phi_{i,\,j}-\phi_{Ci,\,j})n_{i,\,j-1}}{k_{j-1}^2/2}$$

$$-R|_{i,\,j}=0, \tag{8.43b}$$

$$-\mu_p|_{i-1/2,\,j}\frac{B(\phi_{i-1,\,j}-\phi_{Vi-1,\,j}-\phi_{i,\,j}+\phi_{Vi,\,j})p_{i,\,j}-B(\phi_{i,\,j}-\phi_{Vi,\,j}-\phi_{i-1,\,j}+\phi_{Vi-1,\,j})p_{i-1,\,j}}{h_{i-1}^2/2}$$

$$-\mu_p|_{i,\,j-1/2}\frac{B(\phi_{i,\,j-1}-\phi_{Vi,\,j-1}-\phi_{i,\,j}+\phi_{Vi,\,j})p_{i,\,j}-B(\phi_{i,\,j}-\phi_{Vi,\,j}-\phi_{i,\,j-1}+\phi_{Vi,\,j-1})p_{i,\,j-1}}{k_{j-1}^2/2}$$

$$-R|_{i,\,j}=0. \tag{8.43c}$$

8.1.5 Solution of non-linear algebraic equations

After discretization, we obtain equations (8.16) and (8.32a&b) for the inner points, equations (8.33a–c) for the points along the top boundary with the Ohmic contact, equations (8.38a–c) for the points along the top boundary with the infinity insulator contact and for the points along the bottom artificial boundary that truncates the computation domain, equations (8.42a–c) for the points along the vertical artificial boundaries that truncate the computation domain, and equations (8.43a–c) for the four corner points. To be solved, these equations can generally be written as

$$\hat{F}(\hat{w})=0, \tag{8.44}$$

where

$$\hat{F}=\left[\hat{f}_\phi(\hat{w}),\quad \hat{f}_n(\hat{w}),\quad \hat{f}_p(\hat{w})\right]^T, \tag{8.45}$$

$$\hat{w}=\left[\hat{\phi},\quad \hat{n},\quad \hat{p}\right]^T. \tag{8.46}$$

In accordance with the classical Newton's method, we have the following iteration scheme in looking for the solution

$$\hat{w}^{k+1}=\hat{w}^k-B^{-1}(\hat{w}^k)F(\hat{w}^k), \tag{8.47}$$

or

$$B(\hat{w}^k)(\hat{w}^{k+1}-\hat{w}^k)=-F(\hat{w}^k), \tag{8.48}$$

to avoid an expensive inversion of matrix $B(\hat{w}^k)$, where $B(\hat{w}^k) = F'(\hat{w}^k)$ is the Jacobian matrix.

In application of the classical Newton's method to our equations, there is a general tendency to overestimate the length of the actual correction step for the iteration, known as the overshoot problem. To solve this problem, a few modified schemes can be considered [15–17].

Modification A.

Let

$$B(\hat{w}^k) = \frac{1}{t_k} F'(\hat{w}^k), \tag{8.49}$$

where the parameter t_k is chosen as the largest of $1/2^i$ with $i = 1, 2, 3, \ldots$, which makes

$$||F'^{-1}(\hat{w}^k) F[\hat{w}^k - t_k F'^{-1}(\hat{w}^k) F(\hat{w}^k)]|| < ||F'^{-1}(\hat{w}^k) F(\hat{w}^k)||, \tag{8.50}$$

or

$$||D^{-1}(\hat{w}^k) F[\hat{w}^k - t_k F'^{-1}(\hat{w}^k) F(\hat{w}^k)]|| < ||D^{-1}(\hat{w}^k) F(\hat{w}^k)||, \tag{8.51}$$

with $D(\hat{w}^k)$ denoted as the main diagonal of $F'(\hat{w}^k)$.

The parameter t_k can also be selected as

$$t_k = 1/[1 + \kappa_k ||F(\hat{w}^k)||], \tag{8.52}$$

with κ_k such that

$$1 - \frac{||F(\hat{w}^{k+1})||}{||F(\hat{w}^k)||} < \delta t_k \quad \delta \in (0, 1). \tag{8.53}$$

Modification B.

Let

$$B(\hat{w}^k) = s_k I + F'(\hat{w}^k), \tag{8.54}$$

where the parameter s_k is selected as

$$s_k = \sigma_k ||F(\hat{w}^k)|| \quad \sigma_k > 0, \tag{8.55}$$

with σ_k such that

$$||F(\hat{w}^{k+1})|| < ||F(\hat{w}^k)||. \tag{8.56}$$

Modification C.

As a combination of the two previous modifications, we let

$$B(\hat{w}^k) = \frac{1}{t_k}[s_k I + F'(\hat{w}^k)]. \tag{8.57}$$

The quality of $B(\hat{w}^k)$ can be evaluated by

$$\lim_{\alpha\to 0}\frac{1}{\alpha}||F[\hat{w}^k-\alpha B^{-1}(\hat{w}^k)F(\hat{w}^k)]-F(\hat{w}^k)||=C>0, \tag{8.58}$$

where C is a positive constant.

From equation (8.48), we find that a solution of a system of linear algebraic equations at each iteration step is required. The result of the linear algebraic equations obtained here is just an incremental correction to the intermediate approximation of the solution, and therefore the accuracy required is only to preserve the convergence of Newton's algorithm. Hence, to obtain the overall solution to the non-linear system more efficiently, we may consider bringing in some modifications to solving the system of linear algebraic equations in order to reduce the computation cost, although the iteration steps in solving the non-linear system might increase because of such modifications. As a result, a successive over-relaxation (SOR) Newton's method is introduced [12].

Actually, the linear equation system (8.48) at the kth iteration step can be written as

$$\begin{bmatrix} \frac{\partial \hat{f}_\phi}{\partial \hat{\phi}} & \frac{\partial \hat{f}_\phi}{\partial \hat{n}} & \frac{\partial \hat{f}_\phi}{\partial \hat{p}} \\ \frac{\partial \hat{f}_n}{\partial \hat{\phi}} & \frac{\partial \hat{f}_n}{\partial \hat{n}} & \frac{\partial \hat{f}_n}{\partial \hat{p}} \\ \frac{\partial \hat{f}_p}{\partial \hat{\phi}} & \frac{\partial \hat{f}_p}{\partial \hat{n}} & \frac{\partial \hat{f}_p}{\partial \hat{p}} \end{bmatrix}_k \begin{bmatrix} \delta\hat{\phi}_k \\ \delta\hat{n}_k \\ \delta\hat{p}_k \end{bmatrix} = -\begin{bmatrix} \hat{f}_\phi(\hat{\phi}_k,\hat{n}_k,\hat{p}_k) \\ \hat{f}_n(\hat{\phi}_k,\hat{n}_k,\hat{p}_k) \\ \hat{f}_p(\hat{\phi}_k,\hat{n}_k,\hat{p}_k) \end{bmatrix}, \tag{8.59}$$

where we have defined

$$\delta\hat{\phi}_k=\hat{\phi}^{k+1}-\hat{\phi}^k,\quad \delta\hat{n}_k=\hat{n}^{k+1}-\hat{n}^k,\ \text{and}\ \delta\hat{p}_k=\hat{p}^{k+1}-\hat{p}^k. \tag{8.60}$$

Under the assumption that the Jacobian matrix is definite and that all blocks in the main diagonal of equation (8.59) are non-singular, we can use a classical block iteration scheme for the solution of the kth iteration step

$$\begin{bmatrix} \frac{\partial \hat{f}_\phi}{\partial \hat{\phi}} & 0 & 0 \\ \frac{\partial \hat{f}_n}{\partial \hat{\phi}} & \frac{\partial \hat{f}_n}{\partial \hat{n}} & 0 \\ \frac{\partial \hat{f}_p}{\partial \hat{\phi}} & \frac{\partial \hat{f}_p}{\partial \hat{n}} & \frac{\partial \hat{f}_p}{\partial \hat{p}} \end{bmatrix}_k \begin{bmatrix} \delta\hat{\phi}_k \\ \delta\hat{n}_k \\ \delta\hat{p}_k \end{bmatrix}_{m+1} = -\begin{bmatrix} \hat{f}_\phi(\hat{\phi}_k,\hat{n}_k,\hat{p}_k) \\ \hat{f}_n(\hat{\phi}_k,\hat{n}_k,\hat{p}_k) \\ \hat{f}_p(\hat{\phi}_k,\hat{n}_k,\hat{p}_k) \end{bmatrix}$$

$$-\begin{bmatrix} 0 & \frac{\partial \hat{f}_\phi}{\partial \hat{n}} & \frac{\partial \hat{f}_\phi}{\partial \hat{p}} \\ 0 & 0 & \frac{\partial \hat{f}_n}{\partial \hat{p}} \\ 0 & 0 & 0 \end{bmatrix}_k \begin{bmatrix} \delta\hat{\phi}_k \\ \delta\hat{n}_k \\ \delta\hat{p}_k \end{bmatrix}_m. \tag{8.61}$$

Equation (8.61) can therefore be solved as three decoupled linear equation systems sequentially

$$\begin{aligned}\frac{\partial \hat{f}_{\phi k}}{\partial \hat{\phi}}\delta\hat{\phi}_{km+1} &= -\hat{f}_\phi(\hat{\phi}_k, \hat{n}_k, \hat{p}_k) - \frac{\partial \hat{f}_{\phi k}}{\partial \hat{n}}\delta\hat{n}_{km} - \frac{\partial \hat{f}_{\phi k}}{\partial \hat{p}}\delta\hat{p}_{km} \\ &= -\omega\hat{f}_\phi(\hat{\phi}_k, \hat{n}_k + \delta\hat{n}_{km}, \hat{p}_k + \delta\hat{p}_{km}),\end{aligned} \tag{8.62a}$$

$$\begin{aligned}\frac{\partial \hat{f}_{nk}}{\partial \hat{n}}\delta\hat{n}_{km+1} &= -\hat{f}_n(\hat{\phi}_k, \hat{n}_k, \hat{p}_k) - \frac{\partial \hat{f}_{nk}}{\partial \hat{\phi}}\delta\hat{\phi}_{km+1} - \frac{\partial \hat{f}_{nk}}{\partial \hat{p}}\delta\hat{p}_{km} \\ &= -\omega\hat{f}_n(\hat{\phi}_k + \delta\hat{\phi}_{km+1}, \hat{n}_k, \hat{p}_k + \delta\hat{p}_{km}),\end{aligned} \tag{8.62b}$$

$$\begin{aligned}\frac{\partial \hat{f}_{pk}}{\partial \hat{p}}\delta\hat{p}_{km+1} &= -\hat{f}_p(\hat{\phi}_k, \hat{n}_k, \hat{p}_k) - \frac{\partial \hat{f}_{pk}}{\partial \hat{\phi}}\delta\hat{\phi}_{km+1} - \frac{\partial \hat{f}_{pk}}{\partial \hat{n}}\delta\hat{n}_{km+1} \\ &= -\omega\hat{f}_p(\hat{\phi}_k + \delta\hat{\phi}_{km+1}, \hat{n}_k + \delta\hat{n}_{km+1}, \hat{p}_k),\end{aligned} \tag{8.62c}$$

with $\omega < 1$ introduced as a relaxation parameter.

Note that for the reduced problem under the charge neutral assumption, equation (8.62a) can hardly converge since $\partial\hat{f}_\phi/\partial\hat{\phi} = 0$.

Finally, we come to the step of solving the system of linear equations. As a result of using the FD discretization scheme, the linear equation system involved is usually sparse. Therefore, a number of special numerical techniques can be invoked to solve these sparse linear equations efficiently. Since these techniques have little physics involved, we list them in Appendix D and skip further discussions.

8.2 The transient carrier transport equation

Solving the transient carrier transport equation through a numerical approach is similar to the time domain marching techniques introduced in dealing with the optical traveling wave equations. The similarity comes from the fact that both optical traveling wave equations and carrier transport equations are PDEs with first order time derivatives, hence they are both of the parabolic type. However, the optical traveling wave equations are linear and have first order spatial derivatives along the propagation direction, although they may have the second order spatial derivatives along other directions such as in a horn waveguide. On the contrary, although the carrier transport equations have second order spatial derivatives which bring in a damping effect to some extent through the diffusion, they are explicitly non-linear. As a result, in contrast to solving the optical traveling wave equation, where we still manage to find an explicit marching scheme (i.e., the upwind or FT–BS scheme), there seems to be no stable explicit scheme available in dealing with the carrier transport equations.

By introducing $d_m = t_{m+1} - t_m$ and using the symbol u_m to represent $u(x, y, t_m)$, we have the following three stable discretization schemes.

(1) Full backward time difference method (fully implicit).

In this method

$$\hat{f}_\phi(\hat{\phi}_{m+1}, \hat{n}_{m+1}, \hat{p}_{m+1}) = 0, \tag{8.63a}$$

$$\frac{\hat{n}_{m+1} - \hat{n}_m}{d_m} - \hat{f}_n(\hat{\phi}_{m+1}, \hat{n}_{m+1}, \hat{p}_{m+1}) = 0, \tag{8.63b}$$

$$\frac{\hat{p}_{m+1} - \hat{p}_m}{d_m} - \hat{f}_p(\hat{\phi}_{m+1}, \hat{n}_{m+1}, \hat{p}_{m+1}) = 0. \tag{8.63c}$$

In the fully explicit discretization scheme, the stable condition requires the time step $d_m = O(h^2 + k^2)$, which is not feasible in practice. Therefore, this scheme is of little use in dealing with our problem.

(2) Mock's method (semi-implicit) [18–20].

Noting that the Poisson equation can be converted to

$$\nabla \cdot \left(\lambda^2 \nabla \frac{\partial \phi}{\partial t}\right) - \frac{\partial n}{\partial t} + \frac{\partial p}{\partial t} = 0, \tag{8.64}$$

we have

$$\nabla \cdot \left(\lambda^2 \nabla \frac{\hat{\phi}_{m+1} - \hat{\phi}_m}{d_m}\right) - \hat{f}_n(\hat{\phi}_{m+1}, \hat{n}_m, \hat{p}_m) + \hat{f}_p(\hat{\phi}_{m+1}, \hat{n}_m, \hat{p}_m) = 0, \tag{8.65a}$$

$$\frac{\hat{n}_{m+1} - \hat{n}_m}{d_m} - \hat{f}_n(\hat{\phi}_{m+1}, \hat{n}_{m+1}, \hat{p}_m) = 0, \tag{8.65b}$$

$$\frac{\hat{p}_{m+1} - \hat{p}_m}{d_m} - \hat{f}_p(\hat{\phi}_{m+1}, \hat{n}_{m+1}, \hat{p}_{m+1}) = 0. \tag{8.65c}$$

(3) Modified Mock's method (semi-implicit).

In this method

$$\frac{\hat{n}_{m+1} - \hat{n}_m}{d_m} - \hat{f}_n(\hat{\phi}_m, \hat{n}_{m+1}, \hat{p}_m) = 0, \tag{8.66a}$$

$$\frac{\hat{p}_{m+1} - \hat{p}_m}{d_m} - \hat{f}_p(\hat{\phi}_m, \hat{n}_{m+1}, \hat{p}_{m+1}) = 0, \tag{8.66b}$$

$$\nabla \cdot (\lambda^2 \nabla \hat{\phi}_{m+1}) - r(\hat{\phi}_{m+1} - \hat{\phi}_m) - (\hat{n}_{m+1} - \hat{p}_{m+1} - \hat{C}) = 0, \tag{8.66c}$$

where r is an appropriate positive and bounded damping function.

8.3 The carrier rate equation

Since the carrier rate equation (5.41) is posed as an ODE initial value problem, the solution technique is similar to that which has been discussed in Section 7.2.3. Equation (5.41) is normally solved in conjunction with the optical traveling wave or standing wave equations, where the time marching step is usually dictated by the stability requirement of the optical equations. With such a time step, the lower order Runge–Kutta or even Euler method would work as well for the carrier rate equation. Hence, we will not go into detail.

8.4 The thermal diffusion equation

In device performance simulations, the thermal diffusion equation is normally solved in every cross-sectional 2D sheet (i.e., in the xy plane) consistently with the carrier transport equations. As a typical linear diffusion equation, (5.47) is posed as a mixed initial-boundary value problem of parabolic type PDE. Since the spatial derivatives in such a diffusive equation are in the second order, the FT–ECS scheme is stable.

Actually, we can write equation (5.47) in the form

$$\frac{\partial T(x, y, t)}{\partial t} = b(x, y)\nabla_t \cdot d(x, y)\nabla_t T(x, y, t) + f(x, y, t). \tag{8.67}$$

Under the FT–ECS scheme, we can discretize equation (8.67) to obtain

$$\begin{aligned}\frac{T_{i,j}^{k+1} - T_{i,j}^{k}}{\Delta t} &= b_{i,j}\frac{d_{i+1/2,j}(T_{i+1,j}^{k} - T_{i,j}^{k}) - d_{i-1/2,j}(T_{i,j}^{k} - T_{i-1,j}^{k})}{(\Delta x)^2} \\ &+ b_{i,j}\frac{d_{i,j+1/2}(T_{i,j+1}^{k} - T_{i,j}^{k}) - d_{i,j-1/2}(T_{i,j}^{k} - T_{i,j-1}^{k})}{(\Delta y)^2} + f_{i,j}^{k},\end{aligned} \tag{8.68}$$

where the symbol $u_{i,j}^{k}$ represents $u(x = i\Delta x, y = j\Delta y, t = k\Delta t)$. Following the method introduced in Section 6.2.1, we can prove that the scheme, equation (8.68), is consistent and stable once

$$\Delta t \leq \min_{i,j}\left\{\frac{1}{2b_{i,j}\left[\frac{d_{j+1/2,i}}{(\Delta x)^2} + \frac{d_{i,j+1/2}}{(\Delta y)^2}\right]}\right\}. \tag{8.69}$$

Therefore, the explicit scheme (8.68) gives a convergent solution to equation (8.67) subject to condition (8.69).

In DC analysis, the explicit scheme is not necessarily efficient since we need to march many time steps before we can reach the steady state. Therefore, we prefer the explicit scheme which allows us to use a large time step in such applications.

Actually, either the FT–ICS or the Crank–Nicholson scheme would fulfill this task. For example, the Crank–Nicholson scheme can be implemented as

$$\begin{aligned}\frac{T_{i,j}^{k+1}-T_{i,j}^{k}}{\Delta t}&=\frac{b_{i,j}}{2}\frac{d_{i+1/2,j}(T_{i+1,j}^{k+1}-T_{i,j}^{k+1})-d_{i-1/2,j}(T_{i,j}^{k+1}-T_{i-1,j}^{k+1})}{(\Delta x)^2}\\&+\frac{b_{i,j}}{2}\frac{d_{i+1/2,j}(T_{i+1,j}^{k}-T_{i,j}^{k})-d_{i-1/2,j}(T_{i,j}^{k}-T_{i-1,j}^{k})}{(\Delta x)^2}\\&+\frac{b_{i,j}}{2}\frac{d_{i,j+1/2}(T_{i,j+1}^{k+1}-T_{i,j}^{k+1})-d_{i,j-1/2}(T_{i,j}^{k+1}-T_{i,j-1}^{k+1})}{(\Delta y)^2}\\&+\frac{b_{i,j}}{2}\frac{d_{i,j+1/2}(T_{i,j+1}^{k}-T_{i,j}^{k})-d_{i,j-1/2}(T_{i,j}^{k}-T_{i,j-1}^{k})}{(\Delta y)^2}\\&+\frac{1}{2}(f_{i,j}^{k+1}+f_{i,j}^{k}).\end{aligned}\tag{8.70}$$

It is straightforward to prove that this scheme is consistent and unconditionally stable. Hence equation (8.70) gives the convergent solution to equation (8.67) under any time step. The drawback is the implicit nature of this scheme, which forces us to solve a system of linear equations at every time step.

Concerning an efficient solution of the linear equation system at each time step in equation (8.70), we can further take an ADI approach by splitting each time step into two halves; in each half we deal only with one space dimension. The advantage of this Crank–Nicholson plus ADI scheme lies in the fact that, at each time sub-step, only the solution of a linear tridiagonal system is required. As such, the linear equation system will be solved in the most efficient way. The following scheme modifies equation (8.70) through the ADI approach

$$\begin{aligned}\frac{T_{i,j}^{k+1/2}-T_{i,j}^{k}}{\Delta t/2}&=b_{i,j}\frac{d_{i+1/2,j}(T_{i+1,j}^{k+1/2}-T_{i,j}^{k+1/2})-d_{i-1/2,j}(T_{i,j}^{k+1/2}-T_{i-1,j}^{k+1/2})}{(\Delta x)^2}\\&+b_{i,j}\frac{d_{i,j+1/2}(T_{i,j+1}^{k}-T_{i,j}^{k})-d_{i,j-1/2}(T_{i,j}^{k}-T_{i,j-1}^{k})}{(\Delta y)^2}\\&+\frac{1}{2}(f_{i,j}^{k+1/2}+f_{i,j}^{k}),\\\frac{T_{i,j}^{k+1}-T_{i,j}^{k+1/2}}{\Delta t/2}&=b_{i,j}\frac{d_{i+1/2,j}(T_{i+1,j}^{k+1/2}-T_{i,j}^{k+1/2})-d_{i-1/2,j}(T_{i,j}^{k+1/2}-T_{i-1,j}^{k+1/2})}{(\Delta x)^2}\\&+b_{i,j}\frac{d_{i,j+1/2}(T_{i,j+1}^{k+1}-T_{i,j}^{k+1})-d_{i,j-1/2}(T_{i,j}^{k+1}-T_{i,j-1}^{k+1})}{(\Delta y)^2}\\&+\frac{1}{2}(f_{i,j}^{k+1}+f_{i,j}^{k+1/2}).\end{aligned}\tag{8.71}$$

Equation (8.71) is consistent and unconditionally stable, hence it gives the convergent solution to equation (8.67). It is of second order accuracy in terms of time and space steps Δt, Δx, and Δy. In addition, we need only to solve a tridiagonal linear equation system at each half time step. Therefore, we can may conclude that equation (8.71) is a superior discretization scheme.

References

[1] J. Buus, Principles of semiconductor laser modeling. *IEE Proc. J*, **132** (1985), 42–51.

[2] T. Ohtoshi, K. Yamaguchi, C. Nagaoka, *et al.*, A two-dimensional device simulator of semiconductor lasers. *Solid-State Electron.*, **30**:6 (1987), 627–38.

[3] K. B. Kahen, Two-dimensional simulation of laser diodes in the steady state. *IEEE Journal of Quantum Electron.*, **QE-24**:4 (1988), 641–51.

[4] M. Ueno, S. Asada, and S. Kumashiro, Two-dimensional numerical analysis of lasing characteristics for self-aligned structure semiconductor lasers. *IEEE Journal of Quantum Electron.*, **QE-26**:6 (1990), 972–81.

[5] G. L. Tan, N. Bewtra, K. Lee, and J. M. Xu, A two-dimensional finite element simulation of laser diodes. *IEEE Journal of Quantum Electron.*, **QE-29**:3 (1993), 822–35.

[6] M. Grupen and K. Hess, Simulation of carrier transport and nonlinearities in quantum-well laser diodes. *IEEE Journal of Quantum Electron.*, **QE-34**:1 (1998), 120–40.

[7] B. Witzigmann, A. Witzig, and W. Fichtner, A multidimensional laser simulator for edge-emitters including quantum carrier capture. *IEEE Trans. on Electron. Dev.*, **ED-47**:10 (2000), 1926–34.

[8] M. A. Alam, M. S. Hybertsen, R. K. Smith, and G. Baraff, Simulation of semiconductor quantum well lasers. *IEEE Trans. on Electron. Dev.*, **ED-47**:10 (2000), 1917–25.

[9] G. W. Taylor and P. R. Claisse, Transport solutions for the SCH quantum-well laser diode. *IEEE Journal of Quantum Electron.*, **QE-31**:12 (1995), 2133–41.

[10] D. Ahn and S. L. Chuang, Model of the field-effect quantum-well laser with free-carrier screening and valence band mixing. *Journal of Appl. Phys.*, **64**:11 (1988), 6143–49.

[11] J. Yokoyama, T. Yamanaka, and S. Seki, Two-dimensional numerical simulator for multi-electrode distributed feedback laser diodes. *IEEE Journal of Quantum Electron.*, **QE-29**:3 (1993), 856–63.

[12] S. Selberherr, *Analysis and Simulation of Semiconductor Devices*, 1st edn (New York: Springer-Verlag, 1984).

[13] M. S. Lundstrom and R. J. Schuelke, Numerical analysis of heterostructure semiconductor devices. *IEEE Trans. on Electron. Dev.*, **ED-30**:9 (1983), 1151–9.

[14] D. L. Scharfetter and H. K. Gummel, Large-signal analysis of a silicon read diode oscillator. *IEEE Trans. on Electron. Dev.*, **ED-16**:1 (1969), 64–77.

[15] R. E. Bank, D. J. Rose, and W. Fichtner, Numerical methods for semiconductor device simulation. *IEEE Trans. on Electron. Dev.*, **ED-30**:9 (1983), 1031–41.

[16] G. Engeln-Mullges and F. Uhlig, *Numerical Algorithms with Fortran*, 1st edn (Berlin: Springer-Verlag, 1996).

[17] J. Stoer and R. Bulirsch, *Introduction to Numerical Analysis*, 1st edn (Harrisonburg, VA: Springer-Verlag, 1993).

[18] M. S. Mock, Time discretization of a nonlinear initial value problem. *Journal of Comp. Phys.*, **21** (1976), 20–37.

[19] M. S. Mock, The stability problem for time-dependent models. In *An Introduction to the Numerical Analysis of Semiconductor Devices and Integrated Circuits*, 1st edn (Dublin: Boole Press, 1981).

[20] M. S. Mock, *Analysis of Mathematical Models of Semiconductor Devices*. 1st edn (Dublin: Boole Press, 1983).

9 Numerical analysis of device performance

9.1 A general approach

9.1.1 The material gain treatment

As explained in Section 7.3, it is neither feasible nor necessary to compute the material's optical properties through the physics based gain model in an "online" manner. It is not feasible because of the huge number of times that the gain model has to be invoked. It is not necessary because on many of the times that the model is called up it provides exactly the same results.

For this reason, we take an "offline" approach by calling up the physics based gain model at a coarse mesh grid constructed by multiple variables (i.e., the electron and hole densities, the temperature and the frequency) which covers the entire device operation range, and by establishing a set of analytical formulas which reproduce the required material optical property (i.e., the stimulated and spontaneous emission gains and the refractive index change) in the device operation range. By assuming a universal polynomial, exponential and rational dependence on the carrier density, temperature and frequency, respectively, such formulas are therefore parameterized with the unknown parameters obtained from searching for the best fit between the results from the formulas and from the rigorous calculation on the mesh grid points. Interpolations might be necessary to refine the mesh grid before such a fitting.

Once the analytical formulas are extracted, they will be used as the material model to replace the rigorous model for calculation of gains and refractive index change in an "online" manner.

Figure 9.1 (a) and (b) show comparisons of the stimulated emission gain and refractive index change calculated by the rigorous gain model and analytical formulas.

For a given active region structure with adjacent layers, the rigorous gain is obtained through the following procedure.

Firstly, the single electron band structure is solved for the active region comprising the bulk semiconductor, equation (3.76), or the QW structure, equations (3.129) and (3.134). If the strained layer is involved, equations (3.76), (3.129), and (3.134) should be solved with the modified energies in the Luttinger–Kohn Hamiltonian matrix elements (3.154a–d), (3.155), and (3.156a–d), for the bulk semiconductor and QW structure, respectively.

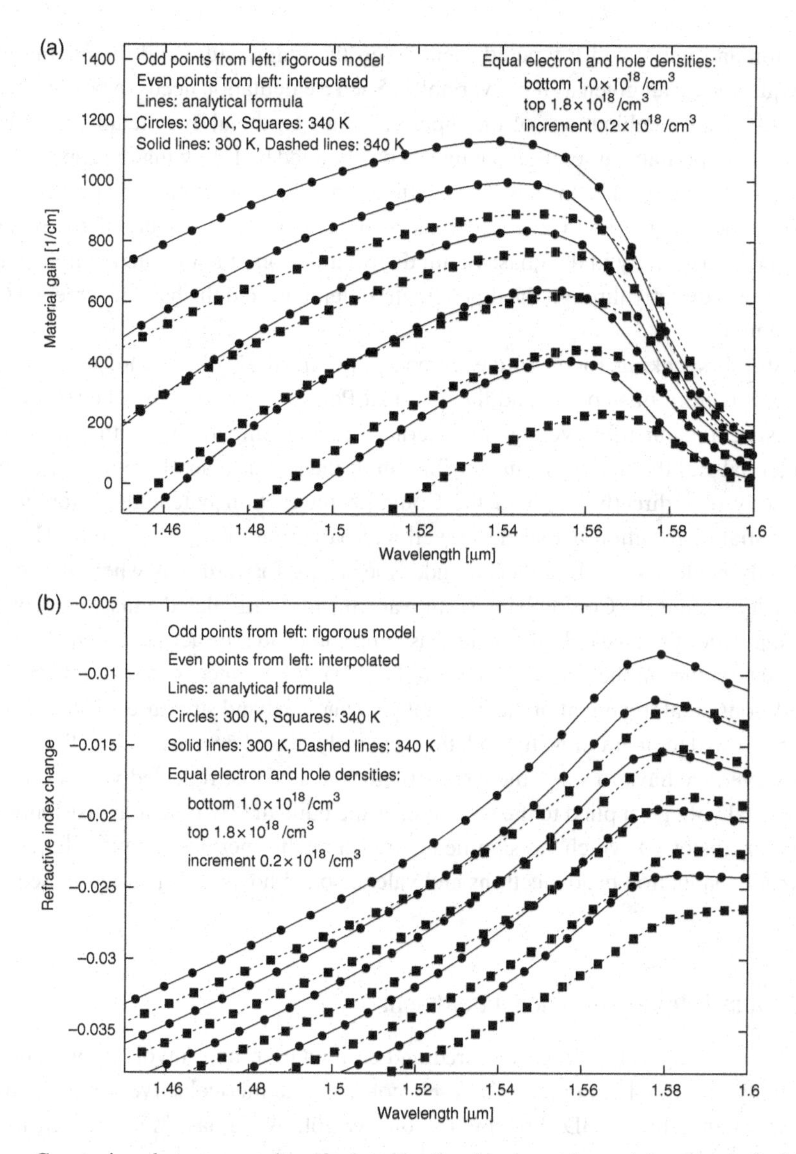

Fig. 9.1. Comparison between results calculated by the rigorous gain model and analytical formulas. (a) The stimulated emission gain. (b) The refractive index change. The rigorous gain is obtained from an active region comprising five 5 nm thickness $In_{0.748}Al_{0.070}Ga_{0.182}As$ QWs separated by six 10 nm thickness $In_{0.529}Al_{0.193}Ga_{0.278}As$ barrier layers.

Secondly, the stimulated emission gain and refractive index change are calculated directly through the free-carrier model, equations (4.65a&b), or the screened Coulomb interaction model, equations (4.109a&b), (4.129a&b), or the many-body model, equations (4.144a&b), or indirectly though solving the governing equations (4.156a–c) in the many-body gain model. In each case, the conduction band electron and valence band hole carrier densities at their respective boundaries between the active region and the N and P type adjacent layer, the temperature and the operating optical frequency are all given as input variables. Such a calculation needs to be done repeatedly for different combinations of the input variables on the above mentioned coarse mesh grid. The spontaneous emission gain can be obtained by following the approach we have used in Section 7.2.1. In these calculations, the conduction band electron and valence band hole energies and wave functions, as the eigensolutions found in the first step, are either used directly in the quasi-Fermi distributions and the Lorentzian line-shape functions or indirectly through the dipole matrix elements and the bare or screened Coulomb potential.

Strictly speaking, for the active region comprising the QW structure, the single electron band structure in step one and the classical Poisson equation must be solved in a self-consistent manner for every given external bias and temperature, as the varying external potential and the non-uniform distribution of the electron and hole wave functions will directly or indirectly (through the Coulomb interaction potential) distort the built-in potential distribution between the well and barrier in the QW structure. However, we usually neglect such dependence under the strong forward bias where the high carrier density screens the Coulomb interaction and the high stimulated emission recombination rate justifies the quasi-charge neutral assumption inside the active region. Therefore, the band structure in the first step needs to be solved only once. One can refer to [1, 2] for self-consistent treatment of the material gain and the band structure under a forward bias for elevated temperatures. In modeling reversely biased devices such as PDs and EAMs, however, we have to solve the band structure in step one repeatedly for every different external voltage applied to the structure, as the built-in QW potential distribution will be heavily distorted, which has significant impact on the energies and wave functions of the bound conduction band electrons and valence band holes as explained in Section 7.1.

9.1.2 The quasi-three-dimensional treatment

In considering the memory required and computation time involved, we find that it is usually impossible to implement a self-consistent numerical solver to model the device performance in full 3D. For this reason, we follow a quasi-3D approach to treat the equations [3].

The 3D device structure is firstly divided into a number of subsections along the wave propagation, or the longitudinal direction z. Under a given external bias, in each subsection, equation (2.34) for the optical field distribution (transverse optical mode), equations (5.29a–c), (5.33a&b) for the electrostatic potential distribution and carrier densities, and equation (5.47) for the temperature distribution are solved in the 2D cross-sectional xy plane using the FD approach, where different computation window sizes and mesh grids are adopted for the optical equation, the carrier equations, and the temperature equation in order to treat different distributions efficiently. A local interpolation technique is used to obtain the values of the different variables at an arbitrary mesh grid point.

For the 2D carrier and temperature distributions obtained in each subsection, the material gains and refractive index change are calculated through a set of analytical formulas extracted beforehand. For the transverse optical mode and effective index obtained in each subsection, the confinement factor and various coupling coefficients are calculated. In most applications, the transverse optical mode, the effective index and other optical modal parameters need to be calculated only once at the very beginning as they normally experience very little change over the entire device operation range. A perturbation approach can be used if significant changes in the transverse optical mode and effective index are found, due to the carrier or temperature induced material refractive index change. Other parameters such as the optical loss, due to non-interband processes such as the free carrier absorption, can also be calculated as long as the transverse optical mode, the carrier density and temperature distributions are all found in each subsection.

Up to this stage, all the material and optical modal parameters that appear in the optical wave equations (either in traveling wave or in standing wave equations) are obtained in each subsection along the longitudinal direction. These material and optical modal parameters are usually different from subsection to subsection (i.e., z dependent).

In summary, by a subdivision of the 3D device structure along the optical wave propagation direction into N 2D sheets, under every bias or time step, the 2D optical, carrier transport and thermal diffusion equations are solved in every sheet, plus the 1D optical (traveling or standing) wave equation is solved along the wave propagation direction, where the material gains and refractive index change are calculated through a set of analytical formulas obtained by an "offline" pre-extraction as explained in Section 9.1.1. Depending on the device to be modeled and the performance to be calculated, these 2D and 1D equations may need to be solved just once or iteratively until a global convergence between the optical field distribution (in each 2D sheet and along the longitudinal 1D direction), the carrier density distributions (in each 2D sheet), and the temperature distribution (in each 2D sheet) is achieved. Those device characteristics can therefore be extracted through a post-processor based on the solutions to those equations.

For example, the output optical power from the left and right facet can be obtained by

$$P_\mathrm{l} = \frac{n_\mathrm{eff}}{2}\sqrt{\frac{\varepsilon_0}{\mu_0}}(1 - |R_\mathrm{l}|^2)|e(0)|^2,$$

and

$$P_\mathrm{r} = \frac{n_\mathrm{eff}}{2}\sqrt{\frac{\varepsilon_0}{\mu_0}}(1 - |R_\mathrm{r}|^2)|e(L)|^2, \tag{9.1}$$

with $e(0)$ and $e(L)$ denoted as the lasing mode optical field slow-varying envelope function in V at the left and right facet, respectively.

The lasing mode frequency is computed through

$$\overline{\omega} = \omega_0 + \frac{\mathrm{d}}{\mathrm{d}t}\left[\frac{e(z,t)}{|e(z,t)|}\right]|_{t\to\infty}, \tag{9.2}$$

where the phase of the lasing mode, i.e., $e(z,t)/|e(z,t)|$ should not have z dependence as $t \to \infty$. While equation (9.2) strictly follows the definition, it cannot be conveniently used for the lasing frequency calculation if the laser has multiple longitudinal modes or if the laser works at an unstable (i.e., self-pulsation) mode. A more comprehensive approach is to take the fast Fourier transform (FFT) of the optical field slow-varying envelope function to convert it from the time domain to the frequency domain with its DC component aligned with the reference frequency ω_0. From the peak positions on such an obtained optical spectrum, we find the lasing mode frequencies.

9.2 Device performance analysis

9.2.1 The steady state analysis

At the steady state, the time derivatives in the governing equations should be set to zero except for the traveling wave equation for the laser, in which the time derivatives of the slow-varying envelope functions should be replaced by $\partial e^{\mathrm{f,b}}/\partial t = -\mathrm{j}\Delta\omega e^{\mathrm{f,b}}$, since the laser will generally oscillate at frequency $\omega_0 + \mathrm{Re}[\Delta\omega]$ with its linewidth proportional to $|\mathrm{Im}[\Delta\omega]|$. That is to say, at the steady state, the laser generates a time harmonic or more precisely, a quasi-harmonic (i.e., with a non-zero linewidth) optical wave as its output. Since the reference frequency is fixed at ω_0, if the lasing oscillation happens at a detuned frequency away from ω_0, the slow-varying envelope functions must contain a phase factor in the form of $\mathrm{e}^{-\mathrm{j}\Delta\omega t}$ at the steady state rather than a time-independent constant, as in the latter case the lasing frequency would be set right at the reference frequency, which conflicts with our presumption following the fact that lasing may happen at a detuned frequency. As for the standing wave equation, any possible detuning has been included in the eigenvalue of the longitudinal mode equation (2.135).

Thus, the traveling wave equation will reduce to an eigenvalue problem with the complex frequency deviation $\Delta\omega$ as the eigenvalue and the slow-varying envelopes $e^{\mathrm{f,b}}$ as the eigenfunctions. For example, equation (2.84) becomes

$$\begin{aligned}\frac{\partial e^{\mathrm{f}}(z)}{\partial z} &= \left[\mathrm{j}\frac{\Delta\omega}{v_{\mathrm{g}}} + \mathrm{j}\delta + \mathrm{j}k_0\Gamma\Delta n(z,\omega_0) + \frac{1}{2}\Gamma g(z,\omega_0) - \frac{1}{2}\bar{\alpha}(z) + \mathrm{j}\kappa_r^{\mathrm{ff}}\right] e^{\mathrm{f}}(z) \\ &\quad + \mathrm{j}(\kappa_M + \kappa_r^{\mathrm{fb}})e^{\mathrm{b}}(z) - \frac{\partial e^{\mathrm{b}}(z)}{\partial z} \\ &= \left[\mathrm{j}\frac{\Delta\omega}{v_{\mathrm{g}}} + \mathrm{j}\delta + \mathrm{j}k_0\Gamma\Delta n(z,\omega_0) + \frac{1}{2}\Gamma g(z,\omega_0) - \frac{1}{2}\bar{\alpha}(z) + \mathrm{j}\kappa_r^{\mathrm{bb}}\right] e^{\mathrm{b}}(z) \\ &\quad + \mathrm{j}(\kappa_{-M} + \kappa_r^{\mathrm{bf}})e^{\mathrm{f}}(z), \end{aligned} \tag{9.3}$$

subject to the boundary conditions (2.137). In equation (9.3), the inhomogeneous spontaneous emission contributions to the forward and backward propagating (i.e., the right- and left-going) waves are ignored.

Comparing with the standing wave longitudinal mode equation (6.79), we find that the two equations (9.3) and (6.79) are identical if

$$\Delta\omega = -\frac{v_g \beta_V^2}{2\beta_0}. \tag{9.4}$$

We can follow the general procedure introduced in Section 9.1.2 to solve the steady state equations in a self-consistent manner. Once the transverse optical mode, the carrier density distributions, and the temperature distributions are found in each cross-sectional sheet, and the material and optical modal parameters are obtained, the static optical field distribution along the cavity known as the optical standing wave pattern, or the longitudinal optical mode, as well as the complex frequency deviation can readily be obtained under the given facet and operating conditions through the transfer matrix method discussed in Section 6.3. They are obtained as the eigenfunction and eigenvalue of equation (9.3), respectively.

As the eigenvalue of the device, the complex frequency deviation takes a clear physical meaning: its real part is the detuning of the lasing frequency from the reference and its imaginary part is the net gain known as the difference between the modal gain provided by the injection and the total loss scaled by the group velocity of the optical wave. The total loss comprises the modal losses through various non-interband absorption processes seen by the optical wave propagating along the waveguide inside the laser cavity and the terminal loss due to the optical power's escape from both ends of the cavity. (It is the escaped optical power that forms the laser output.) As the net gain of the laser, the imaginary part of the complex frequency deviation is also a measure of the laser coherence: it is the linewidth of the lasing mode on the spectrum, while its reciprocal gives the laser coherent time.

Once the longitudinal optical mode is obtained, the stimulated emission recombination rate has to be updated and the carrier transport equations and the thermal equation in each cross-sectional sheet have to be solved again. Under a strong injection, even the transverse optical mode in each cross-sectional sheet needs to be updated for any significant change in the refractive index induced by the varying carrier density or temperature. This iteration will continue until no further changes can be detected on the longitudinal optical mode and complex frequency deviation.

Under a different bias, we have to let such an iteration loop start all over again. The full procedure for laser DC analysis is summarized below.

(1) 3D geometrical structure input.
(2) Material constant input.
(3) Material gain calculation on sampled points in variable space.
(4) Analytical material model extraction.
(5) Longitudinal subdivision (in z direction).
(6) Mesh set-up for cross-sectional sheets (in xy plane).
(7) Solver initialization.
(8) Operating condition input (possible looping starts here).
(9) Variable scaling (physical to numerical).

(10) 2D-1D- iteration loop starts.
(11) Call 2D transverse optical mode solver.
(12) Call 2D carrier transport solver.
(13) Call 2D temperature solver.
(14) Calculate the material and optical modal parameters.
(15) Call longitudinal optical mode solver.
(16) Go to the iteration starting point (step 10) if not converged, otherwise continue.
(17) Variable scaling (numerical to physical).
(18) Post processing for required output assembly.
(19) Go to step 8 for operating condition (bias and temperature) looping, until the maximum settings are reached, otherwise continue.
(20) Stop.

In the subdivision along the device's longitudinal direction, a subsection length in 50–100 wavelength periods is normally sufficient to capture the non-uniform distribution of the optical slow-varying envelope function caused by the LSHB. This requires 10–40 subsections in total for modeling a typical semiconductor laser operated in the C-band and with a cavity length around 0.25–1 mm. In the DC analysis, there is no stability concern as there is no time domain evolution involved. Therefore, there is no other constraint on setting up the subsection length, hence computational effort would be greatly reduced if an adaptive longitudinal subdivision scheme were introduced for lasers with long cavity lengths. Such an adaptive scheme can be established through, for example, repeatedly doubling the subsection number (starting from one) until a converged longitudinal optical mode is obtained. In the process of searching for the optimized subsection number, the carrier rate equation (5.41) and the 1D thermal diffusion equation (5.48) can be used instead of the rigorous 2D carrier transport equations and thermal diffusion equations. The rigorous 2D models will be invoked only after the subsections are set up along the longitudinal direction. Solving the 2D carrier transport equations usually needs more time than completing all other processes, hence the overhead spent on minimizing the total subsection number required is well justified. A more advanced approach is to allow the subsections to take non-uniform lengths. For a given longitudinal structure, we can obtain an initial longitudinal optical field distribution by using, e.g., only one subsection. A non-uniform subsection division strategy can therefore be selected to make the subsection length inversely proportional to the derivative of the total optical field intensity with respect to z. As a result, each subsection length, hence the total subsection number, will be optimized and the number of times that the 2D carrier transport solver is invoked will be reduced.

In many applications, the carrier rate equation (5.41) can be solved instead of invoking the 2D carrier transport solver for those intermediate steps in the iteration. That is to say, the rigorous 2D carrier transport solver will be invoked only at the beginning and end of the iteration. It is called up at the beginning because we need a set of rather "close" initial values. It is called up again at the end because we need the final result to be accurate. The intermediate values only help us to reach the final convergence. Hence their accuracy is not pursued as these values will be dropped anyway.

9.2.2 The small-signal dynamic analysis

The small-signal dynamic analysis is straightforwardly based on the DC analysis [4–6]. By assuming the time-dependent external bias to be in the form

$$I(t) = I_0 + \Delta I(t) = I_0 + I_1 \mathrm{e}^{-\mathrm{j}\omega_1 t} + I_2 \mathrm{e}^{-\mathrm{j}\omega_2 t}, \tag{9.5}$$

or

$$V(t) = V_0 + \Delta V(t) = V_0 + V_1 \mathrm{e}^{-\mathrm{j}\omega_1 t} + V_2 \mathrm{e}^{-\mathrm{j}\omega_2 t}, \tag{9.6}$$

with $|\Delta I(t)| \ll I_0$ and $|\Delta V(t)| \ll V_0$, we can expand all other variables such as the optical field slow-varying envelope functions (both amplitudes and phases), the carrier density distributions, and the temperature distributions in the time-harmonic forms

$$\begin{aligned} u(t) &= u_0 + \Delta u(t) \\ &= u_0 + A_1^1 e^{-j\omega_1 t} + A_2^1 e^{-j\omega_2 t} \\ &\quad + A_{1,1}^2 \mathrm{e}^{-\mathrm{j}2\omega_1 t} + A_{2,2}^2 \mathrm{e}^{-\mathrm{j}2\omega_2 t} + A_{2,1}^2 \mathrm{e}^{-\mathrm{j}(\omega_2+\omega_1)t} + A_{2,-1}^2 \mathrm{e}^{-\mathrm{j}(\omega_2-\omega_1)t} \\ &\quad + A_{1,1,1}^3 \mathrm{e}^{-\mathrm{j}3\omega_1 t} + A_{2,2,2}^3 \mathrm{e}^{-\mathrm{j}3\omega_2 t} + A_{2,1,1}^3 \mathrm{e}^{-\mathrm{j}(\omega_2+2\omega_1)t} \\ &\quad + A_{2,-1,-1}^3 \mathrm{e}^{-\mathrm{j}(\omega_2-2\omega_1)t} + A_{2,2,1}^3 \mathrm{e}^{-\mathrm{j}(2\omega_2+\omega_1)t} + A_{2,2,-1}^3 \mathrm{e}^{-\mathrm{j}(2\omega_2-\omega_1)t} \\ &\quad + \cdots, \end{aligned} \tag{9.7}$$

with $|\Delta u(t)| \ll u_0$. In equation (9.7), $A_{1/2}^1$ indicates the first order (linear) modulation response, $A_{(1,1)/(2,2)}^2$ the second order harmonic distortion response, $A_{(2,1)/(2,-1)}^2$ the second order inter-modulation distortion response, $A_{(1,1,1)/(2,2,2)}^3$ the third order harmonic distortion response, $A_{(2,1,1)/(2,-1,-1)/(2,2,1)/(2,2,-1)}^3$ the third order inter-modulation distortion response, etc. Those higher order harmonic or inter-modulation distortions come from the non-linearity dependence of the variables on the external bias embedded in the governing equations.

Substituting expressions (9.5) or (9.6) and (9.7) for every variable into the governing equations with the time derivatives directly performed on those harmonic terms, we will obtain a set of equations which balances the terms in the same harmonic order. For example, by balancing the time-independent terms involving I_0 or V_0, and u_0's, we obtain the steady state equations again. By balancing the terms with the same harmonic factor $\mathrm{e}^{-\mathrm{j}\omega_1 t}$ or $\mathrm{e}^{-\mathrm{j}\omega_2 t}$ (not both), we obtain the small-signal linear modulation response equations. By following this approach, we can obtain the response equations for all the second, third, or even higher order harmonic and inter-modulation distortions.

Each set of equations becomes time independent and hence can be respectively treated as the DC governing equations with the corresponding frequency taken as a parameter.

These corresponding frequencies are: ω_1 or ω_2 for the linear modulation response, $2\omega_1$ or $2\omega_2$ for the second order harmonic distortion response, $\omega_2 + \omega_1$ and $\omega_2 - \omega_1$ for the second order inter-modulation distortion response, $3\omega_1$ or $3\omega_2$ for the third order harmonic distortion response, $\omega_2 + 2\omega_1$, $\omega_2 - 2\omega_1$, $2\omega_2 + \omega_1$, and $2\omega_2 - \omega_1$ for the third order inter-modulation distortion response, etc. We can vary these frequencies and solve the equations at each frequency to obtain the variable small-signal modulation response over a frequency bandwidth.

In most applications, we need to calculate the optical power modulation response, also known as the intensity modulation (IM) response, and the accompanying phase modulation response, also known as the parasitic frequency modulation (FM) responses, at various orders. These modulation responses are normally scaled by the external bias small-signal modulation amplitude and are defined as

$\frac{A_1^1}{I_1}$ or $\frac{A_2^1}{I_2}$, the first order optical power or frequency modulation responses,
$\frac{A_{1,1}^2}{(I_1)^2}$ or $\frac{A_{2,2}^2}{(I_2)^2}$, the second order optical power or frequency harmonic distortion responses,
$\frac{A_{2,1}^2}{I_2 I_1}$ and $\frac{A_{2,-1}^2}{I_2 I_1}$, the second order optical power or frequency inter-modulation distortion responses,
$\frac{A_{1,1,1}^3}{(I_1)^3}$ or $\frac{A_{2,2,2}^3}{(I_2)^3}$, the third order optical power or frequency harmonic distortion responses,
$\frac{A_{2,1,1}^3}{I_2 (I_1)^2}$, $\frac{A_{2,-1,-1}^3}{I_2 (I_1)^2}$, $\frac{A_{2,2,1}^3}{(I_2)^2 I_1}$, and $\frac{A_{2,2,-1}^3}{(I_2)^2 I_1}$, the third order optical power or frequency inter- modulation distortion responses.

These small-signal responses actually provide the derivative information of the output optical power and lasing frequency in the neighborhood of their DC values corresponding to the DC bias I_0. Therefore, they should be independent of the input (i.e., the external bias) small-signal modulation amplitude.

This method is usually applied to the optical standing wave model, since the variable harmonic expansions can be implemented through the amplitude rate equation (2.126) analytically. If the carrier rate equation model is also used, such an expansion can also be analytically performed on these equations. As a result, after the DC calculation at the reference bias, the small-signal modulation responses can be obtained through a set of linear algebraic equations, hence their values can be expressed in closed forms where only numerical quadratures are involved.

Without this advantage, we can follow the large-signal analysis approach by taking equation (9.5) or (9.6) as the time-dependent external bias to compute directly the output variables in the time domain. Once the time domain variables are obtained, we can find their frequency domain spectra through the FFT. In these optical spectra, we should find a set of peaks at those harmonic frequencies. The heights of the peaks, therefore, give the modulation response. By scanning the small-signal input (i.e., the external bias) modulation frequency, we will be able to obtain the modulation response over a frequency bandwidth.

9.2.3 The large-signal dynamic analysis

The large-signal dynamic analysis is performed by solving all the governing equations in the time domain directly. The following sequential scheme is usually adopted for large-signal analysis.

(1) 3D geometrical structure input.
(2) Material constant input.
(3) Material gain calculation on sampled points in variable space.
(4) Analytical material model extraction.
(5) Longitudinal subdivision (in z direction).
(6) Mesh set-up for cross-sectional sheets (in xy plane).
(7) Solver initialization.
(8) Variable scaling (physical to numerical).
(9) Time domain progression starts (by setting $\Delta t = \eta \Delta z / v_g$).
(10) Operating condition input (read in bias as function of time).
(11) Call 2D transverse optical mode solver.
(12) Call 2D carrier transport solver.
(13) Call 2D temperature solver.
(14) Calculate the material and optical modal parameters.
(15) Call longitudinal optical (traveling or standing) wave equation solver.
(16) Go to the progression starting point (step 9) if the maximum time is not reached, otherwise continue.
(17) Variable scaling (numerical to physical).
(18) Post processing for required output assembly.
(19) Stop.

The major difference between the time domain large-signal approach and the steady state approach introduced in Section 9.2.1 is that there is no iteration required for solution convergence in the time domain approach once the explicit scheme is used. However, this is usually at the cost of using small sequential steps. Before reaching a steady state, we may have to experience many such steps.

Ideally, the device subsection length Δz and the time progression step Δt should be independently selected. The former should be considered to catch up to the spatial non-uniformity of the variables along the device longitudinal direction due to the LSHB, whereas the latter should be designed to follow changing of the variables due to the time-dependent bias. Therefore, we may have two possibilities if the two steps are chosen from different considerations. One makes the progression faster, $\Delta z/\Delta t > v_g$ and the other makes it slower, $\Delta z/\Delta t < v_g$, both in comparison with the physical wave propagation at v_g. In order to make the marching algorithm stable, in the former case we have to use the implicit scheme, whereas in the latter case we can take the explicit scheme.

Without the constraint from the physical problem, we may use either an explicit or implicit marching algorithm depending on which carrier transport solver will be used. If the carrier rate equation solver is called up, we can use any explicit scheme as the evaluation of the carrier density distribution is very efficient. On the contrary, if the rigorous carrier transport solver has to be invoked, we should pick an implicit scheme which allows us to use larger time steps; hence we would call up the rigorous carrier transport solver fewer times.

9.3 Model calibration and validation

Although strictly speaking the physics based model takes only the structural geometrical dimensions and the basic material constants as input parameters, we may still miss some important parameters because rigorous computation of these parameters may not be feasible or cannot justify the effort spent on the calculation, such as the non-radiative recombination constants and the non-linear gain saturation coefficient in equation (7.32), as the former needs knowledge of the full band structure (i.e., the E–$\vec{k}$ relation), not just the band structure in the neighborhood of $\vec{k} = 0$ [7–9], whereas the latter needs to have the Boltzmann transport equation solved to count in the transient carrier–phonon scattering process [10–13]. Some parameters are almost unobtainable due to their randomness, such as the optical scattering loss incurred as the wave propagates along the cavity. Finally, some material constants (such as the background refractive indices, the material compositions, the doping concentrations, and the strain quantities, etc.) and geometrical dimensions (such as the QW thickness, the grating periods, and the grating phase shifts at section ends, etc.) vary quite significantly for different device fabrication technologies, different fabrication facilities, or even different batches fabricated at different times. This makes simulation results disagree with experimentally measured data. Therefore, the numerical solver needs to be calibrated and validated.

Taking the semiconductor laser as an example, different sets of parameters can be extracted depending on different models. For example, effective parameters in the rate equation model have been successfully extracted [14–16]. However, in physics-based laser modeling, we need to extract those intrinsic laser parameters rather than effective parameters. The emission spectra [17], various small-signal modulation responses [18] and far-field patterns can be used for this purpose.

Parameter extraction and solver calibration usually follow such a sequence. We firstly use the measured optical far-field pattern and optical spectrum to extract the background refractive indices, the cross-sectional waveguide structure, and the longitudinal cavity structure [19]. Figure 9.2 shows the calculated optical spectra before and after calibration in comparison with the measured spectrum. The nominal input parameters used for the initial spectrum calculation and the extracted parameters through fitting our calculated spectrum to the measured spectrum are listed in Table 9.1. The spectrum is calculated based on the standing wave transfer matrix approach [20]. The fitting approach is based on a genetic algorithm (GA) [21, 22] that does not need any derivative evaluation of the target function and offers global optimization. In the fitting process, we let the input parameters vary until the best fitting spectrum is obtained. The corresponding input parameter set is therefore taken as the parameters after calibration.

Table 9.1. A comparative list of nominal and extracted parameters

Parameter	Nominal value	Extracted value
Straight waveguide section length, μm	100	100
Grating section length, μm	150	150
Effective index	3.2172	3.2172
Group index	3.6031	3.6031
Confinement factor	1.7998%	1.7998%
Grating period, nm	238.0	238.0
Normalized index coupling coefficient	2.0000	1.6726
Normalized gain/loss coupling coefficient	0.0	0.0
Rear facet amplitude reflectivity (HR coated)	0.75	0.81
Rear facet phase	0.0	35°
Front facet amplitude reflectivity (AR coated)	0.05	0.16
Front facet phase	0.0	143°

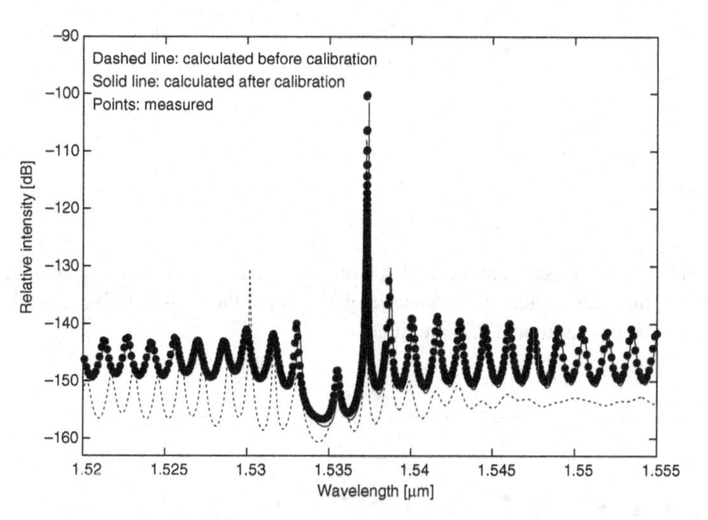

Fig. 9.2. Comparison between the measured semiconductor DFB laser spectrum and the calculated spectra before and after calibration. The semiconductor DFB laser has a two-sectional partially corrugated structure with the rear (the straight waveguide section end) and front (the grating section end) facets high reflection (HR) and anti-reflection (AR) coated, respectively. The nominal parameters (i.e., the parameters before the calibration) and the extracted parameters (i.e., the parameters after calibration) are listed in Table 9.1.

At the second step, the measured I–V (bias current – bias voltage) and $(\mathrm{d}I/\mathrm{d}V)/I$ curves can be used to calibrate those parameters in the carrier transport model.

At the third step, the measured L–I (output optical power – bias current) and small-signal modulation response curves under different ambient temperatures can be used to calibrate the non-radiative recombination constants, the optical modal loss, the non-linear gain saturation coefficient, and the thermal parameters.

Up to this stage, all the input parameters invoked by the numerical model have been fixed. We will use these parameters to simulate, e.g., the time domain large-signal output optical power waveform under a given modulation pattern. Once the simulated result agrees with the measured waveform of the same device, we confirm the consistency of this model.

Shown in Figs. 9.3 and 9.4 are the calibrated results in terms of the measured L–I and small-signal IM response.

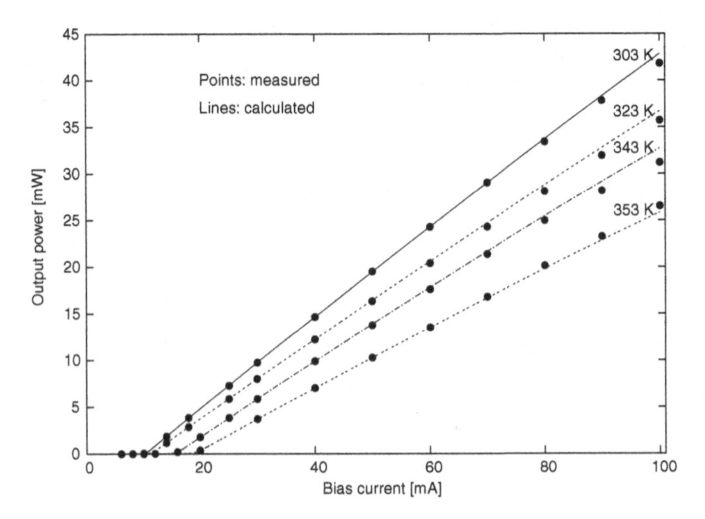

Fig. 9.3. Comparison between measured and calculated output optical power–bias current dependence curves. The thermal parameters in the material gain model, the non-radiative recombination constants, and the optical modal loss are fitted.

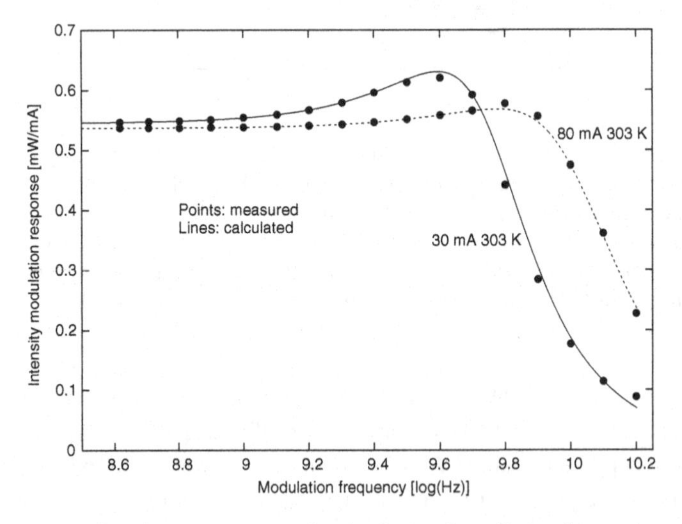

Fig. 9.4. Comparison between measured and calculated small-signal intensity modulation response curves. The non-linear gain saturation factor is fitted.

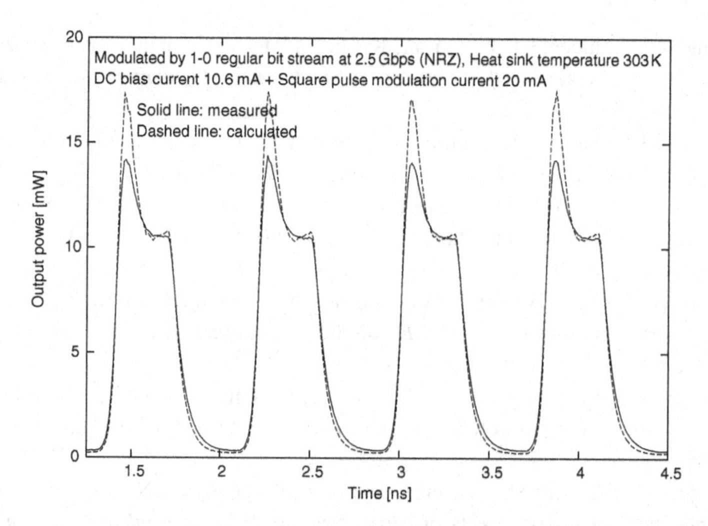

Fig. 9.5. Comparison between measured and calculated large-signal output optical power waveforms under the modulation of a binary bit stream. The bias current comprises a 10.6 mA DC component plus 20 mA peak-to-peak AC square pulses with an equal on and off duration in 400 ps, which is equivalent to a non-return to zero (NRZ) 1-0 regular bit stream at 2.5 Gbps. In this simulation, the laser parameters are all fixed either through self-consistent calculations or through the above fitting processes. As such, no flexible fitting parameter is used.

Figure 9.5 shows a comparison between the simulated and measured large-signal output optical power waveforms. A reasonable agreement between these results indicates the consistency of the model framework and confirms the accuracy of the numerical solution techniques, since there is no flexible fitting parameter used in this step.

References

[1] S. Seki and K. Yokoyama, Electrostatic deformation in band profiles of InP-based strained-layer quantum-well lasers. *Journal of Appl. Phys.*, **77**:10 (1995), 5180–4.

[2] S. Seki and K. Yokoyama, Theoretical analysis of temperature sensitivity of differential gain in 1.55 μm InGaAsP-InP quantum-well lasers. *IEEE Photon. Tech. Lett.*, **7**:3 (1995), 251–3.

[3] X. Li, Distributed feedback lasers: quasi-3D static and dynamic model. Chapter 4 in *Optoelectronic Devices: Advanced Simulation and Analysis*, ed. J. Piprek. 1st edn (New York: Springer Science+Business Media, 2005).

[4] R. H. Wentworth, Small-signal resonance distortion of semiconductor lasers. *IEEE Journal of Quantum Electron.*, **QE-30**:3 (1994), 680–4.

[5] J. L. Bihan and G. Yabre, FM and IM intermodulation distortions in directly modulated single-mode semiconductor lasers. *IEEE Journal of Quantum Electron.*, **QE-30**:4 (1994), 899–904.

[6] W.-P. Huang, X. Li, and T. Makino, Analytical formulas for modulation responses of semiconductor DFB lasers. *IEEE Journal of Quantum Electron.*, **QE-31**:5 (1995), 842–51.

[7] T. Kamiya, Radiative and non-radiative recombination in semiconductors. Chapter 4 in *Handbook of Semiconductor Lasers and Photonic Integrated Circuits*, ed. Y. Suematsu and A. R. Adams (London: Chapman & Hall, 1994).

[8] J. Wang, P. von Allmen, J. Leburton, and K. Linden, Auger recombination in long-wavelength strained-layer quantum-well structures. *IEEE Journal of Quantum Electron.*, **QE-31**:5 (1995), 864–75.

[9] A. Tomita, Free carrier effect on the refractive index change in quantum-well structures. *IEEE Journal of Quantum Electron.*, **QE-30**:12 (1994), 2798–802.

[10] B. N. Gomatam and A. P. Defonzo, Theory of hot carrier effect on nonlinear gain in GaAs-GaAlAs lasers and amplifiers. *IEEE Journal of Quantum Electron.*, **QE-26**:10 (1990), 1689–704.

[11] Y. Lam and J. Singh, Monte Carlo simulation of gain compression effect in GRINSCH quantum well laser structures. *IEEE Journal of Quantum Electron.*, **QE-30**:11 (1994), 2435–42.

[12] A. Uskov, J. Mork, and J. Mark, Wave mixing in semiconductor laser amplifiers due to carrier heating and spectral-hole burning. *IEEE Journal of Quantum Electron.*, **QE-30**:8 (1994), 1769–81.

[13] C.-Y. Tsai, C.-Y. Tsai, R. M. Spencer, Y. -H. Lo, and L. F. Eastman, Nonlinear gain coefficients in semiconductor lasers: effects of carrier heating. *IEEE Journal of Quantum Electron.*, **QE-32**:2 (1996), 201–12.

[14] J. C. Cartledge and R. C. Srinivasan, Extraction of DFB laser rate equation parameters for system simulation purposes. *IEEE/OSA Journal of Lightwave Tech.*, **LT-15**:5 (1997), 852–60.

[15] H. Yasaka, K. Takahata, and M. Naganuma, Measurement of gain saturation coefficients in strained-layer multiple quantum-well distributed feedback lasers. *IEEE Journal of Quantum Electron.*, **QE-28**:5 (1992), 1294–304.

[16] L. Bjerkan, A. Royset, L. Hafskjer, and D. Myhre, Measurement of laser parameters for simulation of high-speed fiberoptic systems. *IEEE/OSA Journal of Lightwave Tech.*, **LT-14**:5 (1996), 839–50.

[17] D. Hofstetter and R. L. Thornton, Measurement of optical cavity properties in semiconductor lasers by Fourier analysis of the emission spectrum. *IEEE Journal of Quantum Electron.*, **QE-34**:10 (1998), 1914–23.

[18] H. M. Salgado, J. M. Ferreira, and J. J. O'Reilly, Extraction of semiconductor intrinsic laser parameters by intermodulation distortion analysis. *IEEE Photon. Tech. Lett.*, **9**:10 (1997), 1331–3.

[19] R. Schatz, E. Berglind, and L. Gillner, Parameter extraction from DFB lasers by means of a simple expression for the spontaneous emission spectrum. *IEEE Photon. Tech. Lett.*, **6**:10 (1994), 1182–4.

[20] T. Makino and J. Glinski, Transfer matrix analysis of the amplified spontaneous emission of DFB semiconductor laser amplifiers. *IEEE Journal of Quantum Electron.*, **QE-24**:8 (1988), 1507–18.

[21] X. Li and W.-P. Huang, Parameter extraction for DFB laser diodes, *Symposium on Advanced Optoelectronic Devices for Optical Fiber Communication Systems, IEICE*, Kofu, Yamanashi, Japan, 1998.

[22] L. Chambers, (Ed.) *The Practical Handbook of Genetic Algorithms Applications*, 1st edn (Boca Raton, FL: Chapman & Hall/CRC, 2001).

10 Design and modeling examples of semiconductor laser diodes

10.1 Design and modeling of the active region for optical gain

10.1.1 The active region material

Specification of the lasing wavelength and substrate material availability usually dictate the active region material selection. However, in some wavelength bands, the selection of the material is not unique. For example, either one of the following systems: InGaAsP on InP substrate [1], InAlGaAs on InP substrate [2], InGaAsN on GaAs substrate [3], and InAsPN on GaAs substrate [4], can be chosen as the active region material for semiconductor laser diodes emitting in the 1300 nm band. In this section, however, we will compare only the first two systems, which seem to be more mature in production. Table 10.1 summarizes the features of the two active region structures made of the two different material systems.

Unlike the InGaAsP/InP heterojunction with a smaller conduction band offset ($\sim$36%) compared to the valence band offset ($\sim$64%), the InAlGaAs system has a larger offset ($\sim$71%) on the conduction band side. As such, the InAlGaAs system has a better confining effect on the injected electrons. Hence it has better temperature characteristics as a result of the effective reduction of electron current leakage, which is the dominant form of leakage due to the high electron mobility [2].

In accordance with the active region designs in Table 10.1, the calculated transverse electric (TE) mode material gain profiles are shown in Fig. 10.1. Since we have used the same well thickness and the same number of QWs in these two material systems, the sheet carrier density in the InAlGaAs and InGaAsP QWs is the same at each carrier injection level. From this result we find that, at the low carrier density, the peak gain of the InAlGaAs system is higher. However, as the carrier density increases, the gain of the InAlGaAs system saturates and the gain of the InGaAsP system surpasses it. Particularly at low carrier densities, the InAlGaAs system has a much broader gain bandwidth. These observations indicate that the InAlGaAs system has a lower transparency carrier density as opposed to the InGaAsP system. Therefore, to take full advantage of the InAlGaAs system, we should always retain lower carrier densities inside the InAlGaAs QWs to obtain a higher differential gain. In this case, we will be able to obtain a lower threshold in DC operation and a better device dynamic performance, as will be discussed in Section 10.1.2. Using the InAlGaAs system with a low carrier density gives us an extra advantage in conjunction with the DFB cavity design. Actually, as the ambient

Table 10.1. Designs of laser active regions with different material systems and different strains at 1310nm

All layers are undoped.

InGaAsP/InP system	Thickness (nm)	Composition of $In_{(1-x)}Ga_xAs_yP_{(1-y)}$ (x, y)	Remarks λ_g (nm) and strain*
Barrier	10	(0.181, 0.395)	1150, 0.0
Well × 6	5	(0.081, 0.547)	1355, CS1.2%
Barrier × 6	10	(0.181, 0.395)	1150, 0.0
Barrier	10	(0.350, 0.130)	1000, 0.0
InAlGaAs/InP system	**Thickness (nm)**	**Composition of $In_{(1-x-y)}Al_xGa_yAs$(x, y)**	**Remarks λ_g (nm) and strain***
Well × 6	5	(0.151, 0.145)	1440, CS1.2%
Barrier × 6	10	(0.350, 0.130)	1000, 0.0
Bulk	–	(0.265, 0.573)	1275, 0.0
Barrier	10	(0.181, 0.395)	1150, 0.0
Well × 6	5	(0.636, 0.987)	1450, TS1.2%
Barrier × 6	10	(0.181, 0.395)	1150, 0.0
Barrier	10	(0.350, 0.130)	1000, 0.0
Well × 6	5	(0.000, 0.642)	1460, TS1.2%
Barrier × 6	10	(0.350, 0.130)	1000, 0.0

* CS = compressive strain, TS = tensile strain.

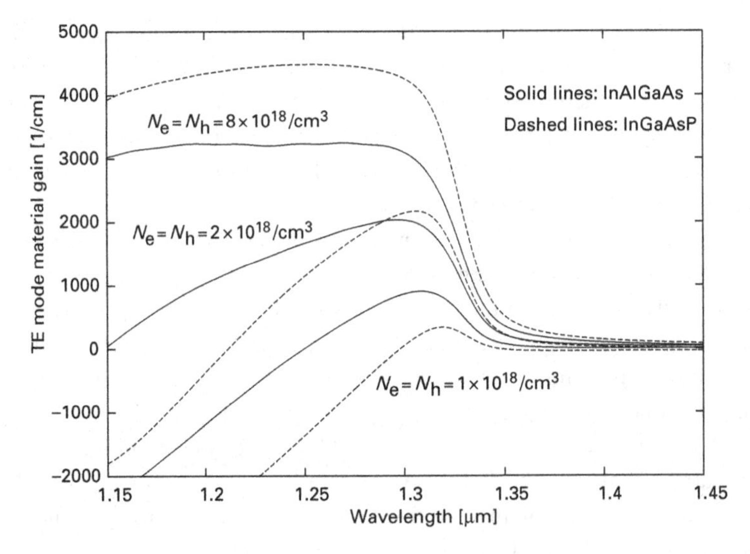

Fig. 10.1. TE mode material gain profiles for compressively strained InAlGaAs and InGaAsP QW structures.

temperature rises, the gain profile exhibits a red-shift at a rate of about 0.25 nm/K as shown in Fig. 9.1(a). It is still higher than the rate of the Bragg wavelength red-shift due to the effective index increase with temperature. As a result, the lasing wavelength moves off the gain peak towards the blue side. In the InGaAsP system, a significant gain drop at the lasing wavelength will occur. In the InAlGaAs system, however, the lasing wavelength sees little change in gain.

At a high carrier density, however, both systems have broad gain bandwidth but the InGaAsP system has significantly higher gain, as this system has relatively high transparent carrier density and hence it saturates at a higher carrier injection level. This observation suggests that in device applications where the cavity has to retain high carrier densities, such as in SOAs and SLEDs, the advantage of less current leakage in the InAlGaAs system could be offset by the disadvantage of lower gain compared with the InGaAsP system.

Once we cannot switch freely between the two material systems, e.g., we are limited by the fabrication technology, to equalize the performance we should use more QWs or longer cavity length in devices made of the InAlGaAs system and vice versa in devices made of the InGaAsP system, as both measures help to reduce and increase the sheet carrier densities inside the InAlGaAs and InGaAsP QWs, respectively. This will be further discussed in Section 10.1.2.

Figure 10.2 shows a comparison of TE mode gains of these two material systems under different compressive strains. Other than the known effect as shown in Fig. 3.5(b), we find that the compressive strain introduces a pure reduction in the transition energy, as a result of $P_{\mathrm{e}}^{\mathrm{c}} - P_{\mathrm{e}}^{\mathrm{v}} + Q_{\mathrm{e}} > 0$. Moreover, in the InAlGaAs system, the compressive strain seems effectively to broaden the gain profile bandwidth as the expansion on the red (longer wavelength) side is greater than the shrinkage on the blue (shorter wavelength) side. In the InGaAsP system, the compressive strain seems to have an optimum value (1.2%) at which the peak gain is the highest.

For the sake of comparison, we also show in Fig. 10.3(a&b) the calculated transverse magnetic (TM) mode material gains for the QW structures with tensile strain and the

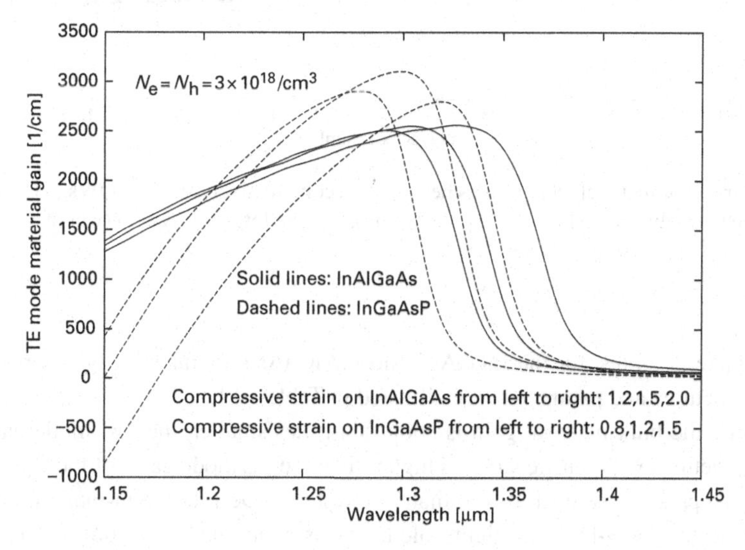

Fig. 10.2. TE mode material gain profiles for InAlGaAs and InGaAsP QW structures under different compressive strains.

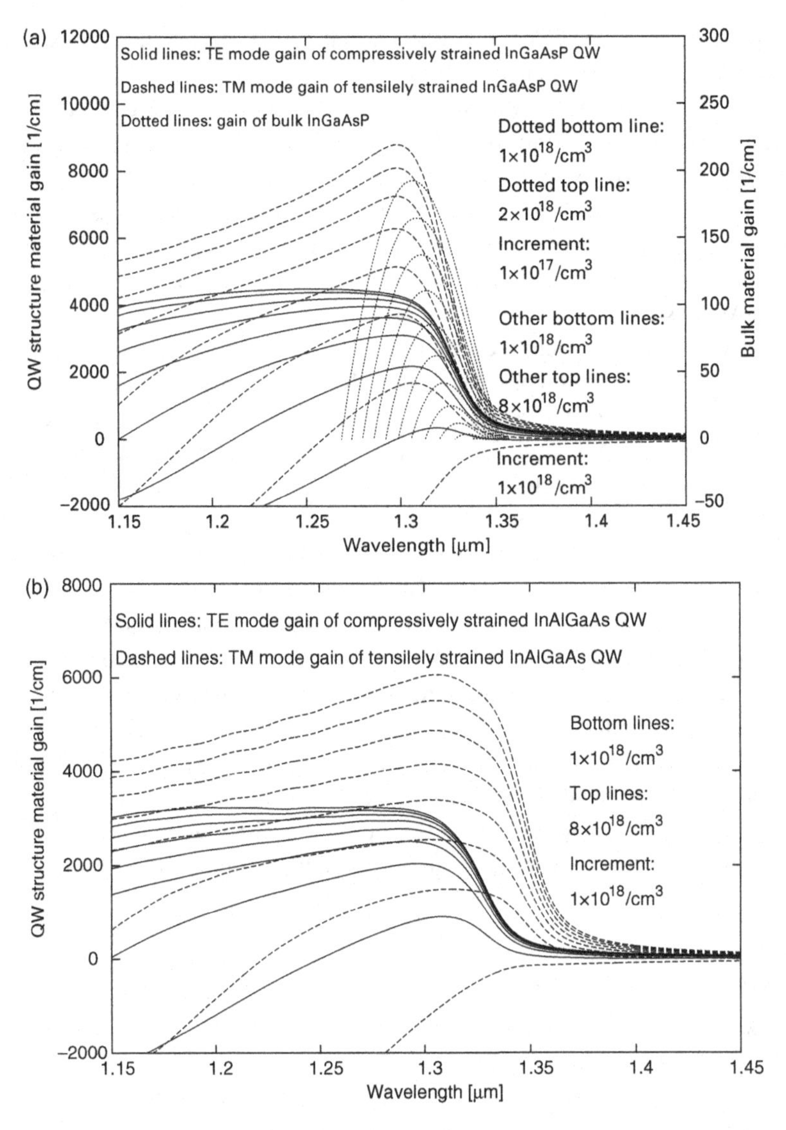

Fig. 10.3. Comparison of material gain profiles between the compressively strained QWs (TE mode gain) and the tensilely strained QWs (TM mode gain). (a) The InGaAsP system (where bulk material gain is also plotted). (b) The InAlGaAs system.

bulk material gains, for the InGaAsP and InAlGaAs systems at 1310 nm, respectively. The structures and parameters are all listed in Table 10.1.

While the bulk material gain is about ten times smaller, the TM mode gains of the QW structures with tensile strain is higher than the TE mode gains of the QW structures with compressive strain for both material systems. Despite the in-plane mixing effect between the heavy-hole and light-hole bands as mentioned in Section 3.2.4, the dipole matrix element between the electron and light-hole band transition in the QW with tensile strain is higher than that between the electron and heavy-hole band transition in the QW with compressive strain because of the larger overlap between the electron and light-hole wave functions. It is also apparent that the QWs with tensile strains have higher transparent carrier density and consequently saturate at high carrier density levels for both material systems. Therefore, to achieve better device performance, we have to retain higher sheet carrier density inside the QWs with tensile strain as opposed to the QWs with compressive strain, which suggests that we use fewer QWs or a shorter cavity length when the tensile strain is incorporated.

10.1.2 The active region structure

As opposed to the bulk material active region, the main advantage of the active region produced by the QW structure lies in its high differential gain at low carrier densities, since the gain in the QW structure grows faster than in bulk as the carrier density increases due to the difference in their density of states [5]. However, if there is no second bound electron state inside the QW, as the carrier density grows to high levels, the QW gain starts to saturate, again due to the constant density of states in the QW. (On the valence band side, usually both heavy-hole and light-hole states are bound and, due to the band mixing effect, there is no selection rule to forbid the transition of electrons to these bands. Thus, saturation is mainly brought by electrons.) Therefore, the differential gain of the QW structure drops as the injected carrier density inside the QW increases. Actually, the gain provided by a QW has a sub-linear dependence on the sheet carrier density due to the constant density of states and the sub-linear quasi-Fermi level dependence on the sheet carrier density according to equations (7.20) and (7.16a). Therefore, the derivative of the gain with respect to the sheet carrier density has to drop as the sheet carrier density increases.

Compared with the bulk material active region, although the transparent carrier density in the QW structure is higher (as the step-function-like density of states becomes non-zero at an energy level higher than the band edge), the threshold current for lasers with a QW active region is still lower, due to the high differential gain at low carrier density. In an FP laser with negligible LSHB effect, the carrier density is fixed and almost uniformly distributed inside the cavity after lasing, hence the differential gain has no effect to the laser output power–bias current dependence when the bias current goes beyond the threshold, as can be seen by the steady state solution of the optical and carrier rate equations. Thus, the differential gain will not affect the FP laser DC performance except for its threshold current. However, the differential gain does have significant impact on the laser dynamic performance, e.g., the 3 dB bandwidth of the small-signal IM response [6, 7]. To achieve a low threshold in DC operation as well as a high modulation bandwidth in alternating current (AC) operation, we have to retain a low carrier density in the entire operation range to maintain a high differential gain in the QW structure.

Therefore, for the design of directly modulated lasers with QW structures, retaining a low carrier density inside the active region can be taken as an important rule to guarantee not only a superior DC, but also a superior AC, performance compared with bulk active region lasers.

We have other applications where high carrier density inside the cavity is preferred, in order, e.g., to maintain a high gain and a high saturation power in SOAs, or to broaden the emission bandwidth in SLEDs. Since these devices are not modulated, low differential gain will not degrade the performance. Hence we can still use the QW structure as the active region for these devices.

In any of the above mentioned applications, we need to control the level of carrier density inside the active region. In a bulk active region, the differential gain is moderate but drops very little as the carrier density increases. In an active region with a QW structure, however, the differential gain is much higher but drops more rapidly as the carrier density increases. Therefore, unless we need to maintain both high carrier density and high differential gain, in which case we would rather select the bulk active region, we should always take a QW structure as the active region. Besides, the implementation of strains in a QW structure is much easier, due to the limitation on the maximum strain–layer thickness product. For this reason, we will mainly focus on QW structures.

In a QW structure design for the active region, there are a number of parameters that need to be determined, including the number of QWs, the well thickness, the well bandgap energy and the strain on the well (i.e., the well material composition), the barrier bandgap energy and the strain on the barrier (i.e., the barrier material composition).

Usually the design goals are:

(1) sufficient gain and differential gain to cover the specified lasing wavelength;
(2) balance on the optical confinement factor and carrier (electron and hole) transport;
(3) higher differential gain and lower transparent carrier density;
(4) control on gain profile bandwidth (broad or narrow);
(5) low temperature dependence.

While every design parameter except the number of QWs can affect the gain peak position, we leave its tuning to the final stage, since there is the least constraint on the alignment of gain peak to the required lasing wavelength (possibly with some detuning).

Other than some special devices where low optical confinement factors are required, such as SOAs for linear amplification, most devices including lasers need larger optical confinement factors. The optical confinement is mainly determined by the number of QWs and the well thickness. To achieve a specific optical confinement factor, we can either select more and thinner QWs or fewer and thicker QWs. On the one hand, we prefer more and thinner QWs as the latter option leads to a lesser quantum effect, and leads to the bulk semiconductor at the extreme case. Besides, to avoid the state filling induced gain saturation [7], we prefer to let the laser operate at moderate sheet carrier densities inside the QWs. The structure with more QWs will therefore raise the bias current at which saturation starts to appear inside the QWs due to the dilution of the sheet carrier density, a similar effect to extending the laser cavity length [6]. On the other hand, a structure with too many QWs increases the threshold current, as the sheet carrier density in each well will be diluted under a fixed bias current. Hence a higher bias current is required to reach the transparent sheet carrier density in each QW. Although this effect can be compensated for by reducing the laser cavity length, unwanted side effects such as poor thermal characteristics and broadened lasing mode linewidth are brought in as well. Moreover, a structure with too many QWs also creates carrier transport problems, as it is getting more difficult for electrons to reach the P side wells and for holes to reach the N side wells with the growing number of QWs. When insufficient injection occurs, population inversion will not happen inside those wells at the two edges. As a result, these wells absorb light rather than emit light. Although this may not be a severe problem in lasers or SOAs with identical wells due to photon assisted carrier transport [8], it may cause insufficient injection to those wells with higher transition energies in lasers or SOAs with non-uniform QWs, or in SLEDs where the photon assisted carrier transport effect is negligible.

For different applications with different specification devices, the optimized number of QWs can be very different and is closely correlated with other device parameters such as the cavity length. Hence a full device simulation is necessary before we can know the best number.

Therefore, at the very beginning, we have to take a trial number of QWs to start the design–simulation–modification loop.

For a fixed trial number of QWs, the well thickness is then fixed in order to meet the required optical confinement factor, and consequently the maximum allowed strain is set, since the total strained QW thickness cannot go beyond a certain critical value at which the strain will be relaxed.

For the fixed number of QWs and well thickness, we can then select the bandgap energy difference between the barrier and the well. This value must be selected to balance carrier injection and confinement, as well as the number of bound states (discrete energy levels) required. Usually we select relatively lower barriers for more wells to facilitate carrier injection. We usually do not apply any strain to the barrier unless we run out of other degrees of freedom. Hence the barrier composition is determined.

Finally, we can jointly tune the well bandgap energy and the strain applied to the wells at the previously determined well thickness and barrier height to align the gain peak to a specific value determined by the lasing wavelength. Hence the well composition is determined.

In the laser design example as shown in Table 10.1, six 5 nm QWs are selected on trial. The barrier bandgap wavelength is selected as 1150 nm for the InGaAsP system. For the InAlGaAs system, however, we usually choose a barrier with a higher bandgap energy, as otherwise the small valence band offset may hardly bind any heavy-hole band. Thus the well bandgap energy and the strain are selected to align the gain peak around 1310 nm.

As shown in Fig. 10.1 and Fig. 10.2, we have already achieved initial designs with reasonable results. Suppose that we have to tune the obtained gain profile towards the red (longer wavelength) side for 35 nm in order, e.g., to acquire a higher differential gain, or to prevent the gain from moving off the lasing wavelength at low ambient temperatures, we can achieve this target by changing a number of parameters, such as reducing the well bandgap energy, or reducing the barrier height (barrier bandgap energy), or increasing the strain, or increasing the well thickness. Figure 10.4 shows the effects of changing these parameters in the InAlGaAs system.

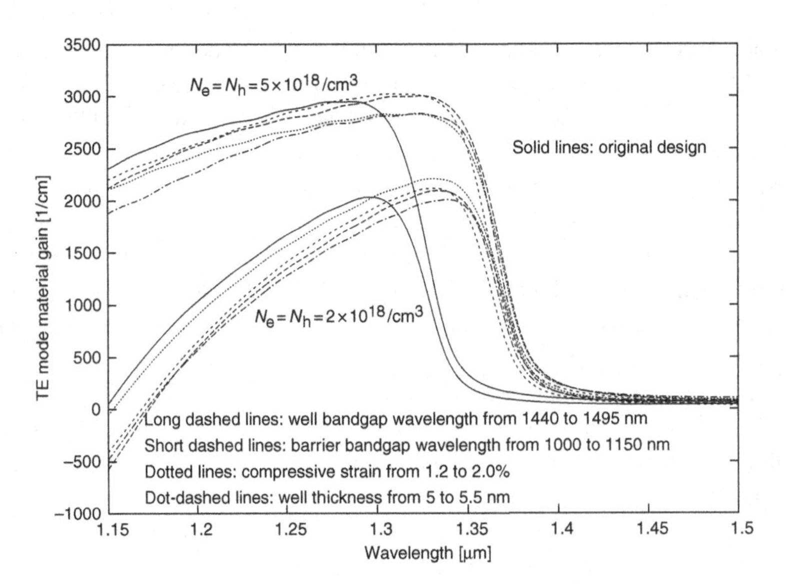

Fig. 10.4. Effects of parameter tuning on gain profiles in the InAlGaAs QW structure.

As can be seen from Fig. 10.4, all these approaches shift the gain peak to a longer wavelength as expected. A common feature is that there is always a gain bandwidth broadening accompanying the gain peak red-shift.

In addition to the common effect, we still make the following observations based on this simulation.

(1) Reducing the well or the barrier bandgap energy has an almost identical effect on the gain profile.

Note that these two approaches change the barrier–well depth (i.e., the bandgap energy difference between the barrier and the well) in an opposite way. The original bandgap energy difference is $1.2405(1.0 - 1.0/1.44) = 379$ meV, the first approach increases the bandgap energy difference to $1.2405(1.0 - 1.0/1.495) = 411$ meV, whereas the second approach reduces it to $1.2405(1.0/1.15 - 1.0/1.44) = 217$ meV. Therefore, we prefer to have the low barrier height design to facilitate carrier transport. This conclusion holds only in this particular case since only the light-hole band becomes unbound in the second approach, whereas there is no extra bound state found in the first approach. With respect to the TE gain, exclusion of the light-hole band has little effect.

(2) Increasing the compressive strain increases the gain bandwidth.

While it offers a relatively higher gain at a low carrier density, increased compressive strain indeed reduces the gain at high carrier densities, which indicates a deterioration of the differential gain at high injection level for the QW structure with high compressive strain. This suggests using relatively high strain for laser design but relatively low strain for SOA or SLED design.

(3) Increasing the well thickness provides lower gain and slightly narrower bandwidth.

This can be attributed to the equivalent reduction of the quantum effect. Therefore, we should always avoid taking this approach in practice.

In the design process, normally we need only to invoke the free-carrier gain model, as computational efficiency is the major concern. Only in the final simulation of a designed structure can we invoke the many-body gain model in order to obtain more accurate results. This strategy works well in practice since the free-carrier gain model does not seem to give any wrong prediction of gain dependence on design parameters, although its accuracy could be poor [9–11].

10.2 Design and modeling of the cross-sectional structure for optical and carrier confinement

10.2.1 General considerations in the layer stack design

The design of the cross-sectional structure is required to provide both optical and carrier confinement. Despite the many designs proposed historically, only buried heterostructure (BH) and ridge waveguide (RW) structures, as shown in Fig. 10.5(a&b), are used practically for device production currently. We will then compare these two structures on their ability to confine the optical field and to prevent the current leakage. To focus on the cross-sectional structures, we will take the previously designed InAlGaAs system as the active region and use a simple Fabry–Perot (FP) structure with a length of 250 μm and as-cleaved facets with equal amplitude reflectivities at 0.565 as the laser cavity.

The parameters in the BH and RW cross-sectional structures are summarized in Table 10.2.

As shown by Table 10.2, the barriers and QWs in the previously designed active region are sandwiched between two graded index separate confinement heterojunction (GRINSCH) regions made of InAlGaAs layers. This well-known design provides us sufficient degrees of freedom simultaneously to optimize confinement of electrons and holes in a small region near the QWs and of the optical wave along the vertical direction [12]. Without this design, it is hard to prevent a deep penetration of the evanescent tail of the optical wave outside the QWs into the heavily doped cladding layers, which leads to strong free-carrier absorptions. Without sufficient doping in the P and N side cladding layers, however, we cannot achieve efficient hole and electron injections into the active region [13]. Moreover, this design also flattens the optical mode profile inside the active region, hence the contribution of the material gain from each individual QW will be more balanced. Two wide bandgap InAlAs layers have also been inserted outside the GRINSCH regions on both the P and N sides, to block electron leakage further and to confine the optical mode better.

An exponentially growing doping profile is used in the P type InP cladding layer from the bottom (closer to the active region) to the top (closer to the cap), which is the best compromise between the conflicting requirements on increasing the hole injection

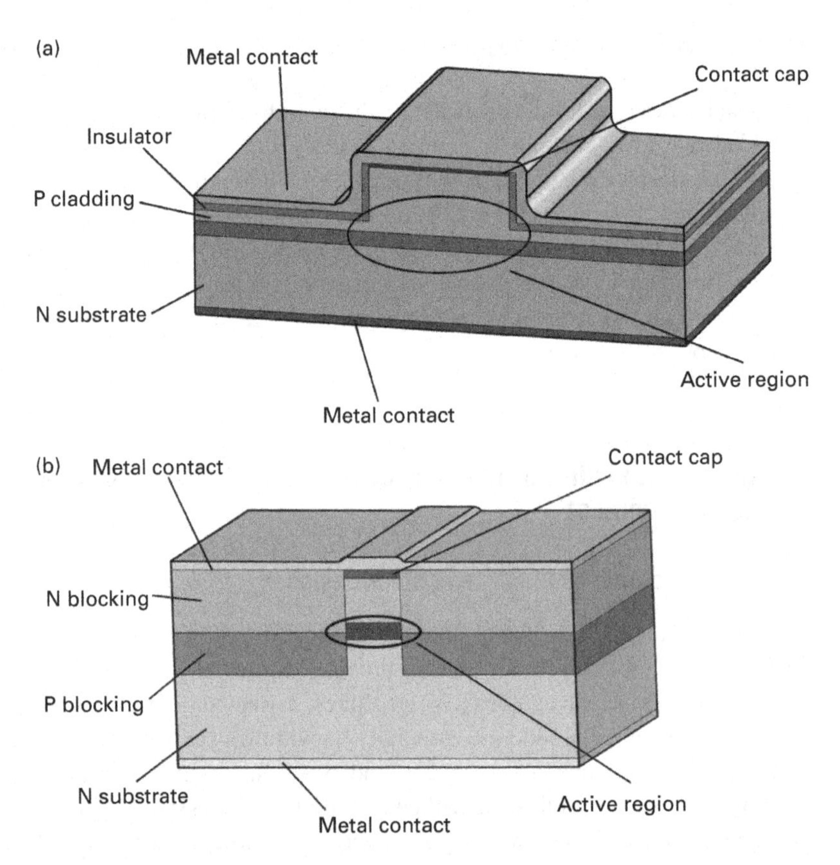

Fig. 10.5. Device design. (a) The ridge waveguide. (b) The buried heterostructure.

efficiency and reducing the optical free-carrier absorption. Actually, as the evanescent tail of the optical wave decays exponentially inside the cladding layer, an exponentially growing doping profile would make a minimum effective overlap between the optical field and the doping concentration, under the constraint that a given averaged doping concentration must be achieved in the cladding layer [13].

Due to the many constraints, there are actually not too many choices of design for the layer stack along the vertical direction. While following the above mentioned general guidance, we have to fine tune the top and bottom GRINSCH regions and to optimize the doping profiles on both P and N sides according to different device specifications for different applications. Since the fine tuning of the vertical layer stack is usually problem dependent, which results in scattered optimized structure parameters in different applications, we will not go into further details on the vertical layer design but shift our focus to two seemingly more common problems in the RW and BH structure, one for each.

10.2.2 The ridge waveguide structure

The main advantage of the RW structure is its ease of fabrication. However, it has drawbacks in performance. Due to concerns about reliability, we have to let the ditch

Table 10.2. Laser cross-sectional structures

The table sets out designs for the InAIGaAs/InP system at 1310 nm.

Layer	Thickness (nm)	Composition of $In_{(1-x-y)}Al_xGa_yAs$ or $In_{(1-x)}Ga_xAs_yP_{(1-y)}$ (x, y)	Doping concentration ($10^{18}/cm^3$)	Remarks λ_g (nm) and strain
Cap P-InGaAs	200	(0.468, 1.000)	10.0 (P)	1654, 0.0
Graded doped P-InP	1800	(0.000, 0.000)	1.0-0.5 (P)	918.6, 0.0
Etching stop InGaAsP	10	(0.108, 0.236)	0.5 (P)	1050, 0.0
Spacer InP	50	(0.000, 0.000)	0.3 (P)	918.6, 0.0
Blocking P-InAlAs	50	(0.479, 0.000)	0.3 (P)	829.2, 0.0
GRINSCH InAlGaAs	100	(0.380, 0.090)-(0.350, 0.130)	undoped	950–1000, 0.0
Barrier InAlGaAs	10	(0.350, 0.130)	undoped	1000, 0.0
Well InAlGaAs × 6	5	(0.155, 0.145)	undoped	1440, CS1.2%
Barrier InAlGaAs × 6	10	(0.350, 0.130)	undoped	1000, 0.0
GRINSCH InAlGaAs	100	(0.350, 0.130)-(0.380, 0.090)	undoped	1000–950, 0.0
Blocking N-InAlAs	50	(0.479, 0.000)	2.0 (N)	829.2, 0.0
Buffer N-InP	650	(0.000, 0.000)	1.0 (N)	918.6, 0.0
Etching stop InGaAsP	10	(0.108, 0.236)	1.0 (N)	1050, 0.0
Substrate N-InP	100 000	(0.000, 0.000)	3.0 (N)	918.6, 0.0
RW width 2000 nm	etching 2000 nm to the first etching stop on P side			
BH width 2000 nm	etching 3110 nm to the second etching stop on N side and regrowth			
Blocking N-InP	2265	(0.000, 0.000)	0.5 (N)	918.6, 0.0
Blocking P-InP	845	(0.000, 0.000)	0.5 (P)	918.6, 0.0

that forms the ridge stop above the active region. In this case, the optical wave overlaps the unpumped active region along the horizontal direction. Therefore, we face a problem in selection of the spacer layer thickness measured from the active region top to the ditch bottom beside the ridge. If we leave a thin or zero spacer, the hole current leakage is minimum and the ridge has the strongest control on the optical mode horizontally. However, the active region outside the ridge becomes totally absorptive due to the zero injection of the hole. Besides, the ridge guidance could become so strong that the second transverse optical mode is brought closer to the cut-off edge. As a result, the second mode also gets a chance to lase under higher injection levels as its threshold is greatly reduced. A well-known consequence of the higher order transverse mode excitation is a kink in the output optical power–bias current curve and the beam steering [14]. On the contrary, a thick spacer may cause an overflow of holes into the area that even the optical field does not reach, which leads to a complete waste, as illustrated by Fig. 10.6. Besides, the ridge will have little effect on the optical mode profile, which leads to an elliptical shaped far-field pattern with poor aspect ratio, and consequently causes a poor coupling to the optical fiber.

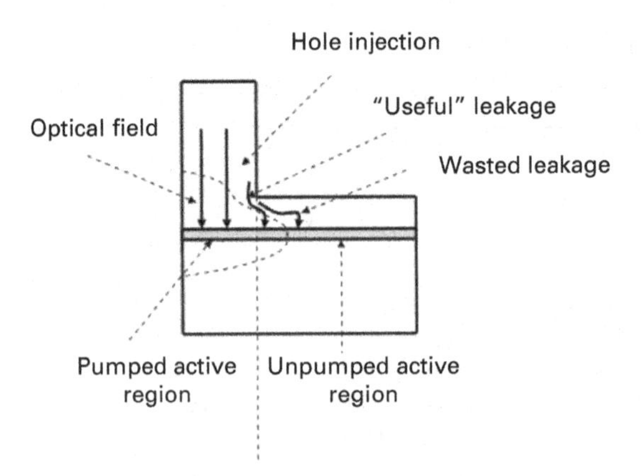

Fig. 10.6. Hole current leakage and the unpumped active region absorption of the optical field.

Therefore, optimization of the spacer thickness is a unique problem in the RW structure and has a significant impact on the laser performance. We calculated the RW structure with four different spacer layer thicknesses and show the results in Figs. 10.7 to 10.10. In these figures, the horizontal axis shows the spacer layer thickness measured from the top barrier boundary. As such, the first structure marked with the spacer thickness 0 is the structure with the ditch beside the ridge etched all the way down to the top of the first barrier layer. This is the thinnest possible spacer if the well is not allowed to be etched. The second structure marked with the spacer thickness 150 nm is the structure with the ditch beside the ridge etched to the top of the P side InAlAs layer. It is obvious that the thickness of the spacer comes from the P side GRINSCH region thickness (100 nm) plus the InAlAs layer thickness (50 nm). The third structure marked with the spacer thickness 200 nm is the structure shown in Table 10.2, in which we leave an extra InP spacer layer 50 nm in addition to the GRINSCH region and the InAlAs layer underneath the ditch beside the ridge. The last structure marked with the spacer thickness 300 nm is the same as the third one, except that the thickness of the InP spacer layer now is 150 nm instead of 50 nm. These structures correspond to the (closed and open) circles in the figures.

Figure 10.7 shows the calculated effective indices and confinement factors for these structures. As the spacer thickness grows, both effective index and confinement factor increase slightly. From these curves we find that the structure with a thicker spacer will have slightly higher modal gain.

Figure 10.8 shows the calculated horizontal and vertical divergence angles in the far-field pattern of these structures. The vertical divergence angle only increases slightly as the spacer gets thicker, as the optical field distribution in the vertical direction is mainly determined by the vertical layer stack. The horizontal divergence angle, however, reduces rapidly as the spacer thickness increases, which indicates a rapid optical field expanding along the horizontal direction. The aspect ratio takes almost a perfect value (1.03:1 for vertical:horizontal) for the structure with the thinnest spacer (at the very left of the figure), and deteriorates rapidly as the spacer thickness increases. It becomes 1.97:1 at the very right of the figure, which indicates that the far-field pattern of the structure with the 300 nm spacer becomes elliptical.

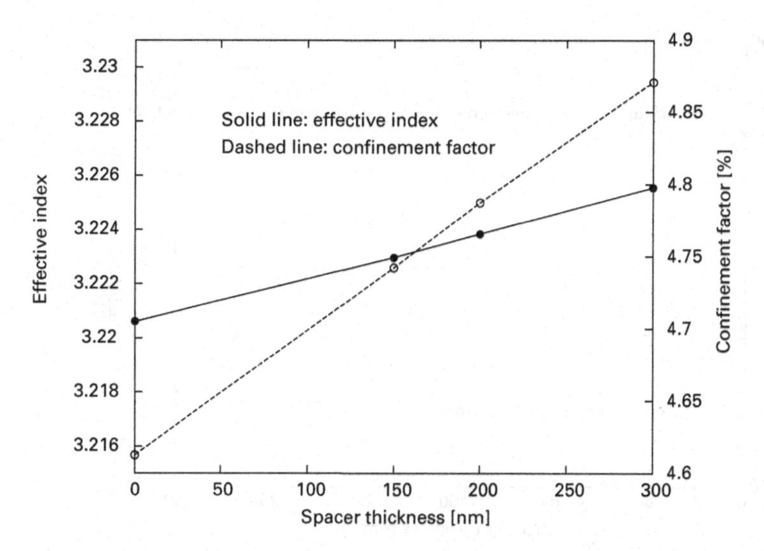

Fig. 10.7. The effective index and confinement factor as functions of the spacer thickness in the RW FP laser.

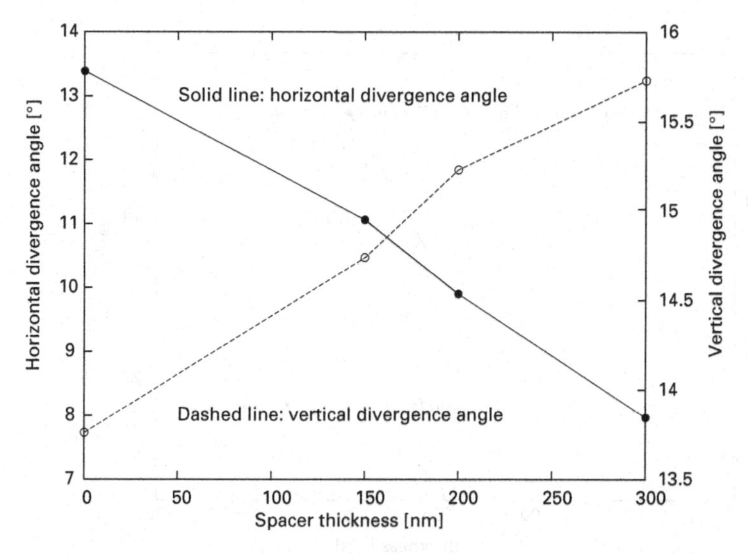

Fig. 10.8. The divergence angles in the far-field pattern as functions of the spacer thickness in the RW FP laser.

Figure 10.9 shows the calculated threshold currents and the injection efficiencies at the threshold of the FP lasers made in cross-sectional designs with the identical cavity specified in Section 10.2.1. The injection efficiency drops as the spacer thickness increases due to hole leakage as expected. Such a drop is minor when the spacer thickness increases from 0 nm to 150 nm, but gains momentum after the spacer goes beyond 150 nm. This is because the first 150 nm thick spacer comprising the GRINSCH and InAlAs layer is undoped, whereas the part of the spacer thicker than 150 nm is made of the P doped InP

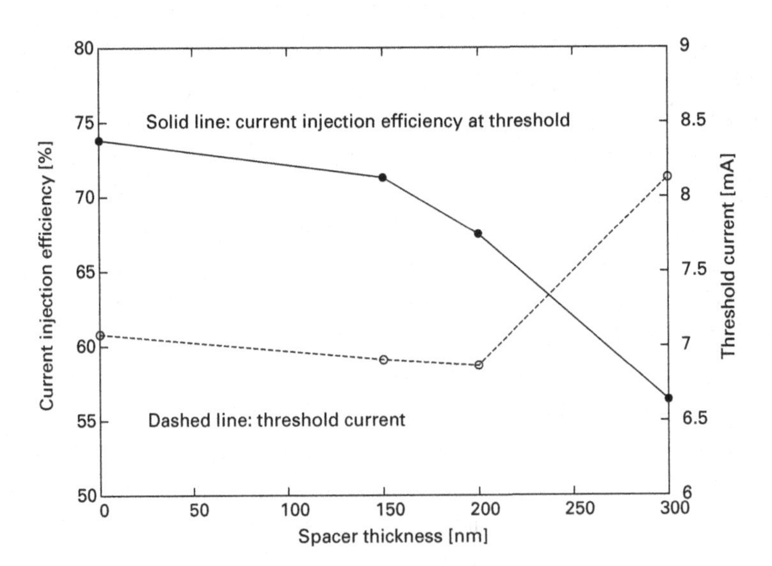

Fig. 10.9. The current injection efficiency and threshold current as functions of the spacer thickness in the RW FP laser.

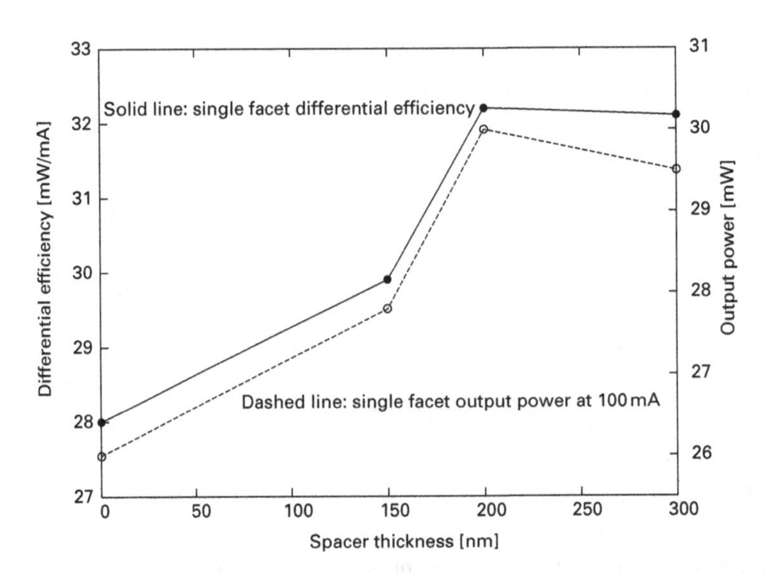

Fig. 10.10. The differential efficiency and output power at 100 mA as functions of the spacer thickness in the RW FP laser.

layer. The threshold current, however, presents a minimum around our standard design shown in Table 10.2. This can be understood from Fig. 10.10, which shows the calculated differential efficiencies and the output powers at 100 mA bias for these structures. Since the differential efficiency is inversely proportional to the optical modal loss, a structure with higher differential efficiency means it has lower optical modal loss. Figure 10.10 shows that a thicker spacer has lower modal loss although the optical field extends more into the active region outside the ridge, as shown by the reduction of the horizontal divergence angle in the far-field pattern. Due to the hole leakage in structures with a thicker spacer, the unpumped active region outside the ridge is actually pumped by the leaked holes from the spacer and electrons from the bottom N side layer. As a result, the absorption in this region disappears, hence the modal loss reduces. Since there is no spatial synchronization between the extended optical field and the leaked holes, if the field extends more, the modal loss increases, if the field extends less, the leaked holes are wasted. While modal loss affects both differential efficiency and threshold, the hole leakage affects only the threshold. As a combination, there must be an optimum spacer thickness at which the device has the lowest threshold: a thinner spacer results in high modal loss as no leaked holes can be consumed to pump the active region outside the ridge; a thicker spacer may still cause the same problem if the field extended farther than the leaked holes, or may cause the leaked holes to be wasted if the field extended nearer than the leaked holes. In either case, the device threshold increases.

Figure 10.9 shows that in this particular case the optimum spacer is the design in Table 10.2. Structures with a thinner spacer have higher injection efficiency but lower differential efficiency, which means that in such structures the leakage is low but the optical modal loss is high. Therefore, the optical field in such structures extends farther than the hole leakage in the active region outside the ridge. The structure with a thicker spacer, however, has lower injection efficiency but almost the same differential efficiency, which means that in this structure the leakage is high but the optical modal loss is the same. Therefore, the optical field in this structure extends nearer than the hole leakage. In the active region outside the ridge, the extended optical field and the leaked holes have the best match in the structure shown in Table 10.2.

10.2.3 The buried heterostructure

The BH solves the hole leakage and optical wave evanescent tail absorption problems simultaneously, as the entire active region is pumped and surrounded by materials with wider bandgap energy. Any horizontal leakage path is blocked either by the depletion region of a reversely biased P–N junction or by semi-insulating semiconductors. However, regrowth is required to fabricate this structure, which increases the cost.

From the device performance point of view, an obvious question is how to select the width of the active region for low threshold current and high differential efficiency.

As the width of the active region grows, the optical confinement factor increases and hence the modal gain increases whereas the modal loss decreases, which results in low threshold current and high differential efficiency. However, the wide active region dilutes the carrier (electron and hole) densities and leads to high threshold current. Lateral spatial hole burning may also occur, which reduces the effective modal gain [15]. Besides, the increased width of the waveguide may lead to multiple transverse optical mode operation and the kink may show up again in the device output power–bias current curve.

We have calculated the BH with three different widths and show the results in Figs. 10.11 to 10.14. In these figures, the horizontal axis shows the width of the active region and each structure corresponds to a pair of closed and open circles.

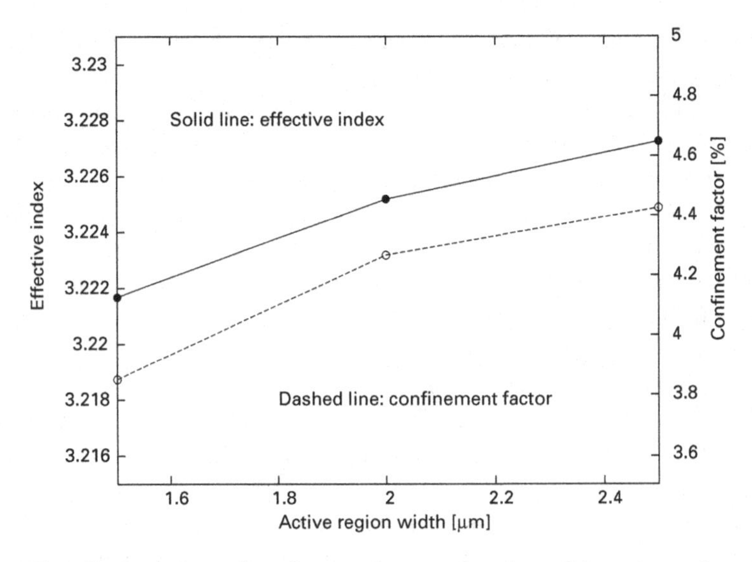

Fig. 10.11. The effective index and confinement factor as functions of the active region width in the BH FP laser.

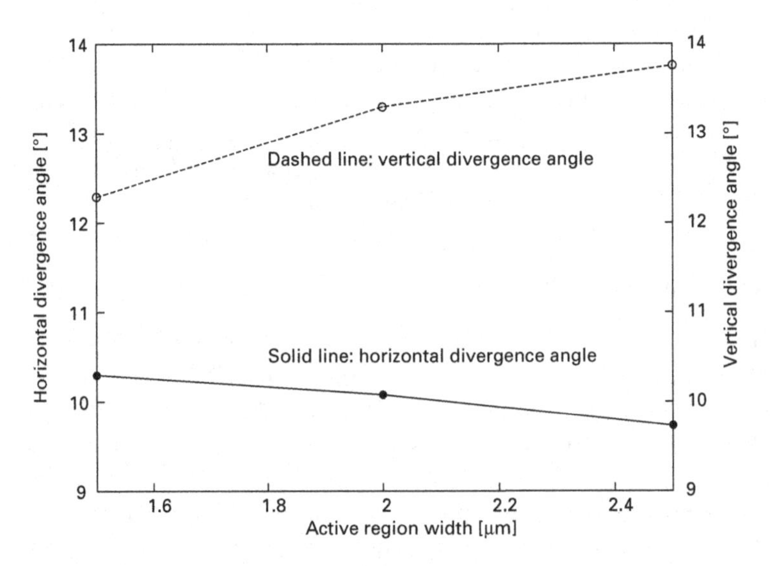

Fig. 10.12. The divergence angles in the far-field pattern as functions of the active region width in the BH FP laser.

Figure 10.11 shows the calculated effective indices and confinement factors for these structures. As the width increases, both effective index and confinement factor increase as expected. Hence the structure with a wider active region has a higher modal gain.

Figure 10.12 shows the calculated horizontal and vertical divergence angles in the far-field pattern of these structures. While the horizontal divergence angles are all slightly smaller than the vertical divergence angles for all structures, the structure with the narrowest width has the best aspect ratio. To reach a perfect aspect ratio, not only do we

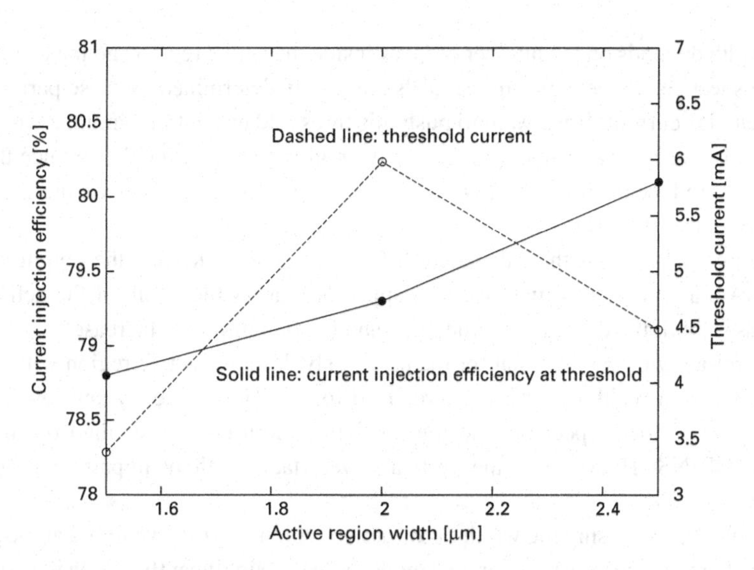

Fig. 10.13. The current injection efficiency and threshold current as functions of the active region width in the BH FP laser.

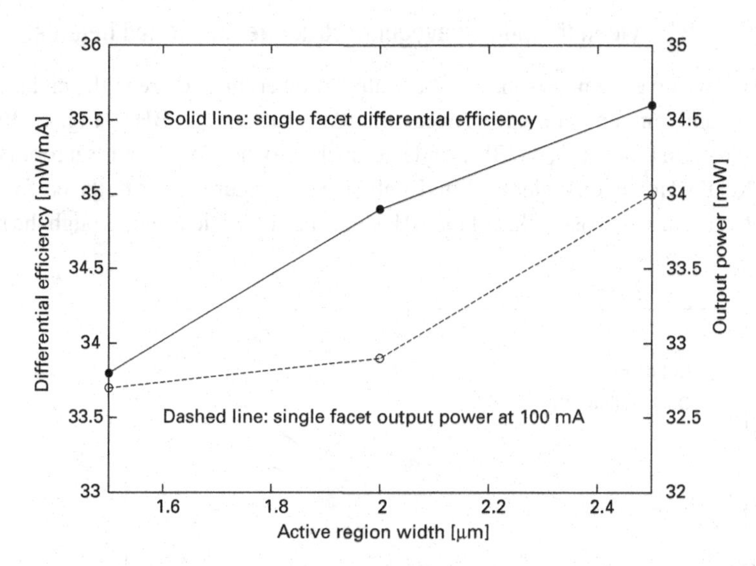

Fig. 10.14. The differential efficiency and output power at 100 mA as functions of the active region width in the BH FP laser.

have to adjust the width of the active region, but also the GRINSCH region design in the vertical layer stack.

Figure 10.13 shows the calculated threshold currents and the injection efficiencies at the threshold of FP lasers of cross-sectional design with an identical cavity to that specified in Section 10.2.1. The injection efficiency increases as the width increases. The threshold current, however, has a maximum when the width is 2.0 μm. Note that this width depends on the number of QWs inside the active region and the cavity length, as the sheet carrier densities in the QWs are jointly determined by these parameters for a given bias current. If we want to push this threshold current maximum away from the width that we normally select for the active region, we should either reduce the cavity length, or reduce the number of QWs, as these approaches will increase the sheet carrier density in the QWs for a fixed bias current.

Figure 10.14 shows the calculated differential efficiencies and the output powers at 100 mA bias for these structures. It is apparent that, as the width of the active region increases, both the differential efficiency and the output power increase.

Therefore, we conclude that for the previously designed active region with six QWs, the BH with a width of 2.5 μm is preferred for its high efficiency and low threshold. Its relatively poor aspect ratio in the far-field pattern can be rectified by fine tuning of the GRINSCH region in the vertical layer stack, without impairing performance otherwise.

By further increasing the width of the active region to 3 μm, we find that the effective index of the second transverse optical mode increases significantly, which is an indication that this mode could also have a chance of lasing under a high injection level.

10.2.4 Comparison between the ridge waveguide structure and buried heterostructure

Finally, we give comparisons of the output power–bias current dependence under different ambient temperatures between the RW structure and BH in Fig. 10.15.

As shown in this figure, the BH is indeed superior to the RW structure at room temperature (300 K). However, under the lifted ambient temperature at 358 K, the performance of the RW structure surpasses that of the BH, since the current leakage through the reversely

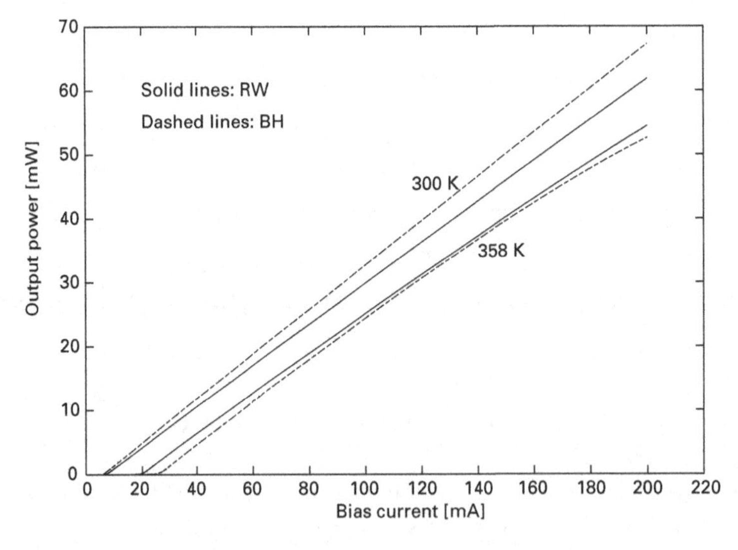

Fig. 10.15. Comparison of the output power–bias current dependences between RW and BH FP lasers at different temperatures.

biased P–N junction grows rapidly with temperature in the BH, whereas the current leakage has little change with ambient temperature in the RW structure. This result is consistent with experimental observation [16], which suggests that, for high temperature operations, either an RW structure or a BH with a semi-insulating blocking layer instead of a reversely biased P–N junction should be used.

10.3 Design and modeling of the cavity for lasing oscillation

10.3.1 The Fabry–Perot laser

By assuming that the FP laser facets are as-cleaved, we have the cavity length as the only design parameter in performance optimization for FP lasers. However, in the previous active region design, we have left the number of QWs yet to be determined, as it is a "global" parameter that impacts the device performance. In this section, we will jointly vary the number of QWs and the cavity length to find the dependence of the device performance on these changes.

As discussed in Section 10.1.2, to take advantage of the QW structures in the laser design we need to retain a low carrier density inside the QWs, hence a high differential gain is obtainable. At a fixed bias current, if we use more QWs in the active region, the sheet carrier density in each QW is lower, hence the gain contributed by each QW is lower. However, the confinement factor must be higher for an active region with more QWs. For optimized GRINSCH regions and fixed well thickness, it is reasonable to expect that the confinement factor will increase linearly with the number of QWs, once the total thickness of QWs (i.e., the product of the well thickness and the total QW number) does not go beyond, e.g., a quarter of the optical wavelength inside the active region [6]. Due to the sub-linear dependence of the gain provided by each QW on the sheet carrier density, as discussed in Section 10.1.2, the total modal gain increases with the number of QWs. To make this argument clear, we assume that $g_{SQW} = \Gamma f(N_{2D})$, where g_{SQW} represents the modal gain of a single QW, Γ the confinement factor of a single QW, N_{2D} the sheet carrier density, and $f(x)$ a sub-linear function for which we have $f(x)/a < f(x/a)$ for $x > x_0$, $f(x_0) = 0$, $f(x) > 0$, and $a > 1$. Consequently, the modal gain of an active region with M QWs will be $g_{MQW} = M\Gamma f(N_{2D}/M)$, where the sheet carrier density has to be diluted M times for a fixed bias current. It is apparent that $g_{MQW} > g_{SQW}$. In lasing operation, the gain is fixed by the total loss if we ignore the spatial and spectral hole burning effects. Therefore, the structure with M QWs just needs to retain lower sheet carrier densities inside the QWs to reach the same gain. Consequently, it will have a higher differential gain. In this sense, we prefer to have more QWs. However, there is a fixed transparent sheet carrier density for each QW; only beyond this will the gain be positive. As M increases, a higher bias current is required to reach this transparent sheet carrier density in each QW. Therefore, compared with the single QW structure, the multiple QW structure has higher differential gain and higher transparent carrier density. We may conclude that there must be an optimum M at which the threshold current will be the lowest.

Alternatively, the sheet carrier density can be adjusted through the cavity length for a given number of QWs. Therefore, for a given number of QWs, there should be an optimum cavity length at which the threshold will be the lowest.

This conclusion, also known as the QW scaling rule [17], is usually treated as a major guideline in designing QW lasers. The fact that the number of QWs and the cavity length can be jointly selected to achieve the lowest threshold current brings us great convenience in laser design. For example, if the cavity length has to be chosen to satisfy other device specifications, we can select the number of QWs to achieve the minimum threshold. Or we can adjust the cavity length to achieve the lowest threshold if the number of QWs is fixed by other conditions.

To confirm this conclusion, we will model RW FP lasers with different numbers of QWs and different cavity lengths by taking the previously designed active region made of the InGaAsP system as an example. The cross-sectional layer structure is given in Table 10.3. The ridge is 2 μm wide and 2 μm deep.

Figure 10.16 shows calculated threshold current as a function of the FP laser cavity length for structures with six and three QWs. This figure gives a consistent result with [6], which confirms the existence of the lowest threshold for any combination of the number of QWs and cavity length. In dealing with real world design problems, the cavity length is normally selected to meet the specifications for the differential efficiency or output power, the small-signal IM 3 dB bandwidth, or the thermal characteristics. Once the cavity length is set, we should follow Fig. 10.16, or results from similar calculations, or even simple estimations [18], to determine the number of QWs. It is worth mentioning that there is a maximum number of QWs beyond which the scaling rule is no longer valid. This can easily be understood if we rewrite the scaled gain in the form $g_{MQW} = G(M)f(N_{2D}/M)$, where $G(M) \approx M\Gamma$ for small M's but becomes a sub-linear function of M as well when M goes beyond a certain integer number. Once M is so large that $g_{MQW} \leq g_{SQW} = \Gamma f(N_{2D})$, the scaling rule fails. If the threshold current is the major concern, we prefer to choose more QWs combined with its optimum cavity length, as the optimum threshold current in the structure with more QWs is still slightly lower than that in the structure with fewer QWs.

Figure 10.17 gives the calculated output power–bias current curves for FP lasers with six and three QWs at different cavity lengths. It shows that the shortest cavity has the largest differential efficiency as expected. As can be seen from the figure, while threshold current depends on both the number of QWs and the cavity length, differential efficiency has little to do with the former. This suggests that, in FP laser design, we should select the cavity length by fitting the specified differential efficiency first, and then select the optimized number of QWs to achieve the lowest possible threshold current.

Up to this stage, we have completed a loop of FP laser design with the help of the numerical simulation tools. At the beginning, we have started with a trial QW number. If the specifications are met, we complete our design loop. Otherwise, we have to adjust the number of QWs and start all over again. By always taking the optimum number of QWs to start the next round of design, we may expect a rapid convergence of this trial-and-error approach, as the selection of the number of QWs is no longer blind from the second round onwards.

Table 10.3. Laser cross-sectional structures

Layer	Thickness (nm)	Composition of $In_{(1-x)}Ga_xAs_y P_{(1-y)}$ (x, y)	Doping concentration ($10^{18}/cm^3$)	Remarks λ_g (nm) and strain
Cap P-InGaAs	200	(0.468, 1.000)	10.0 (P)	1654, 0.0
Graded doped P-InP	1 800	(0.000, 0.000)	1.0-0.5 (P)	918.6, 0.0
Etching stop InGaAsP	10	(0.108, 0.236)	0.5 (P)	1050, 0.0
Spacer InP	50	(0.000, 0.000)	0.3 (P)	918.6, 0.0
GRINSCH InGaAsP	50	(0.108, 0.236)	undoped	1050, 0.0
GRINSCH InGaAsP	50	(0.145, 0.317)	undoped	1100, 0.0
Barrier InGaAsP	10	(0.181, 0.395)	undoped	1150, 0.0
Well InGaAsP × 3 or 6	5	(0.081, 0.547)	undoped	1355, CS1.2%
Barrier InGaAsP × 3 or 6	10	(0.181, 0.395)	undoped	1150, 0.0
GRINSCH InGaAsP	75	(0.145, 0.317)	undoped	1100, 0.0
GRINSCH InGaAsP	75	(0.108, 0.236)	undoped	1050, 0.0
Buffer N-InP	650	(0.000, 0.000)	1.0 (N)	918.6, 0.0
Substrate N-InP	100 000	(0.000, 0.000)	3.0 (N)	918.6, 0.0

The table sets out design for the InGaAsP/InP system at 1310 nm.

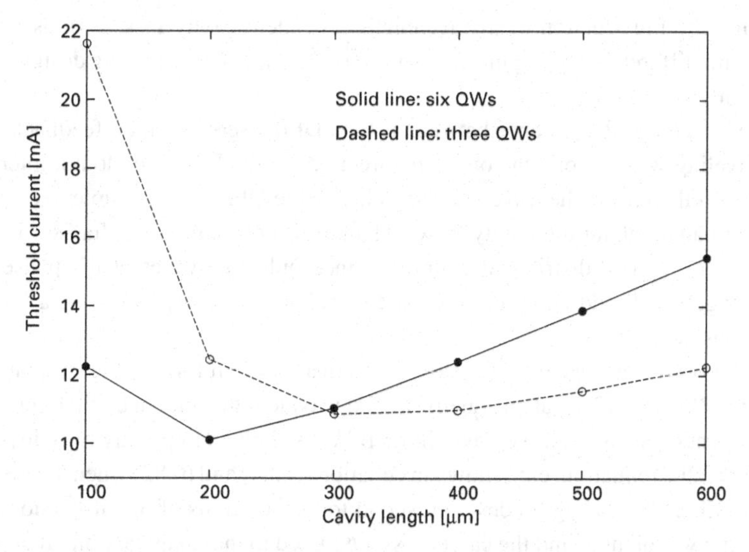

Fig. 10.16. The threshold current as a function of cavity length in an RW FP laser with different numbers of QWs.

10.3.2 Distributed feedback lasers in different coupling mechanisms through grating design

A major problem of the FP laser is its poor control of the lasing wavelength due to the multiple longitudinal mode operation brought about by the narrow spacing between the wavelengths of these modes on a spectrum compared with the gain bandwidth.

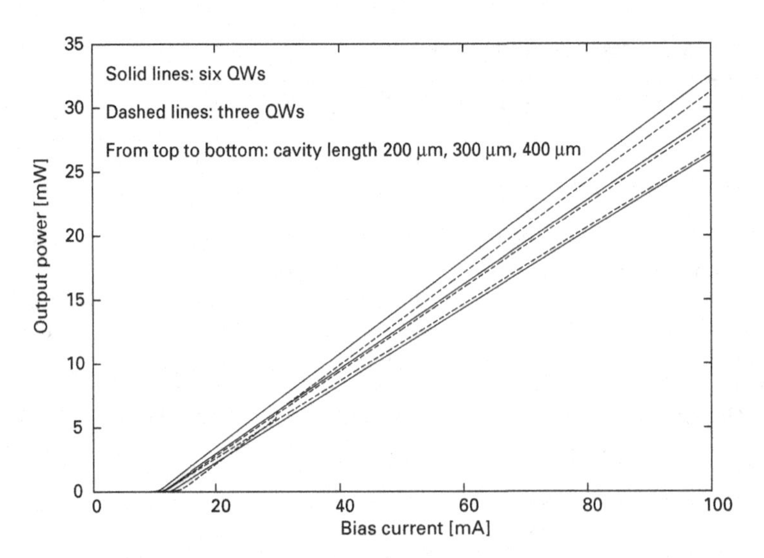

Fig. 10.17. The output power–bias current dependences of a RW FP laser with different cavity lengths and different numbers of QWs.

Through the introduction of a wavelength dependent cavity loss, various DFB structures with different coupling mechanisms realized by different grating designs solve this problem.

The fundamental difference between FP and DFB lasers is in their feedback, for while the traveling waves along the opposite directions in the FP cavity do not interfere until they reach the facets, the traveling waves inside the DFB cavity couple to each other as they propagate along the cavity. A wavelength discrimination mechanism is therefore brought in by such distributed coupling, since only the waves at the phase matched wavelengths will add constructively and establish a standing wave, or a longitudinal mode, inside the cavity.

As opposed to a constant cavity loss given in an explicit form of the structure parameters $|\ln(|R_l||R_r|)/(2L)|$ at an equally spaced wavelength comb in an FP laser, the cavity loss dependence on the wavelength in DFB lasers is given implicitly by a dispersion (or an eigenvalue) equation, e.g., shown in equations (6.81) and (6.82) when the LSHB effect is negligible. The lasing modes correspond to the solutions of the dispersion equation: the lasing wavelength and the cavity loss are linked to the imaginary and real part of the solution. Actually, if we split the complex dispersion equation into two real equations, we will find the latter to be a set of coupled amplitude and phase resonance conditions. However, in contrast to the FP cavity, the amplitude and phase resonance conditions are coupled in the DFB cavity. Therefore, we do not take the real form of the dispersion equation as it neither provides any clearer physical picture nor simplifies the solution technique.

Despite the complexity in solving the dispersion equation (6.81) when the LSHB is negligible, or the original rigorous eigenvalue problem, equation (6.65), we may still state generally that a common feature in DFB lasers is that cavity loss at the lasing wavelength is inversely proportional to the normalized coupling factors $\kappa_{\pm M} L$, with the coupling coefficients $\kappa_{\pm M}$ given in equation (2.79).

Therefore, the design guidelines that have been summarized for FP lasers are still valid for DFB lasers if we use $1/(|\kappa_{\pm M}|L)$ to replace $|\ln(|R_{\mathrm{l}}||R_{\mathrm{r}}|)/(2L)|$.

An extra design problem in the DFB laser, however, is its lasing wavelength control. First of all, the lasing wavelength must be linked to the Bragg wavelength

$$\lambda_{\mathrm{B}} = 2n_{\mathrm{eff}}\Lambda, \tag{10.1}$$

at which $\beta_0 = 2\pi n_{\mathrm{eff}}/\lambda = \pi/\Lambda$ or $\delta = 0$. This is the wavelength at which the wave sees the maximum reflection as it travels through the cold cavity (passive waveguide). It is apparent that, for a DBR laser, lasing will take place right at this wavelength. Since the DBR laser can be viewed as an FP laser with its mirror reflectors at the facets replaced by the wavelength selective grating reflectors, the cavity loss at the Bragg wavelength becomes the smallest due to the highest reflections presented by the gratings at this wavelength. In DFB lasers, however, this is not a generally valid conclusion as discussed in Section 6.3. For example, in the most commonly used first order purely index-coupled gratings, $\kappa_M\kappa_{-M} = \kappa_{-1}\kappa_1$ is real and we find that $\delta = 0$ can never be a solution since equation (6.81) has no real solution for γ. Physically, this is also understandable: due to the maximum reflection, the wave at the Bragg wavelength must have the shortest penetration into the grating region; since the grating region is also the gain region in DFB lasers, the Bragg wave then sees the least gain and hence it cannot be the lasing mode. On the contrary, at the zero-reflection wavelengths, maximum transmission through the grating region is obtained. Although the waves at these particular wavelengths experience the highest gain, they cannot be the lasing modes either since no reflection happens and hence no resonance can be established. In this sense, lasing should occur somewhere in between, i.e., the traveling waves at the lasing wavelength should be partially reflected by the grating to establish a resonance, they should be partially transmitted in order to experience the gain in the grating region. Thus, the phase matching and amplitude sustaining (i.e., the calculation of the round trip gain and the loss, or the unitary net gain) conditions are mixed, in a form of equation (6.81). Only when the required threshold gain for lasing approaches zero do the lasing wavelengths approach the zero-reflection wavelengths, as the waves may acquire sufficient gain to reach their thresholds through only a single pass. Therefore, we may conclude that lasing will occur neither at the Bragg wavelength, nor at the zero-reflection wavelengths of the corresponding grating without gain (i.e., under zero bias). As the threshold gain approaches zero, however, lasing will occur near these zero-reflection wavelengths of the unbiased grating. Since the cavity loss of the DFB laser is proportional to $1/(|\kappa_{\pm M}|L)$, for gratings with higher $|\kappa_{\pm M}|L$, the required threshold gain must be lower. Hence we find that the lasing wavelength of DFB lasers with high $|\kappa_{\pm M}|L$ should be very close to the zero-reflection wavelengths of the unbiased grating.

The zero-reflection wavelengths of the unbiased uniform grating without facet reflections can be found analytically through launching a field into the input port of the grating and calculating the resulting reflected and transmitted fields. Actually, according to equation (6.69), for a given input field F_1, the reflected and transmitted field can be obtained through

$$\begin{bmatrix} F_{N+1} \\ 0 \end{bmatrix} = \begin{bmatrix} a^{11} & a^{12} \\ a^{21} & a^{22} \end{bmatrix} \begin{bmatrix} F_1 \\ R_1 \end{bmatrix}, \tag{10.2}$$

or

$$R_1 = -(a^{21}/a^{22})F_1, \tag{10.3}$$

and

$$F_{N+1} = [(a^{11}a^{22} - a^{12}a^{21})/a^{22}]F_1. \tag{10.4}$$

Once the LSHB effect is ignored, we obtain the reflectivity and transmissivity as

$$R \equiv \frac{R_1}{F_1} = -\frac{a^{21}}{a^{22}} = \frac{\mathrm{j}\kappa_{-1}\sinh(\gamma L)}{\gamma\cosh(\gamma L) - (\alpha + \mathrm{j}\beta)\sinh(\gamma L)}, \tag{10.5}$$

and

$$T \equiv \frac{F_{N+1}}{F_1} = \frac{a^{11}a^{22} - a^{12}a^{21}}{a^{22}} = \frac{\gamma}{\gamma\cos\mathrm{h}(\gamma L) - (a + \mathrm{j}\beta)\sin\mathrm{h}(\gamma L)}, \tag{10.6}$$

where equations (6.75) to (6.77), (6.80), and (6.82) have been used with $\gamma^2 = (\alpha + \mathrm{j}\beta)^2 + \kappa_{-1}\kappa_1$.

For a passive waveguide with either the symmetric or the anti-symmetric purely index-coupled gratings $\alpha = 0$ and $\kappa_{-1}\kappa_1 = |\kappa|^2$, and we have

$$\gamma = \pm\mathrm{j}\sqrt{\left(\beta^2 - |\kappa|\right)^2} \equiv \pm\mathrm{j}u, \tag{10.7}$$

and

$$R = \frac{\kappa_{-1}\sin(uL)}{\beta\sin(uL) - \mathrm{j}u\cos(uL)}, \tag{10.8}$$

and

$$T = \frac{\mathrm{j}u}{\beta\sin(uL) + \mathrm{j}u\cos(uL)}. \tag{10.9}$$

For either real (if $\beta \geq |\kappa|$) or imaginary (if $\beta < |\kappa|$)u, we have

$$|R|^2 = \frac{\mathrm{j}u}{\beta^2|\sin(uL)|^2 + |u|^2|\cos(uL)|^2}, \tag{10.10}$$

and

$$|T|^2 = \frac{|u|^2}{\beta^2|\sin(uL)|^2 + |u|^2|\cos(uL)|^2}. \tag{10.11}$$

Therefore, at $uL = k\pi$, or

$$\beta L = \pm\sqrt{\left(|\kappa L|^2 + (k\pi)^2\right)}, \tag{10.12}$$

$k = \pm 1, \pm 2, \ldots$, we have $|R| = 0$ and $|T| = 1$.

At $\beta L = 0$, or

$$uL = \mathrm{j}|\kappa L|, \tag{10.13}$$

we have $|R| = |\tanh(\kappa L)|$ (maximum) and $|T| = 1/|\cosh(\kappa L)|$ (minimum).

Hence, for the uniform grating on the passive waveguide, we find the maximum reflection at the Bragg wavelength where $\beta L = 0$, and zero-reflection, or full transmission at multiple wavelengths where $\beta L = \pm\sqrt{(|\kappa L|^2 + (k\pi)^2)}$, $k = \pm 1, \pm 2, \ldots$ Normally, we define the Bragg stop-band width as the difference between the two smallest βL's that reach the zero-reflection around the Bragg condition ($\beta L = 0$), i.e.,

$$\Delta\beta L \equiv \sqrt{\left(|\kappa L|^2 + \pi^2\right)} - \left(-\sqrt{\left(|\kappa L|^2 + \pi^2\right)}\right) = 2\sqrt{\left(|\kappa L|^2 + \pi^2\right)}. \tag{10.14}$$

According to equations (6.83) and (2.69), we find the corresponding Bragg stop-band wavelength as

$$\Delta\lambda_{\mathrm{B}} \approx \frac{\lambda_{\mathrm{B}}^2}{2\pi n_{\mathrm{eff}}}\Delta\delta = \frac{\lambda_{\mathrm{B}}^2}{2\pi n_{\mathrm{eff}}}\Delta\beta = \frac{\lambda_{\mathrm{B}}^2\sqrt{(|\kappa L|^2 + \pi^2)}}{\pi n_{\mathrm{eff}} L}. \tag{10.15}$$

Note that λ_{B} in equation (10.1) and $\Delta\lambda_{\mathrm{B}}$ in equation (10.15) can therefore be employed to estimate the lasing wavelengths of DFB lasers, as we know that the lasing wavelengths will be in the neighborhood of λ_{B} and bounded by $\Delta\lambda_{\mathrm{B}}$, as can be seen by comparing equation (10.14) to Fig. 2 in [19].

For a purely index-coupled uniform grating, as the grating normalized coupling factor (κL) increases, the lasing wavelengths approach $\lambda_{\mathrm{B}} \pm \Delta\lambda_{\mathrm{B}}/2$.

For a purely gain-coupled uniform grating, however, the lasing wavelength is at the Bragg wavelength λ_{B}, as discussed in Section 6.3.

For a complex-coupled grating in general, we know that the lasing wavelength falls in a range from $\lambda_{\mathrm{B}} - \Delta\lambda_{\mathrm{B}}/2$ to $\lambda_{\mathrm{B}} + \Delta\lambda_{\mathrm{B}}/2$.

A complete solution of the lasing wavelength and the threshold gain has to be found numerically by solving equation (6.65) through, e.g., the transfer matrix method. For uniform gratings with negligible LSHB effect and zero facet reflections, one can find the solution from equation (6.81) through, e.g., a root searching routine based on Muller's algorithm. The solutions for some specific structures have also been in published [19].

Since there have been numerous works published on the DFB laser design and simulation [20–22], we will not go into every detail in this book. Rather, through a few simulation examples, we will show the main features of some DFB lasers with different coupling mechanisms realized by different gratings.

In these examples, we will still take the previous active region and cross-sectional layer structure design, but with the purely index-coupled, in-phase partially gain-coupled, and loss-coupled grating structures inserted, respectively, as summarized in Table 10.4. All these DFB lasers are assumed to have ridge waveguides with identical ridge width at 2.5 μm and identical ridge height at 2.5 μm. While the gratings are formed at different layers with different etching depths, all these structures have the same grating period at 242 nm with duty cycle at 50%. They also have the same cavity length at 300 μm.

Table 10.4. Laser cross-sectional structures

The table sets out designs with different gratings in the InGaAsP/InP system at 1550 nm.

Layer		Thickness (nm)	Composition of $In_{(1-x)}Ga_x As_yP_{(1-y)}$ (x, y)	Doping concentration ($10^{18}/cm^3$)	Remarks λ_g (nm) and strain
Cap P-InGaAs		200	(0.468, 1.000)	10.0 (P)	1654, 0.0
Graded doped P-InP		2300	(0.000, 0.000)	1.0–0.5 (P)	918.6, 0.0
Etching stop InGaAsP		10	(0.108, 0.236)	0.5 (P)	1050, 0.0
Spacer InP		120	(0.000, 0.000)	0.3 (P)	918.6, 0.0
Grating	index-coupled	40	(0.280, 0.606)	undoped	1300, 0.0
	loss-coupled	40	(0.443, 0.949)	undoped	1600, 0.0
	gain-coupled	0	–	–	–
Spacer InP		10	(0.000, 0.000)	undoped	918.6, 0.0
*GRINSCH InGaAsP		15	(0.145, 0.317)	undoped	1100, 0.0
GRINSCH InGaAsP		15	(0.216, 0.469)	undoped	1200, 0.0
Barrier InGaAsP		10	(0.255, 0.553)	undoped	1260, 0.0
Well InGaAsP × 6		5	(0.189, 0.778)	undoped	1591, CS1.2%
Barrier InGaAsP × 6		10	(0.255, 0.553)	undoped	1260, 0.0
GRINSCH InGaAsP		20	(0.242, 0.525)	undoped	1240, 0.0
GRINSCH InGaAsP		20	(0.216, 0.469)	undoped	1200, 0.0
GRINSCH InGaAsP		20	(0.181, 0.395)	undoped	1150, 0.0
GRINSCH InGaAsP		20	(0.145, 0.317)	undoped	1100, 0.0
GRINSCH InGaAsP		20	(0.108, 0.236)	undoped	1050, 0.0
Buffer N-InP		500	(0.000, 0.000)	1.0 (N)	918.6, 0.0
Substrate N-InP		100 000	(0.000, 0.000)	3.0 (N)	918.6, 0.0

Purely index-coupled grating using the 1300 nm grating layer in 40 nm and etching through
loss-coupled grating using the 1600 nm grating layer in 40 nm and etching through
in-phase partially gain-coupled grating etching from the top GRINSCH layer on P side (marked with *) down to 80 nm until the center barrier layer and removing the 1300 nm grating layer.

To prevent a poor SMSR and mode hopping due to the dual mode operation, the rear and front facets of the purely index-coupled DFB laser are high-reflective (HR) and anti-reflective (AR) coated with amplitude reflectivities at 0.87 and 0.01, respectively. In the two complex-coupled DFB lasers, however, the facets are all as-cleaved with the amplitude reflectivity at 0.565, since they intrinsically work at a single mode, as can be seen from the solution to equation (6.81).

The purely index-coupled grating is made on the InGaAsP grating layer outside the GRINSCH region on the P side, with the bandgap wavelength at 1300 nm. The grating depth is the layer thickness at 40 nm.

The in-phase partially gain-coupled grating is made on top of the GRINSCH region on the P side, and etching is assumed down to 80 nm, hence half of the QWs (three) are removed periodically along the cavity. Since the modal gain in the sections with six QWs is higher than that in the sections with three QWs, in accordance with the argument in the beginning of Section 6.3.1, other than the index-coupling, a gain-coupling mechanism also appears. Since the section with a higher modal gain also has a higher effective index, the grating is in-phase partially gain-coupled.

The loss-coupled grating is the same as the purely index-coupled grating, except that the bandgap wavelength of the InGaAsP grating layer is set at 1600 nm. Due to its narrower bandgap compared with the lasing wavelength, the grating layer is also absorptive. Therefore, other than the index-coupling, a loss-coupling mechanism also exists, which makes this structure loss-coupled.

The calculated coupling coefficients are, for the purely index-coupled structure, 80.0 /cm, for the in-phase partially gain-coupled structure, 125 /cm with a gain coupling ratio of 25%, and for the loss-coupled structure, 128 /cm with a loss coupling ratio of 20%. Thus, the normalized coupling factor for the purely index-coupled structure is $\kappa L = 2.4$. For the in-phase partially gain-coupled structure, the normalized coupling factor is $\kappa L = 3.75 + 0.25(\mathrm{j} + \alpha_{\mathrm{LEF}})g_{\mathrm{m}}/\kappa$, where α_{LEF} indicates the linewidth enhancement factor [23], and g_{m} the modal gain, both bias dependent. For this particular structure, after lasing, the calculated modal gain is approximately fixed at 42.5 /cm, and consequently the fixed α_{LEF} is around -1.39. Thus, we approximately have $\kappa L = 3.75 + 0.25(\mathrm{j} + \alpha_{\mathrm{LEF}})g_{\mathrm{m}}/\kappa = 3.63 + \mathrm{j}0.085$ when the bias current is above the threshold. Actually, as the bias current increases, due to the growing LSHB effect, the averaged modal gain has to increase slightly to compensate its inefficient overlapping with the optical field along the cavity, which causes the real and imaginary parts of the normalized coupling factor to decrease slightly (due to the contribution of the negative α_{LEF}) and increase, respectively. Therefore, the ratio between the gain and index coupling increases with the bias current, which, as a well-known feature of the in-phase gain-coupled grating, helps the structure to maintain a high SMSR despite the increase of LSHB. Finally, for the loss-coupled structure, the normalized coupling factor is $\kappa L = 3.84 - \mathrm{j}0.029$. A major difference between the gain- and loss-coupled gratings is that the loss coupling is not bias dependent [24]. As a result, the loss-coupled grating only provides a mechanism that eliminates the lasing of the red edge mode on the Bragg stop-band without any self-tuning ability to adapt to a changed bias condition. Note that the loss-coupled DFB laser is different from the anti-phase partially gain-coupled DFB laser. The latter not only provides the index-coupling and out-of-phase gain-coupling mechanisms simultaneously, but also adapts its gain-to-index coupling ratio to the changed bias condition, as the imaginary part of its coupling coefficient is bias dependent [25–27]. This structure can be realized, e.g., through a periodically etched reversely doped grating layer near the active region.

Figure 10.18 shows the calculated output power–bias current dependences for these DFB lasers. The purely index-coupled DFB laser has higher differential efficiency due to its smaller $|\kappa L|$. Its output powers from the front and rear facets are different due to the asymmetric coating. Compared with the loss-coupled DFB laser, the in-phase partially gain-coupled DFB laser has slightly higher differential efficiency as its $|\kappa L|$ is slightly higher; besides, its modal loss is slightly lower. Finally, we find that, as the bias current increases, the differential efficiency of the in-phase partially gain-coupled DFB laser slightly increases. This can be attributed to the minor reduction of its $|\kappa L|$. Actually, since α_{LEF} is negative, as the bias current increases, the reduction in the real part of κL cannot be offset by the increase in its imaginary part. Therefore, $|\kappa L|$ still decreases as the bias increases, which makes the output power a superlinear function of the bias current. This

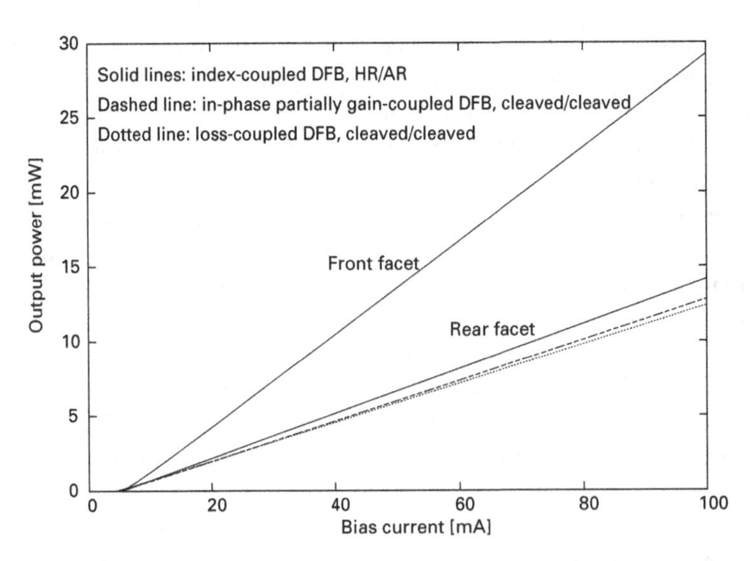

Fig. 10.18. Comparison of the output power–bias current dependences between DFB lasers with different coupling mechanisms.

effect will be more pronounced if $|\alpha_{\mathrm{LEF}}|$ is larger or the LSHB is more strongly dependent on the bias. It is apparent that we will reach the opposite conclusion if $\alpha_{\mathrm{LEF}} > 0$, or we have an anti-phase partially gain-coupled structure with $\alpha_{\mathrm{LEF}} < 0$. Knowing this feature, we must rule out the possibility of using complex-coupled DFB lasers in applications where direct analog modulations are employed, such as in community antenna television (CATV) or wireless signal fiber-optic transmission networks, due to their poor linearity in the output power–bias current dependence.

Figure 10.19 shows the calculated side mode suppression ration (SMSR)–bias current dependences for these DFB lasers. We find that, while complex-coupled DFB lasers have excellent SMSR, the purely index-coupled DFB laser's SMSR is marginal, although its facets are HR and AR coated. This could be partially due to its relatively low $|\kappa L|$. However, as we raise the normalized coupling factor, the LSHB also becomes severe. Under a random grating phase at the HR coated rear facet, the yield distribution would be more random. The fact that the in-phase partially gain-coupled DFB laser has an even higher SMSR than the loss-coupled DFB laser, although the former has a slightly lower $|\kappa L|$, can be attributed to its higher gain-to-index coupling ratio, as we know that the threshold gain difference between the first and second modes increases with this ratio [19].

Figure 10.20 shows the calculated optical spectra for these DFB lasers with their bias currents all set to 50 mA. Due to the InP filled, deeply-etched grating and the removal of the top 1300 nm InGaAsP grating layer, the in-phase partially gain-coupled DFB structure has the lowest effective index. Hence it has the shortest Bragg wavelength, as we have set the same grating period for all these structures. Regarding the loss-coupled DFB laser, it has the longest Bragg wavelength as its effective index is the highest, because the original InGaAsP grating layer in the purely index-coupled structure is replaced by a

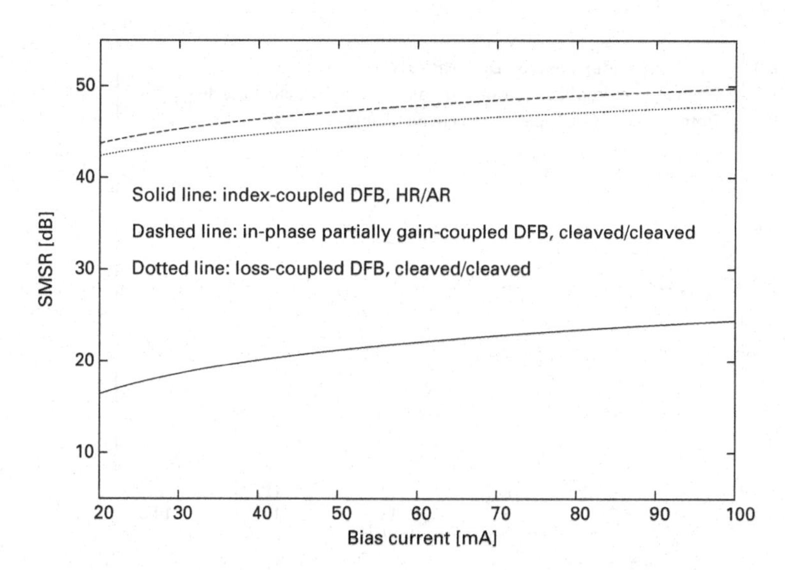

Fig. 10.19. Comparison of the SMSR–bias current dependences between DFB lasers with different coupling mechanisms.

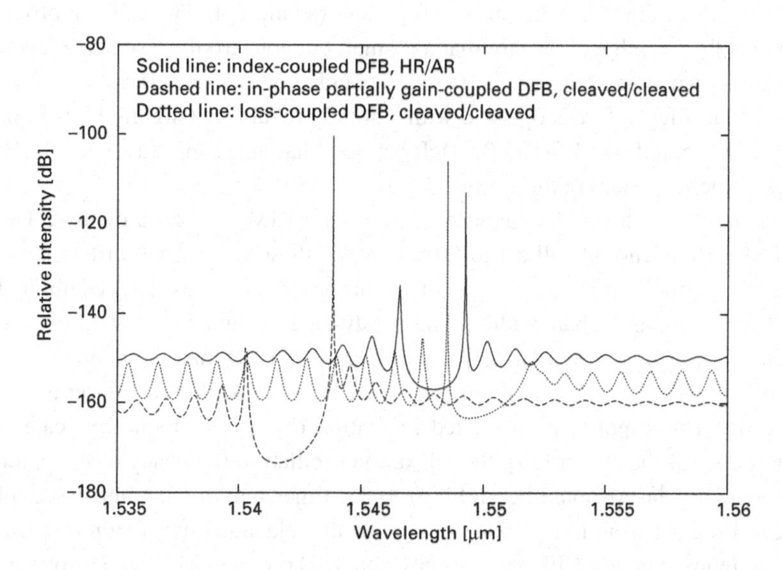

Fig. 10.20. Comparison of the optical spectra between DFB lasers with different coupling mechanisms.

1600 nm InGaAsP layer, which has a higher refractive index. We also find that the Bragg stop-band (not the unbiased waveguide Bragg stop-band, but the stop-band appearing in the spectrum with bias) widths of these DFB lasers are proportional to their $|\kappa L|$ values. Finally, it is again confirmed that the in-phase partially gain-coupled DFB laser has its lasing wavelength at the red edge of the stop-band (not 100%, as the probability of lasing at the blue edge of the stop-band is small but not zero if we scan the facet phases [28]), whereas the loss-coupled (or anti-phase partially gain-coupled) DFB laser has its lasing

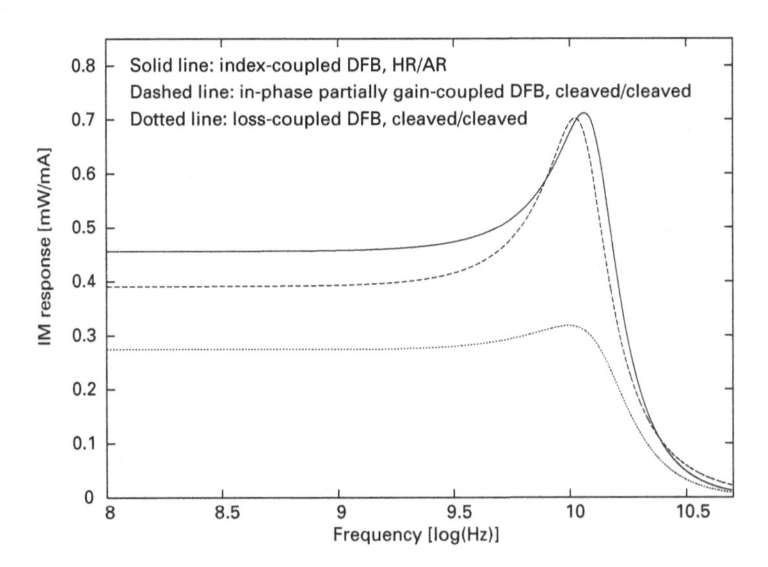

Fig. 10.21. Comparison of the small-signal IM responses between DFB lasers with different coupling mechanisms.

wavelength at the blue edge of the stop-band (again, not 100%, as the probability of lasing at the red edge of the stop-band is small but not zero if we scan the facet phases). The lasing wavelength of the purely index-coupled DFB laser is randomly located at the red or blue edges of the stop-band with a 50:50 probability (slightly higher on the blue edge if we include the LSHB effect). It happens that the lasing wavelength is on the red edge of the stop-band in this particular case.

Figure 10.21 shows the calculated small-signal IM responses of these DFB lasers with their bias currents all set to 50 mA. Since these lasers have different differential efficiencies, their output powers are different for the same bias. The seemingly broadest 3 dB small-signal IM bandwidth in the purely index-coupled DFB laser comes from its highest power output at 50 mA. We find from Fig. 10.18 that the power ratio between the purely index-coupled DFB laser and the complex-coupled DFB laser is about 14/6. Since the IM response peak related relaxation oscillation frequency scales with the square root of the power [29], the relaxation oscillation frequency of the purely index-coupled DFB laser should be $\sqrt{(14/6)} \approx 1.5$ times that of the complex-coupled DFB laser. However, from Fig. 10.21, we find that the relaxation oscillation frequency of the purely index-coupled DFB laser is only about 1.26 times that of the in-phase partially gain-coupled DFB laser (i.e., 12.6 GHz to 10 GHz). We conclude that the relaxation oscillation frequency of the in-phase partially gain-coupled DFB laser is highest if the IM responses are measured under the same output power. Consequently, the in-phase partially gain-coupled DFB laser has the highest 3 dB small-signal IM bandwidth. Since the two complex-coupled DFB lasers have about the same output power at 50 mA according to Fig. 10.18, we find from Fig. 10.21 that the in-phase partially gain-coupled DFB laser has a higher 3 dB small-signal IM bandwidth than the loss-coupled DFB laser. While the two complex-coupled DFB lasers have about the same relaxation oscillation frequency, the loss-coupled DFB laser has an obviously higher damping rate.

10.3.3 Lasers with multiple section designs

Despite the many advantages of DFB lasers compared with the FP laser, the uniform grating purely index-coupled DFB laser intrinsically operates at dual modes, due to the imbedded symmetry in its structure as discussed in Section 6.3. Although this rarely happens in reality, due to the symmetry breaking by the asymmetric facet reflections or grating phases and the LSHB, the SMSR is poor and mode hopping may appear at the red and blue edges of the Bragg stop-band. Asymmetric AR and HR coatings at the front and rear facets help to raise the SMSR and solve the mode hopping problem, but there is still no precise control on the lasing wavelength. The probability of lasing at the red and blue edges of the Bragg stop-band is 50:50 due to the difficulty of cleaving the device facet at a deterministic grating phase. In applications where precise lasing wavelengths are required, e.g., in dense wavelength division multiplexing (DWDM) systems and in monolithic integrations, the yield becomes a major issue. Although complex-coupled DFB lasers have this problem solved, their reliability could be a concern in making products. Therefore, the quarter-wavelength phase-shifted grating DFB structure [30] and a variety of its derived versions, such as the single [31] or dual octant-wavelength phase-shifted grating DFB structure, the partially corrugated DFB structure [32], and the quarter-wavelength phase-shifted asymmetric grating DFB structure, etc., have been proposed with intrinsic single mode operation, high SMSR and precise control on the lasing wavelength.

Here we consider a few examples of such DFB lasers with multiple sections. Their active region and cross-sectional layer structure designs are the same as the purely index-coupled single section (uniform grating) DFB laser detailed in Table 10.4. All these multiple section DFB lasers are still assumed to have ridge waveguides with identical ridge width at 2.5 μm and identical ridge height at 2.5 μm. All the gratings are identical to the one used for the single section purely index-coupled DFB laser, i.e., with a period of 242 nm and a duty cycle of 50%. All these DFB lasers have identical total cavity length at 400 μm. In the quarter-wavelength phased-shifted DFB laser, there are two identical sections at an equal length of 200 μm with a 180° grating phase shift in between. The device rear and front facets are AR and AR coated with identical reflections at 1%. In the dual octant-wavelength phase-shifted DFB laser, there are three identical sections in different lengths. The two equal-length edge sections are 150 μm, the intermediate section is 100 μm. There are two identical 90° grating phase shifts from section to section. Again, the device rear and front facets are AR and AR coated with identical reflections at 1%. The partially corrugated DFB laser comprises two sections, a conventional 100 μm long FP section with identical cross-sectional structure but without the grating, and a 300 μm long DFB section. The device rear (FP side) and front (DFB side) facets are HR and AR coated with amplitude reflectivities at 0.87 and 0.01, respectively. This device is actually the previous purely index-coupled DFB laser with an extended FP section of 100 μm inside the cavity.

Figure 10.22 shows the calculated output power–bias current dependences for these DFB lasers. The partially corrugated DFB laser has the smallest κL (2.4), hence it has the highest differential gain. Moreover, almost doubled output power can be collected

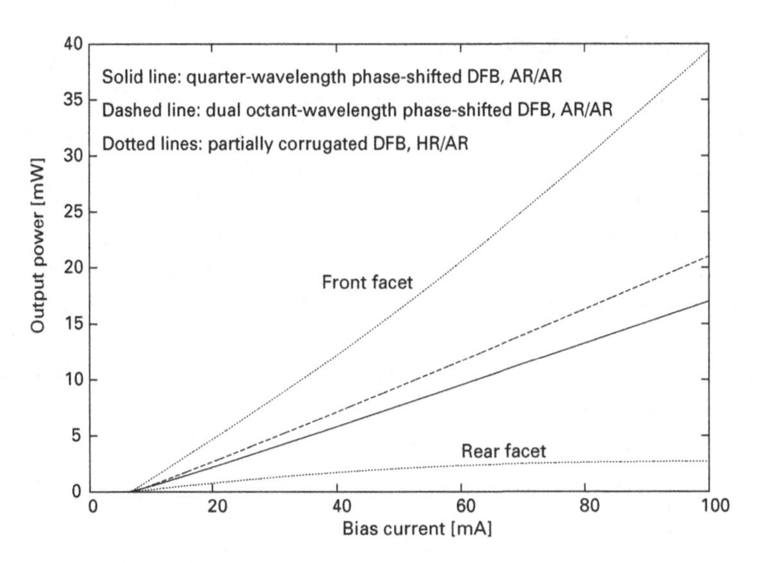

Fig. 10.22. Comparison of the output power–bias current dependences between DFB lasers with different multiple section designs.

from its front facet as the device is HR and AR coated. The facets of the other two DFB lasers, however, have to be AR and AR coated. As a result, half of their powers are simply wasted, as the monitoring PD usually does not need such high power. This is a major drawback of the multiple section DFB lasers with symmetric phase-shifted gratings.

Figure 10.23 shows the calculated side mode suppression ration (SMSR)–bias current dependences for these DFB lasers. All these structure have excellent SMSR without significant difference.

Figure 10.24 shows the calculated optical spectra for these DFB lasers. While the quarter-wavelength phase-shifted DFB laser has its lasing wavelength precisely positioned at the Bragg wavelength, the dual octant-wavelength phase-shifted DFB laser has its lasing wavelength located inside the Bragg stop-band, with its position fixed once the device structure (i.e., the coupling coefficient and each section length) is fixed. Although in both structures their lasing wavelengths still shift with the operating condition, i.e., the bias and the ambient temperature, it is the best we can achieve: with symmetric phase-shifted DFB structures, the lasing wavelength is deterministic at given operating conditions, hence there is not a yield problem from the device design point of view. The lasing wavelength of the partially corrugated DFB laser, however, may appear at the blue or red edge, or even inside the Bragg stop-band. It is jointly determined by the section length and the grating phase between the DFB and FP sections. While the grating phase between the DFB and FP sections is probably controllable, the length of the FP section is still random. An FP length change in a quarter-wavelength, i.e., $\lambda/(4n_{\text{eff}})$, or in the sub-micron range is sufficient to bring in a field reflection with opposite phase and change the oscillation condition. Thus, small fluctuations in the FP section length still cause a yield problem in production.

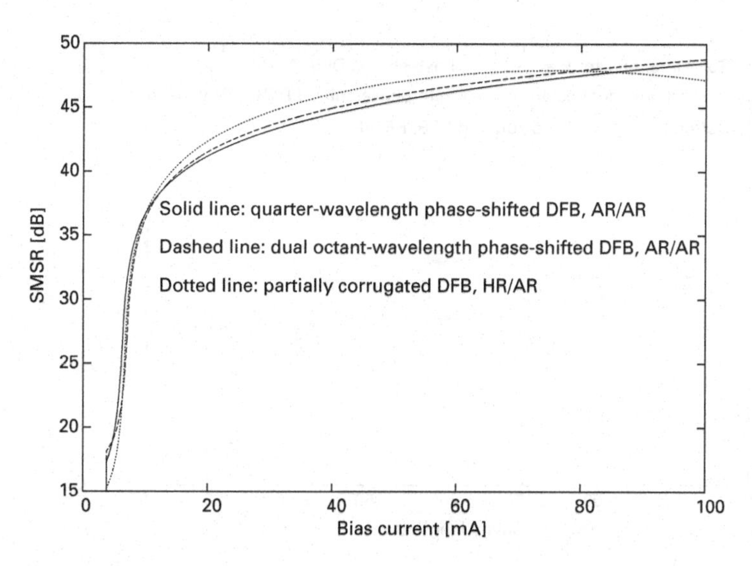

Fig. 10.23. Comparison of the SMSR–bias current dependences between DFB lasers with different multiple section designs.

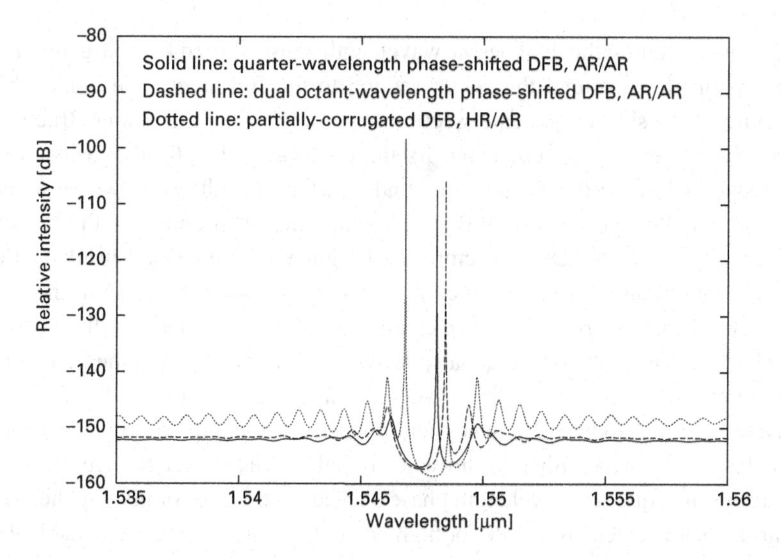

Fig. 10.24. Comparison of the optical spectra between DFB lasers with different multiple section designs.

Figure 10.25 shows the calculated optical field distributions along the cavity for these DFB lasers. It is apparent that the quarter-wavelength phase-shifted DFB laser has a strong LSHB near the center region where the grating has the phase shift. For this reason, the dual octant-wavelength phase-shifted DFB laser is proposed to reduce this effect. Actually, as can be seen from this figure, the field distribution in the dual octant-wavelength phase-shifted DFB laser is indeed flatter.

In summary, we conclude that, if we view the purely index-coupled single section (uniform grating) DFB structure and the quarter-wavelength phase-shifted DFB structure

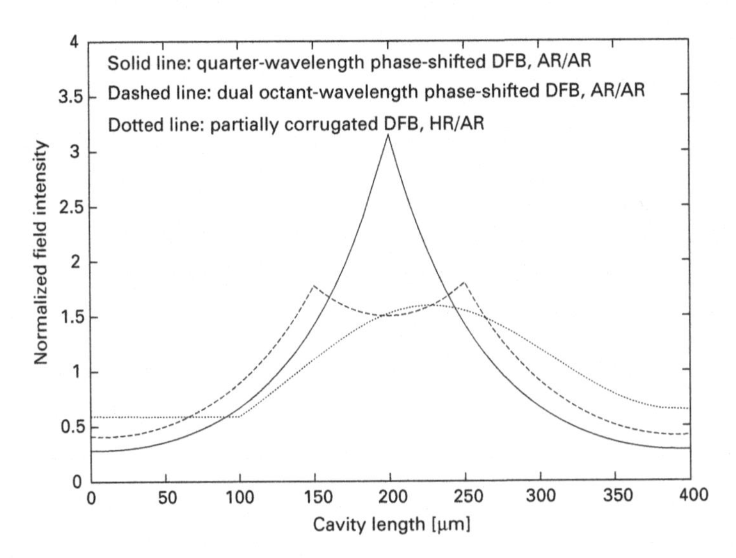

Fig. 10.25. Comparison of the optical field distributions along the cavity between DFB lasers with different multiple section designs.

as the two extremes, the dual octant-wavelength phase-shifted DFB structure is a design in between: if we extend the center section to the full cavity, we obtain the former structure; if we shrink the center section to zero, we obtain the latter structure. As we move the center section length but fix the total cavity length, the lasing wavelength must sweep through the Bragg stop-band. On the other hand, if we view the purely index-coupled single section DFB structure and the FP structure as the two extremes, the partially corrugated DFB structure is a design in between: if we shrink the FP section to zero, we obtain the former structure; if we extend the FP section to the full cavity, we obtain the latter structure. As we extend the FP section length but fix the total cavity length, the Bragg stop-band gradually shrinks and submerges into the FP comb.

Without the complex coupling involved, there seems to be a trend that we cannot achieve both high power efficiency and precise wavelength control simultaneously in DFB lasers. We have either to throw away half of the power for precise wavelength control (as in a quarter-wavelength phase-shifted DFB laser), or give up the wavelength control to some extent as a trade for high power (as in a partially corrugated DFB laser). Whether it is possible to realize a design with both precise wavelength control at 100% yield and full power efficiency is still an open problem.

References

[1] G. P. Agrawal and N. K. Dutta, *Long-wavelength Semiconductor Lasers*, 1st edn (New York: Van Nostrand Reinhold, 1986).

[2] C.-E. Zah, R. B. Bhat, B. N. Pathak, *et al.*, High-performance uncooled 1.3-μm $Al_xGa_yIn_{(1-x-y)}As/InP$ strained-layer quantum-well lasers for subscriber loop applications. *IEEE Journal of Quantum Electron.*, **QE-30**:2 (1994), 511–23.

[3] M. Kondow, K. Uomi, A. Niwa, *et al.*, GaInNAs: a novel material for long-wavelength-range laser diodes with excellent high-temperature performance. *Journal of Appl. Phys. Jpn.*, **35** (1996), 1273–5.

[4] C. W. Tu, W. G. Bi, Y. Ma, *et al.*, A novel material for long-wavelength lasers: InNAsP. *IEEE Journal of Select. Topics in Quantum Electron.*, **4**:3 (1998), 510–3.

[5] L. A. Coldren and S. W. Corzine, *Diode Laser and Photonic Integrated Circuits*, 1st edn (New York: John Wiley & Sons, 1995).

[6] P. Zory, (Ed.) *Quantum Well Lasers*, 1st edn (London: Academic Press, 1993).

[7] E. Kapon, (Ed.) *Semiconductor Laser I: Fundamentals*, 1st edn (London: Academic Press, 1993).

[8] N. Tessler and G. Eisenstein, Transient carrier dynamics and photon-assisted transport in multiple-quantum-well lasers. *IEEE Photon. Tech. Lett.*, **5**:3 (1993), 291–3.

[9] S. W. Koch, J. Hader, A. Tranhardt, and J. V. Moloney, Gain and absorption: many-body effects. Chapter 1 in *Optoelectronic Devices: Advanced Simulation and Analysis*, ed. J. Piprek. 1st edn (New York: Springer Science + Business Media, 2005).

[10] J. Hader, J. V. Moloney, and S. E. Koch, Microscopic theory of gain, absorption, and refractive index in semiconductor laser materials – influence of conduction-band nonparabolicity and Coulomb-induced intersubband coupling. *IEEE Journal of Quantum Electron.*, **QE-35**:11 (1999), 1878–86.

[11] J. Hader, A. R. Zakharian, J. V. Moloney, *et al.*, Quantitative prediction of semiconductor laser characteristics based on low intensity photoluminescence measurements. *IEEE Photon. Tech. Lett.*, **14**:6 (2002), 762–4.

[12] S. R. Chinn, P. S. Zory, and A. R. Reisinger, A model for GRIN-SCH-SQW diode lasers. *IEEE Journal of Quantum Electron.*, **QE-24**:11 (1988), 2191–214.

[13] I. K. Han, S. H. Cho, P. J. S. Heim, *et al.*, Dependence of the light-current characteristics of 1.55μm broad-area lasers on different p-doping profiles. *IEEE Photon. Tech. Lett.*, **12**:3 (2000), 251–3.

[14] G. -L. Tan, R. S. Mand, and J. M. Xu, Self-consistent modeling of beam instability in 980 nm fiber pump lasers. *IEEE Journal of Quantum Electron.*, **QE-33**:8 (1997), 1384–95.

[15] X. Li, Distributed feedback lasers: quasi-3D static and dynamic model. Chapter 4 in *Optoelectronic Devices: Advanced Simulation and Analysis*, ed. J. Piprek. 1st edn (New York: Springer Science + Business Media, 2005).

[16] M. Aoki, M. Komori, T. Tsuchiya, *et al.*, InP-based reverse-mesa ridge-waveguide structure for high-performance long-wavelength laser diodes. *IEEE Journal of Select. Topics in Quantum Electron.*, **3**:2 (1997), 672–83.

[17] Y. Arakawa and A. Yariv, Theory of gain, modulation response, and spectral linewidth in AlGaAs quantum well lasers. *IEEE Journal of Quantum Electron.*, **QE-21**:10 (1985), 1666–74.

[18] S. L. Chuang, *Physics of Optoelectronic Devices*, 1st edn (New York: John Wiley & Sons, 1995).

[19] E. Kapon, A. Hardy, and A. Katzir, The effect of complex coupling coefficients on distributed feedback lasers. *IEEE Journal of Quantum Electron.*, **QE-18**:1 (1982), 66–71.

[20] J. Carroll, J. Whiteaway, and D. Plumb, *Distributed Feedback Semiconductor Lasers*, 1st edn (London: IEE, 1998).

[21] G. Morthier and P. Vankwikelberge, *Handbook of Distributed Feedback Laser Diodes*, 1st edn (Norwood, MA: Artech House, 1997).

[22] H. Ghafouri-Shiraz and B. S. K. Lo, *Distributed Feedback Laser Diodes: Principles and Physical Modeling*, 1st edn (New York: John Wiley & Sons, 1996).

[23] C. Henry, Theory of the linewidth of semiconductor lasers. *IEEE Journal of Quantum Electron.*, **QE-18**:2 (1982), 259–64.

[24] S. Kuen and C. Liew, Above-threshold analysis of loss-coupled DFB lasers: threshold current and power efficiency. *IEEE Photon. Tech. Lett.*, **7**:12 (1995), 1400–2.

[25] K. Kudo, J. I. Shim, K. Komori, and S. Arai, Reduction of effective linewidth enhancement factor α_{eff} of DFB lasers with complex coupling coefficients. *IEEE Photon. Tech. Lett.*, **4**:6 (1992), 531–4.

[26] A. J. Lowery and D. Novak, Enhanced maximum intrinsic modulation bandwidth of complex-coupled DFB semiconductor lasers. *Electron. Lett.*, **29**:5 (1993), 461–2.

[27] J. Hong, T. Makino, H. Lu, and G. P. Li, Effect of in-phase and antiphase gain coupling on high-speed properties of MQW DFB lasers. *IEEE Photon. Tech. Lett.*, **7**:9 (1995), 956–8.

[28] J. Hong, K. W. Leong, T. Makino, X. Li, and W.-P. Huang, Impact of random facet phases on modal properties of partly gain-coupled distributed-feedback lasers. *IEEE Journal of Select. Topics in Quantum Electron.*, **3**:2 (1997), 555–68.

[29] K. Petermann, *Laser Diode Modulation and Noise*, 1st edn (Dordrecht, The Netherlands: Kluwer Academic Publishers, 1991).

[30] K. Utaka, S. Akiba, K. Sakai, and Y. Matsushima, Analysis of quarter-wave-shifted DFB laser, *Electron. Lett.*, **20**:8 (1984), 326–7.

[31] Y. Huang, K. Sato, T. Okuda, *et al.*, Low-chirp and external optical feedback resistant characteristics in $\lambda/8$ phase-shifted distributed feedback laser diodes under direct modulation. *IEEE Journal of Quantum Electron.*, **QE-38**:11 (2002), 1479–84.

[32] Y. Huang, T. Okuda, K. Shiba, *et al.*, External optical feedback resistant 2.5 Gb/s transmission of partially corrugated waveguide laser diodes over a $-40\,°C$ to $80\,°C$ temperature range. *IEEE Photon. Tech. Lett.*, **11**:11 (1999), 1482–4.

Further reading

On laser design

[1] P. J. A. Thijs, L. F. Tiemeijer, J. J. M. Binsma, and Teus van Dongen, Progress in long-wavelength strained-layer InGaAs(P) quantum-well semiconductor lasers and amplifiers. *IEEE Journal of Quantum Electron.*, **QE-30**:2 (1994), 477–99.

[2] E. P. O'Reilly and A. R. Adams, Band-structure engineering in strained semiconductor lasers. *IEEE Journal of Quantum Electron.*, **QE-30**:2 (1994), 366–79.

[3] M. Silver and E. P. O'Reilly, Optimization of long wavelength InGaAsP strained quantum-well lasers. *IEEE Journal of Quantum Electron.*, **QE-31**:7 (1995), 1193–200.

[4] S. R. Selmic, T. -M. Chou, J. P. Sih, *et al.*, Design and characterization of 1.3 μm AlGaInAs-InP multiple-quantum-well lasers. *IEEE Journal of Select. Topics in Quantum Electron.*, **7**:2 (2001), 340–9.

[5] J. Piprek, J. K. White, and A. J. Spring thorpe, What limits the maximum output power of long-wavelength AlGaInAs/InP laser diodes? *IEEE Journal of Quantum Electron.*, **QE-38**:9 (2002), 1253–9.

[6] G. L. Belenky, C. L. Reybolds, Jr., R. F. Kazarinov, *et al.*, Effect of p-doping profile on performance of strained multi-quantum-well InGaAsP-InP lasers. *IEEE Journal of Quantum Electron.*, **QE-32**:8 (1996), 1450–5.

[7] T. Namegaya, N. Matsumoto, N. Yamanaka, *et al.*, Effects of well number in 1.3 μm GaInAsP/InP GRIN-SCH strained-layer quantum-well lasers. *IEEE Journal of Quantum Electron.*, **QE-30**:2 (1994), 578–84.

[8] M. J. Hamp, D. T. Cassidy, B. J. Robinson, Q. C. Zhao, and D. A. Thomson, Nonuniform carrier distribution in asymmetric multiple-quantum-well InGaAsP laser structure with different numbers of quantum wells. *Appl. Phys. Lett.*, **74**:5 (1999), 744–6.

[9] M. J. Hamp, D. T. Cassidy, B. J. Robinson, *et al.*, Effect of barrier height on the uneven carrier distribution in asymmetric multiple-quantum-well InGaAsP lasers. *IEEE Photon. Tech. Lett.*, **10**:10 (1998), 1380–2.

[10] X. Zhu, D. T. Cassidy, M. J. Hamp, *et al.*, 1.4 μm InGaAsP-InP strained multiple-quantum-well laser for broad-wavelength tenability. *IEEE Photon. Tech. Lett.*, **9**:9 (1997), 1202–4.

[11] M. Kito, N. Otsuka, M. Ishino, and Y. Matsui, Barrier composition dependence of differential gain and external differential quantum efficiency in 1.3 μm strained-layer multiquantum-well lasers. *IEEE Journal of Quantum Electron.*, **QE-32**:1 (1996), 38–42.

[12] G. Morthier, Design and optimization of strained-layer-multiquantum-well lasers for high-speed analog communications. *IEEE Journal of Quantum Electron.*, **QE-30**:7 (1994), 1520–8.

[13] Y. Lam, J. P. Loehr, and J. Singh, Effects of strain on the high speed modulation of GaAs- and InP-based quantum-well lasers. *IEEE Journal of Quantum Electron.*, **QE-29**:1 (1993), 42–50.

[14] H. Hirayama, Y. Miyake, and M. Asada, Analysis of current injection efficiency of separate-confinement-heterostructure quantum-film lasers. *IEEE Journal of Quantum Electron.*, **QE-28**:1 (1992), 68–74.

[15] S. Y. Hu, D. B. Young, A. C. Gossard, and L. A. Coldren, The effect of lateral leakage current on the experimental gain/current-density curve in quantum-well ridge-waveguide lasers. *IEEE Journal of Quantum Electron.*, **QE-30**:10 (1994), 2245–50.

[16] M. Amann and W. Thulke, Current confinement and leakage currents in planar buried-ridge-structure laser diode on n-substrate. *IEEE Journal of Quantum Electron.*, **QE-25**:7 (1989), 1595–602.

[17] S. F. Yu and E. H. Li, Effects of lateral modes on the static and dynamic behavior of buried heterostructure DFB lasers. *IEE Proc. Optoelectron.*, **142**:2 (1995), 97–102.

[18] T. Ohtoshi, K. Yamaguchi, and N. Chinone, Analysis of current leakage in InGaAsP/InP buried heterostructure lasers. *IEEE Journal of Quantum Electron.*, **QE-25**:6 (1989), 1369–74.

11 Design and modeling examples of other solitary optoelectronic devices

11.1 The electro-absorption modulator

11.1.1 The device structure

The schematic structure of electro-absorption modulators (EAMs) with shallowly-etched (left) and deeply-etched (right) ridge waveguides is shown in Fig. 11.1.

The cross-sectional layer structure of a typical EAM operated in the 1550 nm wavelength band is shown in Table 11.1.

The device ridge waveguide has width 1.5 μm and height 2.0 μm and 3.0 μm in the shallow and deep ridge design, respectively. Thus, the etched ditch stops on top of the active region in the shallow ridge design but goes through the active region in the deep ridge design, as shown schematically in Fig. 11.1. The length of the EAM is set to be 200 μm in device performance modeling.

11.1.2 Simulated material properties and device performance

Figure 11.2 shows the calculated energy band edge diagrams of the device active region under different reverse bias voltages. Due to the compressive strain applied to the QWs, the heavy-hole and light-hole band edges in the QW material split, with the heavy-hole band edge on top.

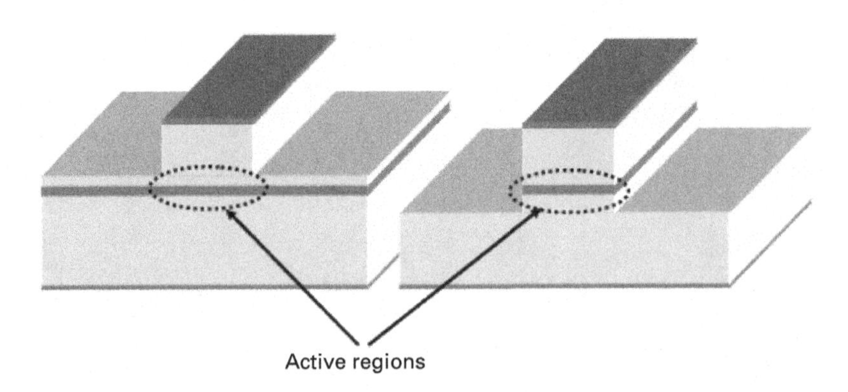

Fig. 11.1. Shallowly etched (left) and deeply etched (right) ridge waveguide structures in EAMs.

Table 11.1. EAM cross-sectional structure

The table sets out a design in the InGaAsP/InP system at 1550 nm.

Layer	Thickness (nm)	Composition of $In_{(1-x)}Ga_xAs_yP_{(1-y)}$ (x, y)	Doping concentration ($10^{18}/cm^3$)	Remarks λ_g (nm) and strain
Cap P-InGaAs	200	(0.468, 1.000)	10.0 (P)	1654, 0.0
Cladding P-InP	1800	(0.000, 0.000)	0.5 (P)	918.6, 0.0
Etching stop	10	(0.108, 0.236)	0.5 (P)	1050, 0.0
Spacer InP	50	(0.000, 0.000)	undoped	918.6, 0.0
SCH InGaAsP	100	(0.216, 0.469)	undoped	1200, 0.0
Barrier	6	(0.196, 0.343)	undoped	1100, TS0.27%
Well × 10	9	(0.356, 0.871)	undoped	1574, CS0.35%
Barrier × 10	6	(0.196, 0.343)	undoped	1100, TS0.27%
SCH InGaAsP	80	(0.216, 0.469)	undoped	1200, 0.0
Buffer N-InP	600	(0.000, 0.000)	1.0 (N)	918.6, 0.0
Etching stop	10	(0.108, 0.236)	0.5 (P)	1050, 0.0
Substrate N-InP	100 000	(0.000, 0.000)	3.0 (N)	918.6, 0.0
Modified well	9	(0.318, 0.826)	undoped	1545, CS0.46%

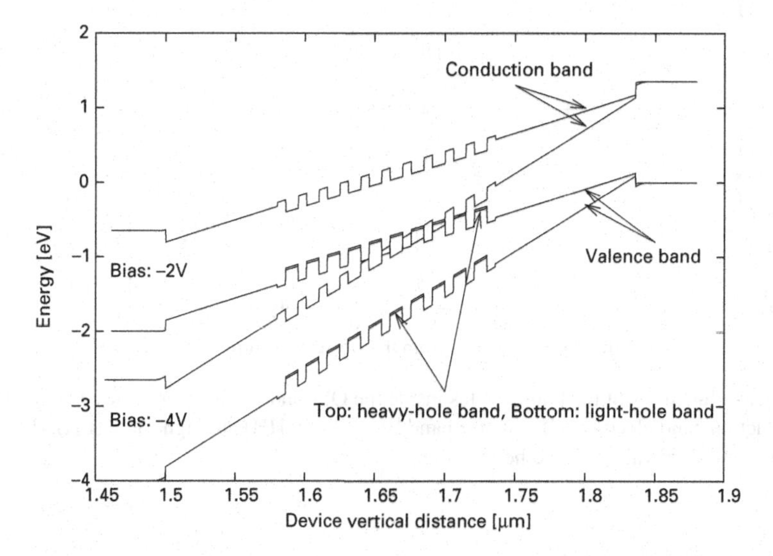

Fig. 11.2. Energy band diagrams along the device vertical direction under different reverse bias voltages.

Figures 11.3(a&b) show the calculated conduction band electron and valence band heavy-hole and light-hole wave functions under 1 V reverse bias.

Figures 11.4(a&b) give the calculated material absorptions for the TE and TM modes under different reverse bias voltages, respectively.

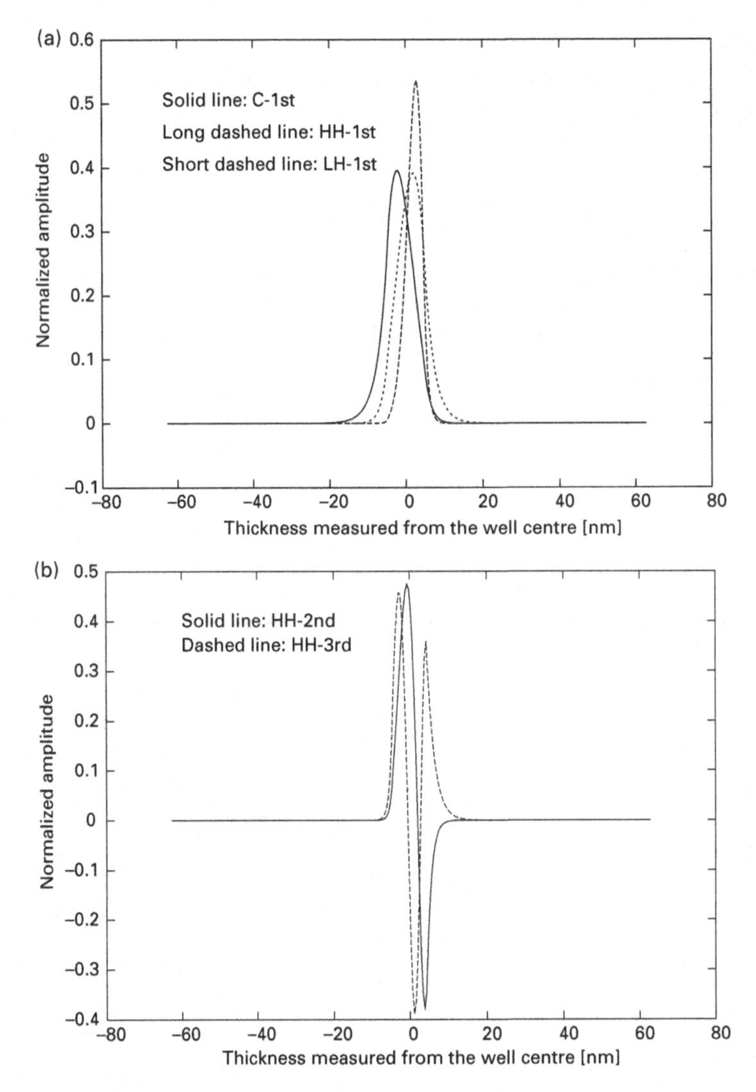

Fig. 11.3. The wave functions of the bound states inside the QW under 1 V reverse bias. (a) The first conduction band electron (C), valence band heavy-hole (HH) and light-hole (LH). (b) The second and third valence band heavy-hole.

Figures 11.5(a&b) give the calculated refractive index changes for the TE and TM modes under different reverse bias voltages, respectively, with the refractive index change at zero bias taken as the reference.

Figures 11.6(a&b) show the calculated shallow ridge EAM insertion loss under zero bias, and the extinction ratios under different reverse bias voltages for the TE and TM modes, respectively.

Figures 11.7(a&b) show the calculated shallow ridge EAM chirp parameters under different reverse bias voltages for the TE and TM modes, respectively. The chirp parameter

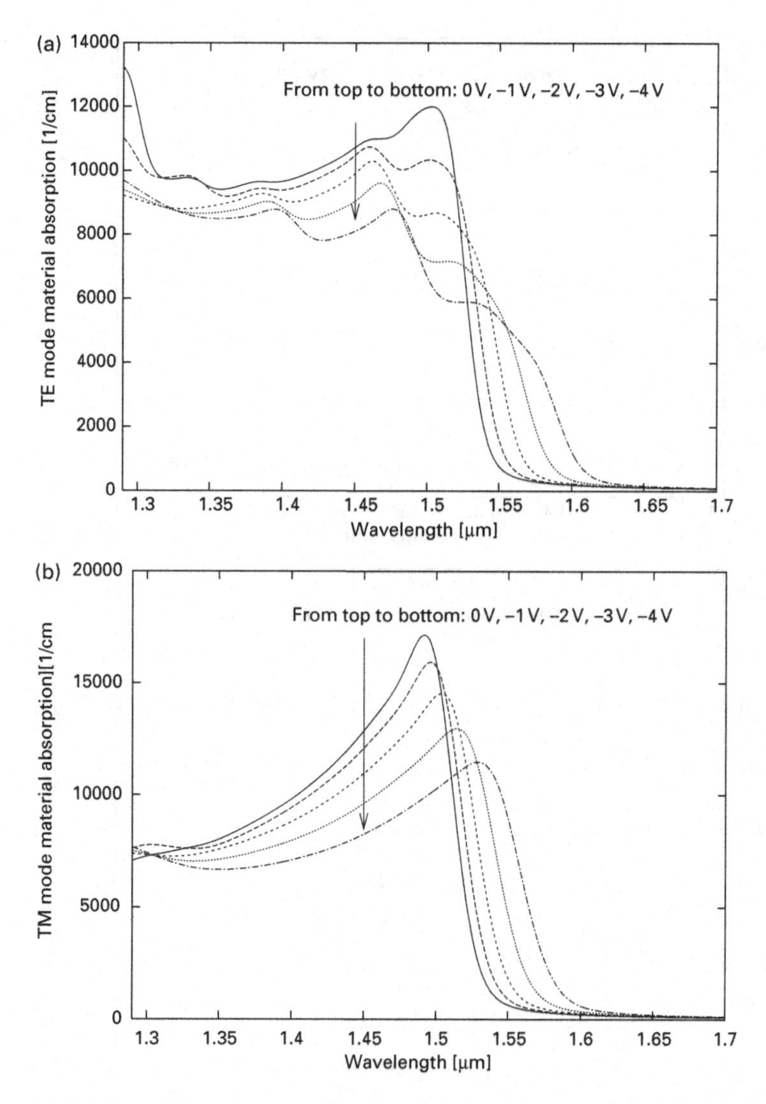

Fig. 11.4. Material absorption spectra under different reverse bias voltages. (a) For the TE mode. (b) For the TM mode.

is defined as

$$C(\lambda) \equiv \frac{4\pi}{\lambda} \frac{\Delta n(\lambda, V)}{\alpha(\lambda, V) - \alpha(\lambda, 0)}$$

with $\Delta n(\lambda, V)$ and $\alpha(\lambda, V)$ indicating the device modal refractive index change and modal absorption under bias voltage V at wavelength λ. Through these curves, we can clearly recognize the exciton absorption peak positions, as sharp turns of the chirp parameter appear at the peak wavelengths. These curves also show that considerable chirps occur at the transparent edge of the EAM (i.e., for wavelengths longer than 1.65 μm), which suggests that an efficient phase modulator can be built in this region.

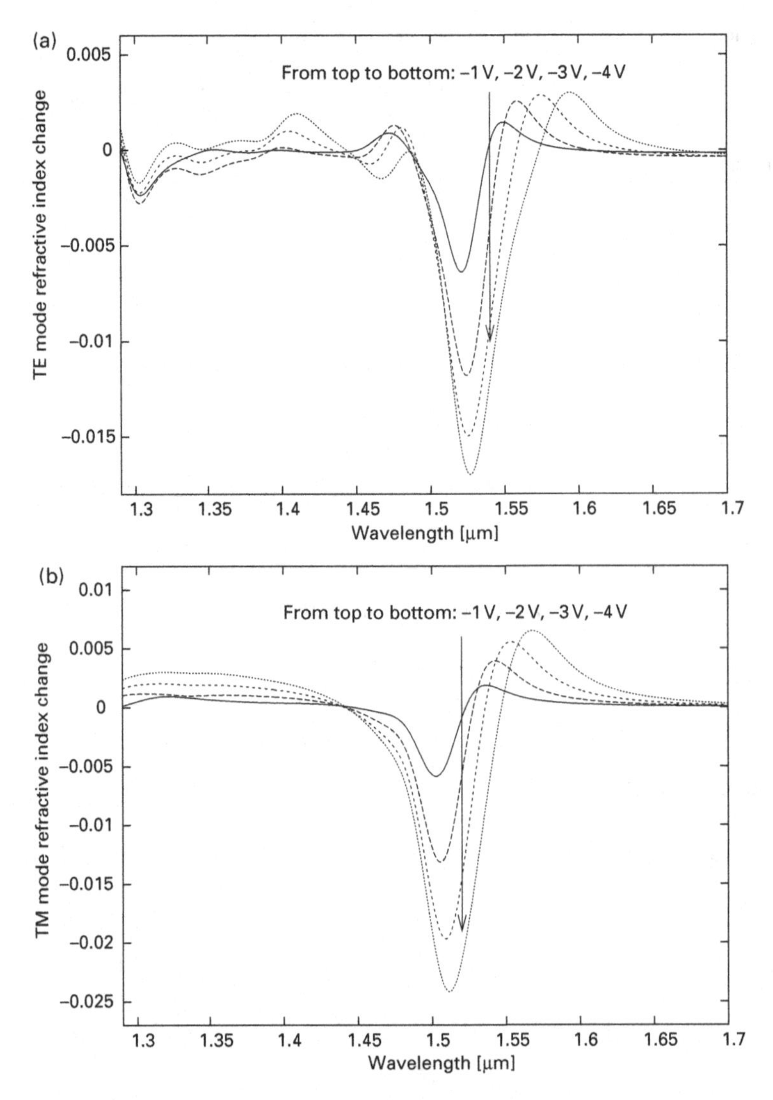

Fig. 11.5. Refractive index change spectra under different reverse bias voltages. (a) For the TE mode. (b) For the TM mode.

11.1.3 Design for high extinction ratio and low insertion loss

In the shallow ridge waveguide design, the light extended outside the ridge along the horizontal direction overlaps with the unbiased active region, since the active region is not etched through the outside of the ridge. Therefore, as light passes through the EAM, it has a poor confinement and suffers constant absorption in the unbiased active region outside the ridge. The former reduces the extinction ratio while the latter increases the insertion loss.

Since the EAM always works under reverse bias, the carrier density will never be high inside the active region. Thus, exposure of the active region side walls will not

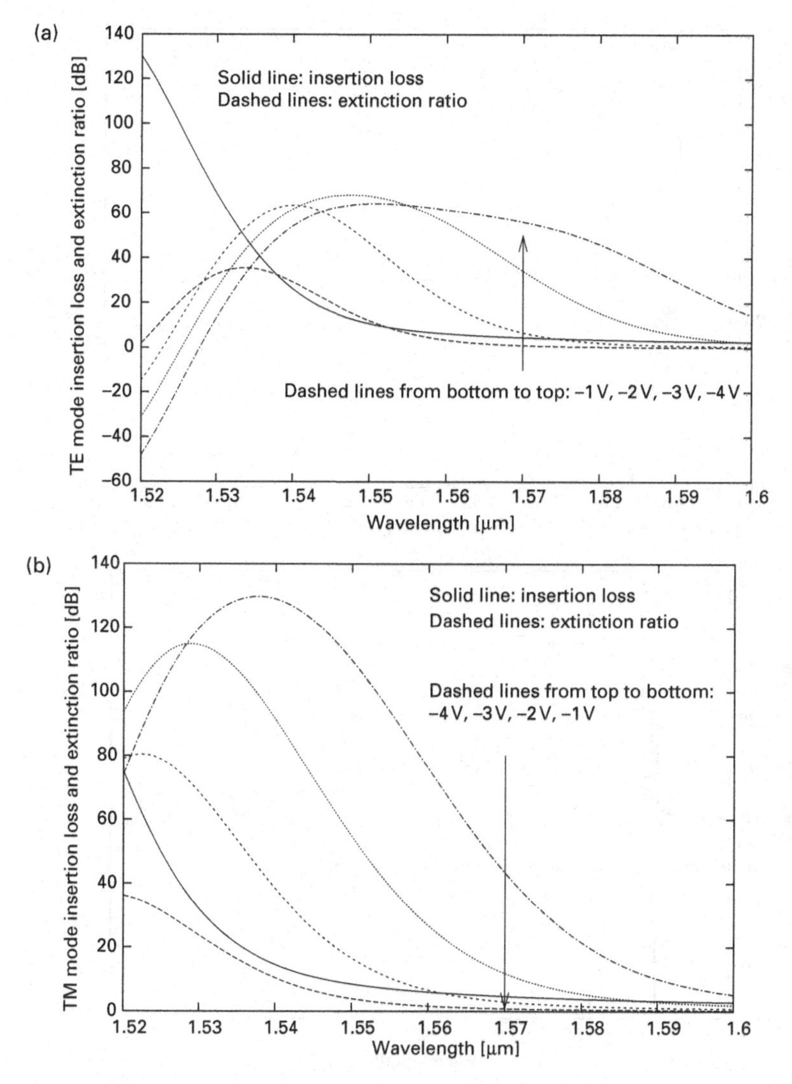

Fig. 11.6. Insertion loss under zero bias and extinction ratios under different reverse bias voltages in the EAM with a shallow RW structure. (a) For the TE mode. (b) For the TM mode.

normally give rise to any appreciable reliability issue. Hence we can follow the deep ridge waveguide design by etching through the active region as shown in Fig. 11.1.

The calculated cross-sectional optical modes in shallow and deep ridge waveguides are shown in Figs. 11.8(a&b) respectively. The extracted effective indices, the group indices, and the confinement factors for these two structures are summarized in Table 11.2. We find that the total confinement factor (for the ten QWs) is indeed higher in the deep ridge waveguide although its effective index becomes lower, when compared to the shallow ridge waveguide.

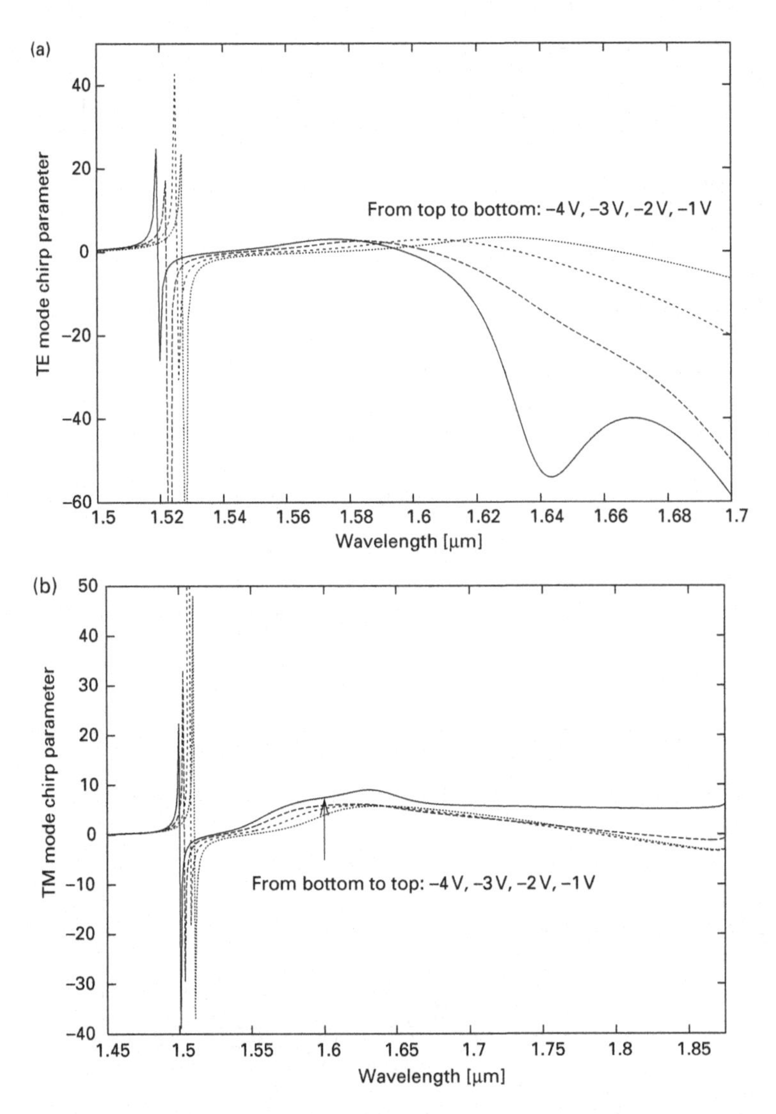

Fig. 11.7. Chirp parameters under different reverse bias voltages in an EAM with a shallow RW structure. (a) For the TE mode. (b) For the TM mode.

Consequently, we can calculate the insertion loss and extinction ratios under different reverse bias voltages for the deep ridge EAM and show the comparative results in Figs. 11.9(a&b) for the TE and TM modes, respectively. From these calculations, we find that both the extinction ratio and insertion loss are improved, although the latter is not significant. This is because the confinement increase raises the device modal absorption, which offsets the effect of removing the constant absorption from the active region outside the ridge. As a result, the reduction in the insertion loss is hardly appreciable unless we reduce the confinement factor in the deep ridge EAM as well. This can actually

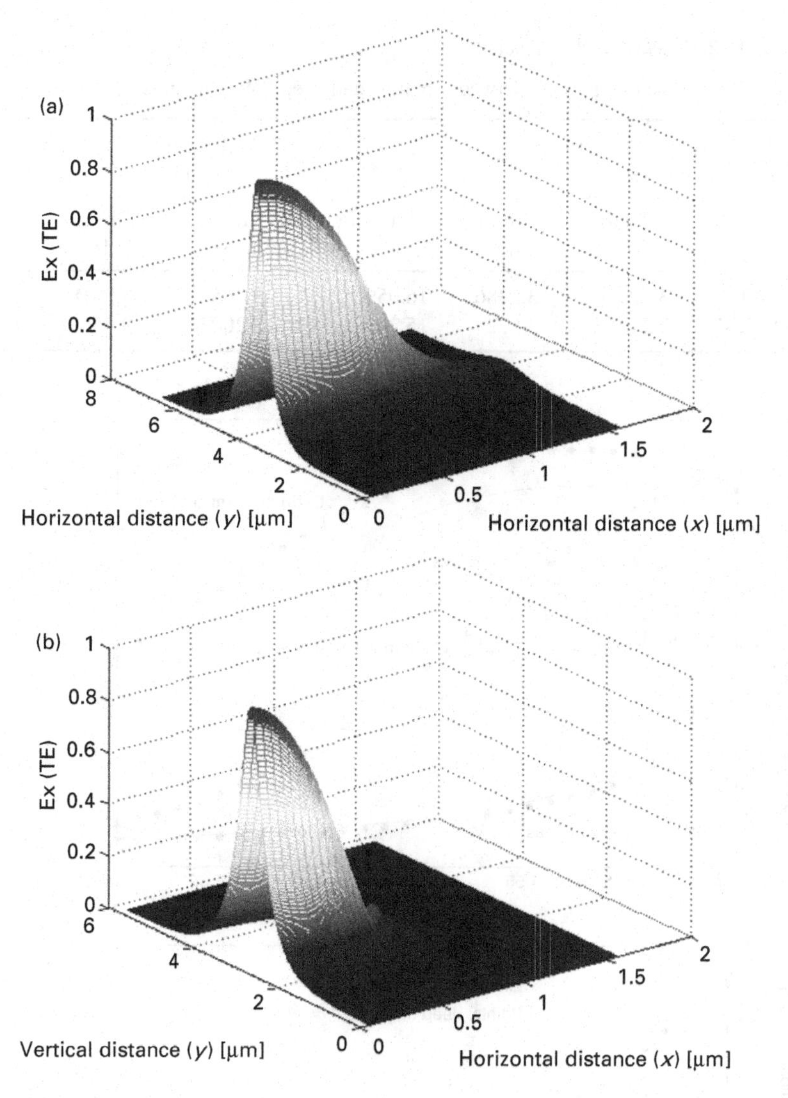

Fig. 11.8. Cross-sectional optical field distributions in EAMs. (a) With a shallow RW structure. (b) With a deep RW structure.

be realized by redesigning the cross-section layer structure. However, it will be at a cost of spoiling the extinction ratio.

While it is obvious that the insertion loss and the extinction ratio are scaled similarly by the device length, the operating conditions (bias and wavelength) are left to us as the last few degrees of freedom in device design work unless these conditions are pre-specified. However, we will leave them for the more challenging task of equalizing the TE and TM absorptions, as it is crucial for solitary EAMs when they are not monolithically integrated with semiconductor lasers. To complete this section, we show the dependences of the

Table 11.2. Optical modal parameters

The table gives a comparison between shallow and deep RW structures.

	Shallow ridge waveguide			Deep ridge waveguide		
	Effective index	Group index	Confinement factor	Effective index	Group index	Confinement factor
Total TE	3.2247	3.5386	16.05%	3.2056	3.5818	17.55%
Total TM	3.2145	3.5229	15.24%	3.2003	3.5569	16.83%

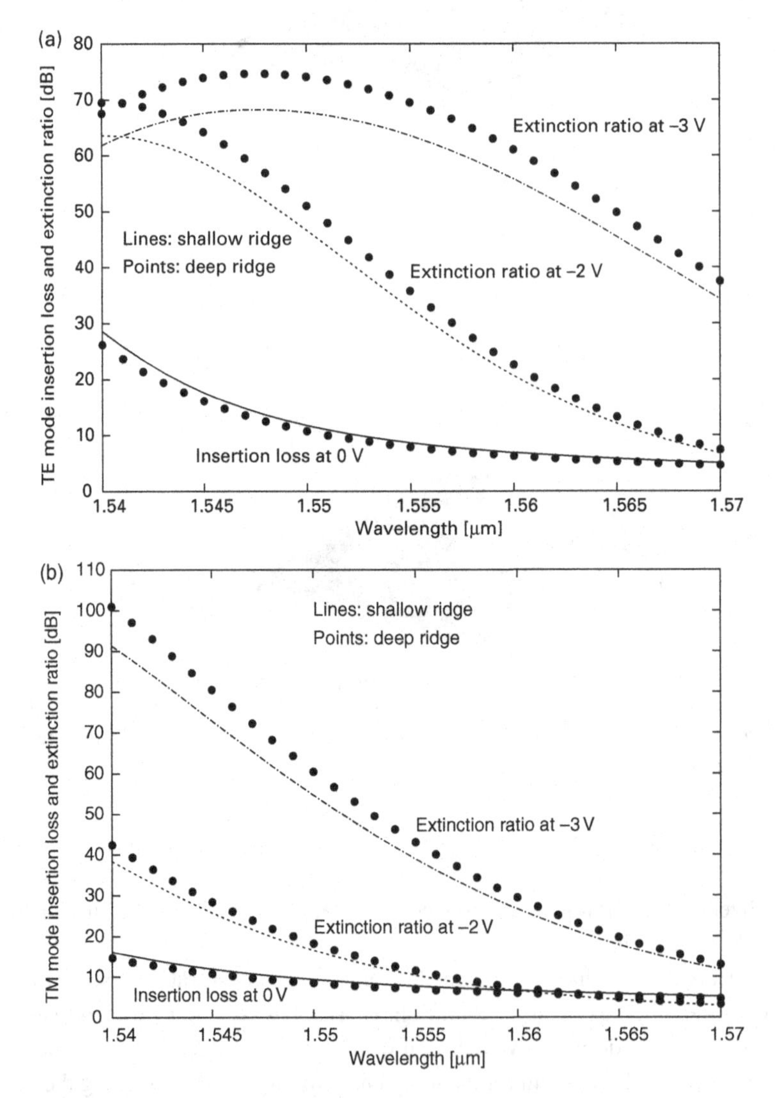

Fig. 11.9. Comparisons of insertion loss and extinction ratio under different reverse bias voltages between EAMs with a shallow or deep RW structure. (a) For the TE mode. (b) For the TM mode.

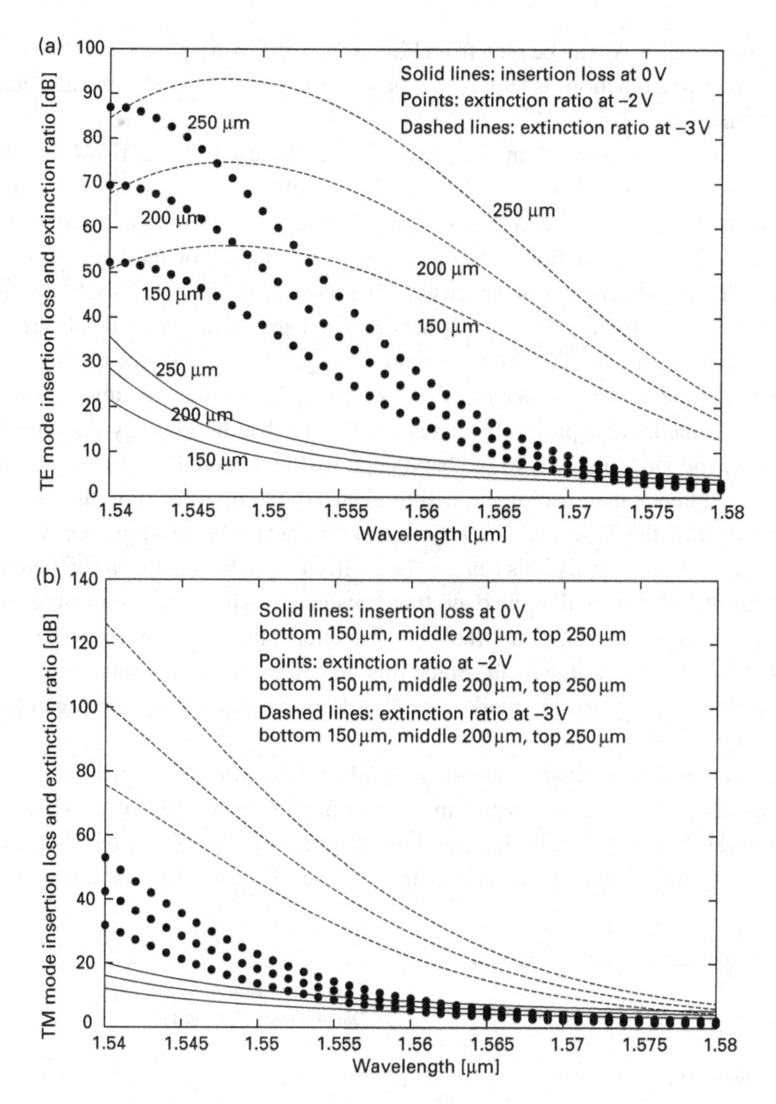

Fig. 11.10. Insertion loss and extinction ratios under different reverse bias voltages in EAMs with different lengths. (a) The TE mode. (b) The TM mode.

insertion loss and extinction ratio on cavity length under a few different reverse bias voltages in Figs. 11.10(a&b) for the TE and TM modes, respectively.

11.1.4 Design for polarization independent absorption

From the simulation results in Sections 11.1.2 and 11.1.3, we find that the absorption peak value of the TM mode is higher than that of the TE mode. This is attributed to the larger dipole matrix element between the (conduction band) electron and (valence band) light-hole compared to that between the (conduction band) electron and (valence band) heavy-hole. As can be seen from Fig. 11.3(a), the overlap between the electron and light-hole wave functions is larger than the overlap between the electron and heavy-hole wave functions.

However, at the absorption spectrum edge that is used by the EAM, the TE mode absorption is usually larger, as the wavelength of the TM mode absorption peak is shorter because the light-hole band edge is lower, which results in a higher transition energy from the conduction band to the light-hole band. For this reason, to align the TE and TM mode absorption spectrum edges, we have either to use the tensile strain to lift the light-hole band edge and hence to red-shift the TM mode absorption peak (towards a longer wavelength), or to reduce the gap between the heavy-hole and light-hole band edge by, e.g., weakening the QW effect. To facilitate the implementation, we will follow the latter approach by increasing the QW bandgap energy. As a result, on the valence band side, the heavy-hole band edge will be lowered more than the light-hole band edge, hence their difference will be reduced. On the spectrum, this is effectively a blue-shift of the TE mode absorption peak to reach a better alignment with the TM mode peak. Theoretically, this can also be realized by reducing the barrier height, which gives us the advantage that the operating wavelength will change very little. However, this approach needs to have the whole cross-sectional layer structure redesigned for both optical and carrier confinement. Since this book's focus is on explaining ideas rather than on tedious engineering work, we will follow the easier approach regardless of the operating wavelength shift.

We must also notice that the confinement of the ridge waveguide to these two polarization modes is also different, which may offset part of the material absorption difference in one way or another, and bring us totally different modal absorptions. Since the ridge waveguide always has a better confinement to the TE than to the TM mode, we have to

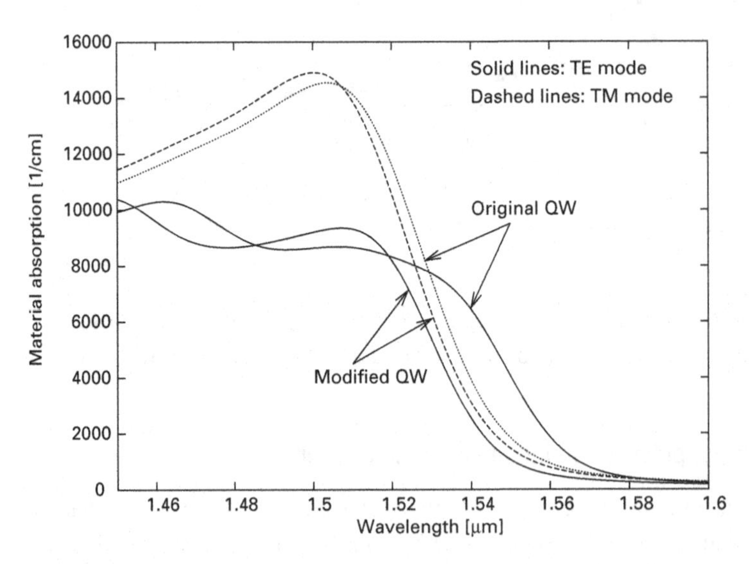

Fig. 11.11. Comparison of the TE and TM mode material absorptions under a 2 V reverse bias for deep ridge EAMs with the original and modified QW structures.

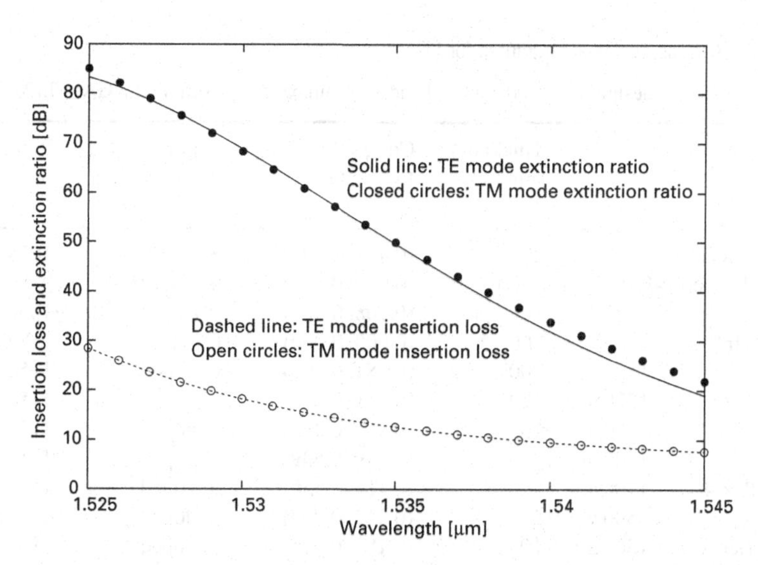

Fig. 11.12. The TE and TM mode insertion loss and extinction ratio under a 2 V reverse bias for the deep ridge EAM with a modified QW structure.

maintain a slightly higher TM absorption in an effort to bring the material TE and TM absorptions closer, to leave room for the final modal absorption balance.

A modified QW design is given in the last line in Table 11.1. The calculated material absorptions for the TE and TM modes under a 2 V reverse bias are shown in Fig. 11.11, where the results from the original structure are also plotted for the sake of comparison. We find that the TE absorption peak indeed has a blue-shift, whereas the TM absorption peak is almost unaffected. As a result, the two absorption edges are well aligned. We have also slightly increased the compressive strain to the QW in order to get the same slope for the TE and TM absorption edges, with TM absorption slightly above TE absorption.

Finally, Fig. 11.12 shows the TE and TM mode insertion losses and extinction ratios under a 2 V reverse bias for the EAM with a deeply-etched ridge waveguide and with the modified QWs, from which we find that the absorptions for the TE and TM polarization modes are almost perfectly aligned.

11.2 The semiconductor optical amplifier

11.2.1 The device structure

Two traveling wave SOAs, one in the 1300 nm band and the other in the 1500 nm band, are modeled in this section. Both SOAs have ridge waveguide structures with the same ridge height of 2.5 μm but with different ridge widths of 1.4 μm for the 1300 nm SOA and 1.6 μm for the 1550 nm SOA, respectively. These SOAs have the same length, namely 1 mm, with all facets assumed to be perfectly AR coated with zero reflection. Their cross-sectional layer structures are given in Table 11.3.

Table 11.3. Cross-sectional structures for SOAs

The table sets out designs for 1300 nm band and 1500 nm band SOAs in the InGaAsP/InP system.

Layer		Thickness (nm)	Composition of $In_{(1-x)}Ga_xAs_y P_{(1-y)}$ (x, y)	Doping concentration ($10^{18}/cm^3$)	Remarks λ_g (nm) and strain
Cap P-InGaAs		200	(0.468, 1.000)	10.0 (P)	1654, 0.0
Graded doped P-InP		2300	(0.000, 0.000)	0.7–0.1 (P)	918.6, 0.0
Etching stop		10	(0.108, 0.236)	0.1 (P)	1050, 0.0
Spacer InP		100	(0.000, 0.000)	0.1 (P)	918.6, 0.0
SCH	1300 nm	100	(0.181, 0.395)	undoped	1150, 0.0
InGaAsP	1500 nm	120	(0.181, 0.395)	undoped	1150, 0.0
Barrier	1300 nm	10	(0.181, 0.395)	undoped	1150, 0.0
	1500 nm	10	(0.249, 0.539)	undoped	1250, 0.0
Well × 4	1300 nm	5	(0.114, 0.556)	undoped	1347, CS1.0%
	1500 nm	5	(0.212, 0.797)	undoped	1600, CS1.1%
Barrier × 4	1300 nm	10	(0.181, 0.395)	undoped	1150, 0.0
	1500 nm	10	(0.249, 0.539)	undoped	1250, 0.0
SCH InGaAsP	1300 nm	100	(0.181, 0.395)	0.5 (N)	1150, 0.0
	1500 nm	120	(0.181, 0.395)	0.5 (N)	1150, 0.0
Buffer N-InP		500	(0.000, 0.000)	1.0 (N)	918.6, 0.0
Substrate N-InP		100 000	(0.000, 0.000)	3.0 (N)	918.6, 0.0

In both structures, relatively low compressive strains of 1.0% and 1.1% for the 1300 nm and 1500 nm band SOAs, respectively, are applied to the QWs. Since SOAs are normally operated at high carrier density levels inside the active region, we use a relatively low compressive strain to maintain a high material gain in operation, by following the conclusion obtained in Section 10.1.2. An exponentially growing doping profile is again adopted in the P type InP cladding to achieve a compromise between hole injection efficiency and optical free-carrier absorption. This design is more crucial to SOAs with a small confinement factor design for linear in-line amplifications, as there is usually a stronger evanescent tail of the optical wave overlapping with the P cladding layer if the confinement factor is reduced.

11.2.2 Simulated semiconductor optical amplifier performance

The calculated stimulated and spontaneous emission gain spectra for the two SOAs are shown in Figs. 11.13(a&b), and Figs. 11.14 (a&b), respectively.

The simulated time domain optical output power waveforms are shown in Fig. 11.15, where both SOAs are under a DC bias of 120 mA. The input optical signals have their channel wavelengths set at 1300 nm (for the first SOA designed for the 1300 nm band) and 1510 nm (for the second SOA designed for the 1500 nm band), respectively. Their waveforms, however, are identical, comprising an ideal binary square pulse stream

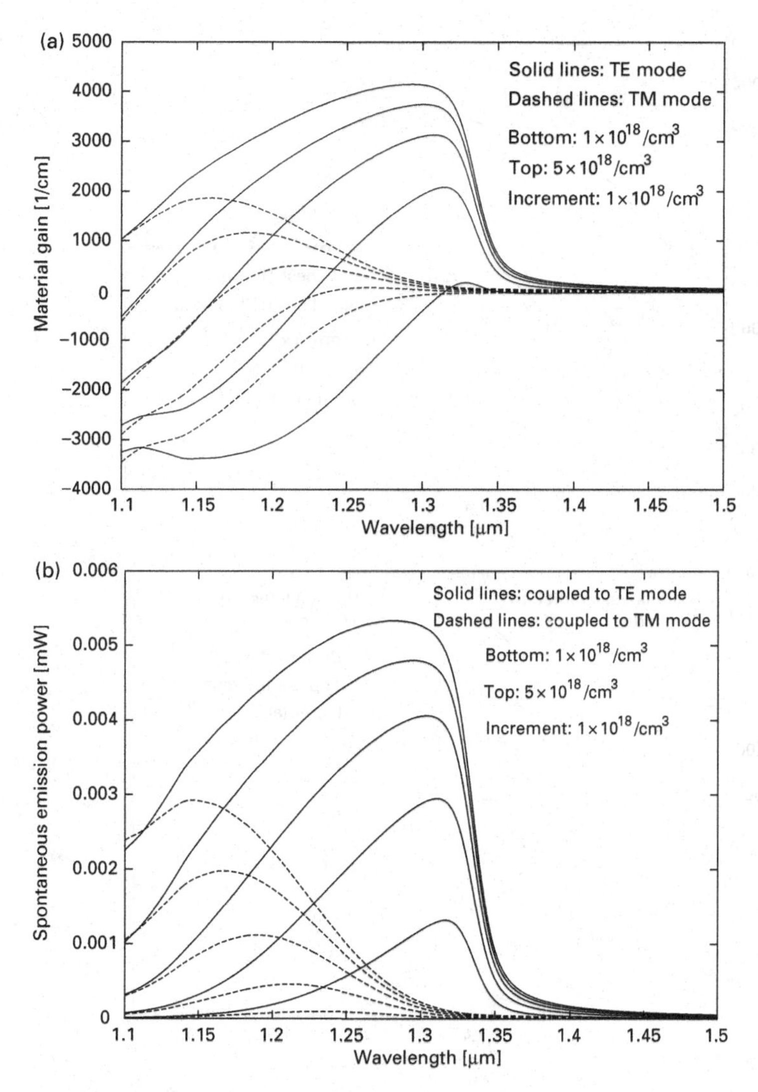

Fig. 11.13. (a) Results for the material (stimulated emission) TE and TM gain profiles. (b) Results for the TE and TM spontaneous emission power spectra; different carrier densities in the 1300 nm band SOA.

in a sequence 1011100101, with symbols 0 and 1 power readings at −20 dBm and 0 dBm, respectively, and with identical symbols 0 and 1 time durations at 400 ps. In this case, the input signal is equivalent to a non-return-to-zero (NRZ) binary bit stream at 2.5 Gbps.

Figure 11.16 gives the calculated optical gain dependences on the input optical power for the two SOAs when both are biased under 120 mA.

Finally, Fig. 11.17 shows the calculated noise figures for the two SOAs under different bias currents.

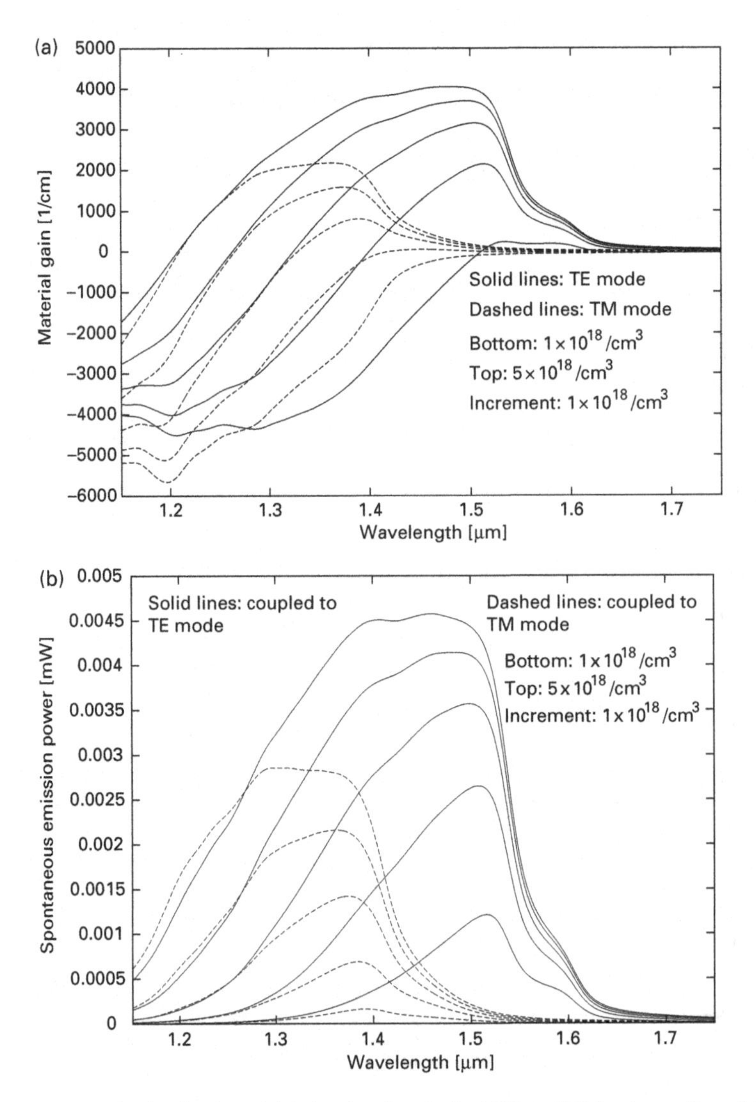

Fig. 11.14. (a) Results for the material (stimulated emission) TE and TM gain profiles. (b) Results for the TE and TM spontaneous emission power spectra; different carrier densities in the 1500 nm band SOA.

11.2.3 Design for performance enhancement

In applications such as linear in-line amplification, SOAs need to be designed with a high saturation power or a broad linear gain region. It then becomes crucial to reduce the device confinement factor [1, 2]. By reducing the number of QWs, we manage to reduce the confinement factor and obtain higher saturation powers, as shown in Fig. 11.18, in which simulated optical gains are plotted as functions of input optical power under a constant bias current of 120 mA for SOAs with their active regions comprising different numbers of QWs. Because of the reduction in the confinement factor, the small-signal optical gain drops as a side effect, despite the increased carrier density inside the QWs

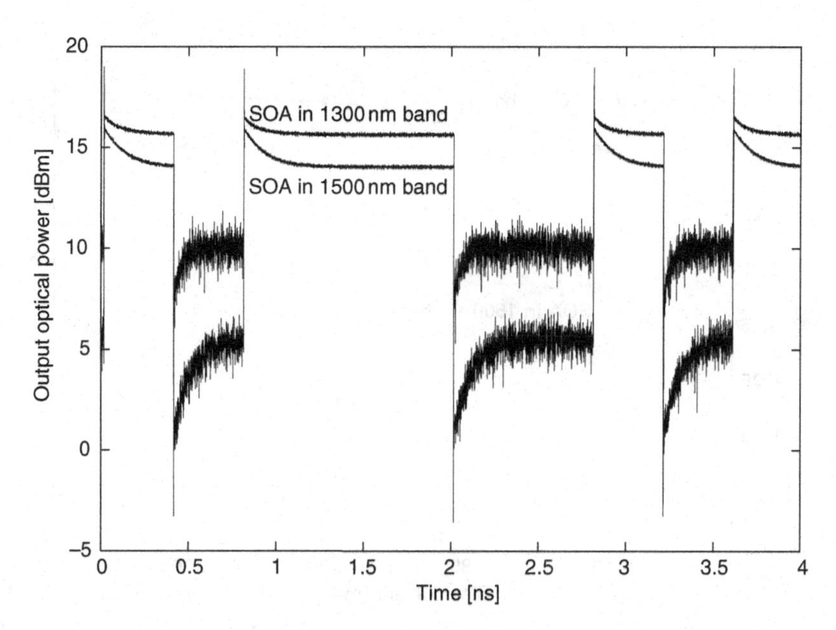

Fig. 11.15. Time domain optical output power waveforms from the 1300 nm band and 1500 nm band SOAs under the same bias current of 120 mA and with the same input optical signal waveform carried by the 1300 nm and 1510 nm channels, respectively.

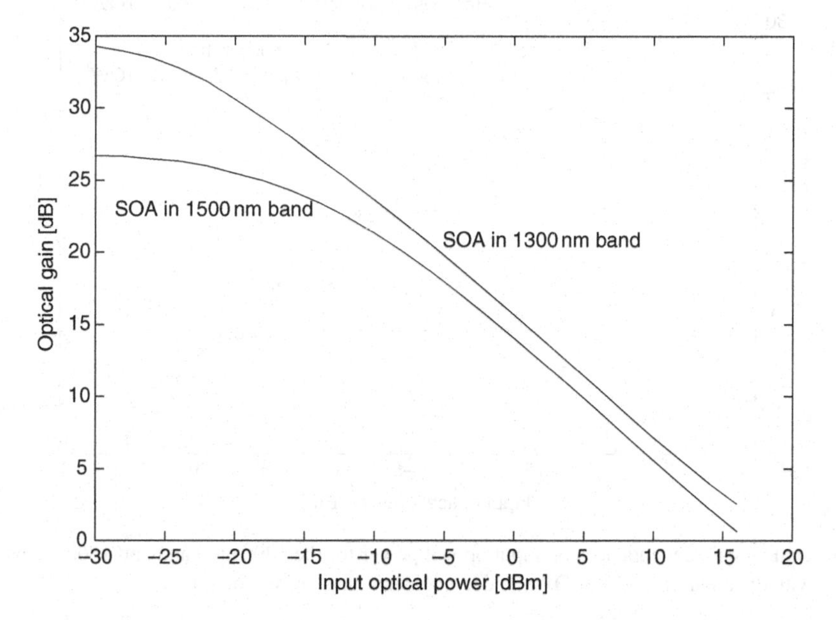

Fig. 11.16. Optical gain dependences on the input optical power for the 1300 nm band and 1500 nm band SOAs when both are biased under 120 mA.

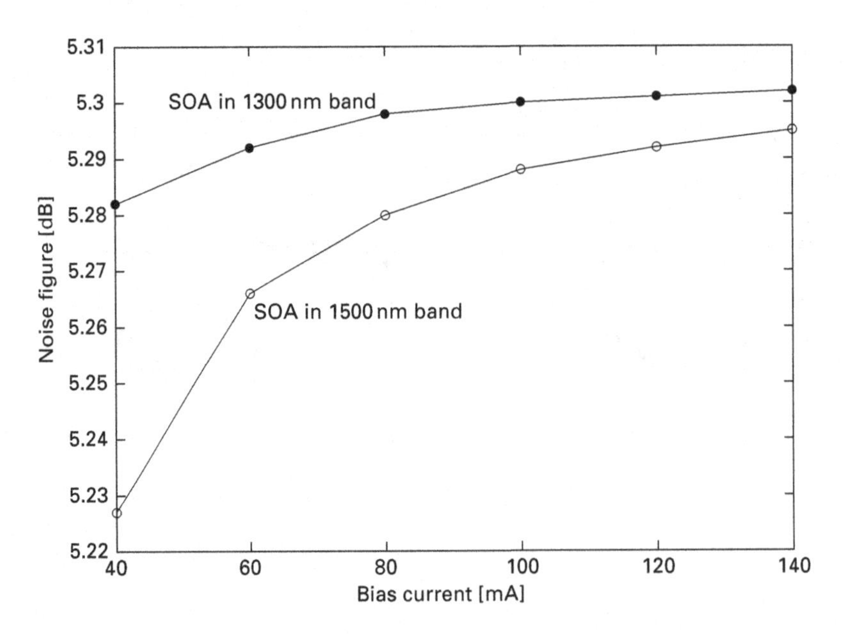

Fig. 11.17. Noise figures of 1300 nm band and 1500 nm band SOAs as functions of the bias current.

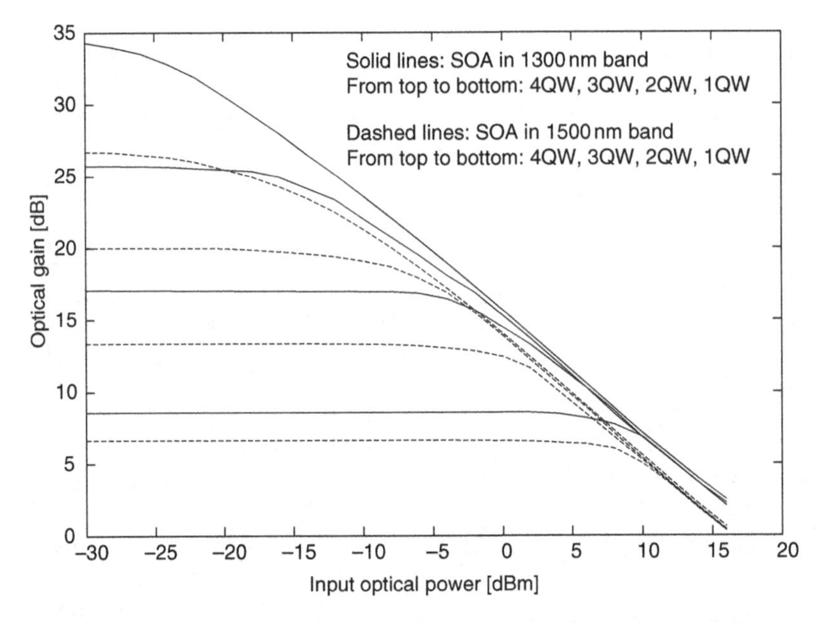

Fig. 11.18. Optical gain dependences on input optical power for the 1300 nm band and 1500 nm band SOAs with different numbers of QWs under a constant bias current of 120 mA.

Table 11.4. Cross-sectional structure for SLED

The table sets out a design for an 820 nm band SLED in the GaAs/AlGaAs system.

Layer	Thickness (nm)	Composition of $Al_xGa_{(1-x)}As$ (x)	Doping concentration (10^{18} /cm^3)	Remarks λ_g (nm) and strain
Cap P-GaAs	200	(0.000)	10.0 (P)	870.2, 0.0
Cladding P-AlGaAs	1300	(0.500)	0.7 (P)	N/A, 0.0
GRINSCH AlGaAs	100	(0.300)	0.2 (P)	685.7, 0.0
GRINSCH AlGaAs	100	(0.225)	undoped	726.7, 0.0
Barrier	10	(0.150)	undoped	771.0, 0.0
Well × 3	5–8	(0.000)	undoped	870.2, 0.0
Barrier × 3	10	(0.150)	undoped	771.0, 0.0
GRINSCH AlGaAs	100	(0.225)	undoped	726.7, 0.0
GRINSCH AlGaAs	100	(0.300)	0.8 (N)	685.7, 0.0
Buffer N-AlGaAs	1500	(0.500)	1.0 (N)	N/A, 0.0
Substrate N-GaAs	100 000	(0.000)	3.0 (N)	870.2, 0.0

for invariant bias current. The reason has been explained in Section 10.3.1. However, we can always extend the SOA length to obtain higher gains without significant reduction in the saturation power. That is to say, we should use a low confinement factor with a long cavity design in a SOA to achieve high gain and high saturation power.

11.3 The superluminescent light emitting diode

11.3.1 The device structure

The device to be modeled is an 820 nm band SLED with the cross-sectional layer structure summarized in Table 11.4. The device has a ridge waveguide structure with the ridge width and height 2.0 μm and 1.5 μm, respectively. The device length is 800 μm with the front facet perfectly AR coated with zero reflection and the rear facet either perfectly AR coated or HR coated with an amplitude reflection of 0.9.

11.3.2 Simulated superluminescent light emitting diode performance

The calculated stimulated and spontaneous emission gain spectra for the TE and TM modes with different QW widths are shown in Figs. 11.19(a&b) and Figs. 11.20(a&b), respectively.

It is found that, as expected, the gain profile has a blue-shift as the QW thickness reduces. However, such a shift is not linearly proportional to the thickness reduction.

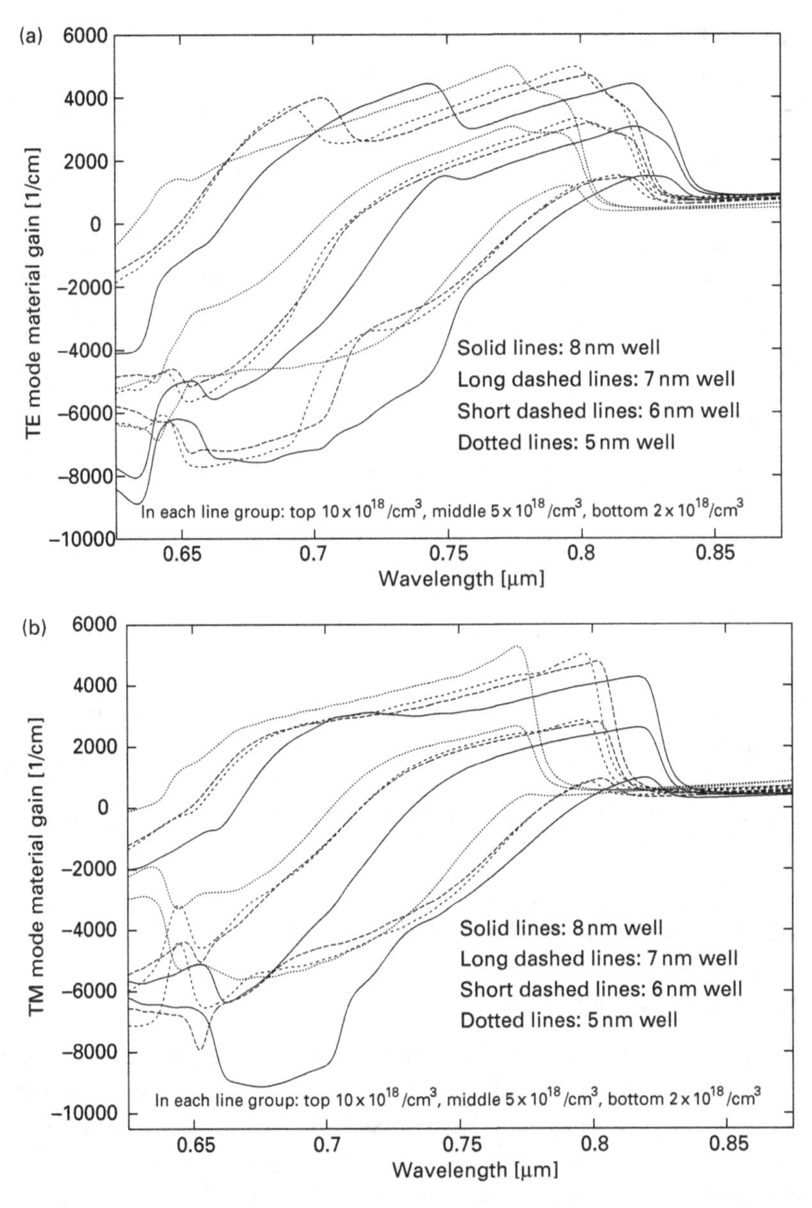

Fig. 11.19. Material (stimulated emission) gain spectra for different QW widths under different carrier densities. (a) For the TE mode. (b) For the TM mode.

11.3.3 Design for performance enhancement

By combining QWs with different widths, we may flatten and broaden both the stimulated emission and spontaneous emission gain spectra. For example, a combination of three QWs with thicknesses of 8 nm, 5 nm, and 6 nm from the P to N side would

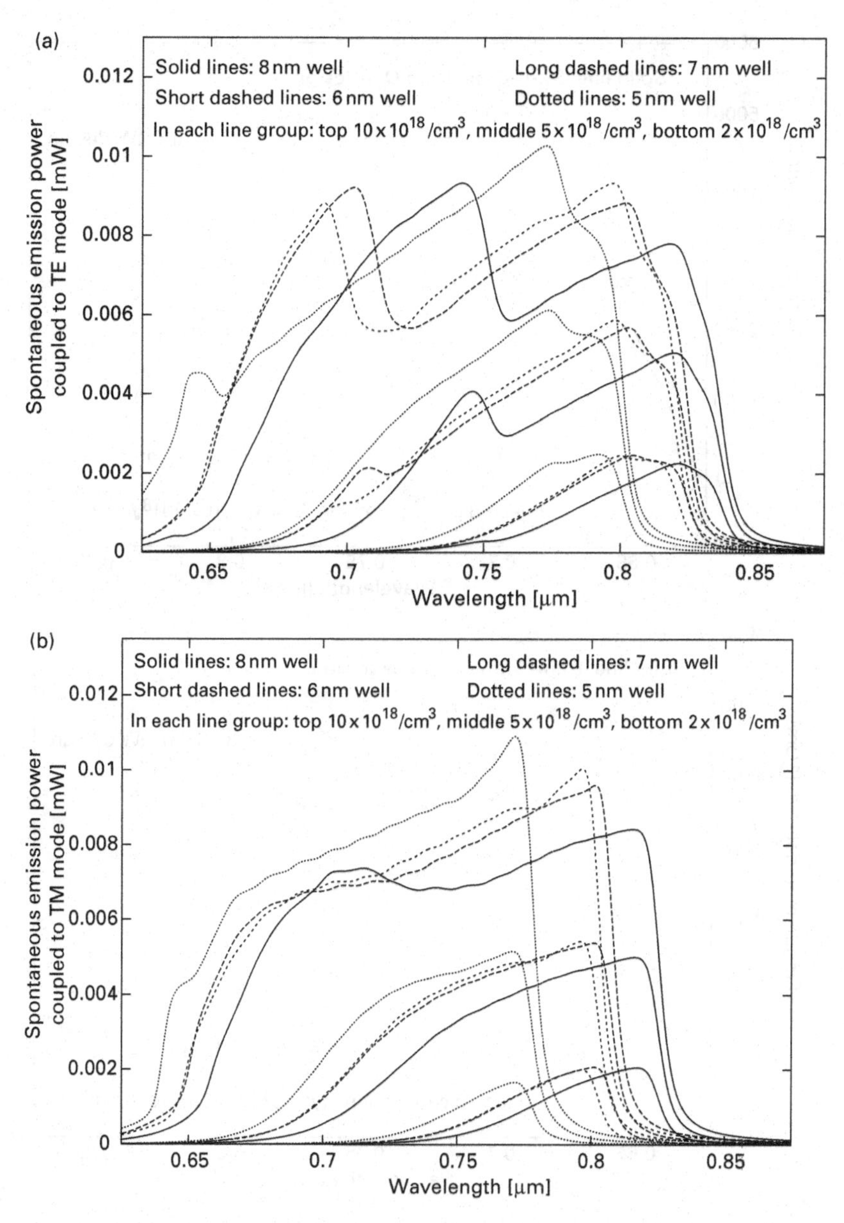

Fig. 11.20. Spontaneous emission power spectra for different QW widths under different carrier densities. (a) For the TE mode. (b) For the TM mode.

give the calculated stimulated emission and spontaneous emission spectra shown in Figs. 11.21(a&b) and Figs. 11.22(a&b) for the TE and TM modes.

With the expanded gain profile design, the calculated output powers with the rear facet either AR or HR coated are shown in Fig. 11.23 as functions of the bias current.

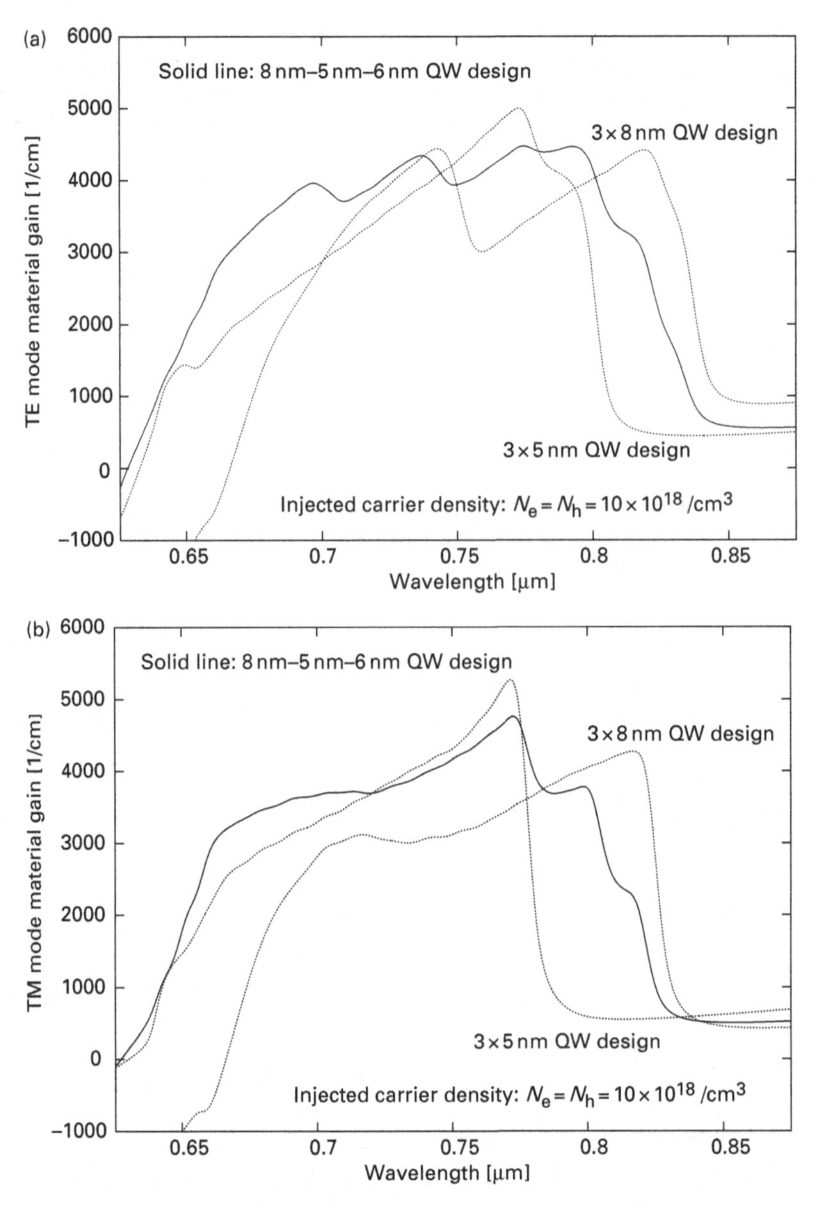

Fig. 11.21. Material (stimulated emission) gain spectra for a combination of three QWs in different widths under different carrier densities. (a) For the TE mode. (b) For the TM mode.

Finally, Fig. 11.24 shows the calculated output optical power spectra under different bias currents. The ripples appearing in the spectra are attributed to transitions between discrete energy levels inside the QWs. Fine tuning of the thicknesses of the QWs and of the heights of the barriers (i.e., the composition of the barrier AlGaAs) might be necessary in order to obtain broad and smooth spectral profiles.

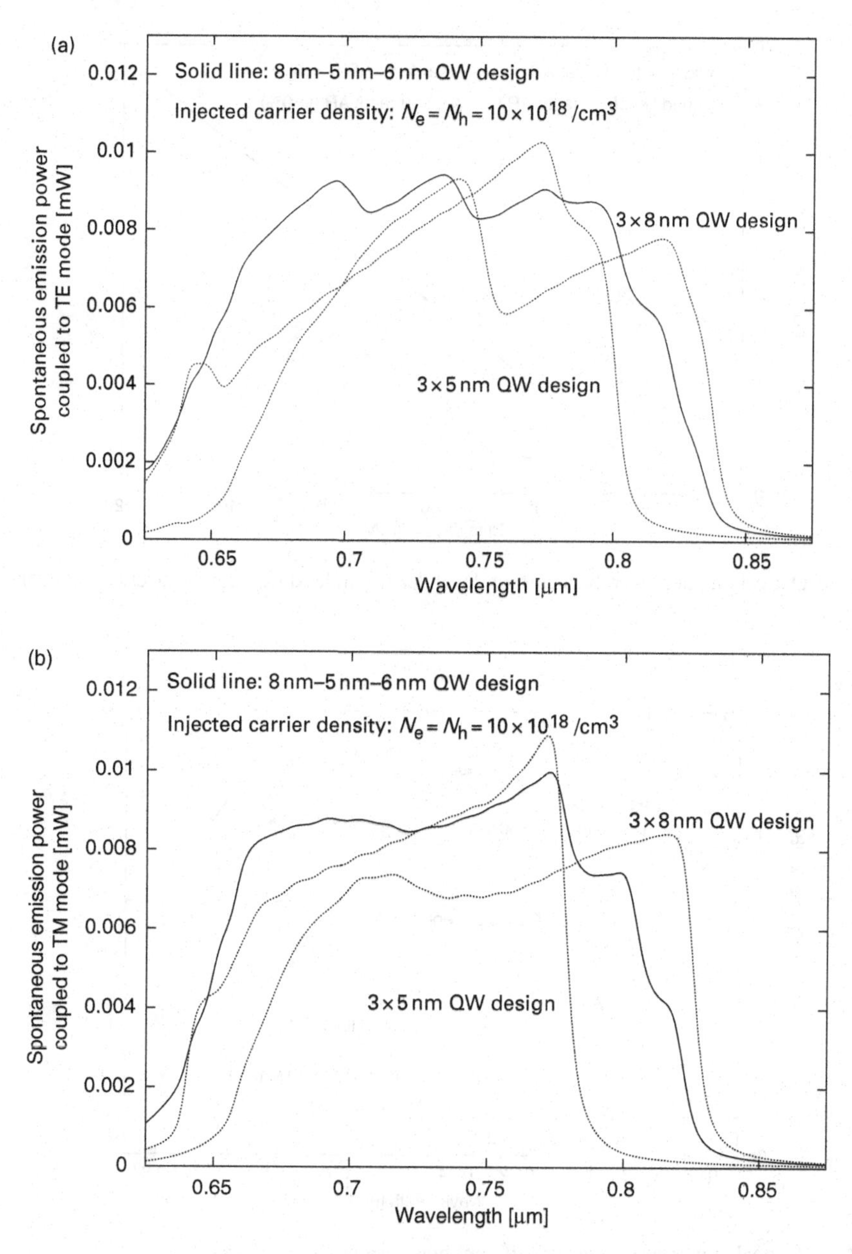

Fig. 11.22. Spontaneous emission power spectra for a combination of three QWs in different widths under different carrier densities. (a) For the TE mode. (b) For the TM mode.

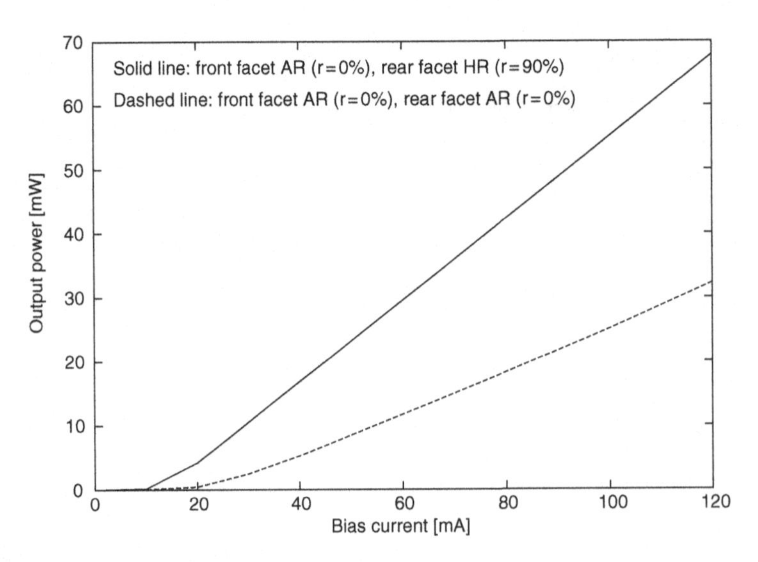

Fig. 11.23. Optical output power–bias current dependences under different rear facet coating conditions.

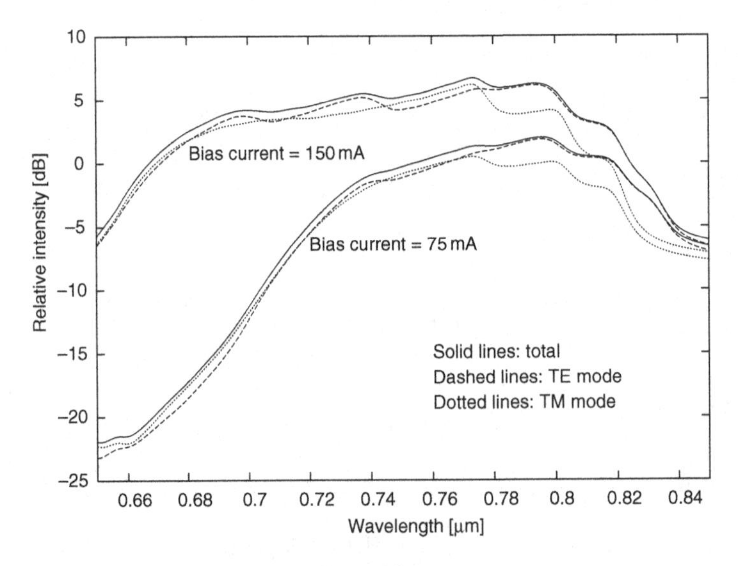

Fig. 11.24. Optical power spectra under different bias currents.

References

[1] T. D. Visser, H. Blok, B. Demeulenaere, and D. Lenstra, Confinement factors and gain in optical amplifiers. *IEEE Journal of Quantum Electron.*, **QE-33**:10 (1997), 1763–6.

[2] I. Kim, K. Uppal, and P. D. Dapkus, Gain saturation in traveling-wave semiconductor optical amplifiers. *IEEE Journal of Quantum Electron.*, **QE-34**:10 (1998), 1949–52.

Further reading

On solitary EAM

[1] F. Devaux, F. Dorgeuille, A. Ougazzaden, *et al.*, 20 Gb/s operation of a high-efficiency InGaAsP/InGaAsP MQW electroabsorption modulator with 1.2 V drive voltage. *IEEE Photon. Tech. Lett.*, **5**:11 (1993), 1288–90.

[2] T. Aizawa, K. G. Ravikumar, S. Suzuki, T. Watanabe, and R. Yamauchi, Polarization-independent quantum-confined Stark effect in an InGaAs/InP tensile-strained quantum well. *IEEE Journal of Quantum Electron.*, **QE-30**:2 (1994), 585–92.

[3] G. L. Li and P. K. L. Yu, Optical intensity modulators for digital and analog applications. *IEEE/OSA Journal of Lightwave Tech.*, **LT-21**:9 (2003), 2010–30.

On SOA

[1] K. Morito, M. Ekawa, T. Watanabe, and Y. Kotaki, High-output-power polarization-insensitive semiconductor optical amplifier. *IEEE/OSA Journal of Lightwave Tech.*, **LT-21**:1 (2003), 176–81.

[2] T. Kakitsuka, Y. Shibata, M. Itoh, *et al.*, Influence of buried structure on polarization sensitivity in strained bulk semiconductor optical amplifier. *IEEE Journal of Quantum Electron.*, **QE-38**:1 (2002), 85–91.

[3] M. Silver, A. F. Phillips, A. R. Adams, P. D. Greene, and A. J. Collar, Design and ASE characteristics of 1550 nm polarization-insensitive semiconductor optical amplifiers containing tensile and compressive well. *IEEE Journal of Quantum Electron.*, **QE-36**:1 (2000), 118–22.

[4] M. A. Newkirk, B. I. Miller, U. Koren, *et al.*, 1.5 μm multiquantum-well semiconductor optical amplifier with tensile and compressively strained wells for polarization-independent gain. *IEEE Photon. Tech. Lett.*, **5**:4 (1993), 406–8.

[5] P. Koonath, S. Kim, W.-J. Cho, and Anand Gopinath, Polarization-insensitive optical amplifiers in AlInGaAs. *IEEE Photon. Tech. Lett.*, **13**:8 (2001), 779–81.

[6] M. J. Connelly, *Semiconductor Optical Amplifier*, 1st edn (Dordrecht, The Netherlands: Kluwer Academic Publishers, 2002).

[7] X. Li and J.-W. Park, Time-domain simulation of channel crosstalk and inter-modulation distortion in gain clamped semiconductor optical amplifiers. *Optics Comm.*, **263** (2006), 219–28.

[8] J.-W. Park, Y. Kawakami, X. Li, and W.-P. Huang, Comparative analysis of the effects of internal lasing oscillation and external light injection on semiconductor optical amplifier performance. *Optics Comm.*, **267** (2006), 379–87.

[9] J.-W. Park, X. Li, and W.-P. Huang, Comparative study of mixed frequency-time-domain models of semiconductor laser optical amplifiers. *IEE Proc. Optoelectron.*, **152** (2005), 151–9.

[10] J.-W. Park, X. Li, and W.-P. Huang, Gain clamping in semiconductor optical amplifiers with second-order index-coupled DFB grating. *IEEE Journal of Quantum Electron.*, **41**:3 (2005), 366–75.

[11] J.-W. Park, W.-P. Huang, and X. Li, Investigation of semiconductor optical amplifier integrated with DBR laser for high saturation power and fast gain dynamics. *IEEE Journal of Quantum Electron.*, **40**:11 (2004), 1540–7.

[12] J.-W. Park, X. Li, and W.-P. Huang, Performance simulation and design optimization of gain-clamped semiconductor optical amplifiers based on distributed Bragg reflectors. *IEEE Journal of Quantum Electron.*, **39**:11 (2003), 1415–23.

On SLED

[1] J. H. Song, S. H. Cho, I. K. Han, *et al.*, High-power broad-band superluminescent diode with low spectral modulation at 1.5 μm wavelength. *IEEE Photon. Tech. Lett.*, **12**:7 (2000), 783–5.
[2] G. A. Alphonse, D. B. Gilbert, M. G. Harvey, and M. Ettenberg, High-power superluminescent diodes. *IEEE Journal of Quantum Electron.*, **QE-24**:12 (1988), 2454–7.
[3] L. Fu, H. Schweizer, Y. Zhang, *et al.*, Design and realization of high-power ripple-free superluminescent diodes at 1300 nm. *IEEE Journal of Quantum Electron.*, **QE-40**:9 (2004), 1270–4.
[4] N. S. K. Kwong, K. Y. Lau, and N. Bar-Chaim, High-power high-efficiency GaAlAs superluminescent diodes with an internal absorber for lasing suppression. *IEEE Journal of Quantum Electron.*, **QE-25**:4 (1989), 696–704.
[5] I. P. Kaminow, G. Eisenstein, L. W. Stulz, and A. G. Dentai, Lateral confinement InGaAsP superluminescent diode at 1.3 μm. *IEEE Journal of Quantum Electron.*, **QE-19**:1 (1983), 78–82.
[6] J.-W. Park and X. Li, Theoretical and numerical analysis of superluminescent diodes, *IEEE/OSA Journal of Lightwave Tech.*, **24**:6 (2006), 2473–80.
[7] D. Labukhin and X. Li, Polarization insensitive asymmetric ridge waveguide design for semiconductor optical amplifiers and super-luminescent light emitting diodes. *IEEE Journal of Quantum Electron.*, **42**:11 (2006), 1137–43.

12 Design and modeling examples of integrated optoelectronic devices

12.1 The integrated semiconductor distributed feedback laser and electro-absorption modulator

12.1.1 The device structure

A schematic structure of an integrated semiconductor DFB laser and EAM is shown in Fig. 12.1.

The cross-sectional layer structures in the DFB and EAM sections are shown in Tables 12.1 and 12.2, respectively.

Both sections of this device are made of a ridge waveguide with the same ridge width of 2.0 μm, but with different ridge heights of 2.0 μm in the DFB laser section and 3.0 μm in the EAM section, respectively. Thus, the etched ditch stops on top of the laser active region but goes through the modulator active region. This design makes a better match of the effective indices between the DFB laser and EAM sections and hence the reflection from the interface between these two sections will be effectively reduced. The deep etching in the modulator section also improves the far-field pattern emitted from the front facet. Finally, by removing the EAM active region outside the ridge, the EAM insertion loss will reduce and its extinction ratio will increase as shown in Section 11.1.3.

The laser section and the modulator section are 300 μm and 200 μm long, respectively, the electrodes are isolated by a 10 μm wide and 1.0 μm deep trench etched across the ridge. The DFB laser has a uniform purely index-coupled grating with a normalized coupling coefficient of 4. This design is a balance between the output optical power, immunity to feedback from the modulator, and the lasing wavelength uncertainty in a range roughly equal to the Bragg stop-band width. The rear (i.e., the DFB laser side) and the front (i.e., the EAM side) facets are HR and AR coated, respectively, with amplitude reflectivities of 0.87 and 0.01, respectively.

Before the growth of the buffer N type InP layer, an InGaAsP etching stop layer is pre-buried to facilitate the final ridge etching in the EAM section. Different GRINSCH structures are designed for the DFB laser and EAM sections in such a way that the optical mode profiles in these two sections will be best matched. In the DFB laser section, a floating grating design is adopted to control the uniformity as the yield in the integrated optoelectronic device is usually a major concern. Again, an exponentially growing doping profile is used in the P type InP cladding layer in the DFB laser section for the reason

Table 12.1. Cross-sectional structure for DFB laser

The table sets out a design for the DFB laser section in the InGaAsP/InP system at 1550 nm.

Layer	Thickness (nm)	Composition of $In_{(1-x)}Ga_xAs_y P_{(1-y)}$ (x, y)	Doping concentration ($10^{18}/cm^3$)	Remarks λ_g (nm) and strain
Cap P-InGaAs	200	(0.468, 1.000)	10.0 (P)	1654, 0.0
Graded doped P-InP	1800	(0.000, 0.000)	0.7–0.2 (P)	918.6, 0.0
Etching stop	10	(0.108, 0.236)	0.2 (P)	1050, 0.0
Spacer InP	50	(0.000, 0.000)	undoped	918.6, 0.0
Grating InGaAsP	80	(0.249, 0.539)	undoped	1250, 0.0
Spacer InP	20	(0.000, 0.000)	undoped	918.6, 0.0
SCH InGaAsP	10	(0.249, 0.539)	undoped	1250, 0.0
Barrier	10	(0.249, 0.539)	undoped	1250, 0.0
Well × 5	5.5	(0.122, 0.728)	undoped	1571, CS1.50%
Barrier × 5	10	(0.249, 0.539)	undoped	1250, 0.0
GRINSCH InGaAsP	10	(0.249, 0.539)	0.5 (N)	1250, 0.0
GRINSCH InGaAsP	20	(0.216, 0.469)	0.5 (N)	1200, 0.0
GRINSCH InGaAsP	20	(0.181, 0.395)	0.5 (N)	1150, 0.0
GRINSCH InGaAsP	20	(0.145, 0.317)	0.5 (N)	1100, 0.0
GRINSCH InGaAsP	20	(0.108, 0.236)	0.5 (N)	1050, 0.0
Buffer N-InP	650	(0.000, 0.000)	1.0 (N)	918.6, 0.0
Etching stop	10	(0.108, 0.236)	3.0 (N)	1050, 0.0
Substrate N-InP	100 000	(0.000, 0.000)	3.0 (N)	918.6, 0.0

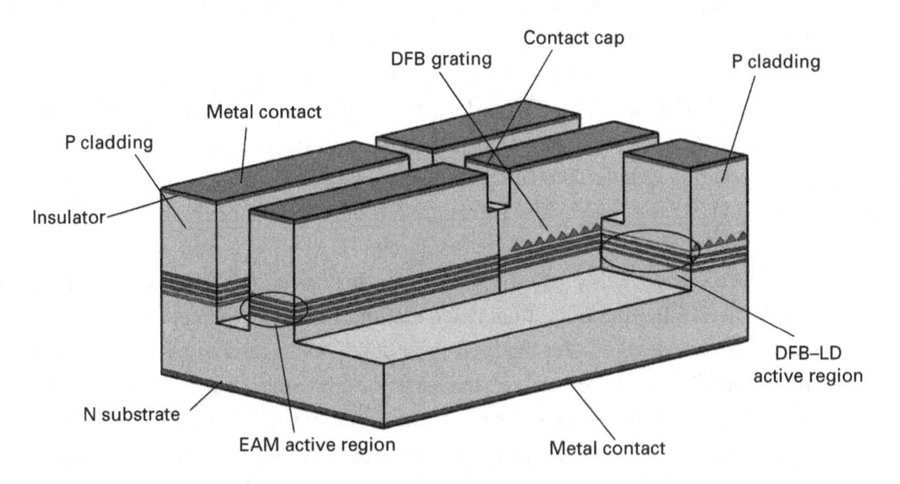

Fig. 12.1. An integrated DFB LD and EAM structure.

Table 12.2. Cross-sectional structure for EAM section

The table sets out a design for the EAM section in the InGaAsP/InP system at 1550 nm.

Layer	Thickness (nm)	Composition of $In_{(1-x)}Ga_xAs_y$ $P_{(1-y)}$ (x, y)	Doping concentration ($10^{18}/cm^3$)	Remarks λ_g (nm) and strain
Cap P-InGaAs	200	(0.468, 1.000)	10.0 (P)	1654, 0.0
Cladding P-InP	1850	(0.000, 0.000)	0.5 (P)	918.6, 0.0
GRINSCH InGaAsP	50	(0.108, 0.236)	undoped	1050, 0.0
GRINSCH InGaAsP	50	(0.181, 0.395)	undoped	1150, 0.0
Barrier	6	(0.196, 0.343)	undoped	1100, TS0.27%
Well × 10	9	(0.336, 0.861)	undoped	1580, CS0.45%
Barrier × 10	6	(0.196, 0.343)	undoped	1100, TS0.27%
GRINSCH InGaAsP	20	(0.181, 0.395)	undoped	1150, 0.0
GRINSCH InGaAsP	25	(0.108, 0.236)	undoped	1050, 0.0
Buffer N-InP	650	(0.000, 0.000)	1.0 (N)	918.6, 0.0
Etching stop	10	(0.108, 0.236)	3.0 (N)	1050, 0.0
Substrate N-InP	100 000	(0.000, 0.000)	3.0 (N)	918.6, 0.0

explained in Section 10.2.1. Compressive strains are also applied to both QWs in the DFB laser and EAM sections to eliminate TM mode emission in the laser and to enhance the extinction ratio in the modulator. In the DFB laser section, we have used a slightly higher strain level of 1.5% in order to achieve a relatively high gain under the laser operating condition, as revealed in Section 10.1.2.

12.1.2 The interface

The calculated optical modes in the DFB laser and EAM sections are shown in Figs. 12.2(a&b), respectively. The calculated effective indices for the TE mode in the DFB laser and EAM sections are 3.177 and 3.206, respectively. If using the same shallow ridge height in the DFB laser and EAM sections, i.e., without the deep etching (through the active region) in the EAM section, the calculated TE mode effective index in the EAM section would be 3.225.

The calculated far-field pattern emitted from the front facet is shown in Fig. 12.3, an aspect ratio of 1.36:1 (vertical:horizontal) is achieved, as can be seen.

12.1.3 Simulated distributed feedback laser performance

Figures 12.4, 12.5, and 12.6 show the calculated DFB laser spectrum, the output power – bias current dependence, and the side mode suppression ratio (SMSR) – bias current dependence under different grating phases at the HR coated rear facet.

From Fig. 12.4, we find that the lasing wavelengths are within a range from 1559.5 nm to 1563.5 nm. The spacing is around 4 nm as expected, which equals the Bragg stop-band

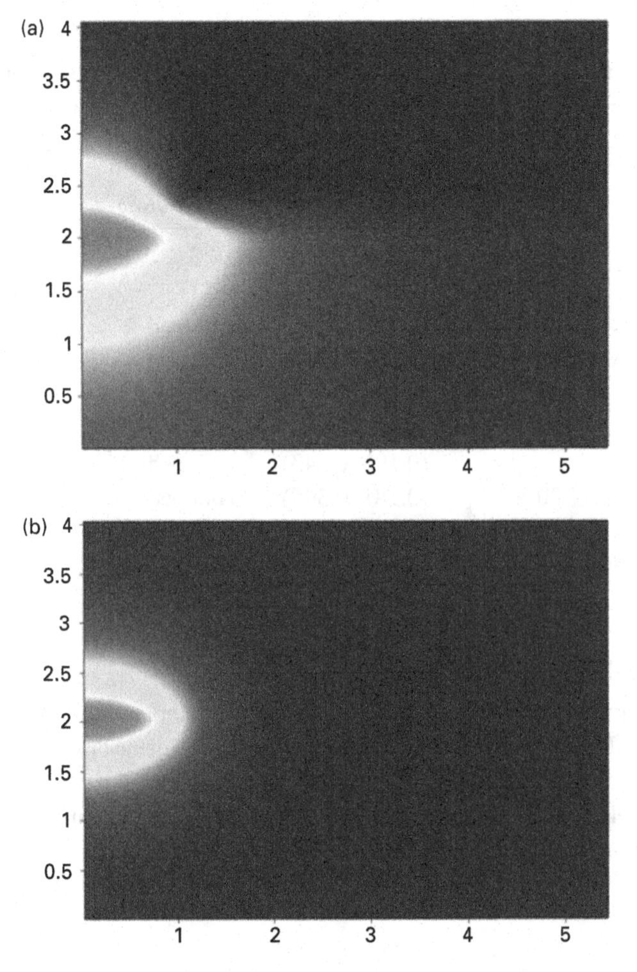

Fig. 12.2. Cross sectional optical field distributions. (a) In the DFB laser section. (b) In the EAM section.

width, as the lasing mode must be on either side of the Bragg stop-band edge for an index-coupled uniform grating DFB laser as discussed in Section 10.3.2.

Both Figs. 12.5 and 12.6 show that mode hopping happens at a specific grating phase among the equally sampled eight different grating phases from 0° to 360°. This is actually a mode switching from one Bragg stop-band edge to the other. Figure 12.5 also shows that at a different specific grating phase the output power is extremely low. Actually, the majority of the power goes to the other direction with this grating phase. Hence the output power from the rear facet is higher despite the HR coating, due to the favorable grating phase condition. Therefore, we may conclude that the yield due to the random grating phase at the HR coated rear facet is roughly 75%, since there are two counts of failures among the eight uniformly sampled grating phases. Provided that other fabrication related issues on the yield can be controlled, this structure-dependent non-uniformity gives the inherent limit to the yield of such an integrated device.

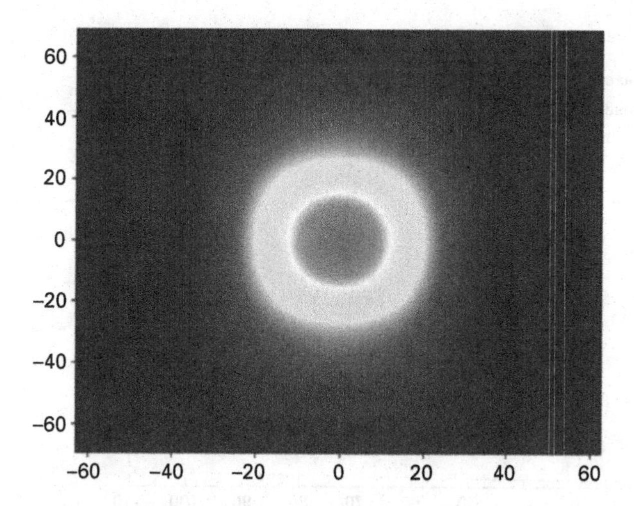

Fig. 12.3. Optical far-field pattern from the EAM front facet.

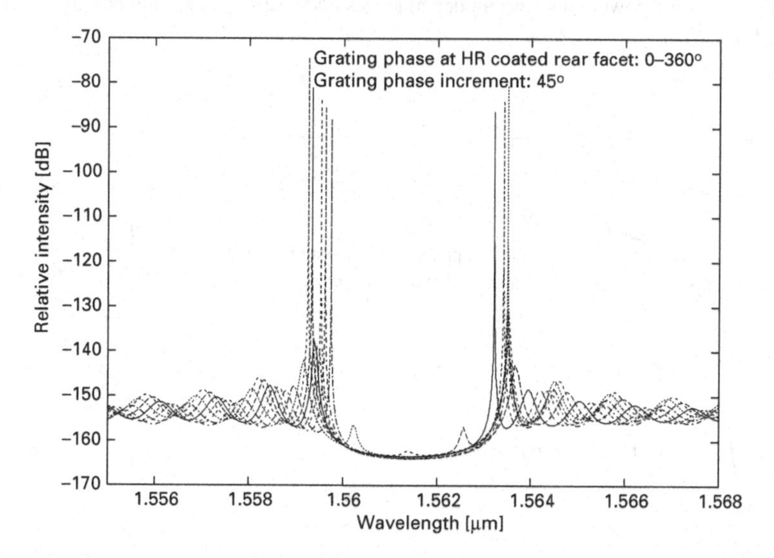

Fig. 12.4. DFB laser spectra under different grating phases at the HR coated rear facet.

12.1.4 Simulated electro-absorption modulator performance

Figure 12.7 shows the calculated modal absorption spectra under different reverse bias voltages. We have only plotted the absorptions of the TE mode since a DFB laser with compressively strained QW structure emits the TE mode only.

Figure 12.8 gives the calculated TE mode extinction ratios under different reverse bias voltages.

Figure 12.9(a) shows the calculated TE mode insertion loss under zero bias and the extinction ratio under 2 V reverse bias in the wavelength range of interest. As shown in the

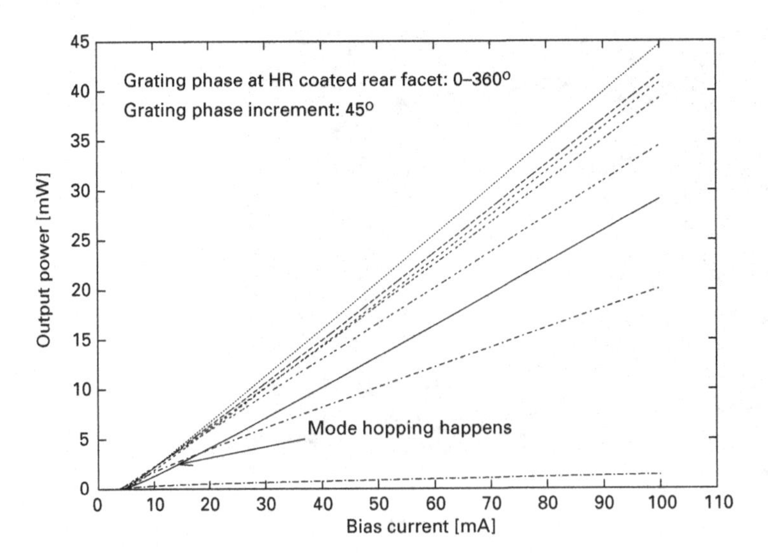

Fig. 12.5. DFB laser output power–bias current dependences under different grating phases at the HR coated rear facet.

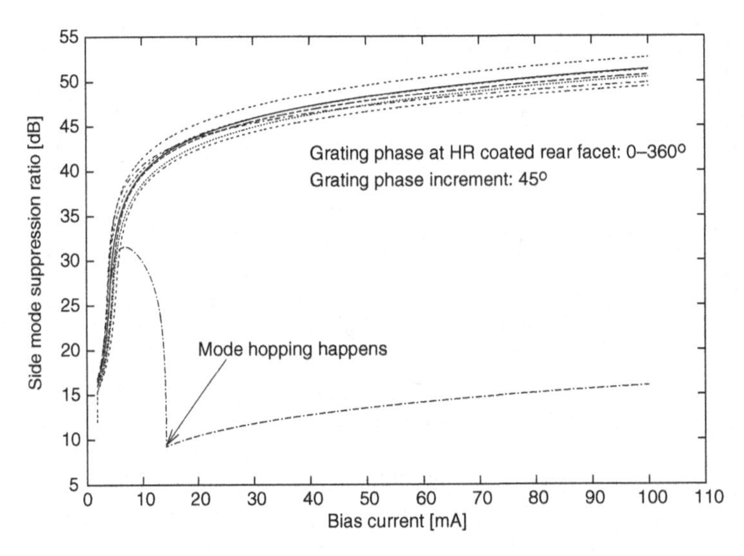

Fig. 12.6. DFB laser SMSR–bias current dependences under different grating phases at the HR coated rear facet.

figure, within the possible lasing wavelength range, i.e., from 1559.5 nm to 1563.5 nm, a 2 V reverse bias voltage will provide an extinction ratio over 44 dB. An insertion loss below 15.7 dB is also achievable over the same wavelength range. This design is certainly far from an optimized one. In dealing with real world problems, however, we can take this design as an initial step from which we proceed towards an optimized structure. For

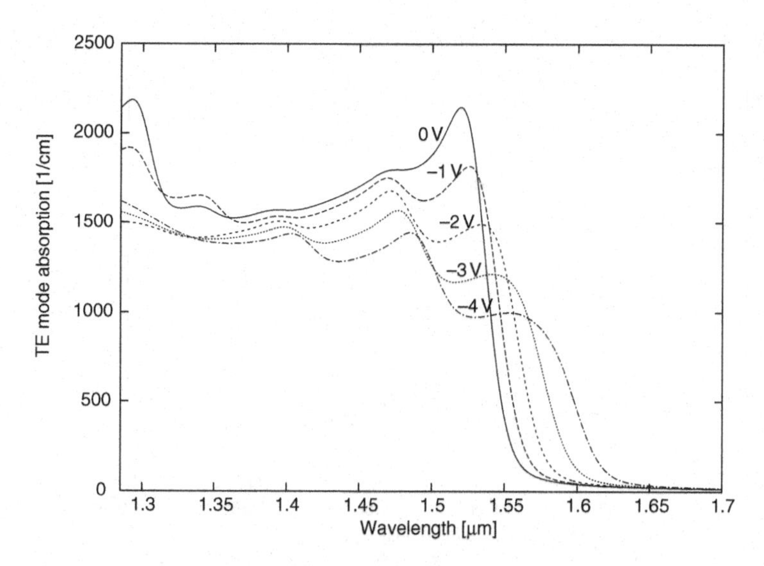

Fig. 12.7. EAM modal absorption spectra for the TE mode under different reverse bias voltages.

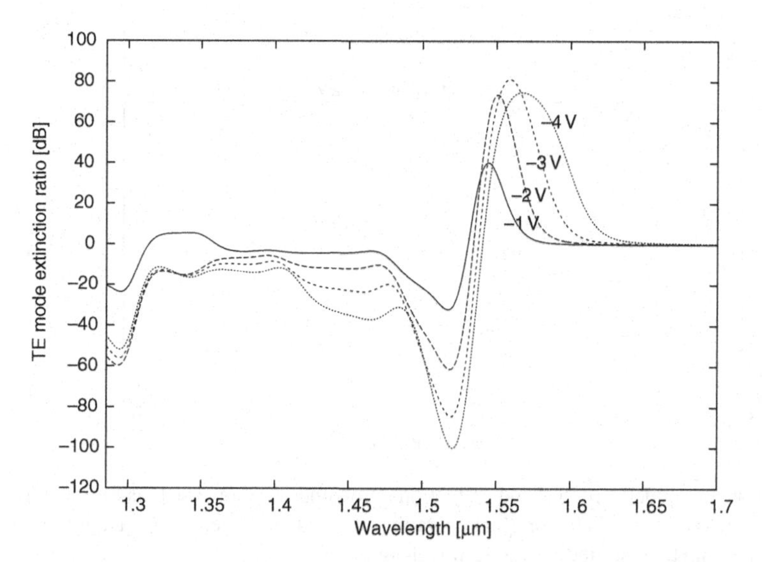

Fig. 12.8. EAM extinction ratios for the TE mode under different reverse bias voltages.

example, Fig. 12.9(a) gives us the hint that if we move our operating wavelength towards the blue side, we may reach a better performance as the insertion loss does not increase as fast as the extinction ratio. If we cannot afford any increase in the insertion loss, we could cut down the EAM length to scale down the loss. Although the extinction ratio will be scaled down as well, we still obtain an increased gap between the insertion loss and the extinction ratio. For example, by blue-shifting the entire operating wavelength range by 3 nm and cutting the EAM section length by half (i.e., to 100 µm), we obtain

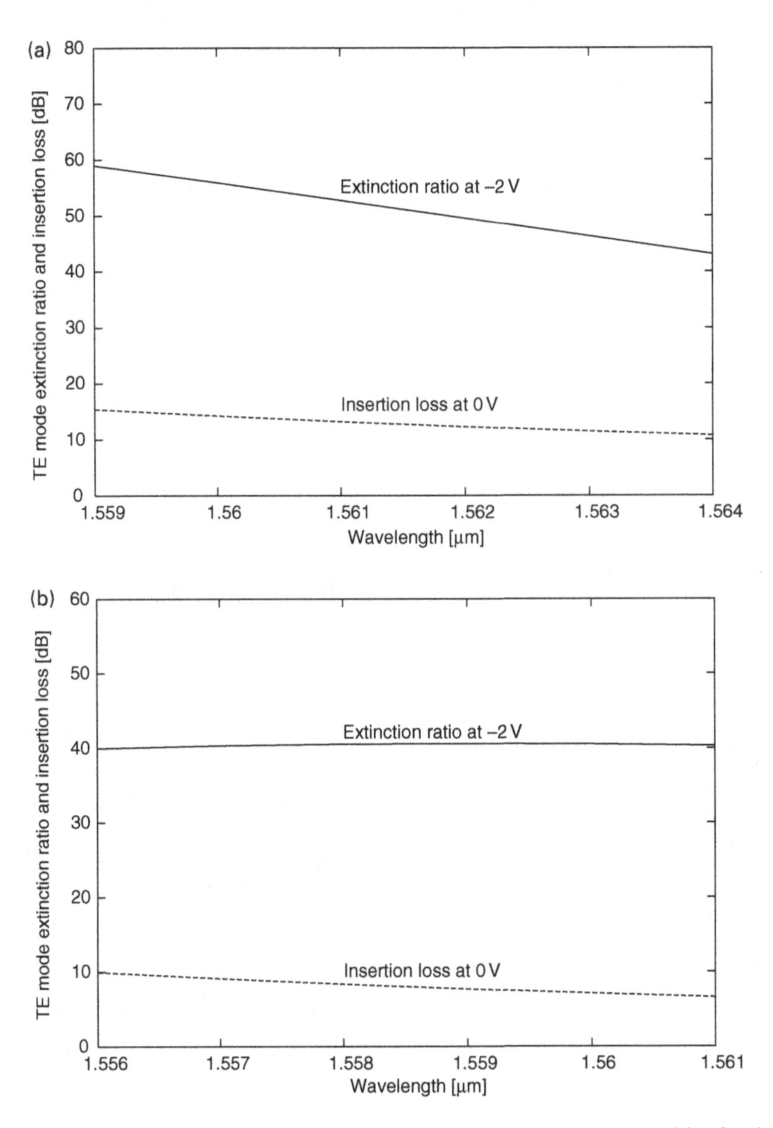

Fig. 12.9. Insertion loss under zero bias and extinction ration under 2 V reverse bias for the EAMs. (a) In a length of 200 μm within the original operating wavelength range. (b) In a length of 100 μm within a 3 nm blue-shifted operating wavelength range.

an almost flat extinction ratio around 40 dB under a 2 V reverse bias, with a reduced insertion loss below 10 dB in the wavelength range from 1556.5 nm to 1560.5 nm, as shown in Fig. 12.9(b). Therefore, at the cost of dropping the extinction ratio by 4 dB, we manage to cut the insertion loss by 5.7 dB.

We may also consider reducing the normalized coupling coefficient in the DFB laser to increase the laser output power and reduce the uncertainty range of the lasing wavelength, since the Bragg stop-band width shrinks as the normalized coupling coefficient decreases.

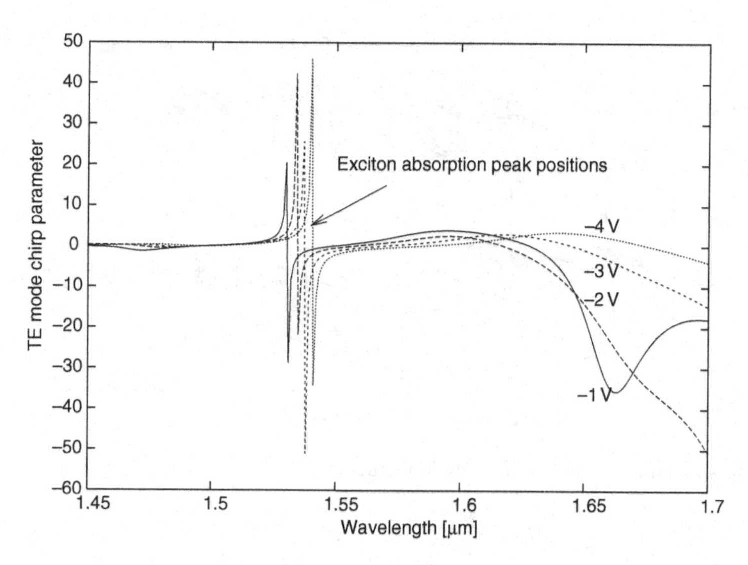

Fig. 12.10. Chirp parameter spectra under different reverse bias voltages.

However, this will reduce the laser's immunity to reflection from the modulator, which may adversely affect the SMSR or cause lasing mode instability [1, 2].

Finally, Fig. 12.10 shows the calculated chirp parameter spectra under different reverse bias voltages.

12.2 The integrated semiconductor distributed feedback laser and monitoring photodetector

12.2.1 The device structure

A schematic structure of an integrated semiconductor DFB laser and monitoring PD is shown in Fig. 12.11. Less popular than the DFB laser and EAM integration that has found many applications in long haul fiber-optic communication backbone networks, this integrated device has been proposed as a component in in-line bidirectional optical diplexer transceivers as the optical network unit (ONU) for passive optical network (PON) based access networks such as fiber-to-the-home (FTTH) systems. As shown in Fig. 12.12, this integrated device is used as a directly modulated laser source for upstream signal transmission through a single mode fiber (SMF), and as a passive waveguide to receive the incoming downstream optical signal from the SMF and to couple the light to a conventional surface incident type PD positioned at the rear end of the device through hybrid integration. The monolithically integrated PD with the DFB laser is for monitoring laser output power and for absorbing the back-traveling laser beam. Knowing that the upstream signal carrier (at 1310 nm) is on the blue (i.e., shorter wavelength) side of the downstream signal carrier (at 1490 nm) according to the ITU standard [3], and by utilizing the fact that an active region designed for lasing at 1310 nm can be made transparent

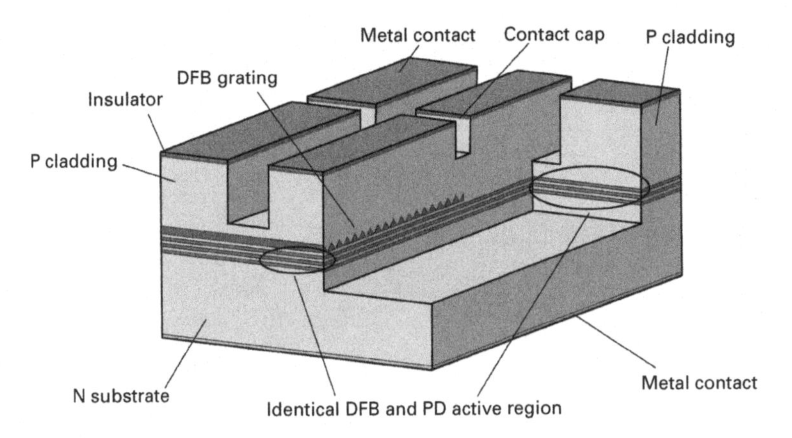

Fig. 12.11. An integrated DFB LD and monitoring PD structure.

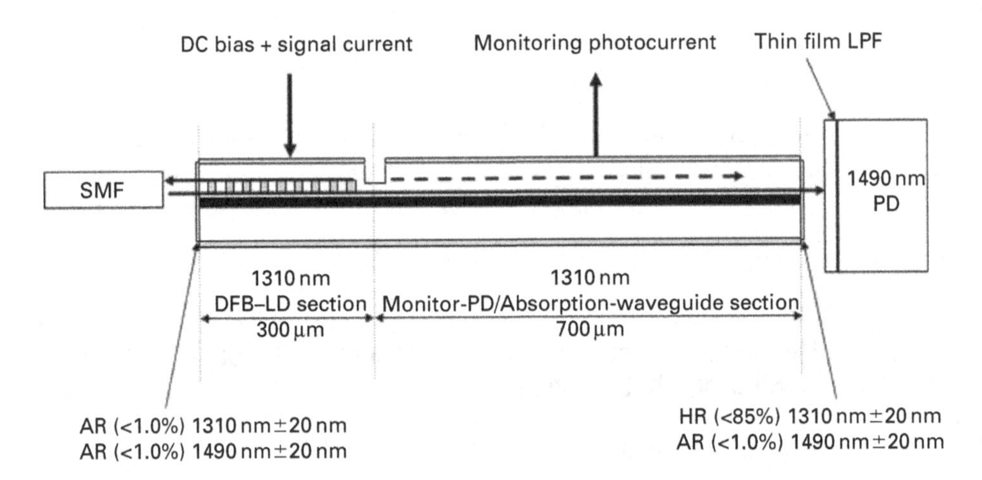

Fig. 12.12. Structure of the in-line bidirectional optical diplexer transceiver.

to light at 1490 nm, we take this integration design to provide the functionality of a wavelength demultiplexer, which has to be used in bidirectional transceivers to separate the upstream and downstream signals carried by different wavelength channels but shared by the same space channel during the transmission.

In this design, the DFB laser section and the monitoring PD section share an identical structure but separated electrode, with the active region gain peak as well as the DFB grating Bragg wavelength designed near 1310 nm. The DC bias and modulation signal currents are applied to the DFB laser section only. As the lasing happens near 1490 nm, the output optical power from the front facet will be coupled to the SMF. The output power from the rear end enters into the unbiased or reversely biased monitoring PD section and gets absorbed. The photocurrent can therefore be picked up at the monitoring PD electrode for laser power control. The incoming downstream signal at 1490 nm is coupled to the device waveguide through the SMF simultaneously. Since the waveguide in both sections is designed to be transparent at 1490 nm, the incoming light will simply pass through the waveguide and reach the rear end hybrid PD for signal detection. The residual light near 1310 nm escaping from the rear facet will be rejected by a thin film low pass filter (LPF) mounted or directly coated on the surface of the hybrid PD to avoid direct optical crosstalk through the detection PD at 1490 nm which responds to both channels. The device front facet is AR coated for both channels, whereas its rear facet is HR and AR coated for the 1310 nm and 1490 nm channels, respectively, for the obvious reason of giving further discrimination on power in these two channels to avoid crosstalk in the detection PD.

The cross-sectional layer structure of this device is shown in Table 12.3. A non-uniform QW design is used for the active region to flatten the material gain in the neighborhood of the operating wavelength (1310 nm). As a result, the laser will perform more uniformly within the operating temperature range of 233 K to 358 K specified for ONUs in PON based FTTH networks [3].

Both sections of this device have the same ridge with a width of 2.0 μm and a height of 2.0 μm. The DFB laser section and monitoring PD section are 300 μm and 700 μm long, respectively, the electrodes are isolated by a 10 μm wide and 1.0 μm deep trench etched across the ridge. The DFB laser has a uniform purely index-coupled grating with a normalized coupling coefficient of 2.1, which is optimized in terms of the laser overall DC and dynamic performance. This design is a balance between threshold current, output optical power, SMSR and mode stability, immunity to external feedback, and small-signal intensity modulation bandwidth. In the neighborhood of 1310 nm, the front (i.e., the DFB laser side) and the rear (i.e., the monitoring PD side) facets are AR and HR coated, respectively, with amplitude reflectivities at 0.01 and 0.87, respectively.

In the DFB laser section, again, a floating grating design is adopted to control the uniformity of the device. An exponentially growing doping profile is also used in the P type InP cladding layer. A slightly higher compressive strain of 1.5% and a well thickness of 5 nm are applied to every QW regardless of material composition. The barriers are uniform and unstrained, with their bandgap wavelength at 1000 nm. The material composition of the QWs and the well thickness are jointly fine-tuned to set the gain peak detuned from the lasing wavelength for about 10 nm on the red side. Since the required lasing wavelength is at 1310 nm, and we have to set the Bragg grating period in accordance with this wavelength, the gain peak, therefore, must be aligned with 1320 nm. Again, this is a balanced design dictated by the required device DC and dynamic performance over the required temperature operating range. Actually, without this detuning the device will have a low small-signal modulation bandwidth due to the low differential gain. Besides, the device DC performance drops at low temperatures under high injection, since the carrier induced gain peak blue-shift makes the gain drop rapidly at the Bragg wavelength. With further detuning, however, the device DC performance deteriorates at higher ambient temperatures, due to the faster red-shift of the gain peak compared to the Bragg wavelength shift, which makes the lasing wavelength move further away from the gain peak. Consequently, there is not sufficient gain at the lasing wavelength again [4].

Table 12.3. Cross-sectional structure for integrated DFB laser

The table sets out design for the integrated DFB laser and monitoring PD in the InAlGaAs/InP system at 1310 nm.

Layer	Thickness (nm)	Composition of $In_{(1-x)}Ga_xAs_y P_{(1-y)}$ (x, y)	Doping concentration ($10^{18}/cm^3$)	Remarks λ_g (nm) and strain
Cap P-InGaAs	200	(0.468, 1.000)	10.0 (P)	1654, 0.0
Graded doped P-InP	1800	(0.000, 0.000)	1.0–0.5 (P)	918.6, 0.0
Etching stop InGaAsP	10	(0.108, 0.236)	0.5 (P)	1050, 0.0
Spacer InP	140	(0.000, 0.000)	0.5 (P)	918.6, 0.0
Grating InGaAsP	40	(0.249, 0.539)	0.3 (P)	1250, 0.0
Spacer InP	40	(0.000, 0.000)	0.3 (P)	918.6, 0.0
Blocking P-InAlAs	50	(0.479, 0.000)	0.3 (P)	829.2, 0.0
GRINSCH InAlGaAs	50	(0.380, 0.090) –(0.350, 0.130)	undoped	950–1000, 0.0
Barrier InAlGaAs	8.5	(0.350, 0.130)	undoped	1000, 0.0
Well InAlGaAs	5	(0.137, 0.116)	undoped	1501, CS1.50%
Barrier InAlGaAs	8.5	(0.350, 0.130)	undoped	1000, 0.0
Well InAlGaAs	5	(0.190, 0.063)	undoped	1359, CS1.50%
Barrier InAlGaAs	8.5	(0.350, 0.130)	undoped	1000, 0.0
Well InAlGaAs	5	(0.155, 0.098)	undoped	1451, CS1.50%
Barrier InAlGaAs	8.5	(0.350, 0.130)	undoped	1000, 0.0
Well InAlGaAs	5	(0.137, 0.116)	undoped	1501, CS1.50%
Barrier InAlGaAs	8.5	(0.350, 0.130)	undoped	1000, 0.0
Well InAlGaAs	5	(0.190, 0.063)	undoped	1359, CS1.50%
Barrier InAlGaAs	8.5	(0.350, 0.130)	undoped	1000, 0.0
Well InAlGaAs	5	(0.155, 0.098)	undoped	1451, CS1.50%
Barrier InAlGaAs	8.5	(0.350, 0.130)	undoped	1000, 0.0
GRINSCH InAlGaAs	100	(0.350, 0.130) –(0.380, 0.090)	undoped	1000–950, 0.0
Blocking N-InAlAs	30	(0.479, 0.000)	2.0 (N)	829.2, 0.0
Buffer N-InP	500	(0.000, 0.000)	1.0 (N)	918.6, 0.0
Substrate N-InP	100 000	(0.000, 0.000)	3.0 (N)	918.6, 0.0

12.2.2 Simulated distributed feedback laser performance

Figures 12.13 and 12.14 give the calculated 1310 nm channel output optical power and SMSR dependences on the bias current under different ambient temperatures. It shows that such a DFB laser design performs quite uniformly over the temperature range from 233 K to 358 K. Without the red-shift detuning of the gain peak from the

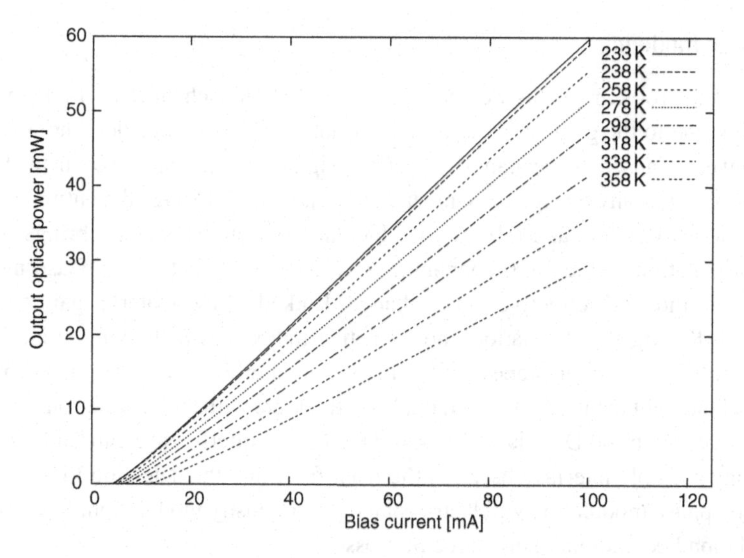

Fig. 12.13. The output power–bias current dependences of the integrated DFB laser under different ambient temperatures.

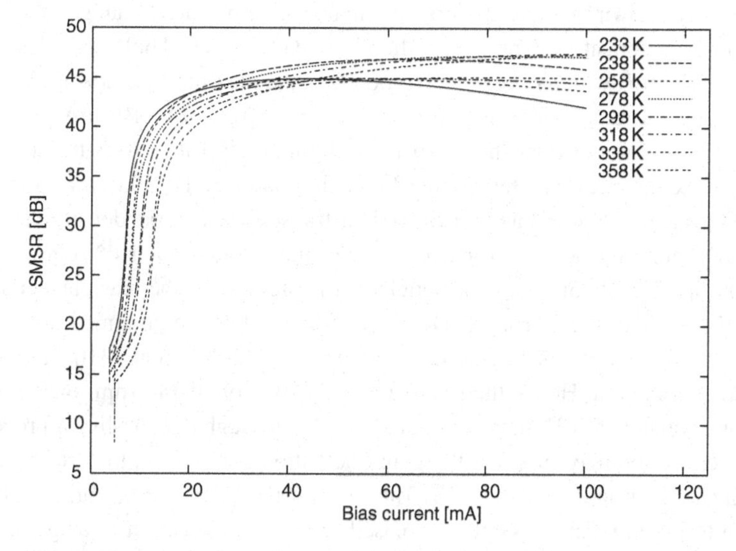

Fig. 12.14. SMSR–bias current dependences of the integrated DFB laser under different ambient temperatures.

Bragg wavelength as mentioned in Section 12.2.1, the device SMSR drops at low ambient temperature under high injection, due to the blue-shift of the gain peak. A sign of such an effect emerges even in the present design, in which a 10 nm red-shift of the gain peak from the Bragg wavelength has been introduced, as can be seen from Fig. 12.14, in which the SMSR under the low ambient temperature drops continually as the bias current increases.

12.2.3 Crosstalk modeling

A potential problem of this device lies in the 1310 nm channel to 1490 nm channel crosstalk due to the gain and phase modulations. Due to the imbedded homogeneity of material gain caused by the many-body effect, signal switches on and off in the DFB laser section will not only change the gain at 1310 nm, but will change the entire gain profile. Thus, the gain or loss at 1490 nm will fluctuate as well. This effect brings in a direct gain modulation to the 1490 channel. An even worse effect is the phase modulation due to the induced refractive index change. Linked to the material gain through the Kramers–Kronig transformation, the refractive index sees its maximum change at the gain profile edge, which happens to fall in the neighborhood of 1490 nm. As a result, the phase of the light in the 1490 nm channel will fluctuate with the modulation current when it passes through the DFB laser section. Due to self-interference caused by reflections from any possible interface between the front facet and the detection PD surface, such parasitic phase modulation will be translated into intensity modulation. Hence the phase modulation becomes another source of crosstalk.

Figures 12.15(a–d) show the calculated material gain profile and refractive index change at different carrier density injection levels. Figure 12.16 gives the calculated average carrier density inside the active region as a function of the bias current.

As clearly shown in Fig. 12.15(b), the material gain indeed changes roughly from -75 /cm to 75 /cm at 1490 nm, when the injected electron and hole densities vary from 0 to 3×10^{18} cm^3, which roughly corresponds to the maximum device operating range, i.e., the bias current applied to the device ranging from 0 to 100 mA, as confirmed by Fig. 12.16. Therefore, as the downstream light at 1490 nm passes through the DFB section, it experiences a material gain fluctuation once the DFB laser section is modulated. As the gain roughly changes from 0 at the transparency carrier densities to 2000 /cm at the maximum injected electron and hole densities around 3×10^{18} /cm^3 at 1310 nm, as shown in Fig. 12.15(a), we can roughly estimate that, as the bias current applied to the DFB section changes from its DC value to its peak value (assumed to be 100 mA), the material gain presents a dynamic range from 0 to 2000 cm at 1310 nm, and from 0 to 75 cm at 1490 nm. Hence the relative gain fluctuation at 1490 nm, once normalized by the gain change at 1310 nm, is about -14 dB. Although the non-linear process in the DFB laser section may suppress the gain fluctuation at 1490 nm, the crosstalk induced by such gain correlation between 1310 nm and 1490 nm can still be considerable. Since the monitoring PD section is either unbiased or reversely biased, as the light at 1490 nm passes through this section, it also suffers a loss due to the material absorption around -75 /cm. In the cross-sectional layer structure design, we have considered this issue by using relatively thin InAlAs blocking layers. As a result, the light at 1490 nm has poor confinement and hence the modal loss is effectively reduced. This design also helps to suppress the modal gain fluctuation at 1490 nm.

Figures 12.15(c&d) indicate that the refractive index changes at 1490 nm and at 1310 nm are on a roughly equal scale. Thus, the parasitic phase modulation at 1490 nm due to the bias change in the DFB laser section is not negligible. Either we have to reduce the residual reflections of the light at 1490 nm to prevent conversion of the phase modulation into intensity modulation due to self-interference, or we have to utilize the phase

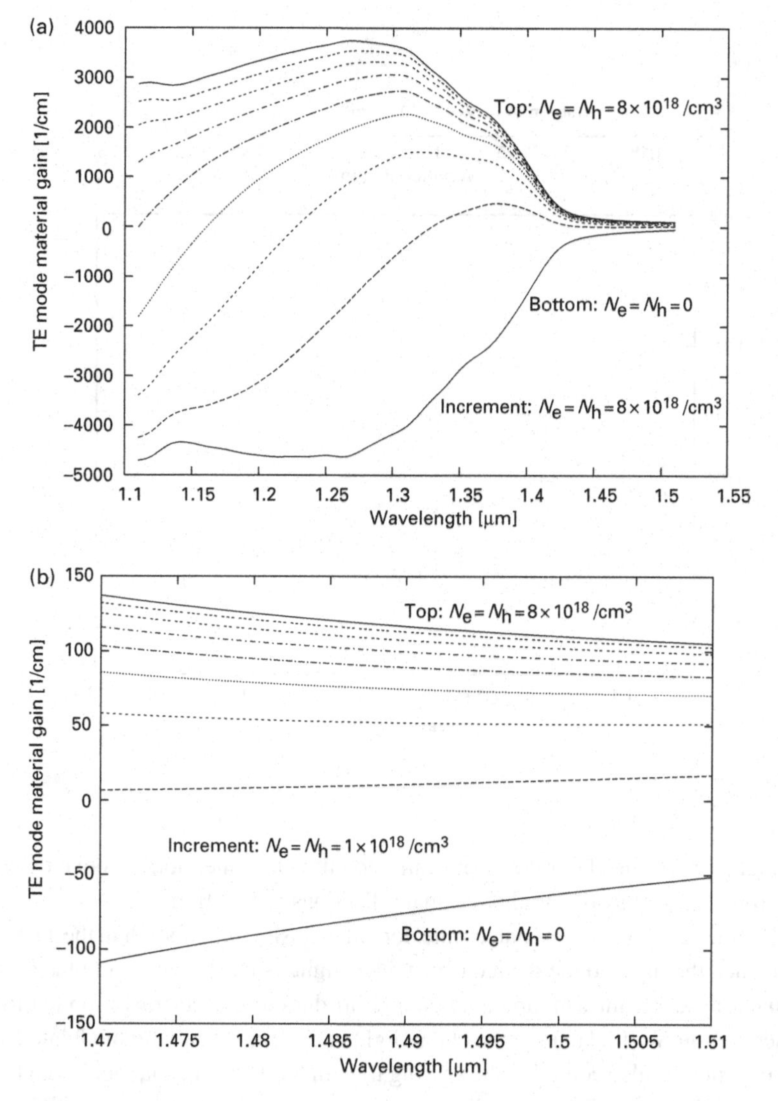

Fig. 12.15. TE mode material gain profiles: (a) full wavelength scale; (b) the neighborhood of 1490 nm. TE mode refractive index change (overleaf); (c) full wavelength scale; (d) the neighborhood of 1490 nm. Each under different carrier densities.

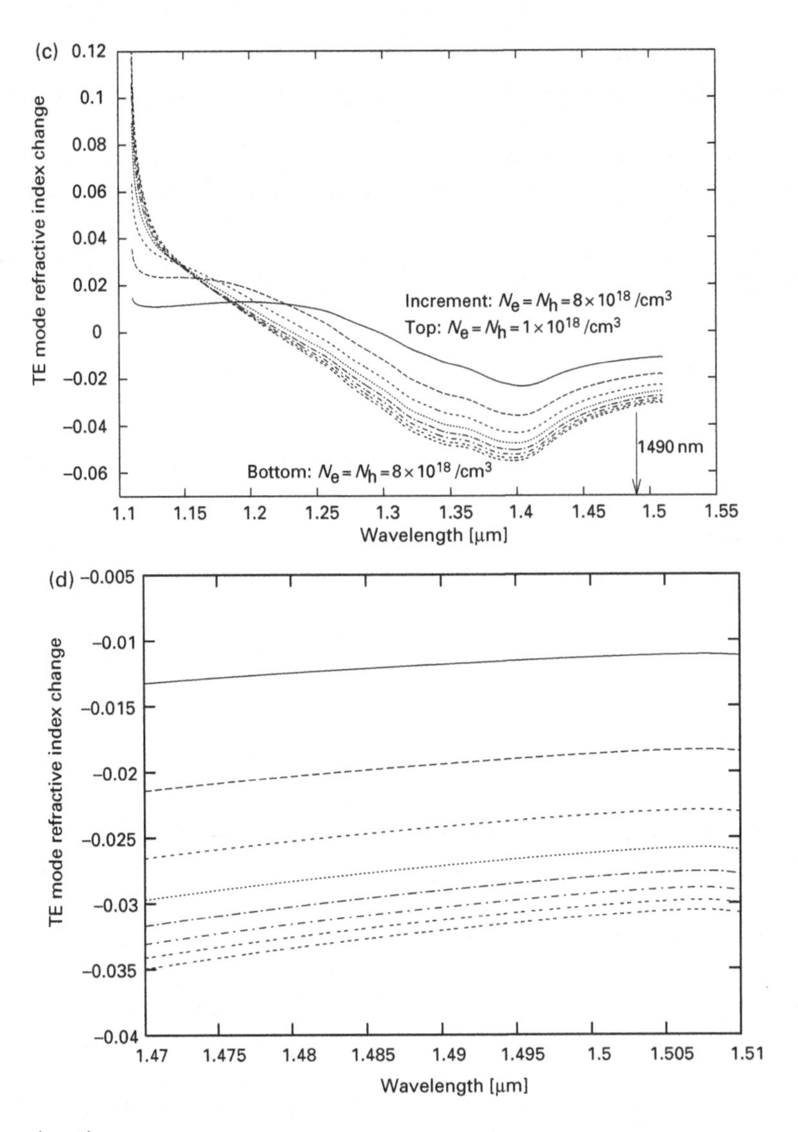

Fig. 12.15. *(cont.)*

modulation to cancel out the direct gain modulation as mentioned above by controlling the round trip delay due to the residual reflections at 1490 nm.

Figure 12.17 shows simulated time domain parasitic modulation in the 1490 nm channel when the input to the device is just a DC light. A DC bias current plus a 1-0 regular square pulse stream with an equal on and off duration of 200 ps is applied to the DFB laser section. We find that, as the output light at 1310 nm is directly modulated, the waveform pattern is also copied to the passing light in the 1490 nm channel through the above mentioned effects. Figure 12.18 shows the corresponding frequency domain base-band spectrum of the light at 1490 nm. It clearly shows that many harmonic frequency lines appear due to the parasitic modulation. If we define the crosstalk as the intensity ratio

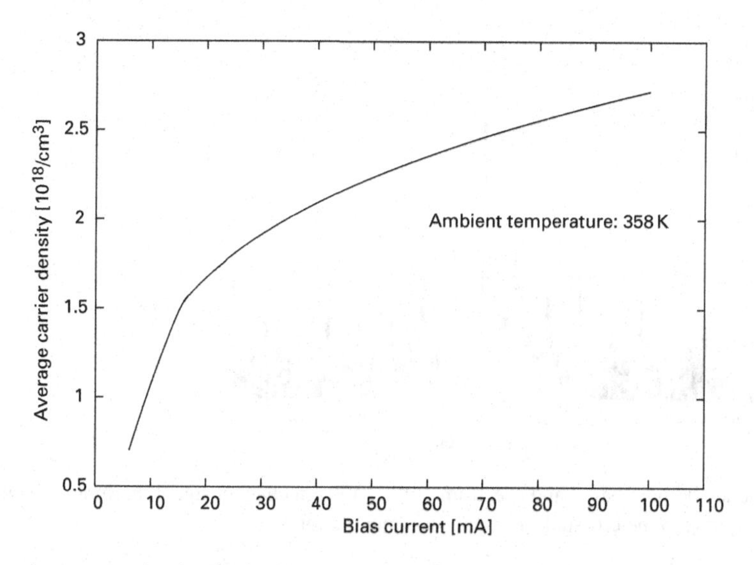

Fig. 12.16. Average carrier density inside the active region as a function of the bias current.

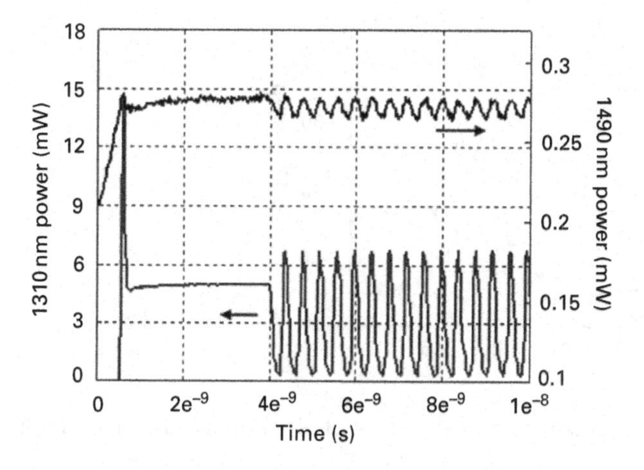

Fig. 12.17. Time domain parasitic modulation in the 1490 nm channel (upper curve) after passing through the device. Its modulation pattern is a duplication of that in the 1310 nm channel (lower curve) when the input (in the 1490 nm channel) is just a DC signal.

between the strongest (which is the first in this case) harmonic component and the DC (undistorted) component, we find from Fig. 12.18 that the crosstalk is around −21 dB.

Figure 12.19 shows the calculated power of the light at 1490 nm and the crosstalk as functions of the change in the round trip phase delay defined as $\Delta\phi \equiv 4\pi \int_0^L \Delta n dz/\lambda$, where $L = 1000\,\mu\text{m}$ indicates the total length of the DFB laser and monitoring PD sections, $\lambda = 1490$ nm the downstream channel wavelength, and Δn the refractive index change at 1490 nm. If the monitoring PD section is not biased, the change in the round trip phase delay comes from the DFB section only. In Fig. 12.19, the crosstalk lobes correspond to the regions where the direct gain modulation and phase modulation induced

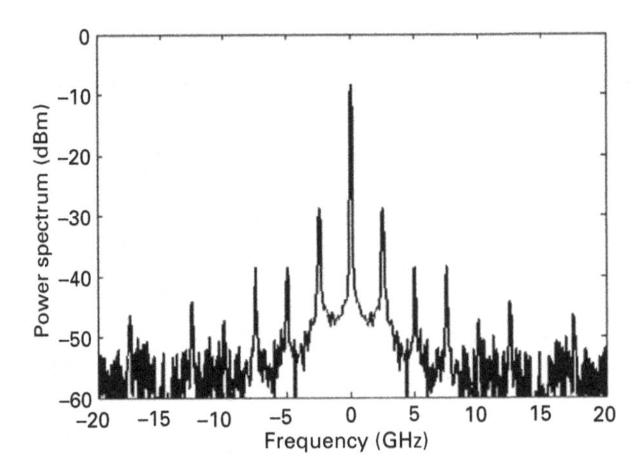

Fig. 12.18. Frequency domain base band spectrum of the 1490 nm channel signal; harmonic frequency lines show up due to the crosstalk (i.e., parasitic modulation).

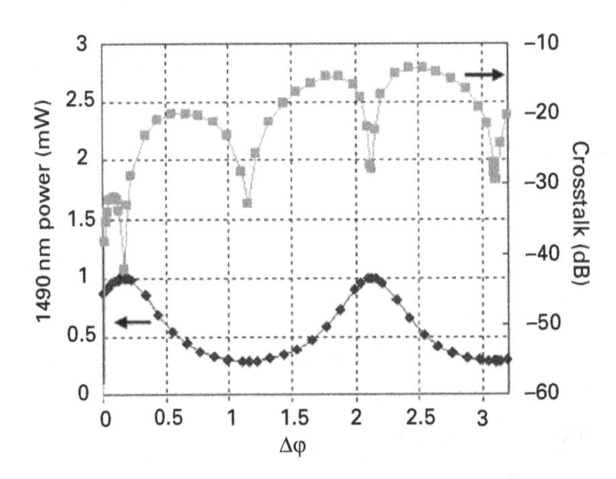

Fig. 12.19. Crosstalk and optical power in the 1490 nm channel as functions of the round trip phase delay inside the device.

intensity changes are added in in-phase and anti-phase fashions alternately. At the boundaries of these regions, crosstalk decreases rapidly due to the cancellation of the two crosstalk mechanisms, which gives us the hint that we should make our device operate at these boundary points so that the crosstalk will be minimized.

References

[1] X. Li and W.-P. Huang, Analysis of frequency chirp in DFB lasers integrated with external modulators. *IEEE Journal of Quantum Electron.*, **QE-30**:12 (1994), 2756–66.

[2] X. Li, W.-P. Huang, D. M. Adams, C. Rolland, and T. Makino, Modeling and design for DFB laser integrated with Mach–Zehnder modulator. *IEEE Journal of Quantum Electron.*, **QE-34**:10 (1998), 1807–15.

[3] W.-P. Huang, X. Li, C.-Q. Xu, *et al.*. Optical transceivers for fiber-to-the-premises applications: system requirements and enabling technologies. *IEEE/OSA Journal of Lightwave Tech.*, **LT-25**:1 (2007), 11–27.

[4] H. Lu, C. Blaauw, and T. Makino, High-temperature single-mode operation of 1.3 μm strained MQW gain-coupled DFB lasers. *IEEE Photon. Tech. Lett.*, **7**:6 (1995), 611–3.

Further reading

Integrated DFB and EAM

[1] T. Watanabe, K. Sato, and H. Soda, Low drive voltage and low chirp modulator integrated DFB laser light source for multiple-gigabit systems. *Fujitsu Sci. Tech. J.*, **28**:1 (1992), 115–21.

[2] P. Ojala, C. Pettersson, B. Stoltz, *et al.*, DFB laser monolithically integrated with an absorption modulator with low residual reflectance and small chirp. *Electron. Lett.*, **29**:10 (1993), 859–60.

[3] G. L. Li and P. K. L. Yu, Optical intensity modulators for digital and analog applications. *IEEE/OSA Journal of Lightwave Tech.*, **LT-21**:9 (2003), 2010–30.

Appendices

A Lowdin's renormalization theory

Using the projection operator $|\varphi_n\rangle\langle\varphi_n|$ we can expand a state $|\phi\rangle$ through base $|\varphi_n\rangle$, $n = 1, 2, 3, \ldots$, i.e., $|\phi\rangle = \sum_{n=1}^{\infty} |\varphi_n\rangle\langle\varphi_n \mid \phi\rangle = \sum_{n=1}^{\infty} a_n |\varphi_n\rangle$, where we have utilized $\sum_{n=1}^{\infty} |\varphi_n\rangle\langle\varphi_n| \equiv 1$ and $a_n \equiv \langle\varphi_n \mid \phi\rangle$. Therefore, multiply both sides of equation $\vec{H}|\phi\rangle = E|\phi\rangle$ by $\langle\varphi_m|$ to yield

$$\sum_{n=1}^{\infty} (H_{mn} - E\delta_{mn}) a_n = 0, \tag{A.1}$$

where we have defined $H_{mn} \equiv \langle\varphi_m| \vec{H} |\phi_n\rangle$ and $\langle\varphi_m \mid \varphi_n\rangle = \delta_{mn}$ is used.

According to Lowdin's renormalization theory, equation (A.1) can also be written as

$$(H_{mm} - E)a_m + \sum_{n\neq m}^{\infty} H_{mn} a_n = 0. \tag{A.2}$$

Now, we partition the complete set of eigensolutions into two groups A and B, with group A containing all those eigenvalues that we are interested in and group B the rest. Our intention is to find the solution for group A states but with the effect of group B states included. If we only put one eigensolution into group A, this method reduces to the normal perturbation approach. However, when there are strongly coupled states, e.g., energy degenerated states, multiple states can be put into group A, which makes this method perfectly suitable for treating the degenerated valence band. Under such a partition, equation (A.2) becomes

$$(H_{mm} - E)a_m + \sum_{n\neq m}^{\mathrm{A}} H_{mn} a_n + \sum_{n\neq m}^{\mathrm{B}} H_{mn} a_n = 0, \tag{A.3}$$

or

$$a_m = \sum_{n\neq m}^{\mathrm{A}} \frac{H_{mn}}{E - H_{mm}} a_n + \sum_{n\neq m}^{\mathrm{B}} \frac{H_{mn}}{E - H_{mm}} a_n. \tag{A.4}$$

Firstly, we use $a_m^{(0)} \approx \sum_{n\neq m}^{A} \frac{H_{mn}}{E-H_{mm}} a_n$ to plug into the second term on the RHS of equation (A.4) to obtain

$$
\begin{aligned}
a_m^{(1)} &\approx \sum_{n\neq m}^{A} \frac{H_{mn}}{E-H_{mm}} a_n + \sum_{n\neq m}^{B} \frac{H_{mn}}{E-H_{mm}} \sum_{\alpha\neq n}^{A} \frac{H_{n\alpha}}{E-H_{nn}} a_\alpha \\
&= \sum_{n\neq m}^{A} \frac{H_{mn}}{E-H_{mm}} a_n + \sum_{n\neq \alpha}^{A} \frac{1}{E-H_{mm}} \sum_{\alpha\neq m}^{B} \frac{H_{m\alpha}H_{\alpha n}}{E-H_{\alpha\alpha}} a_n \\
&= \sum_{n\neq m}^{A} \frac{H_{mn} + \sum_{\alpha\neq m}^{B} \frac{H_{m\alpha}H_{\alpha n}}{E-H_{\alpha\alpha}}}{E-H_{mm}} a_n \\
&= \sum_{n}^{A} \frac{H_{mn} + \sum_{\alpha\neq m}^{B} \frac{H_{m\alpha}H_{\alpha n}}{E-H_{\alpha\alpha}} - H_{mn}\delta_{mn}}{E-H_{mm}} a_n .
\end{aligned}
\tag{A.5}
$$

Such an iteration process can be performed repeatedly by substituting equation (A.5) into the second term on the RHS of (A.4), and hence we obtain

$$
a_m = \sum_{n}^{A} \frac{U_{mn}^{A} - H_{mn}\delta_{mn}}{E-H_{mm}} a_n, \tag{A.6}
$$

with

$$
U_{mn}^{A} \equiv H_{mn} + \sum_{\alpha\neq m}^{B} \frac{H_{m\alpha}H_{\alpha n}}{E-H_{\alpha\alpha}} + \sum_{\substack{\alpha,\beta\neq m,n \\ \alpha\neq\beta}}^{B} \frac{H_{m\alpha}H_{\alpha\beta}H_{\beta n}}{(E-H_{\alpha\alpha})(E-H_{\alpha\alpha})} + \cdots . \tag{A.7}
$$

Equation (A.6) is equivalent to

$$
\sum_{n}^{A} (U_{mn}^{A} - E\delta_{mn}) a_n = 0, \tag{A.8}
$$

for states belonging to group A, or $n \subset$ A and $m \subset$ A.

Once those group A coefficients are obtained by solving equation (A.8), we can also find the group B coefficients according to equation (A.6)

$$
a_\gamma = \sum_{n}^{A} \frac{U_{\gamma n}^{A} - H_{\gamma n}\delta_{\gamma n}}{E-H_{\gamma\gamma}} a_n, \tag{A.9}
$$

for states belonging to group B, or $\gamma \subset$ B.

A necessary condition for the expansion (A.7) to converge is

$$|H_{m\gamma}| \ll |E - H_{\gamma\gamma}|, \tag{A.10}$$

for states m and γ belonging to groups A and B, or $m \subset \text{A}$ and $\gamma \subset \text{B}$, respectively.

In particular, to first order accuracy, equation (A.7) becomes

$$U^{\text{A}}_{mn} = H_{mn} + \sum_{\alpha \neq m}^{\text{B}} \frac{H_{m\alpha} H_{\alpha n}}{E - H_{\alpha\alpha}}. \tag{A.11}$$

B Integrations in the many-body gain model

For quantum well structures, by following equation (7.2a), we convert the wave vector domain summations into integrals in computing the coefficients summarized in Table 4.3. These coefficients appear in the governing ODEs, i.e., equation (4.156a) for the slow-varying envelope function of the polariton number expectation, and equations (4.156b&c) for the conduction band electron and valence band hole number expectations, respectively. In the following derivations, we will also rewrite equation (4.9d) as

$$V_{|\vec{k}-\vec{k}'|} = \frac{e^2 \Phi(|\vec{k} - \vec{k}'|)}{2\varepsilon_0 \Sigma |\vec{k} - \vec{k}'|}. \tag{B.1}$$

The 1st order diagonal summation is

$$\begin{aligned}
I^1_{\vec{k}} &\equiv \sum_{\vec{l} \neq \vec{k}} V_{|\vec{l}-\vec{k}|}(f^{\text{e}}_{\vec{l}} + f^{\text{h}}_{\vec{l}} - 1) = \sum_{\vec{l} \neq \vec{k}} \frac{e^2 \Phi(|\vec{l} - \vec{k}|)}{2\varepsilon_0 \Sigma |\vec{l} - \vec{k}|}(f^{\text{e}}_{\vec{l}} + f^{\text{h}}_{\vec{l}} - 1) \\
&= \sum_{n_z} \frac{e^2}{4\pi^2 \varepsilon_0} \int_0^\infty \int_0^{2\pi} l_r \mathrm{d}l_r \, \mathrm{d}\varphi_l \\
&\quad \times \left\{ \frac{\Phi\left[\sqrt{(k_r^2 + l_r^2 - 2k_r l_r \cos(\varphi_k - \varphi_l))}\right](f^{\text{e}}_{l_r \varphi_l} + f^{\text{h}}_{l_r \varphi_l} - 1)}{\sqrt{(k_r^2 + l_r^2 - 2k_r l_r \cos(\varphi_k - \varphi_l))}} \Big|_{l_r \neq k_r \cup \varphi_l \neq \varphi_k} \right\} \\
&= \sum_{n_z} \left[\int_0^{k_r - \Delta} \mathrm{d}l_r \phi_1(k_r, \varphi_k, l_r) + \int_{k_r + \Delta}^\infty \mathrm{d}l_r \phi_1(k_r, \varphi_k, l_r) \right. \\
&\quad \left. + \int_{k_r - \Delta}^{k_r + \Delta} \mathrm{d}l_r \phi_1(k_r, \varphi_k, l_r)|_{\varphi_l \neq \varphi_k} \right],
\end{aligned} \tag{B.2a}$$

where

$$\phi_1(k_r, \varphi_k, l_r) \equiv \frac{e^2 l_r}{4\pi^2 \varepsilon_0} \int_0^{2\pi} \mathrm{d}\varphi_l \left\{ \frac{\Phi\left[\sqrt{(k_r^2 + l_r^2 - 2k_r l_r \cos(\varphi_k - \varphi_l))}\right]}{\sqrt{(k_r^2 + l_r^2 - 2k_r l_r \cos(\varphi_k - \varphi_l))}} (f^{\text{e}}_{l_r \varphi_l} + f^{\text{h}}_{l_r \varphi_l} - 1) \right\}. \tag{B.2b}$$

The 1st order non-diagonal summation is

$$I_{\vec{k}}^{2} \equiv \sum_{\vec{l} \neq \vec{k}} V_{|\vec{l}-\vec{k}|}\widetilde{p}_{\vec{l}} = \sum_{\vec{l} \neq \vec{k}} \frac{\mathrm{e}^2\Phi(|\vec{l}-\vec{k}|)}{2\varepsilon_0\Sigma|\vec{l}-\vec{k}|}\widetilde{p}_{\vec{l}}$$

$$= \sum_{n_z} \frac{\mathrm{e}^2}{4\pi^2\varepsilon_0} \int_0^\infty \int_0^{2\pi} l_r \mathrm{d}l_r \, \mathrm{d}\varphi_l \frac{\Phi\left[\sqrt{\left(k_r^2 + l_r^2 - 2k_r l_r \cos(\varphi_k - \varphi_l)\right)}\right]\widetilde{p}_{l_r\varphi_l}}{\sqrt{\left(k_r^2 + l_r^2 - 2k_r l_r \cos(\varphi_k - \varphi_l)\right)}}|_{l_r \neq k_r \bigcup \varphi_l \neq \varphi_k}$$

$$= \sum_{n_z} \left[\int_0^{k_r-\Delta} \mathrm{d}l_r \phi_2(k_r, \varphi_k, l_r) + \int_{k_r+\Delta}^{\infty} \mathrm{d}l_r \phi_2(k_r, \varphi_k, l_r)\right.$$

$$\left. + \int_{k_r-\Delta}^{k_r+\Delta} \mathrm{d}l_r \phi_2(k_r, \varphi_k, l_r)|_{\varphi_l \neq \varphi_k}\right], \tag{B.3a}$$

where

$$\phi_2(k_r, \varphi_k, l_r) \equiv \frac{\mathrm{e}^2 l_r}{4\pi^2\varepsilon_0} \int_0^{2\pi} \mathrm{d}\varphi_l \left\{\frac{\Phi\left[\sqrt{\left(k_r^2 + l_r^2 - 2k_r l_r \cos(\varphi_k - \varphi_l)\right)}\right]}{\sqrt{\left(k_r^2 + l_r^2 - 2k_r l_r \cos(\varphi_k - \varphi_l)\right)}}\widetilde{p}_{l_r\varphi_l}\right\}. \tag{B.3b}$$

The 2nd order diagonal summation is

$$J_{\vec{k}}^{1} \equiv \sum_{\alpha,\beta=e,h} \sum_{\vec{k}''} \sum_{\vec{l} \neq 0} \left(2\widetilde{V}_{|\vec{l}|}^2 - \delta_{\alpha\beta}\widetilde{V}_{|\vec{l}|}\widetilde{V}_{|\vec{k}''-\vec{l}-\vec{k}|}\right)$$

$$\times g\left[\varepsilon_{\alpha\vec{k}} + \varepsilon_{\beta\vec{k}''} - \varepsilon_{\alpha(\vec{k}+\vec{l})} - \varepsilon_{\beta(\vec{k}''-\vec{l})}\right]$$

$$\times \left[f_{\vec{k}+\vec{l}}^{\alpha}(1 - f_{\vec{k}''}^{\beta})f_{\vec{k}''-\vec{l}}^{\beta} + (1 - f_{\vec{k}+\vec{l}}^{\alpha})f_{\vec{k}''}^{\beta}(1 - f_{\vec{k}''-\vec{l}}^{\beta})\right]$$

$$= \left(\frac{\mathrm{e}^2}{2\varepsilon_0\Sigma}\right)^2 \sum_{\alpha,\beta=e,h} \sum_{\vec{k}''} \sum_{\vec{l} \neq 0} \left[\frac{2\Phi(|\vec{l}|)}{\varepsilon_{\vec{l}}|\vec{l}|} - \frac{\delta_{\alpha\beta}\Phi(|\vec{k}''-\vec{l}-\vec{k}|)}{\varepsilon_{\vec{k}''-\vec{l}-\vec{k}}|\vec{k}''-\vec{l}-\vec{k}|}\right]\frac{\Phi(|\vec{l}|)}{\varepsilon_{\vec{l}}|\vec{l}|}$$

$$\times \frac{\delta + \mathrm{j}\left[\varepsilon_{\alpha\vec{k}} + \varepsilon_{\beta\vec{k}''} - \varepsilon_{\alpha(\vec{k}+\vec{l})} - \varepsilon_{\beta(\vec{k}''-\vec{l})}\right]}{\delta^2 + \left[\varepsilon_{\alpha\vec{k}} + \varepsilon_{\beta\vec{k}''} - \varepsilon_{\alpha(\vec{k}+\vec{l})} - \varepsilon_{\beta(\vec{k}''-\vec{l})}\right]^2}$$

$$\times \left[f_{\vec{k}+\vec{l}}^{\alpha}(1 - f_{\vec{k}''}^{\beta})f_{\vec{k}''-\vec{l}}^{\beta} + (1 - f_{\vec{k}+\vec{l}}^{\alpha})f_{\vec{k}''}^{\beta}(1 - f_{\vec{k}''-\vec{l}}^{\beta})\right]$$

$$= \left(\frac{\mathrm{e}^2}{4\pi^2\varepsilon_0}\right)^2 \sum_{\alpha,\beta=e,h} \sum_{n_{z1}} \sum_{n_{z2}} \int_0^\infty \int_0^{2\pi} k_r'' \mathrm{d}k_r'' \, \mathrm{d}\varphi_{k''} \int_0^\infty \int_0^{2\pi} \mathrm{d}l_r \, \mathrm{d}\varphi_l \left\{\left\{\frac{2\Phi(l_r)}{\varepsilon(l_r, \varphi_l)l_r}\right.\right.$$

$$\left. - \frac{\delta_{\alpha\beta}\Phi\left[Q(k_r, \varphi_k, k_r'', \varphi_{k''}, l_r, \varphi_l)\right]}{\varepsilon(k_r, \varphi_k, k_r'', \varphi_{k''}, l_r, \varphi_l)Q(k_r, \varphi_k, k_r'', \varphi_{k''}, l_r, \varphi_l)}\right\}\frac{\Phi(l_r)}{\varepsilon(l_r, \varphi_r)}$$

$$\left. \times \frac{\delta + i\varepsilon_{\Delta 1\alpha\beta}(k_r, \varphi_k, k_r'', \varphi_{k''}, l_r, \varphi_l)}{\delta^2 + \varepsilon_{\Delta 1\alpha\beta}^2(k_r, \varphi_k, k_r'', \varphi_{k''}, l_r, \varphi_l)} F_{1\alpha\beta}(k_r, \varphi_k, k_r'', \varphi_{k''}, l_r, \varphi_l)\right\}|_{l_r \neq 0}$$

$$= \sum_{\alpha,\beta=e,h} \sum_{n_{z1}} \sum_{n_{z2}} \int_0^\infty \int_0^{2\pi} k_r'' \mathrm{d}k_r'' \, \mathrm{d}\varphi_{k''} \left[\int_\Delta^\infty \mathrm{d}l_r \phi_{3\alpha\beta}(k_r, \varphi_k, k_r'', \varphi_{k''}, l_r)\right], \tag{B.4a}$$

where we have defined

$$\phi_{3\alpha\beta}(k_r, \varphi_k, k_r^{''}, \varphi_{k''}, l_r) \equiv \phi_{31\alpha\beta}(k_r, \varphi_k, k_r^{''}, \varphi_{k''}, l_r) + \phi_{32\alpha\beta}(k_r, \varphi_k, k_r^{''}, \varphi_{k''}, l_r), \tag{B.4b}$$

$$\begin{aligned}\phi_{31\alpha\beta}(k_r, \varphi_k, k_r^{''}, \varphi_{k''}, l_r) \equiv & \left(\frac{\mathrm{e}^2}{4\pi^2\varepsilon_0}\right)^2 \int_0^{2\pi} \mathrm{d}\varphi_l \left\{\left\{\frac{2\Phi(l_r)}{\varepsilon(l_r, \varphi_l) l_r}\right.\right. \\ & - \left.\frac{\delta_{\alpha\beta}\Phi\left[Q(k_r, \varphi_k, k_r^{''}, \varphi_{k''}, l_r, \varphi_l)\right]}{\varepsilon(k_r, \varphi_k, k_r^{''}, \varphi_{k''}, l_r, \varphi_l) Q(k_r, \varphi_k, k_r^{''}, \varphi_{k''}, l_r, \varphi_l)}\right\} \frac{\Phi(l_r)}{\varepsilon(l_r, \varphi_l)} \\ & \times \left.\frac{\delta + \mathrm{j}\varepsilon_{\Delta 1\alpha\beta}(k_r, \varphi_k, k_r^{''}, \varphi_{k''}, l_r, \varphi_l)}{\delta^2 + \varepsilon^2_{\Delta 1\alpha\beta}(k_r, \varphi_k, k_r^{''}, \varphi_{k''}, l_r, \varphi_l)} F_{11\alpha\beta}(k_r, \varphi_k, k_r^{''}, \varphi_{k''}, l_r, \varphi_l)\right\}\Big|_{l_r \neq 0},\end{aligned} \tag{B.4c}$$

$$\begin{aligned}\phi_{32\alpha\beta}(k_r, \varphi_k, k_r^{''}, \varphi_{k''}, l_r) \equiv & \left(\frac{\mathrm{e}^2}{4\pi^2\varepsilon_0}\right)^2 \int_0^{2\pi} \mathrm{d}\varphi_l \left\{\left\{\frac{2\Phi(l_r)}{\varepsilon(l_r, \varphi_l) l_r}\right.\right. \\ & - \left.\frac{\delta_{\alpha\beta}\Phi\left[Q(k_r, \varphi_k, k_r^{''}, \varphi_{k''}, l_r, \varphi_l)\right]}{\varepsilon(k_r, \varphi_k, k_r^{''}, \varphi_{k''}, l_r, \varphi_l) Q(k_r, \varphi_k, k_r^{''}, \varphi_{k''}, l_r, \varphi_l)}\right\} \frac{\Phi(l_r)}{\varepsilon(l_r, \varphi_l)} \\ & \times \left.\frac{\delta + \mathrm{j}\varepsilon_{\Delta 1\alpha\beta}(k_r, \varphi_k, k_r^{''}, \varphi_{k''}, l_r, \varphi_l)}{\delta^2 + \varepsilon^2_{\Delta 1\alpha\beta}(k_r, \varphi_k, k_r^{''}, \varphi_{k''}, l_r, \varphi_l)} F_{12\alpha\beta}(k_r, \varphi_k, k_r^{''}, \varphi_{k''}, l_r, \varphi_l)\right\}\Big|_{l_r \neq 0}.\end{aligned} \tag{B.4d}$$

We have also defined

$$\begin{aligned}Q(k_r, \varphi_k, k_r^{''}, \varphi_{k''}, l_r, \varphi_l) &\equiv |\vec{k}'' - \vec{l} - \vec{k}| \\ &= \sqrt{\left(k_r^2 + k_r^{''2} + l_r^2 - 2k_r k_r^{''}\cos(\varphi_k - \varphi_{k''}) + 2k_r l_r \cos(\varphi_k - \varphi_l) - 2k_r^{''} l_r \cos(\varphi_{k''} - \varphi_l)\right)},\end{aligned} \tag{B.4e}$$

$$\begin{aligned}\varepsilon(l_r, \varphi_l) \equiv \varepsilon_{\vec{l}} &= 1 - V_{|\vec{l}|} \sum_{\vec{x}} \sum_{\alpha=e,h} \frac{f^{\alpha}_{\vec{x}-\vec{l}} - f^{\alpha}_{\vec{x}}}{\hbar\omega + i\delta + \varepsilon_{\alpha(\vec{x}-\vec{l})} - \varepsilon_{\alpha\vec{x}}} \\ &= 1 - \frac{\mathrm{e}^2\Phi(l_r)}{4\pi^2\varepsilon_0 l_r} \sum_{\alpha=e,h} \sum_{n_z} \int_0^{\infty} \int_0^{2\pi} x_r\, \mathrm{d}x_r\, \mathrm{d}\varphi_x \frac{f_1(l_r, \varphi_l, x_r, \varphi_x)}{\hbar\omega + \mathrm{j}\delta + \varepsilon_{\delta 1}(l_r, \varphi_l, x_r, \varphi_x)},\end{aligned} \tag{B.4f}$$

$$f_1(l_r, \varphi_l, x_r, \varphi_x) \equiv f^{\alpha}_{\vec{x}-\vec{l}} - f^{\alpha}_{\vec{x}} = f^{\alpha}_{d_1\varphi_{d_1}} - f^{\alpha}_{x_r\varphi_x}, \tag{B.4g}$$

$$d_1 = \sqrt{(x_r^2 + l_r^2 - 2x_r l_r \cos(\varphi_x - \varphi_l))}, \tag{B.4h}$$

$$\varphi_{d_1} \equiv \tan^{-1}\left[\frac{x_r \sin\varphi_x - l_r \sin\varphi_l}{x_r \cos\varphi_x - l_r \cos\varphi_l}\right], \tag{B.4i}$$

$$\begin{aligned}\varepsilon_{\delta 1}(l_r, \varphi_l, x_r, \varphi_x) \equiv \varepsilon_{\alpha(\vec{x}-\vec{l})} - \varepsilon_{\alpha\vec{x}} &\approx \frac{\hbar^2}{2m_\alpha}(|\vec{x} - \vec{l}|^2 - |\vec{x}|^2) \\ &= \frac{\hbar^2}{2m_\alpha}\left[l_r^2 - 2l_r x_r \cos(\varphi_l - \varphi_x)\right],\end{aligned} \tag{B.4j}$$

$$\varepsilon(k_r, \varphi_k, k_r^{''}, \varphi_{k''}, l_r, \varphi_l) \equiv \varepsilon_{\vec{k}''-\vec{l}-\vec{k}} = 1 - V_{|\vec{k}''-\vec{l}-\vec{k}|}$$

$$\times \sum_{\vec{x}} \sum_{\alpha=e,h} \frac{f^{\alpha}_{\vec{x}-(\vec{k}''-\vec{l}-\vec{k})} - f^{\alpha}_{\vec{x}}}{\hbar\omega + j\delta + \varepsilon_{\alpha[\vec{x}-(\vec{k}''-\vec{l}-\vec{k})]} - \varepsilon_{\alpha\vec{x}}}$$

$$= 1 - \frac{e^2\Phi\left[Q(k_r, \varphi_k, k_r^{''}, \varphi_{k''}, l_r, \varphi_l)\right]}{\pi\varepsilon Q(k_r, \varphi_k, k_r^{''}, \varphi_{k''}, l_r, \varphi_l)} \sum_{\alpha=e,h} \sum_{n_z} \int_0^{\infty} \int_0^{2\pi} \frac{f_2(l_r, \varphi_l, x_r, \varphi_x) x_r \, dx_r \, d\varphi_x}{\hbar\omega + j\delta + \varepsilon_{\delta 2}(l_r, \varphi_l, x_r, \varphi_x)}, \tag{B.4k}$$

$$f_2(l_r, \varphi_l, x_r, \varphi_x) \equiv f^{\alpha}_{\vec{x}-(\vec{k}''-\vec{l}-\vec{k})} - f^{\alpha}_{\vec{x}} = f^{\alpha}_{d_2\varphi_{d_2}} - f^{\alpha}_{x_r\varphi_x}, \tag{B.4l}$$

$$d_2 \equiv \sqrt{\left(Q^2 + x_r^2 - 2Qx_r\cos(\varphi_q - \varphi_x)\right)}, \tag{B.4m}$$

$$\varphi_q \equiv \tan^{-1}\left[\frac{k_r\sin\varphi_k - k_r^{''}\sin\varphi_{k''} + l_r\sin\varphi_l}{k_r\cos\varphi_k - k_r^{''}\cos\varphi_{k''} + l_r\cos\varphi_l}\right], \tag{B.4n}$$

$$\varphi_{d_2} \equiv \tan^{-1}\left[\frac{x_r\sin\varphi_x - k_r\sin\varphi_k + k_r^{''}\sin\varphi_{k''} - l_r\sin\varphi_l}{x_r\cos\varphi_x - k_r\cos\varphi_k + k_r^{''}\cos\varphi_{k''} - l_r\cos\varphi_l}\right], \tag{B.4o}$$

$$\begin{aligned}\varepsilon_{\delta 2}(l_r, \varphi_l, x_r, \varphi_x) &\equiv \varepsilon_{\alpha[\vec{x}-(\vec{k}''-\vec{l}-\vec{k})]} - \varepsilon_{\alpha\vec{x}} \\ &\approx \frac{\hbar^2}{2m_\alpha}\left[|\vec{x} - (\vec{k}'' - \vec{l} - \vec{k})|^2 - |\vec{x}|^2\right] \\ &= \frac{\hbar^2}{2m_\alpha}\left[Q^2 - 2Qx_r\cos(\varphi_q - \varphi_x)\right],\end{aligned} \tag{B.4p}$$

$$\begin{aligned}\varepsilon_{\Delta 1\alpha\beta}(k_r, \varphi_k, k_r^{''}, \varphi_{k''}, l_r, \varphi_l) &\equiv \varepsilon_{\alpha\vec{k}} + \varepsilon_{\beta\vec{k}''} - \varepsilon_{\alpha(\vec{k}+\vec{l})} - \varepsilon_{\beta(\vec{k}''-\vec{l})} \\ &\approx \frac{\hbar^2}{2m_\alpha}(|\vec{k}|^2 - |\vec{k} + \vec{l}\,|^2) + \frac{\hbar^2}{2m_\beta}(|\vec{k}''|^2 - |\vec{k}'' - \vec{l}\,|^2) \\ &= -\frac{\hbar^2}{2m_\alpha}\left[l_r^2 + 2k_r l_r\cos(\varphi_k - \varphi_l)\right] - \frac{\hbar^2}{2m_\beta}\left[l_r^2 - 2k_r^{''} l_r\cos(\varphi_{k''} - \varphi_l)\right],\end{aligned} \tag{B.4q}$$

$$\begin{aligned}F_{1\alpha\beta}(k_r, \varphi_k, k_r^{''}, \varphi_{k''}, l_r, \varphi_l) &\equiv F_{11\alpha\beta}(k_r, \varphi_k, k_r^{''}, \varphi_{k''}, l_r, \varphi_l) \\ &\quad + F_{12\alpha\beta}(k_r, \varphi_k, k_r^{''}, \varphi_{k''}, l_r, \varphi_l),\end{aligned} \tag{B.4r}$$

$$F_{11\alpha\beta}(k_r, \varphi_k, k_r^{''}, \varphi_{k''}, l_r, \varphi_l) \equiv f^{\alpha}_{\vec{k}+\vec{l}}(1 - f^{\beta}_{\vec{k}''}) f^{\beta}_{\vec{k}''-\vec{l}} = f^{\alpha}_{d_3\varphi_{d_3}}(1 - f^{\beta}_{k_r^{''}\varphi_{k''}}) f^{\beta}_{d_4\varphi_{d_4}}, \tag{B.4s}$$

$$\begin{aligned}F_{12\alpha\beta}(k_r, \varphi_k, k_r^{''}, \varphi_{k''}, l_r, \varphi_l) &\equiv (1 - f^{\alpha}_{\vec{k}+\vec{l}}) f^{\beta}_{\vec{k}''}(1 - f^{\beta}_{\vec{k}''-\vec{l}}) \\ &= (1 - f^{\alpha}_{d_3\varphi_{d_3}}) f^{\beta}_{k_r^{''}\varphi_{k''}}(1 - f^{\beta}_{d_4\varphi_{d_4}}),\end{aligned} \tag{B.4t}$$

$$d_3 \equiv \sqrt{\left(k_r^2 + l_r^2 + 2k_r l_r\cos(\varphi_k - \varphi_l)\right)}, \tag{B.4u}$$

$$\varphi_{d_3} \equiv \tan^{-1}\left[\frac{k_r \sin\varphi_k + l_r \sin\varphi_l}{k_r \cos\varphi_k + l_r \cos\varphi_l}\right], \tag{B.4v}$$

$$d_4 \equiv \surd\left(k_r^{''2} + l_r^2 - 2k_r^{''} l_r \cos(\varphi_{k''} - \varphi_l)\right), \tag{B.4w}$$

$$\varphi_{d_4} \equiv \tan^{-1}\left[\frac{k_r^{''} \sin\varphi_{k''} - l_r \sin\varphi_l}{k_r^{''} \cos\varphi_{k''} - l_r \cos\varphi_l}\right]. \tag{B.4x}$$

The 2nd order non-diagonal summation is

$$\begin{aligned}
J^2_{\vec{k}\alpha\beta} &= \sum_{\vec{k}''}\sum_{\vec{l}\neq\vec{k}} \left(2\widetilde{V}^2_{|\vec{l}-\vec{k}|} - \delta_{\alpha\beta}\widetilde{V}_{|\vec{l}-\vec{k}|}\widetilde{V}_{|\vec{k}''-\vec{l}|}\right) \\
&\quad \times g\left[-\varepsilon_{\alpha\vec{k}} - \varepsilon_{\beta\vec{k}''} + \varepsilon_{\alpha\vec{l}} + \varepsilon_{\beta(\vec{k}''-\vec{l}+\vec{k})}\right]\left[\widetilde{p}_{\vec{l}}(1-f^{\beta}_{\vec{k}''})f^{\beta}_{\vec{k}''-\vec{l}+\vec{k}}\right] \\
&= \left(\frac{e^2}{2\varepsilon_0\Sigma}\right)^2 \sum_{\vec{k}''}\sum_{\vec{l}\neq\vec{k}} \left[\frac{2\Phi(|\vec{l}-\vec{k}|)}{\varepsilon_{\vec{l}-\vec{k}}|\vec{l}-\vec{k}|} - \frac{\delta_{\alpha\beta}\Phi(|\vec{k}''-\vec{l}|)}{\varepsilon_{\vec{k}''-\vec{l}}|\vec{k}''-\vec{l}|}\right]\frac{\Phi(|\vec{l}-\vec{k}|)}{\varepsilon_{\vec{l}-\vec{k}}|\vec{l}-\vec{k}|} \\
&\quad \times \frac{\delta + \mathrm{j}\left[-\varepsilon_{\alpha\vec{k}} - \varepsilon_{\beta\vec{k}''} + \varepsilon_{\alpha\vec{l}} + \varepsilon_{\beta(\vec{k}''-\vec{l}+\vec{k})}\right]}{\delta^2 + \left[-\varepsilon_{\alpha\vec{k}} - \varepsilon_{\beta\vec{k}''} + \varepsilon_{\alpha\vec{l}} + \varepsilon_{\beta(\vec{k}''-\vec{l}+\vec{k})}\right]^2}\left[\widetilde{p}_{\vec{l}}(1-f^{\beta}_{\vec{k}''})f^{\beta}_{\vec{k}''-\vec{l}+\vec{k}}\right] \\
&= \left(\frac{e^2}{4\pi^2\varepsilon_0}\right)^2 \sum_{n_{z1}}\sum_{n_{z2}} \int_0^\infty\int_0^{2\pi} k_r^{''}\,\mathrm{d}k_r^{''}\,\mathrm{d}\varphi_{k''}\int_0^\infty\int_0^{2\pi} l_r\,\mathrm{d}l_r\,\mathrm{d}\varphi_l \\
&\quad \times \left\{\left\{\frac{2\Phi[\surd(k_r^2+l_r^2-2k_rl_r\cos(\varphi_k-\varphi_l))]}{\varepsilon(k_r,\varphi_k,l_r,\varphi_l)\surd(k_r^2+l_r^2-2k_rl_r\cos(\varphi_k-\varphi_l))}\right.\right. \\
&\quad \left. - \frac{\delta_{\alpha\beta}\Phi[\surd\left(k_r^{''2}+l_r^2-2k_r^{''}l_r\cos(\varphi_{k''}-\varphi_l)\right)]}{\varepsilon(l_r,\varphi_r,k_r^{''},\varphi_{k''})\surd(k_r^{''2}+l_r^2-2k_r^{''}l_r\cos(\varphi_{k''}-\varphi_l))}\right\} \\
&\quad \times \frac{\Phi\left(\surd(k_r^2+l_r^2-2k_rl_r\cos(\varphi_k-\varphi_l))\right)}{\varepsilon(k_r,\varphi_k,l_r,\varphi_l)\surd(k_r^2+l_r^2-2k_rl_r\cos(\varphi_k-\varphi_l))} \\
&\quad \left.\times \frac{\delta + \mathrm{j}\varepsilon_{\Delta 2\alpha\beta}(k_r,\varphi_k,k_r^{''},\varphi_{k''},l_r,\varphi_l)}{\delta^2+\varepsilon^2_{\Delta 2\alpha\beta}(k_r,\varphi_k,k_r^{''},\varphi_{k''},l_r,\varphi_l)} F_{2\alpha\beta}(k_r,\varphi_k,k_r^{''},\varphi_{k''},l_r,\varphi_l)\right\}\Big|_{l_r\neq k_r\bigcup\varphi_l\neq\varphi_k} \\
&= \sum_{n_{z1}}\sum_{n_{z2}} \int_0^\infty\int_0^{2\pi} k_r^{''}\,\mathrm{d}k_r^{''}\,\mathrm{d}\varphi_{k''}\left[\int_0^{k_r-\Delta}\mathrm{d}l_r\phi_{4\alpha\beta}(k_r,\varphi_k,k_r^{''},\varphi_{k''},l_r)\right. \\
&\quad + \int_{k_r+\Delta}^\infty \mathrm{d}l_r\phi_{4\alpha\beta}(k_r,\varphi_k,k_r^{''},\varphi_{k''},l_r) \\
&\quad \left.+ \int_{k_r-\Delta}^{k_r+\Delta}\mathrm{d}l_r\phi_{4\alpha\beta}(k_r,\varphi_k,k_r^{''},\varphi_{k''},l_r)|_{\varphi_l\neq\varphi_k}\right],
\end{aligned} \tag{B.5a}$$

where we have defined

$$\phi_{4\alpha\beta}(k_r, \varphi_k, k_r^{''}, \varphi_{k''}, l_r)$$
$$\equiv \left(\frac{\mathrm{e}^2}{4\pi^2\varepsilon_0}\right)^2 l_r \int_0^{2\pi} \mathrm{d}\varphi_l \left\{\left\{\frac{2\Phi\left[\sqrt{\left(k_r^2 + l_r^2 - 2k_r l_r \cos(\varphi_k - \varphi_l)\right)}\right]}{\varepsilon(k_r, \varphi_k, l_r, \varphi_l)\sqrt{\left(k_r^2 + l_r^2 - 2k_r l_r \cos(\varphi_k - \varphi_l)\right)}}\right.\right.$$
$$\left. - \frac{\delta_{\alpha\beta}\Phi\left[\sqrt{\left(k_r^{''2} + l_r^2 - 2k_r^{''} l_r \cos(\varphi_{k''} - \varphi_l)\right)}\right]}{\varepsilon(l_r, \varphi_r, k_r^{''}, \varphi_{k''})\sqrt{\left(k_r^{''2} + l_r^2 - 2k_r^{''} l_r \cos(\varphi_{k''} - \varphi_l)\right)}}\right\}$$
$$\times \frac{\Phi\left(\sqrt{\left(k_r^2 + l_r^2 - 2k_r l_r \cos(\varphi_k - \varphi_l)\right)}\right)}{\varepsilon(k_r, \varphi_k, l_r, \varphi_l)\sqrt{\left(k_r^2 + l_r^2 - 2k_r l_r \cos(\varphi_k - \varphi_l)\right)}}$$
$$\left. \times \frac{\delta + \mathrm{j}\varepsilon_{\Delta 2\alpha\beta}(k_r, \varphi_k, k_r^{''}, \varphi_{k''}, l_r, \varphi_l)}{\delta^2 + \varepsilon^2_{\Delta 2\alpha\beta}(k_r, \varphi_k, k_r^{''}, \varphi_{k''}, l_r, \varphi_l)} F_{2\beta}(k_r, \varphi_k, k_r^{''}, \varphi_{k''}, l_r, \varphi_l)\right\}. \tag{B.5b}$$

We have also defined

$$\varepsilon(k_r, \varphi_k, l_r, \varphi_l) \equiv \varepsilon_{\vec{l}-\vec{k}}$$
$$= 1 - V_{|\vec{l}-\vec{k}|}\sum_{\vec{x}}\sum_{\alpha=e,h}\frac{f^{\alpha}_{\vec{x}-(\vec{l}-\vec{k})} - f^{\alpha}_{\vec{x}}}{\hbar\omega + \mathrm{j}\delta + \varepsilon_{\alpha[\vec{x}-(\vec{l}-\vec{k})]} - \varepsilon_{\alpha\vec{x}}}$$
$$= 1 - \frac{e^2\Phi\left[\sqrt{\left(k_r^2 + l_r^2 - 2k_r l_r \cos(\varphi_k - \varphi_l)\right)}\right]}{4\pi^2\varepsilon_0\sqrt{\left(k_r^2 + l_r^2 - 2k_r l_r \cos(\varphi_k - \varphi_l)\right)}}$$
$$\times \sum_{\alpha=e,h}\sum_{n_z}\int_0^{\infty}\int_0^{2\pi} x_r\,\mathrm{d}x_r\,\mathrm{d}\varphi_x \frac{f_3(k_r, \varphi_k, l_r, \varphi_l, x_r, \varphi_x)}{\hbar\omega + i\delta + \varepsilon_{\delta 3}(k_r, \varphi_k, l_r, \varphi_l, x_r, \varphi_x)}, \tag{B.5c}$$

$$f_3(k_r, \varphi_k, l_r, \varphi_l, x_r, \varphi_x) \equiv f^{\alpha}_{\vec{x}-(\vec{l}-\vec{k})} - f^{\alpha}_{\vec{x}} = f^{\alpha}_{d_5\varphi_{d_5}} - f^{\alpha}_{x_r\varphi_x}, \tag{B.5d}$$

$$d_5 \equiv \sqrt{\left(x_r^2 + k_r^2 + l_r^2 - 2x_r k_r \cos(\varphi_x - \varphi_k) + 2x_r l_r \cos(\varphi_x - \varphi_l)\right.}$$
$$\left. - 2k_r l_r \cos(\varphi_k - \varphi_l)\right), \tag{B.5e}$$

$$\varphi_{d_5} \equiv \tan^{-1}\left[\frac{x_r \sin\varphi_x - k_r \sin\varphi_k + l_r \sin\varphi_l}{x_r \cos\varphi_x - k_r \cos\varphi_k + l_r \cos\varphi_l}\right], \tag{B.5f}$$

$$\varepsilon_{\delta 3}(k_r, \varphi_k, l_r, \varphi_l, x_r, \varphi_x) \equiv \varepsilon_{\alpha[\vec{x}-(\vec{l}-\vec{k})]} - \varepsilon_{\alpha\vec{x}} \approx \frac{\hbar^2}{2m_\alpha}\left[|\vec{x} - (\vec{l} - \vec{k})|^2 - |\vec{x}|^2\right]$$
$$= \frac{\hbar^2}{2m_\alpha}\left[k_r^2 + l_r^2 - 2x_r k_r \cos(\varphi_x - \varphi_k) + 2x_r l_r \cos(\varphi_x - \varphi_l) - 2k_r l_r \cos(\varphi_k - \varphi_l)\right], \tag{B.5g}$$

$$\varepsilon(l_r, \varphi_l, k_r^{''}, \varphi_{k''}) \equiv \varepsilon_{\vec{k}''-\vec{l}} = 1 - V_{|\vec{k}''-\vec{l}|} \sum_{\vec{x}} \sum_{\alpha=e,h} \frac{f^{\alpha}_{\vec{x}-(\vec{k}''-\vec{l})} - f^{\alpha}_{\vec{x}}}{\hbar\omega + i\delta + \varepsilon_{\alpha[\vec{x}-(\vec{k}''-\vec{l})]} - \varepsilon_{\alpha\vec{x}}}$$

$$= 1 - \frac{e^2 \Phi\left[\sqrt{\left(k_r^{''2} + l_r^2 - 2k_r^{''} l_r \cos(\varphi_{k''} - \varphi_l)\right)}\right]}{4\pi^2 \varepsilon_0 \sqrt{\left(k_r^{''2} + l_r^2 - 2k_r^{''} l_r \cos(\varphi_{k''} - \varphi_l)\right)}}$$

$$\times \sum_{\alpha=e,h} \sum_{n_z} \int_0^{\infty} \int_0^{2\pi} x_r \, \mathrm{d}x_r \, \mathrm{d}\varphi_x \frac{f_4(l_r, \varphi_l, k_r^{''}, \varphi_{k''}, x_r, \varphi_x)}{\hbar\omega + i\delta + \varepsilon_{\delta 4}(l_r, \varphi_l, k_r^{''}, \varphi_{k''}, x_r, \varphi_x)}, \tag{B.5h}$$

$$f_4(l_r, \varphi_l, k_r^{''}, \varphi_{k''}, x_r, \varphi_x) \equiv f^{\alpha}_{\vec{x}-(\vec{k}''-\vec{l})} - f^{\alpha}_{\vec{x}} = f^{\alpha}_{d_6 \varphi_{d_6}} - f^{\alpha}_{x_r \varphi_x}, \tag{B.5i}$$

$$d_6 \equiv \sqrt{\left(x_r^2 + l_r^2 + k_r^{''2} - 2x_r l_r \cos(\varphi_x - \varphi_l) + 2x_r k_r^{''} \cos(\varphi_x - \varphi_{k''})\right.}$$
$$\left.-2l_r k_r^{''} \cos(\varphi_l - \varphi_{k''})\right), \tag{B.5j}$$

$$\varphi_{d_6} \equiv \tan^{-1}\left[\frac{x_r \sin\varphi_x - l_r \sin\varphi_l + k_r^{''} \sin\varphi_{k''}}{x_r \cos\varphi_x - l_r \cos\varphi_l + k_r^{''} \cos\varphi_{k''}}\right], \tag{B.5k}$$

$$\varepsilon_{\delta 4}(l_r, \varphi_l, k_r^{''}, \varphi_{k''}, x_r, \varphi_x) \equiv \varepsilon_{\alpha[\vec{x}-(\vec{k}''-\vec{l})]} - \varepsilon_{\alpha\vec{x}} \approx \frac{\hbar^2}{2m_\alpha}\left[|\vec{x} - (\vec{k}'' - \vec{l})|^2 - |\vec{x}|^2\right]$$

$$= \frac{\hbar^2}{2m_\alpha}\left[l_r^2 + k_r^{''2} - 2x_r l_r \cos(\varphi_x - \varphi_l)\right.$$
$$\left.+ 2x_r k_r^{''} \cos(\varphi_x - \varphi_{k''}) - 2l_r k_r^{''} \cos(\varphi_l - \varphi_{k''})\right], \tag{B.5l}$$

$$\varepsilon_{\Delta 2\alpha\beta}(k_r, \varphi_k, k_r^{''}, \varphi_{k''}, l_r, \varphi_l) \equiv -\varepsilon_{\alpha\vec{k}} - \varepsilon_{\beta\vec{k}''} + \varepsilon_{\alpha\vec{l}} + \varepsilon_{\beta(\vec{k}''-\vec{l}+\vec{k})}$$

$$\approx \frac{\hbar^2}{2m_\alpha}(|\vec{l}|^2 - |\vec{k}|^2) + \frac{\hbar^2}{2m_\beta}(|\vec{k}'' - \vec{l} + \vec{k}|^2 - |\vec{k}''|^2)$$

$$= \frac{\hbar^2}{2m_\alpha}(l_r^2 - k_r^2) + \frac{\hbar^2}{2m_\beta}\left[k_r^2 + l_r^2 + 2k_r^{''} k_r \cos(\varphi_k - \varphi_{k''})\right.$$
$$\left.- 2k_r^{''} l_r \cos(\varphi_{k''} - \varphi_l) - 2k_r l_r \cos(\varphi_k - \varphi_l)\right], \tag{B.5m}$$

$$F_{2\beta}(k_r, \varphi_k, k_r^{''}, \varphi_{k''}, l_r, \varphi_l) \equiv \widetilde{p}_{\vec{l}}(1 - f^{\beta}_{\vec{k}''}) f^{\beta}_{\vec{k}''-\vec{l}+\vec{k}} = \widetilde{p}_{l_r \varphi_l}(1 - f^{\beta}_{k_r^{''} \varphi_{k''}}) f^{\beta}_{d_7 \varphi_{d_7}}, \tag{B.5n}$$

$$d_7 \equiv \sqrt{\left(k_r^2 + k_r^{''2} + l_r^2 + 2k_r k_r^{''} \cos(\varphi_k - \varphi_{k''}) - 2k_r l_r \cos(\varphi_k - \varphi_l)\right.}$$
$$\left.-2k_r^{''} l_r \cos(\varphi_{k''} - \varphi_l)\right), \tag{B.5o}$$

$$\varphi_{d_7} \equiv \tan^{-1}\left[\frac{k_r \sin\varphi_k + k_r^{''} \sin\varphi_{k''} - l_r \sin\varphi_l}{k_r \cos\varphi_k + k_r^{''} \cos\varphi_{k''} - l_r \cos\varphi_l}\right]. \tag{B.5p}$$

Similarly, we have

$$\begin{aligned}
J^3_{\vec{k}\alpha\beta} &\equiv \sum_{\vec{k}''}\sum_{\vec{l}\neq\vec{k}} \left(2\widetilde{V}^2_{|\vec{l}-\vec{k}|} - \delta_{\alpha\beta}\widetilde{V}_{|\vec{l}-\vec{k}|}\widetilde{V}_{|\vec{k}''-\vec{l}|}\right) \\
&\quad \times g\left[-\varepsilon_{\alpha\vec{k}} - \varepsilon_{\beta\vec{k}''} + \varepsilon_{\alpha\vec{l}} + \varepsilon_{\beta(\vec{k}''-\vec{l}+\vec{k})}\right]\left[\widetilde{p}_{\vec{l}} f^{\beta}_{\vec{k}''}\left(1 - f^{\beta}_{\vec{k}''-\vec{l}+\vec{k}}\right)\right] \\
&= \sum_{n_{z1}}\sum_{n_{z2}} \int_0^{\infty}\int_0^{2\pi} k''_r\,\mathrm{d}k''_r\,\mathrm{d}\varphi_{k''}\left[\int_0^{k_r-\Delta} \mathrm{d}l_r \phi_{5\alpha\beta}(k_r,\varphi_k,k''_r,\varphi_{k''},l_r)\right. \\
&\quad + \int_{k_r+\Delta}^{\infty} \mathrm{d}l_r \phi_{5\alpha\beta}(k_r,\varphi_k,k''_r,\varphi_{k''},l_r) \\
&\quad \left. + \int_{k_r-\Delta}^{k_r+\Delta} \mathrm{d}l_r \phi_{5\alpha\beta}(k_r,\varphi_k,k''_r,\varphi_{k''},l_r)|_{\varphi_l\neq\varphi_k}\right],
\end{aligned} \tag{B.6a}$$

where we have defined

$$\begin{aligned}
\phi_{5\alpha\beta}(k_r,\varphi_k,k''_r,\varphi_{k''},l_r) &\equiv \left(\frac{\mathrm{e}^2}{4\pi^2\varepsilon_0}\right)^2 l_r \\
&\times \int_0^{2\pi} \mathrm{d}\varphi_l \left\{\left\{\frac{2\Phi\left[\sqrt{\left(k_r^2 + l_r^2 - 2k_r l_r\cos(\varphi_k - \varphi_l)\right)}\right]}{\varepsilon(k_r,\varphi_k,l_r,\varphi_l)\sqrt{\left(k_r^2 + l_r^2 - 2k_r l_r\cos(\varphi_k - \varphi_l)\right)}}\right.\right. \\
&- \left.\frac{\delta_{\alpha\beta}\Phi\left[\sqrt{\left(k''^2_r + l_r^2 - 2k''_r l_r\cos(\varphi_{k''} - \varphi_l)\right)}\right]}{\varepsilon(l_r,\varphi_r,k''_r,\varphi_{k''})\sqrt{\left(k''^2_r + l_r^2 - 2k''_r l_r\cos(\varphi_{k''} - \varphi_l)\right)}}\right\} \\
&\times \frac{\Phi\left[\sqrt{\left(k_r^2 + l_r^2 - 2k_r l_r\cos(\varphi_k - \varphi_l)\right)}\right]}{\varepsilon(k_r,\varphi_k,l_r,\varphi_l)\sqrt{\left(k_r^2 + l_r^2 - 2k_r l_r\cos(\varphi_k - \varphi_l)\right)}} \\
&\times \left.\frac{\delta + \mathrm{j}\varepsilon_{\Delta 2\alpha\beta}(k_r,\varphi_k,k''_r,\varphi_{k''},l_r,\varphi_l)}{\delta^2 + \varepsilon^2_{\Delta 2\alpha\beta}(k_r,\varphi_k,k''_r,\varphi_{k''},l_r,\varphi_l)} F_{3\beta}(k_r,\varphi_k,k''_r,\varphi_{k''},l_r,\varphi_l)\right\},
\end{aligned} \tag{B.6b}$$

$$F_{3\beta}(k_r,\varphi_k,k''_r,\varphi_{k''},l_r,\varphi_l) \equiv \widetilde{p}_{\vec{l}} f^{\beta}_{\vec{k}''}\left(1 - f^{\beta}_{\vec{k}''-\vec{l}+\vec{k}}\right) = \widetilde{p}_{l_r\varphi_l} f^{\beta}_{k''_r\varphi_{k''}}\left(1 - f^{\beta}_{d_7\varphi_{d_7}}\right). \tag{B.6c}$$

The phonon related diagonal contribution is

$$\begin{aligned}
K^1_{\vec{k}} \equiv \hbar^2 \sum_{\alpha=e,h}\sum_{\vec{l}\neq 0} G^2_{|\vec{l}|}&\left\{g\left[\varepsilon_{\alpha\vec{k}} - \varepsilon_{\alpha(\vec{k}-\vec{l})} - \hbar\omega_{\mathrm{LO}}\right]\left[(1 - f^{\alpha}_{\vec{k}-\vec{l}})f^{\mathrm{p}}_{\vec{l}} + f^{\alpha}_{\vec{k}-\vec{l}}(1 + f^{\mathrm{p}}_{\vec{l}})\right]\right. \\
&\left. + g\left[\varepsilon_{\alpha\vec{k}} - \varepsilon_{\alpha(\vec{k}-\vec{l})} + \hbar\omega_{\mathrm{LO}}\right]\left[(1 - f^{\alpha}_{\vec{k}-\vec{l}})(1 + f^{\mathrm{p}}_{\vec{l}}) + f^{\alpha}_{\vec{k}-\vec{l}} f^{\mathrm{p}}_{\vec{l}}\right]\right\}
\end{aligned}$$

$$
\begin{aligned}
&= \frac{e^2\hbar\omega_{LO}}{4\varepsilon_0\Sigma}\left(\frac{1}{\varepsilon_\infty}-\frac{1}{\varepsilon_0}\right)\sum_{\alpha=e,h}\sum_{\vec{l}\neq 0}\frac{\Phi(|\vec{l}\,|)}{\varepsilon_{\vec{l}}|\vec{l}\,|}\left\{\frac{\delta+\mathrm{j}\left[\varepsilon_{\alpha\vec{k}}-\varepsilon_{\alpha(\vec{k}-\vec{l})}-\hbar\omega_{LO}\right]}{\delta^2+\left[\varepsilon_{\alpha\vec{k}}-\varepsilon_{\alpha(\vec{k}-\vec{l})}-\hbar\omega_{LO}\right]^2}\right.\\
&\quad\times\left[(1-f^\alpha_{\vec{k}-\vec{l}})f^p_{\vec{l}}+f^\alpha_{\vec{k}-\vec{l}}(1+f^p_{\vec{l}})\right]+\frac{\delta+\mathrm{j}\left[\varepsilon_{\alpha\vec{k}}-\varepsilon_{\alpha(\vec{k}-\vec{l})}+\hbar\omega_{LO}\right]}{\delta^2+\left[\varepsilon_{\alpha\vec{k}}-\varepsilon_{\alpha(\vec{k}-\vec{l})}+\hbar\omega_{LO}\right]^2}\\
&\quad\left.\times\left[(1-f^\alpha_{\vec{k}-\vec{l}})(1+f^{\mathrm{p}}_{\vec{l}})+f^\alpha_{\vec{k}-\vec{l}}f^{\mathrm{p}}_{\vec{l}}\right]\right\}\\
&= \frac{e^2\hbar\omega_{LO}}{8\pi^2\varepsilon_0}\left(\frac{1}{\varepsilon_\infty}-\frac{1}{\varepsilon_0}\right)\sum_{\alpha=e,h}\sum_{n_z}\int_0^\infty\int_0^{2\pi}\mathrm{d}l_r\,\mathrm{d}\varphi_l\frac{\Phi(l_r)}{\varepsilon(l_r,\varphi_l)}\\
&\quad\times\left\{\frac{\delta+\mathrm{j}\left[\varepsilon_{\Delta 3\alpha}(k_r,\varphi_k,l_r,\varphi_l)-\hbar\omega_{LO}\right]}{\delta^2+\left[\varepsilon_{\Delta 3\alpha}(k_r,\varphi_k,l_r,\varphi_l)-\hbar\omega_{LO}\right]^2}F_{4\alpha}(k_r,\varphi_k,l_r,\varphi_l)\right.\\
&\quad\left.+\frac{\delta+\mathrm{j}\left[\varepsilon_{\Delta 3\alpha}(k_r,\varphi_k,l_r,\varphi_l)+\hbar\omega_{LO}\right]}{\delta^2+\left[\varepsilon_{\Delta 3\alpha\beta}(k_r,\varphi_k,l_r,\varphi_l)+\hbar\omega_{LO}\right]^2}F_{5\alpha}(k_r,\varphi_k,l_r,\varphi_l)\right\}\\
&= \sum_{\alpha=e,h}\sum_{n_z}\int_0^\infty \mathrm{d}l_r\phi_{6\alpha}(k_r,\varphi_k,l_r),
\end{aligned}
\tag{B.7a}
$$

where we have defined

$$
\begin{aligned}
\phi_{6\alpha}(k_r,\varphi_k,l_r) &\equiv \frac{e^2\hbar\omega_{LO}}{8\pi^2\varepsilon_0}\left(\frac{1}{\varepsilon_\infty}-\frac{1}{\varepsilon_0}\right)\int_0^{2\pi}\mathrm{d}\varphi_l\frac{\Phi(l_r)}{\varepsilon(l_r,\varphi_l)}\\
&\quad\times\left\{\frac{\delta+\mathrm{j}\left[\varepsilon_{\Delta 3\alpha}(k_r,\varphi_k,l_r,\varphi_l)-\hbar\omega_{LO}\right]}{\delta^2+\left[\varepsilon_{\Delta 3\alpha}(k_r,\varphi_k,l_r,\varphi_l)-\hbar\omega_{LO}\right]^2}F_{4\alpha}(k_r,\varphi_k,l_r,\varphi_l)\right.\\
&\quad\left.+\frac{\delta+\mathrm{j}\left[\varepsilon_{\Delta 3\alpha}(k_r,\varphi_k,l_r,\varphi_l)+\hbar\omega_{LO}\right]}{\delta^2+\left[\varepsilon_{\Delta 3\alpha}(k_r,\varphi_k,l_r,\varphi_l)+\hbar\omega_{LO}\right]^2}F_{5\alpha}(k_r,\varphi_k,l_r,\varphi_l)\right\}.
\end{aligned}
\tag{B.7b}
$$

We have also defined

$$
\begin{aligned}
\varepsilon_{\Delta 3\alpha}(k_r,\varphi_k,l_r,\varphi_l) &\equiv \varepsilon_{\alpha\vec{k}}-\varepsilon_{\alpha(\vec{k}-\vec{l})}\approx\frac{\hbar^2}{2m_\alpha}(|\vec{k}|^2-|\vec{k}-\vec{l}|^2)\\
&= -\frac{\hbar^2}{2m_\alpha}\left[l_r^2-2k_rl_r\cos(\varphi_k-\varphi_l)\right],
\end{aligned}
\tag{B.7c}
$$

$$
\begin{aligned}
F_{4\alpha}(k_r,\varphi_k,l_r,\varphi_l) &\equiv (1-f^\alpha_{\vec{k}-\vec{l}})f^{\mathrm{p}}_{\vec{l}}+f^\alpha_{\vec{k}-\vec{l}}(1+f^{\mathrm{p}}_{\vec{l}})\\
&= \frac{1-f^\alpha_{d_8\varphi_{d_8}}+f^\alpha_{d_8\varphi_{d_8}}e^{\hbar\omega_{LO}/k_BT}}{e^{\hbar\omega_{LO}/k_BT}-1},
\end{aligned}
\tag{B.7d}
$$

$$F_{5\alpha}(k_r,\varphi_k,l_r,\varphi_l) \equiv (1-f^{\alpha}_{\vec{k}-\vec{l}})(1+f^{\mathrm{p}}_{\vec{l}}) + f^{\alpha}_{\vec{k}-\vec{l}} f^{\mathrm{p}}_{\vec{l}}$$

$$= \frac{(1-f^{\alpha}_{d_8\varphi_{d_8}})\mathrm{e}^{\hbar\omega_{\mathrm{LO}}/\mathrm{k_B}T} + f^{\alpha}_{d_8\varphi_{d_8}}}{\mathrm{e}^{\hbar\omega_{\mathrm{LO}}/\mathrm{k_B}T}-1}, \tag{B.7e}$$

$$d_8 \equiv \sqrt{\left(k_r^2 + l_r^2 - 2k_r l_r \cos(\varphi_k - \varphi_l)\right)}, \tag{B.7f}$$

$$\varphi_{d_8} \equiv \tan^{-1}\left[\frac{k_r \sin\varphi_k - l_r \sin\varphi_l}{k_r \cos\varphi_k - l_r \cos\varphi_l}\right]. \tag{B.7g}$$

The phonon related non-diagonal contribution is

$$K^2_{\vec{k}\alpha} \equiv \hbar^2 \sum_{\vec{l}\neq 0} \widetilde{p}_{\vec{k}+\vec{l}} G^2_{|\vec{l}|} \Big\{ g\left[\varepsilon_{\alpha\vec{k}} - \varepsilon_{\alpha(\vec{k}-\vec{l})} - \hbar\omega_{\mathrm{LO}}\right] f^{\mathrm{p}}_{\vec{l}}$$

$$+ g\left[\varepsilon_{\alpha\vec{k}} - \varepsilon_{\alpha(\vec{k}-\vec{l})} + \hbar\omega_{\mathrm{LO}}\right](1+f^{\mathrm{p}}_{\vec{l}}) \Big\}$$

$$= \frac{\mathrm{e}^2\hbar\omega_{\mathrm{LO}}}{4\varepsilon_0\Sigma}\left(\frac{1}{\varepsilon_\infty} - \frac{1}{\varepsilon_0}\right) \sum_{\vec{l}\neq 0} \frac{\Phi(|\vec{l}\,|)}{\varepsilon_{\vec{l}}\,|\vec{l}\,|} \widetilde{p}_{\vec{k}+\vec{l}} \left\{ \frac{\delta + \mathrm{j}\left[\varepsilon_{\alpha\vec{k}} - \varepsilon_{\alpha(\vec{k}-\vec{l})} - \hbar\omega_{\mathrm{LO}}\right]}{\delta^2 + \left[\varepsilon_{\alpha\vec{k}} - \varepsilon_{\alpha(\vec{k}-\vec{l})} - \hbar\omega_{\mathrm{LO}}\right]^2} f^{\mathrm{p}}_{\vec{l}} \right.$$

$$\left. + \frac{\delta + \mathrm{j}\left[\varepsilon_{\alpha\vec{k}} - \varepsilon_{\alpha(\vec{k}-\vec{l})} + \hbar\omega_{\mathrm{LO}}\right]}{\delta^2 + \left[\varepsilon_{\alpha\vec{k}} - \varepsilon_{\alpha(\vec{k}-\vec{l})} + \hbar\omega_{\mathrm{LO}}\right]^2}(1+f^{\mathrm{p}}_{\vec{l}}) \right\}$$

$$= \frac{\mathrm{e}^2\hbar\omega_{\mathrm{LO}}}{8\pi^2\varepsilon_0}\left(\frac{1}{\varepsilon_\infty} - \frac{1}{\varepsilon_0}\right) \sum_{n_z} \int_0^\infty \int_0^{2\pi} \mathrm{d}l_r\, \mathrm{d}\varphi_l \frac{\Phi(l_r)}{\varepsilon(l_r,\varphi_l)} \widetilde{p}_{d_3\varphi_{d_3}}$$

$$\times \left\{ \frac{\delta + \mathrm{j}\left[\varepsilon_{\Delta 3\alpha}(k_r,\varphi_k,l_r,\varphi_l) - \hbar\omega_{\mathrm{LO}}\right]}{\delta^2 + \left[\varepsilon_{\Delta 3\alpha}(k_r,\varphi_k,l_r,\varphi_l) - \hbar\omega_{\mathrm{LO}}\right]^2} \frac{1}{\mathrm{e}^{\hbar\omega_{\mathrm{LO}}/\mathrm{k_B}T}-1} \right.$$

$$\left. + \frac{\delta + \mathrm{j}\left[\varepsilon_{\Delta 3\alpha}(k_r,\varphi_k,l_r,\varphi_l) + \hbar\omega_{\mathrm{LO}}\right]}{\delta^2 + \left[\varepsilon_{\Delta 3\alpha}(k_r,\varphi_k,l_r,\varphi_l) + \hbar\omega_{\mathrm{LO}}\right]^2} \frac{\mathrm{e}^{\hbar\omega_{\mathrm{LO}}/\mathrm{k_B}T}}{\mathrm{e}^{\hbar\omega_{\mathrm{LO}}/\mathrm{k_B}T}-1} \right\}$$

$$= \sum_{n_z} \int_0^\infty \mathrm{d}l_r \phi_{7\alpha}(k_r,\varphi_k,l_r), \tag{B.8a}$$

where we have defined

$$\phi_{7\alpha}(k_r,\varphi_k,l_r) \equiv \frac{\mathrm{e}^2\hbar\omega_{\mathrm{LO}}}{8\pi^2\varepsilon_0}\left(\frac{1}{\varepsilon_\infty} - \frac{1}{\varepsilon_0}\right) \int_0^{2\pi} \mathrm{d}\varphi_l \frac{\Phi(l_r)}{\varepsilon(l_r,\varphi_l)} \widetilde{p}_{d_3\varphi_{d_3}}$$

$$\times \left\{ \frac{\delta + \mathrm{j}\left[\varepsilon_{\Delta 3\alpha}(k_r,\varphi_k,l_r,\varphi_l) - \hbar\omega_{\mathrm{LO}}\right]}{\delta^2 + \left[\varepsilon_{\Delta 3\alpha}(k_r,\varphi_k,l_r,\varphi_l) - \hbar\omega_{\mathrm{LO}}\right]^2} \frac{1}{\mathrm{e}^{\hbar\omega_{\mathrm{LO}}/\mathrm{k_B}T}-1} \right.$$

$$\left. + \frac{\delta + \mathrm{j}\left[\varepsilon_{\Delta 3\alpha}(k_r,\varphi_k,l_r,\varphi_l) + \hbar\omega_{\mathrm{LO}}\right]}{\delta^2 + \left[\varepsilon_{\Delta 3\alpha}(k_r,\varphi_k,l_r,\varphi_l) + \hbar\omega_{\mathrm{LO}}\right]^2} \frac{\mathrm{e}^{\hbar\omega_{\mathrm{LO}}/\mathrm{k_B}T}}{\mathrm{e}^{\hbar\omega_{\mathrm{LO}}/\mathrm{k_B}T}-1} \right\}. \tag{B.8b}$$

Similarly, we have

$$K^3_{\vec{k}\alpha} \equiv \hbar^2 \sum_{\vec{l}\neq 0} \widetilde{p}_{\vec{k}+\vec{l}} G^2_{|\vec{l}|} \left\{ g\left[\varepsilon_{\alpha\vec{k}} - \varepsilon_{\alpha(\vec{k}-\vec{l})} - \hbar\omega_{\mathrm{LO}}\right](1 + f^{\mathrm{p}}_{\vec{l}}) + g\left[\varepsilon_{\alpha\vec{k}} - \varepsilon_{\alpha(\vec{k}-\vec{l})} + \hbar\omega_{\mathrm{LO}}\right] f^{\mathrm{p}}_{\vec{l}} \right\}$$

$$= \sum_{n_z} \int_0^\infty \mathrm{d}l_r \phi_{8\alpha}(k_r, \varphi_k, l_r), \tag{B.9a}$$

where we have defined

$$\phi_{8\alpha}(k_r, \varphi_k, l_r) \equiv \frac{\mathrm{e}^2\hbar\omega_{\mathrm{LO}}}{8\pi^2\varepsilon_0}\left(\frac{1}{\varepsilon_\infty} - \frac{1}{\varepsilon_0}\right) \int_0^{2\pi} \mathrm{d}\varphi_l \frac{\Phi(l_r)}{\varepsilon(l_r, \varphi_l)} \widetilde{p}_{d_3\varphi_{d_3}}$$
$$\times \left\{ \frac{\delta + \mathrm{j}\left[\varepsilon_{\Delta 3\alpha}(k_r, \varphi_k, l_r, \varphi_l) - \hbar\omega_{\mathrm{LO}}\right]}{\delta^2 + \left[\varepsilon_{\Delta 3\alpha}(k_r, \varphi_k, l_r, \varphi_l) - \hbar\omega_{\mathrm{LO}}\right]^2} \frac{\mathrm{e}^{\hbar\omega_{\mathrm{LO}}/\mathrm{k_B}T}}{\mathrm{e}^{\hbar\omega_{\mathrm{LO}}/\mathrm{k_B}T} - 1} \right.$$
$$\left. + \frac{\delta + \mathrm{j}\left[\varepsilon_{\Delta 3\alpha}(k_r, \varphi_k, l_r, \varphi_l) + \hbar\omega_{\mathrm{LO}}\right]}{\delta^2 + \left[\varepsilon_{\Delta 3\alpha}(k_r, \varphi_k, l_r, \varphi_l) + \hbar\omega_{\mathrm{LO}}\right]^2} \frac{1}{\mathrm{e}^{\hbar\omega_{\mathrm{LO}}/\mathrm{k_B}T} - 1} \right\}. \tag{B.9b}$$

The carrier–carrier collision coefficients are

$$\Sigma^{\mathrm{e/h-out}}_{\vec{k}} \equiv \pi \sum_{\alpha=e,h} \sum_{\vec{l}\neq 0} \sum_{\vec{k}''} \left(2\widetilde{V}_{|\vec{l}|} - \delta_{(\mathrm{e/h})\alpha} \widetilde{V}_{|\vec{k}''-\vec{l}-\vec{k}|}\right) \widetilde{V}_{|\vec{l}|}$$
$$\times \delta\left[\varepsilon_{(\mathrm{e/h})\vec{k}} + \varepsilon_{\alpha\vec{k}''} - \varepsilon_{(\mathrm{e/h})(\vec{k}+\vec{l})} - \varepsilon_{\alpha(\vec{k}''-\vec{l})}\right](1 - f^{\mathrm{e/h}}_{\vec{k}+\vec{l}}) f^{\alpha}_{\vec{k}''}(1 - f^{\alpha}_{\vec{k}''-\vec{l}})$$
$$= \pi \left(\frac{\mathrm{e}^2}{2\varepsilon_0\Sigma}\right)^2 \sum_{\alpha=e,h} \sum_{\vec{l}\neq 0} \sum_{\vec{k}''} \left[\frac{2\Phi(|\vec{l}|)}{\varepsilon_{\vec{l}}|\vec{l}|} - \frac{\delta_{(\mathrm{e/h})\alpha}\Phi(|\vec{k}''-\vec{l}-\vec{k}|)}{\varepsilon_{\vec{k}''-\vec{l}-\vec{k}}|\vec{k}''-\vec{l}-\vec{k}|}\right] \frac{\Phi(|\vec{l}|)}{\varepsilon_{\vec{l}}|\vec{l}|}$$
$$\times \delta\left[\varepsilon_{(\mathrm{e/h})\vec{k}} + \varepsilon_{\alpha\vec{k}''} - \varepsilon_{(\mathrm{e/h})(\vec{k}+\vec{l})} - \varepsilon_{\alpha(\vec{k}''-\vec{l})}\right](1 - f^{\mathrm{e/h}}_{\vec{k}+\vec{l}}) f^{\alpha}_{\vec{k}''}(1 - f^{\alpha}_{\vec{k}''-\vec{l}})$$
$$= \frac{1}{\pi^3}\left(\frac{\mathrm{e}^2}{4\varepsilon_0}\right)^2 \sum_{\alpha=e,h} \sum_{n_{z1}} \sum_{n_{z2}} \int_0^\infty \int_0^{2\pi} k''_r \, \mathrm{d}k''_r \, \mathrm{d}\varphi_{k''} \int_0^\infty \int_0^{2\pi} \mathrm{d}l_r \, \mathrm{d}\varphi_l \left\{ \left\{ \frac{2\Phi(l_r)}{\varepsilon(l_r, \varphi_l) l_r} \right.\right.$$
$$\left. - \frac{\delta_{(\mathrm{e/h})\alpha}\Phi\left[Q(k_r, \varphi_k, k''_r, \varphi_{k''}, l_r, \varphi_l)\right]}{\varepsilon(k_r, \varphi_k, k''_r, \varphi_{k''}, l_r, \varphi_l) Q(k_r, \varphi_k, k''_r, \varphi_{k''}, l_r, \varphi_l)} \right\} \frac{\Phi(l_r)}{\varepsilon(l_r, \varphi_r)}$$
$$\left. \times \delta\left[\varepsilon_{\Delta 1(\mathrm{e/h})\alpha}(k_r, \varphi_k, k''_r, \varphi_{k''}, l_r, \varphi_l)\right] F_{12(\mathrm{e/h})\alpha}(k_r, \varphi_k, k''_r, \varphi_{k''}, l_r, \varphi_l) \right\}\Big|_{l_r\neq 0}$$
$$= \sum_{\alpha=e,h} \sum_{n_{z1}} \sum_{n_{z2}} \int_0^\infty \int_0^{2\pi} k''_r \, \mathrm{d}k''_r \, \mathrm{d}\varphi_{k''} \left[\int_\Delta^\infty \mathrm{d}l_r \phi'_{32(\mathrm{e/h})\alpha}(k_r, \varphi_k, k''_r, \varphi_{k''}, l_r)\right], \tag{B.10a}$$

with

$$\phi'_{32(\mathrm{e/h})\alpha}(k_r,\varphi_k,k_r'',\varphi_{k''},l_r) \equiv \frac{1}{\pi^3}\left(\frac{\mathrm{e}^2}{4\varepsilon_0}\right)^2 \int_0^{2\pi} \mathrm{d}\varphi_l \left\{\left\{\frac{2\Phi(l_r)}{\varepsilon(l_r,\varphi_l)l_r}\right.\right.$$
$$\left. - \frac{\delta_{(\mathrm{e/h})\alpha}\Phi\left[Q(k_r,\varphi_k,k_r'',\varphi_{k''},l_r,\varphi_l)\right]}{\varepsilon(k_r,\varphi_k,k_r'',\varphi_{k''},l_r,\varphi_l)Q(k_r,\varphi_k,k_r'',\varphi_{k''},l_r,\varphi_l)}\right\}\frac{\Phi(l_r)}{\varepsilon(l_r,\varphi_l)}$$
$$\left.\times\,\delta\left[\varepsilon_{\Delta 1(\mathrm{e/h})\alpha}(k_r,\varphi_k,k_r'',\varphi_{k''},l_r,\varphi_l)\right]F_{12(\mathrm{e/h})\alpha}(k_r,\varphi_k,k_r'',\varphi_{k''},l_r,\varphi_l)\right\}|_{l_r\neq 0}, \tag{B.10b}$$

and

$$\Sigma_{\vec{k}}^{\mathrm{e/h-in}} \equiv \pi \sum_{\alpha=e,h}\sum_{\vec{l}\neq 0}\sum_{\vec{k}''}(2\widetilde{V}_{|\vec{l}|} - \delta_{(\mathrm{e/h})\alpha}\widetilde{V}_{|\vec{k}''-\vec{l}-\vec{k}|})\widetilde{V}_{|\vec{l}|}$$
$$\times\,\delta\left[\varepsilon_{(\mathrm{e/h})\vec{k}} + \varepsilon_{\alpha\vec{k}''} - \varepsilon_{(\mathrm{e/h})(\vec{k}+\vec{l})} - \varepsilon_{\alpha(\vec{k}''-\vec{l})}\right] f^{\mathrm{e/h}}_{\vec{k}+\vec{l}}(1-f^{\alpha}_{\vec{k}''})f^{\alpha}_{\vec{k}''-\vec{l}}$$
$$= \pi\left(\frac{\mathrm{e}^2}{2\varepsilon_0\Sigma}\right)^2 \sum_{\alpha=e,h}\sum_{\vec{l}\neq 0}\sum_{\vec{k}''}\left[\frac{2\Phi(|\vec{l}\,|)}{\varepsilon_{\vec{l}}|\vec{l}\,|} - \frac{\delta_{(\mathrm{e/h})\alpha}\Phi(|\vec{k}''-\vec{l}-\vec{k}|)}{\varepsilon_{\vec{k}''-\vec{l}-\vec{k}}|\vec{k}''-\vec{l}-\vec{k}|}\right]\frac{\Phi(|\vec{l}\,|)}{\varepsilon_{\vec{l}}|\vec{l}\,|}$$
$$\times\,\delta\left[\varepsilon_{(\mathrm{e/h})\vec{k}} + \varepsilon_{\alpha\vec{k}''} - \varepsilon_{(\mathrm{e/h})(\vec{k}+\vec{l})} - \varepsilon_{\alpha(\vec{k}''-\vec{l})}\right] f^{\mathrm{e/h}}_{\vec{k}+\vec{l}}(1-f^{\alpha}_{\vec{k}''})f^{\alpha}_{\vec{k}''-\vec{l}}$$
$$= \frac{1}{\pi^3}\left(\frac{\mathrm{e}^2}{4\varepsilon_0}\right)^2 \sum_{\alpha=e,h}\sum_{n_{z1}}\sum_{n_{z2}}\int_0^{\infty}\!\!\int_0^{2\pi} k_r''\,\mathrm{d}k_r''\,\mathrm{d}\varphi_{k''}\int_0^{\infty}\!\!\int_0^{2\pi}\mathrm{d}l_r\,\mathrm{d}\varphi_l\left\{\left\{\frac{2\Phi(l_r)}{\varepsilon(l_r,\varphi_l)l_r}\right.\right.$$
$$\left. - \frac{\delta_{(\mathrm{e/h})\alpha}\Phi\left[Q(k_r,\varphi_k,k_r'',\varphi_{k''},l_r,\varphi_l)\right]}{\varepsilon(k_r,\varphi_k,k_r'',\varphi_{k''},l_r,\varphi_l)Q(k_r,\varphi_k,k_r'',\varphi_{k''},l_r,\varphi_l)}\right\}\frac{\Phi(l_r)}{\varepsilon(l_r,\varphi_r)}$$
$$\left.\times\,\delta\left[\varepsilon_{\Delta 1(\mathrm{e/h})\alpha}(k_r,\varphi_k,k_r'',\varphi_{k''},l_r,\varphi_l)\right]F_{11(\mathrm{e/h})\alpha}(k_r,\varphi_k,k_r'',\varphi_{k''},l_r,\varphi_l)\right\}|_{l_r\neq 0}$$
$$= \sum_{\alpha=e,h}\sum_{n_{z1}}\sum_{n_{z2}}\int_0^{\infty}\int_0^{2\pi} k_r''\,\mathrm{d}k_r''\,\mathrm{d}\varphi_{k''}\left[\int_{\Delta}^{\infty}\mathrm{d}l_r\phi'_{31(\mathrm{e/h})\alpha}(k_r,\varphi_k,k_r'',\varphi_{k''},l_r)\right], \tag{B.11a}$$

with

$$\phi'_{31(\mathrm{e/h})\alpha}(k_r,\varphi_k,k_r'',\varphi_{k''},l_r) \equiv \frac{1}{\pi^3}\left(\frac{\mathrm{e}^2}{4\varepsilon_0}\right)^2 \int_0^{2\pi} \mathrm{d}\varphi_l \left\{\left\{\frac{2\Phi(l_r)}{\varepsilon(l_r,\varphi_l)l_r}\right.\right.$$
$$\left. - \frac{\delta_{(\mathrm{e/h})\alpha}\Phi\left[Q(k_r,\varphi_k,k_r'',\varphi_{k''},l_r,\varphi_l)\right]}{\varepsilon(k_r,\varphi_k,k_r'',\varphi_{k''},l_r,\varphi_l)Q(k_r,\varphi_k,k_r'',\varphi_{k''},l_r,\varphi_l)}\right\}\frac{\Phi(l_r)}{\varepsilon(l_r,\varphi_l)}$$
$$\left.\times\,\delta\left[\varepsilon_{\Delta 1(\mathrm{e/h})\alpha}(k_r,\varphi_k,k_r'',\varphi_{k''},l_r,\varphi_l)\right]F_{11(\mathrm{e/h})\alpha}(k_r,\varphi_k,k_r'',\varphi_{k''},l_r,\varphi_l)\right\}|_{l_r\neq 0}. \tag{B.11b}$$

The carrier–phonon scattering coefficients are

$$
\begin{aligned}
L_{\vec{k}}^{(\mathrm{e/h})-\mathrm{out,p}} &\equiv 2\pi\hbar^2 \sum_{\vec{q}\neq 0,\pm} G_{|\vec{q}|}^2 \delta\left[\varepsilon_{(\mathrm{e/h})\vec{k}} - \varepsilon_{(\mathrm{e/h})(\vec{k}-\vec{q})} \mp \hbar\omega_{\mathrm{LO}}\right] \\
&\quad \times (f_{\vec{q}}^{\mathrm{p}} + \frac{1}{2} \pm \frac{1}{2})(1 - f_{\vec{k}-\vec{q}}^{\mathrm{e/h}}) \\
&= 2\pi\hbar^2 \sum_{\vec{l}\neq 0} G_{|\vec{l}|}^2 (1 - f_{\vec{k}-\vec{l}}^{\mathrm{e/h}}) \Big\{ \delta\left[\varepsilon_{(\mathrm{e/h})\vec{k}} - \varepsilon_{(\mathrm{e/h})(\vec{k}-\vec{l})} - \hbar\omega_{\mathrm{LO}}\right] (f_{\vec{l}}^{\mathrm{p}} + 1) \\
&\qquad + \delta\left[\varepsilon_{(\mathrm{e/h})\vec{k}} - \varepsilon_{(\mathrm{e/h})(\vec{k}-\vec{l})} + \hbar\omega_{\mathrm{LO}}\right] f_{-\vec{l}}^{\mathrm{p}} \Big\} \\
&= \frac{\pi \mathrm{e}^2 \hbar\omega_{\mathrm{LO}}}{2\varepsilon_0 \Sigma} \left(\frac{1}{\varepsilon_\infty} - \frac{1}{\varepsilon_0}\right) \sum_{\vec{l}\neq 0} \frac{\Phi(|\vec{l}\,|)}{\varepsilon_{\vec{l}}\,|\vec{l}\,|} \\
&\quad \times (1 - f_{\vec{k}-\vec{l}}^{\mathrm{e/h}}) \Big\{ \delta\left[\varepsilon_{(\mathrm{e/h})\vec{k}} - \varepsilon_{(\mathrm{e/h})(\vec{k}-\vec{l})} - \hbar\omega_{\mathrm{LO}}\right] (f_{\vec{l}}^{\mathrm{p}} + 1) \\
&\qquad + \delta\left[\varepsilon_{(\mathrm{e/h})\vec{k}} - \varepsilon_{(\mathrm{e/h})(\vec{k}-\vec{l})} + \hbar\omega_{\mathrm{LO}}\right] f_{-\vec{l}}^{\mathrm{p}} \Big\} \\
&= \frac{\mathrm{e}^2 \hbar\omega_{\mathrm{LO}}}{4\pi\varepsilon_0} \left(\frac{1}{\varepsilon_\infty} - \frac{1}{\varepsilon_0}\right) \sum_{n_z} \int_0^\infty \int_0^{2\pi} \mathrm{d}l_r \, \mathrm{d}\varphi_l \frac{\Phi(l_r)}{\varepsilon(l_r, \varphi_l)} (1 - f_{d_8\varphi_{d_8}}^{\mathrm{e/h}}) \\
&\quad \times \Big\{ \delta\left[\varepsilon_{\Delta 3(\mathrm{e/h})}(k_r, \varphi_k, l_r, \varphi_l) - \hbar\omega_{\mathrm{LO}}\right] \frac{\mathrm{e}^{\hbar\omega_{\mathrm{LO}}/\mathrm{k_B}T}}{\mathrm{e}^{\hbar\omega_{\mathrm{LO}}/\mathrm{k_B}T} - 1} \\
&\quad + \delta\left[\varepsilon_{\Delta 3(\mathrm{e/h})}(k_r, \varphi_k, l_r, \varphi_l) + \hbar\omega_{\mathrm{LO}}\right] \frac{1}{\mathrm{e}^{\hbar\omega_{\mathrm{LO}}/\mathrm{k_B}T} - 1} \Big\} \\
&= \sum_{n_z} \int_0^\infty \mathrm{d}l_r \phi_{91(\mathrm{e/h})}(k_r, \varphi_k, l_r),
\end{aligned}
\tag{B.12a}
$$

with

$$
\begin{aligned}
\phi_{91(\mathrm{e/h})}(k_r, \varphi_k, l_r) \equiv{}& \frac{\mathrm{e}^2 \hbar\omega_{\mathrm{LO}}}{4\pi\varepsilon_0} \left(\frac{1}{\varepsilon_\infty} - \frac{1}{\varepsilon_0}\right) \int_0^{2\pi} \mathrm{d}\varphi_l \frac{\Phi(l_r)}{\varepsilon(l_r, \varphi_l)} (1 - f_{d_8\varphi_{d_8}}^{\mathrm{e/h}}) \\
& \times \Big\{ \delta\left[\varepsilon_{\Delta 3(\mathrm{e/h})}(k_r, \varphi_k, l_r, \varphi_l) - \hbar\omega_{\mathrm{LO}}\right] \frac{\mathrm{e}^{\hbar\omega_{\mathrm{LO}}/\mathrm{k_B}T}}{\mathrm{e}^{\hbar\omega_{\mathrm{LO}}/\mathrm{k_B}T} - 1} \\
& + \delta\left[\varepsilon_{\Delta 3(\mathrm{e/h})}(k_r, \varphi_k, l_r, \varphi_l) + \hbar\omega_{\mathrm{LO}}\right] \frac{1}{\mathrm{e}^{\hbar\omega_{\mathrm{LO}}/\mathrm{k_B}T} - 1} \Big\},
\end{aligned}
\tag{B.12b}
$$

and

$$\begin{aligned}
L_{\vec{k}}^{(\mathrm{e/h})-\mathrm{in,p}} &\equiv 2\pi\hbar^2 \sum_{\vec{q}\neq 0,\pm} G_{|\vec{q}|}^2 \delta\left[\varepsilon_{(\mathrm{e/h})\vec{k}} - \varepsilon_{(\mathrm{e/h})(\vec{k}-\vec{q})} \mp \hbar\omega_{\mathrm{LO}}\right]\left(f_{\vec{q}}^{\mathrm{p}} + \frac{1}{2} \mp \frac{1}{2}\right) f_{\vec{k}-\vec{q}}^{\mathrm{e/h}} \\
&= 2\pi\hbar^2 \sum_{\vec{l}\neq 0} G_{|\vec{l}|}^2 f_{\vec{k}-\vec{l}}^{\mathrm{e/h}} \Big\{ \delta\left[\varepsilon_{(\mathrm{e/h})\vec{k}} - \varepsilon_{(\mathrm{e/h})(\vec{k}-\vec{l})} - \hbar\omega_{\mathrm{LO}}\right] f_{\vec{l}}^{\mathrm{p}} \\
&\qquad + \delta\left[\varepsilon_{(\mathrm{e/h})\vec{k}} - \varepsilon_{(\mathrm{e/h})(\vec{k}-\vec{l})} + \hbar\omega_{\mathrm{LO}}\right]\left(f_{-\vec{l}}^{\mathrm{p}} + 1\right)\Big\} \\
&= \frac{\pi \mathrm{e}^2\hbar\omega_{\mathrm{LO}}}{2\varepsilon_0 \Sigma}\left(\frac{1}{\varepsilon_\infty} - \frac{1}{\varepsilon_0}\right) \sum_{\vec{l}\neq 0} \frac{\Phi(|\vec{l}|)}{\varepsilon_{\vec{l}}|\vec{l}|} f_{\vec{k}-\vec{l}}^{\mathrm{e/h}} \Big\{ \delta\left[\varepsilon_{(\mathrm{e/h})\vec{k}} - \varepsilon_{(\mathrm{e/h})(\vec{k}-\vec{l})}\right. \\
&\quad \left. - \hbar\omega_{\mathrm{LO}}\right] f_{\vec{l}}^{\mathrm{p}} + \delta\left[\varepsilon_{(\mathrm{e/h})\vec{k}} - \varepsilon_{(\mathrm{e/h})(\vec{k}-\vec{l})} + \hbar\omega_{\mathrm{LO}}\right]\left(f_{-\vec{l}}^{\mathrm{p}} + 1\right)\Big\} \\
&= \frac{\mathrm{e}^2\hbar\omega_{\mathrm{LO}}}{4\pi\varepsilon_0}\left(\frac{1}{\varepsilon_\infty} - \frac{1}{\varepsilon_0}\right) \sum_{n_z} \int_0^\infty \int_0^{2\pi} \mathrm{d}l_r\, \mathrm{d}\varphi_l \frac{\Phi(l_r)}{\varepsilon(l_r,\varphi_l)} f_{d_8\varphi_{d_8}}^{\mathrm{e/h}} \\
&\quad \times \Big\{ \delta\left[\varepsilon_{\Delta 3(\mathrm{e/h})}(k_r,\varphi_k,l_r,\varphi_l) - \hbar\omega_{\mathrm{LO}}\right] \frac{1}{\mathrm{e}^{\hbar\omega_{\mathrm{LO}}/\mathrm{k_B}T} - 1} \\
&\qquad + \delta\left[\varepsilon_{\Delta 3(\mathrm{e/h})}(k_r,\varphi_k,l_r,\varphi_l) + \hbar\omega_{\mathrm{LO}}\right] \frac{\mathrm{e}^{\hbar\omega_{\mathrm{LO}}/\mathrm{k_B}T}}{\mathrm{e}^{\hbar\omega_{\mathrm{LO}}/\mathrm{k_B}T} - 1}\Big\} \\
&= \sum_{n_z} \int_0^\infty \mathrm{d}l_r \phi_{92(\mathrm{e/h})}(k_r,\varphi_k,l_r),
\end{aligned} \tag{B.13a}$$

with

$$\begin{aligned}
\phi_{92(\mathrm{e/h})}(k_r,\varphi_k,l_r) &\equiv \frac{\mathrm{e}^2\hbar\omega_{\mathrm{LO}}}{4\pi\varepsilon_0}\left(\frac{1}{\varepsilon_\infty} - \frac{1}{\varepsilon_0}\right) \int_0^{2\pi} \mathrm{d}\varphi_l \frac{\Phi(l_r)}{\varepsilon(l_r,\varphi_l)} f_{d_8\varphi_{d_8}}^{\mathrm{e/h}} \\
&\quad \times \Big\{ \delta\left[\varepsilon_{\Delta 3(\mathrm{e/h})}(k_r,\varphi_k,l_r,\varphi_l) - \hbar\omega_{\mathrm{LO}}\right] \frac{1}{\mathrm{e}^{\hbar\omega_{\mathrm{LO}}/\mathrm{k_B}T} - 1} \\
&\qquad + \delta\left[\varepsilon_{\Delta 3(\mathrm{e/h})}(k_r,\varphi_k,l_r,\varphi_l) + \hbar\omega_{\mathrm{LO}}\right] \frac{\mathrm{e}^{\hbar\omega_{\mathrm{LO}}/\mathrm{k_B}T}}{\mathrm{e}^{\hbar\omega_{\mathrm{LO}}/\mathrm{k_B}T} - 1}\Big\}.
\end{aligned} \tag{B.13b}$$

C Cash–Karp's implementation of the fifth order Runge–Kutta method [1]

For a system of ODEs

$$\frac{\mathrm{d}}{\mathrm{d}t'} y_k(t') = F_k\left[y_k(t'), t'\right], \tag{C.1}$$

Table A.1. Cash–Karp coefficients

i	a_i	b_{i1}	b_{i2}	b_{i3}	b_{i4}	b_{i5}	c_i	c_i^*
1	–	–	–	–	–	–	37/378	2825/27648
2	1/5	1/5	–	–	–	–	0	0
3	3/10	3/40	9/40	–	–	–	250/621	18575/48384
4	3/5	3/10	-9/10	6/5	–	–	125/594	13525/55296
5	1	-11/54	5/2	-70/27	35/27	–	0	277/14336
6	7/8	1631/55296	175/512	575/13824	44275/110592	253/4096	512/1771	1/4

the fifth order Runge–Kutta algorithm is implemented as

$$y_k(t'_{j+1}) = y_k(t'_j) + \sum_{i=1}^{6} c_i d_{ki} + O(h^6), \tag{C.2}$$

with the embedded fourth order formula in the form

$$y_k^*(t'_{j+1}) = y_k(t'_j) + \sum_{i=1}^{6} c_i^* d_{ki} + O(h^5). \tag{C.3}$$

The parameters are evaluated through

$$\begin{aligned}
d_{k1} &= hF_k\left[y_k(t'_j), t'_j\right]\\
d_{k2} &= hF_k\left[y_k(t'_j) + b_{21}d_{k1}, t'_j + a_2h\right]\\
d_{k3} &= hF_k\left[y_k(t'_j) + b_{31}d_{k1} + b_{32}d_{k2}, t'_j + a_3h\right]\\
d_{k4} &= hF_k\left[y_k(t'_j) + b_{41}d_{k1} + b_{42}d_{k2} + b_{43}d_{k3}, t'_j + a_4h\right]\\
d_{k5} &= hF_k\left[y_k(t'_j) + b_{51}d_{k1} + b_{52}d_{k2} + b_{53}d_{k3} + b_{54}d_{k4}, t'_j + a_5h\right]\\
d_{k6} &= hF_k\left[y_k(t'_j) + b_{61}d_{k1} + b_{62}d_{k2} + b_{63}d_{k3} + b_{64}d_{k4} + b_{65}d_{k5}, t'_j + a_6h\right].
\end{aligned} \tag{C.4}$$

The coefficients in equations (C.2), (C.3) and (C.4) are given in [2], as in Table A.1.

D The solution of sparse linear equations [3, 4]

To solve a non-linear system of algebraic equations, we need to employ the iteration algorithm developed from Newton's method or its derived versions. At each iteration step, a system of linear algebraic equations in the form

$$Ax = B, \tag{D.1}$$

must be solved. In dealing with such a problem, we generally have two available approaches, known as the direct method and the iterative method.

D.1 The direct method

In seeking for the solution of equation (D.1), we need to factorize the coefficient matrix A to make

$$PAQ = LU, \tag{D.2}$$

where P and Q are permutation matrices and L and U are lower and upper triangular matrices, respectively.

Prior to starting the factorization of a matrix, a diagonal scaling scheme can be introduced to make all non-zero elements of comparable size in order to minimize the round-off error

$$D^{-1}Ax = D^{-1}B, \tag{D.3}$$

$$D^{-1/2}AD^{-1/2}D^{1/2}x = D^{-1/2}B, \tag{D.4}$$

where D is a diagonal matrix formed by the main diagonal of A.

The former is a single row scaling while the latter is a row and column double scaling. Obviously, the double scaling scheme is applicable only if all main diagonal elements of A are positive or negative. A unique feature of the double scaling scheme is that it preserves the symmetry of A.

When a sparse matrix is factorized, the fill normally increases, i.e., the factors of a matrix taken together are usually not as sparse as the matrix itself. A transformation in the form

$$PAP^{-1}Px = PB, \tag{D.5}$$

can be introduced to reduce the fill, where P is the permuting matrix to make the permuted matrix PAP^{-1} exhibit a different fill. For an appropriate choice of P, we can often reduce the fill significantly.

For example, for a rectangular mesh with N_x in the vertical and N_y in the horizontal, we will have a matrix A of the order of $N_x \times N_y$. The number of non-zero elements prior to factorization is usually around 5–9 $\times$ $(N_x \times N_y)$, depending on the discretization scheme. After the factorization, however, the total number of non-zero elements grows significantly. By reordering the mesh points, i.e., performing equation (D.5), we will be able to reduce the increment on the non-zero elements in the factorized matrices. In doing so, we usually take the following principles.

(1) If $N_x > N_y$, order the mesh points along N_y (the shorter one), then along N_x (the longer one). For example, if $N_x = 5$, $N_y = 4$, do not order the mesh point along $N_x = 5$:

$$\begin{matrix} 16 & 17 & 18 & 19 & 20 \\ 11 & 12 & 13 & 14 & 15 \\ 6 & 7 & 8 & 9 & 10 \\ 1 & 2 & 3 & 4 & 5 \end{matrix}, \tag{D.6}$$

but along $N_y = 4$

$$\begin{matrix} 4 & 8 & 12 & 16 & 20 \\ 3 & 7 & 11 & 15 & 19 \\ 2 & 6 & 10 & 14 & 18 \\ 1 & 5 & 9 & 13 & 17 \end{matrix}. \tag{D.7}$$

(2) The total non-zero elements in the factorized matrices will be fewer if we order the mesh points along the diagonals:

$$\begin{matrix} 7 & 11 & 15 & 18 & 20 \\ 4 & 8 & 12 & 16 & 19 \\ 2 & 5 & 9 & 13 & 17 \\ 1 & 3 & 6 & 10 & 14 \end{matrix}. \tag{D.8}$$

There are still better numbering strategies such as reverse diagonal, one-way dissection, nested dissection, and minimum degree methods [5]. However, the implementation is a non-trivial task and the improvement on reducing the number of non-zero elements is not so significant.
Particularly, if the matrix A can be permuted such that

$$PAP^{-1} = \begin{bmatrix} D_R & C_R \\ C_B & D_B \end{bmatrix}, \tag{D.9}$$

where D_R and D_B are diagonal matrices of rank n_R and n_B, $rank(A) = n_R + n_B$, respectively, we can introduce checker-board ordering

$$\begin{matrix} 12 & 4 & 16 & 8 & 20 \\ 2 & 14 & 6 & 18 & 10 \\ 11 & 3 & 15 & 7 & 19 \\ 1 & 13 & 5 & 17 & 9 \end{matrix}, \tag{D.10}$$

or the alternating diagonal ordering

$$\begin{matrix} 13 & 5 & 17 & 9 & 20 \\ 2 & 14 & 6 & 18 & 10 \\ 11 & 3 & 15 & 7 & 19 \\ 1 & 12 & 4 & 16 & 8 \end{matrix}. \tag{D.11}$$

Either one will offer us fewer non-zero elements after factorization.

It should be noted that all the above ordering procedures are designed only to make the fill smaller after the factorization. It is assumed that column and row interchanges to maintain numerical stability are not necessary for the factorization. This is true only for positive (or negative) definite matrices. Fortunately, the linear systems arising from the discretization of the PDEs by various FD schemes usually have this property.

D.2 The iterative method

(1) *Relaxation method.*

From equation (D.1), we have

$$Cx + (A - C)x = B. \tag{D.12}$$

An iterative scheme can be established for equation (D.12) as

$$Cx_{k+1} = (C - A)x_k + B, \tag{D.13}$$

if equation (D.13) can trivially be solved through

$$x_{k+1} = (I - C^{-1}A)x_k + C^{-1}B \equiv Mx_k + C^{-1}B. \tag{D.14}$$

The basis for the application of relaxation methods to a system is splitting of the coefficient matrix A

$$A = D - L - U, \tag{D.15}$$

with D as a non-singular matrix, L as a strict lower triangular matrix and U as a strict upper triangular matrix.

For matrix C, there are usually four selection schemes:

(A) The Jacobi method.

$$C_{\mathrm{J}} = D, \text{ and } M_{\mathrm{J}} = D^{-1}(L + U). \tag{D.16}$$

Advantages:

If D is diagonal, the solution can be vectorized easily, leading to high efficiency for vector computing.

If M_{J} is symmetric (which can easily be achieved if A is symmetric), M_{J} only has real eigenvalues, which allows the application of many convergence acceleration methods.

The major disadvantage is its poor convergence property.

(B) The Gauss–Seidel method.

$$C_{\mathrm{GS}} = D - L, \text{ and } M_{\mathrm{GS}} = (D - L)^{-1}U. \tag{D.17}$$

(C) The successive over-relaxation (SOR) method.

$$C_{\mathrm{SOR}} = \frac{1}{\omega}D - L, \text{ and } M_{\mathrm{SOR}} = (D - \omega L)^{-1}\left[\omega U + (1 - \omega)D\right]. \tag{D.18}$$

When $\omega = 1$, the SOR method reduces to the G–S method.

(D) The symmetric SOR (SSOR) method.

$$C_{\mathrm{SSOR},1} = \frac{1}{\omega}D - L$$
$$C_{\mathrm{SSOR},2} = \frac{1}{\omega}D - U, \tag{D.19a}$$

$$C_{\mathrm{SSOR},1}x_{k+1/2} = (C_{\mathrm{SSOR},1} - A)x_k + B$$
$$C_{\mathrm{SSOR},2}x_{k+1} = (C_{\mathrm{SSOR},2} - A)x_{k+1/2} + B, \tag{D.19b}$$

$$M_{\mathrm{SSOR}} = (D - \omega U)^{-1}\left[\omega L + (1-\omega)D\right] \times (D - \omega L)^{-1}\left[\omega U + (1-\omega)D\right]. \tag{D.19c}$$

Usually the convergence of the SSOR method differs very marginally from that of the SOR method. However, matrix M_{SSOR} preserves the symmetry of A, which enables the application of acceleration methods, whereas M_{SOR} is always non-symmetric. For this reason, the SSOR method is used only in conjunction with acceleration methods.

(2) *The alternating-direction implicit (ADI) method.*

By assuming

$$A = H + V, \tag{D.20}$$

we have

$$(H + \alpha_{k+1}I)x_{k+1/2} = (\alpha_{k+1}I - V)x_k + B$$
$$(V + \alpha_{k+1}I)x_k = (\alpha_{k+1}I - H)x_{k+1/2} + B, \tag{D.21}$$

where α_{k+1} is a positive constant acting as an acceleration parameter. The iteration matrix is therefore given by

$$M_{\mathrm{ADI}} = (V + \alpha_{k+1}I)^{-1}(\alpha_{k+1}I - H)(H + \alpha_{k+1}I)^{-1}(\alpha_{k+1}I - V). \tag{D.22}$$

The ADI method is convergent for arbitrary positive α_{k+1} when at least one of H and V is positive definite and the other is not negative definite. An optimum sequence of α_{k+1}, however, can greatly improve the average rate of convergence.

For coefficient matrix A with bounded diagonal non-zero elements, we usually put the horizontal neighbors in H and vertical neighbors in V with the main diagonal elements of A equally divided and then put into H and V. The coefficient matrices in equation (D.21) are tri-diagonal after proper permutations, hence (D.21) can be solved by direct (implicit) methods.

(3) *The strongly implicit (SI) method.*

From equation (D.13), we find that matrix C should be selected as close to A as possible in order to accelerate the convergence. If we can find

$$A + N = LU, \tag{D.23}$$

with $||N|| \ll ||A||$, we can readily select $C = LU$, hence equation (D.13) becomes

$$LUx_{k+1} = Nx_k + B. \tag{D.24}$$

Obviously, iterations based on equation (D.24) will converge rapidly.

(4) *Convergence acceleration.*

(A) The Lyusternik method.

Following equation (D.14), by assuming $x^* = x_k - e_k = x_{k+1} - e_{k+1}$, where x^* is the exact solution and e_k the error, we have $e_{k+1} = Me_k = \lambda e_k + \delta_k$, with λ as the maximum eigenvalue of M. Therefore

$$x^* = x_k + \frac{x_{k+1} - x_k}{1-\lambda} - \frac{\delta_k}{1-\lambda} \approx x_k + \frac{x_{k+1} - x_k}{1-\lambda}, \tag{D.25}$$

where $||\delta_k|| < 1 - \lambda$. If λ can be evaluated as the spectral radius of the iteration matrix, equation (D.25) can be viewed as the extrapolation of x^*.

All eigenvalues of the iteration matrix are required to be real and positive for this acceleration method.

(B) The Aitken method.

Following equation (D.14), we can also represent the extrapolation scheme given as

$$x_i^* \approx x_k^i + \frac{x_{k+1}^i - x_k^i}{1 - \frac{x_{k+1}^i - x_k^i}{x_k^i - x_{k-1}^i}}. \tag{D.26}$$

In this scheme, there is no need to evaluate λ.

(C) The semi-iterative method.

All iterative methods considered so far, i.e., the relaxation, ADI, and SI methods are one-step stationary approaches that consist of the mapping

$$x_{k+1} = Mx_k + B. \tag{D.27}$$

An optimal method may not follow this form as information obtained from earlier iterations is not utilized as feedback to improve the iteration.

Therefore, in the semi-iterative method, we try multiple-step mapping

$$y_k = N_k(x_k, x_{k-1}, \ldots, x_0), \tag{D.28}$$

with

$$\lim_{k\to\infty} (y_k - x^*) = 0. \tag{D.29}$$

For example, we may use the following linear combination scheme

$$y_k = \sum_{i=0}^{k} c_i^k x_i, \tag{D.30}$$

with

$$\sum_{i=0}^{k} c_i^k = 1. \tag{D.31}$$

This method is equivalent to a scheme in which the iterates x_i pass through an extra finite impulse response (FIR) filter at a length of $k + 1$.

A non-stationary two-step iterative method is given by

$$y_{k+1} = \omega_{k+1}(My_k + B) + (1 - \omega_{k+1})y_{k-1}, \tag{D.32}$$

with

$$\omega_1 = 1, \text{ and } \omega_{k+1} = 1 + \frac{C^{k-1}(1/\rho)}{C^{k+1}(1/\rho)}, \tag{D.33}$$

where $C^k(x) = \cos\left[k\cos^{-1}(x)\right]$, $x \in [-1, 1]$, $k > 0$ is the kth Chebyshev polynomial and $\rho = \lambda_{\max} = -\lambda_{\min}$, with $\lambda_{\max}$ and $\lambda_{\min}$ denoting the maximal and minimal eigenvalues of a symmetric iteration matrix M with real eigenvalues only.

This scheme also shows that it is not necessary to express the accelerated solutions y_k explicitly in terms of the iterates x_k, which is equivalent to letting the iterates x_k pass through an extra infinite impulse response (IIR) filter.

(D) The conjugate gradient method.

This method comes from minimization of the target function

$$\frac{1}{2}(Az - B)^T A^{-1}(Az - B). \tag{D.34}$$

When A is positive definite, the target function is zero (and minimal) only for $z = x^*$, the solution of equation (D.1).

By defining the residual vector as

$$r_k = Ax_k - B, \tag{D.35}$$

we construct

$$x_{k+1} = x_k + \lambda_k d_k, \tag{D.36}$$

where

$$\lambda_k = -\frac{d_k^T r_k}{d_k^T A d_k}, \tag{D.37}$$

$$d_0 = -r_0, \text{ and } d_{k+1} = -r_{k+1} + \frac{r_{k+1}^T A d_k}{d_k^T A d_k} d_k. \tag{D.38}$$

Obviously, this algorithm would terminate theoretically after $rank(A)$ iterations since no more orthogonal search directions exist. However, the existence of round-off errors may need further iterations until the residual is sufficiently small.

This method can also be used to accelerate the general iteration scheme equation (D.13), once C is symmetric and positive definite.

Actually, we establish the following algorithm

$$Cs_k = r_k, \tag{D.39}$$

$$x_{k+1} = x_k + \lambda_k e_k, \tag{D.40}$$

$$e_0 = -s_0, \text{ and } e_{k+1} = -s_{k+1} + \frac{s_{k+1}^T A e_k}{e_k^T A e_k} e_k, \tag{D.41}$$

$$r_0 = Ax_0 - B, \text{ and } r_{k+1} = r_k + \lambda_k A e_k, \tag{D.42}$$

$$\lambda_k = \frac{e_k^T r_k}{e_k^T A e_k}. \tag{D.43}$$

When $\lambda_k = 1$, $s_{k+1}^T A e_k = 0$, this algorithm reduces to equation (D.13).

If $C = I$, this algorithm reduces to the basic conjugate gradient method, equations (D.35)–(D.38). Thus, the basic conjugate gradient method can be understood as the accelerated Jacobi method for the preconditioned system

$$D^{-1/2} A D^{-1/2} D^{1/2} x = D^{-1/2} B, \tag{D.44}$$

where D is the main diagonal of A.

References

[1] W. H. Press, S. A. Teukolsky, W. T. Vetterling, and B. P. Flannery, *Numerical Recipes in Fortran: The Art of Scientific Computing*, 2nd edn (Cambridge, UK: Cambridge University Press, 1992).

[2] J. R. Cash and A. H. Karp, A variable order Runge–Kutta method for initial value problems with rapidly varying right-hand sides. *ACM Trans. on Mathematical Software*, **16** (1990), 201–22.

[3] G. H. Golub and C. F. Van Loan, *Matrix Computations*, 2nd edn (Baltimore: The John Hopkins University Press, 1989).

[4] W. Hackbusch, *Iterative Solution of Large Sparse Systems of Equations*, 1st edn (New York: Springer-Verlag, 1994).

[5] S. Selberherr, *Analysis and Simulation of Semiconductor Devices*, 1st edn (New York: Springer-Verlag, 1984).

Index

absorbing boundary condition (ABC), 172
active region, 157–159
active region boundary, 158, 159
active region volume, 156
ADI approach, 234, 352
ADI method, 182, 352, 353
Aitken method, 353
alternating current (AC), 255
alternating-direction implicit *see* ADI approach
alternating-direction implicit method, *see* ADI method
amplitude-modulated harmonic wave, 17
amplitude-modulated plane wave, 17
anti-reflective (AR) coat, 276, 278, 281, 282, 299, 305, 307, 314, 323
associated refractive index, 19
Auger impact ionization, 163
Auger recombination, 163, 164

background refractive index, 44
backward-time backward-space (BT–BS), 178, 179
backward-time forward-space (BT–FS), 178, 179
backward-time implicit scheme, 179
backward-time scheme, 178, 180
band edge, 156
band structure, 79
Bernoulli function, 223
Bloch equation, 36, 37, 40, 59, 65
Bloch function, 63, 64
Bloch's theorem, 57, 59–61, 75, 81
Bohr radius, 136
Boltzmann equation, 152
Boltzmann transport equation, 246
Born–Oppenheimer approximation, 55
Bose–Einstein distribution, 114
bottom valence band, 157
boundary value, 4
box scheme, 181, 182
Bragg condition, 275
Bragg scattering, 61, 64
Bragg stop-band, 275, 279, 281, 282, 284, 313, 315
Bragg stop-band wavelength, 275
Brillouin zone (BZ), 59–64, 82–84
  Brillouin zone (BZ) centers, 64, 84
broadband optical field, 40
bulk active region, 162
bulk semiconductor, 56–60, 63, 64, 161
  3D, 57
buried heterostructure (BH), 259, 260, 265, 268, 269

carrier density, 158, 159, 163
  2D, 205
carrier and temperature distribution, 2D, 239
carrier distribution, 159
carrier rate equation, 1D, 168
carrier redistribution, 159
carrier transport, 1
  2D, 162, 242
carrier transport equation, 152, 157, 158
  2D, 160, 242
carrier transport model, 2D, 165
carrier–carrier collision, 152, 344
  second order, 146
carrier–phonon scattering, 152, 156
  second order, 146
carrier–phonon scattering coefficient, 346
carrier–phonon scattering contribution, 147
Cash–Karp's implementation, 209
center difference scheme, 179
channel frequency spacing, 39
charge neutrality condition, 164
classical carrier transport equation, 158
classical carrier transport model, 159
classical Drude formula, 132
classical thermal diffusion equation, 166
cold carrier, 156–159
cold carrier densities, 157–159
cold carrier distribution, non-uniform, 159
community antenna television (CATV), 278
complex-coupled DFB laser, 278, 280
conduction, 157
conduction band, 66, 68–70, 75, 81, 84, 85, 88, 93
conduction band hydrostatic energy, 93, 94
conservation law, 2
continuum band, 162
convergence acceleration, 4, 5
cooling and heating damping term, 157
cooling and heating time constant, 159

cooling time constant, 158
Coulomb interaction, 55
Coulomb potential, 61
Coulomb potential energy, 56, 57
coupled ODEs, 148
 1D, 200
coupled traveling wave equation, 28
coupling contribution, 157
coupling matrix, 108
Courant–Friedrichs–Lewy stability criterion, 176
Crank–Nicholson plus ADI scheme, 234
Crank–Nicholson scheme, 181, 182, 234
cross-sectional area, 159, 164
cross-sectional region, 164
cross-sectional structure, 165
current density divergence, 161

degenerated heavy- and light-hole band, 77, 78
dense wavelength division multiplexing (DWDM), 281
density of states (DOS)
 3D, 202
 2D, 202
density of state (DOS) function
 3D, 35
 2D, 36
differential–integral equation, 31
digital filtering, 4
 algorithm, 34
direct current (DC), 3, 24, 151, 179, 255, 323, 328
 analysis, 191, 233, 241–243
 bias, 244, 300, 322, 328
 calculation, 244
 component, 24, 83, 240, 329
 governing equations for, 243
 operation, 251, 255
 performance, 255, 323
 value, 244, 326
direct discretization method, 182
direct injection term, 160
direct method, 4
Dirichlet boundary condition, 224
discrete energy levels, 162
discretization scheme, 172, 180
distributed Bragg reflector (DBR), 8, 21, 29
distributed feedback (DFB), 5, 8, 21, 29, 191, 195, 251, 272, 281, 282, 313, 328
 DFB grating, Bragg wavelength, 321
 DFB laser, 272, 273, 275–284, 313–318, 320–329
  purely index-coupled, 280, 281
  uniform grating, 281, 283, 284
 DFB laser spectrum, 315
 DFB section, 326
 DFB side, 281
drift and diffusion model, 166
dual octant-wavelength phase-shifted DFB laser, 282, 283
dynamic performance, 323

edge emitting device, 43, 159, 168
effective Hamiltonian matrix, 100
effective index, 48
eigenfunction expansion, 182
eigenstate, 66, 75, 78, 81, 82
eigenvalue, 66, 78
eigenvalue equation, 75
 1D, 20
 2D, 16
eigenvalue problem, 4
 1D, 46
 2D, 172
Einstein relation, 153
electro-absorption modulator (EAM), 5, 137, 200, 238, 288, 291, 292, 294, 295, 298, 299, 313–315, 317, 319, 321
 shallow ridge EAM chirp parameter, 290
 shallow ridge EAM insertion loss, 290
electromagnetic wave theory, 1
electron and hole Auger recombination constant, 163
electron and hole capture, 163
electron and hole densities, 163, 164
electron density, 164
electron spin effect, 66
electrons and holes, 161, 163, 164
Elliot formula, 136
elliptical PDE, 9
energy conservation condition, 166
envelope function, 9, 11–13, 15, 17, 18
 2D, 21
Euler method, 233
 exciton effect, 2D, 137

Fabry–Perot (FP) devices, 5, 37, 259, 272, 281, 282
 FP laser, 255, 263, 267, 269–273, 281
 FP section, 281, 282, 284
 FB side, 281
 FP structure, 259, 284
fast Fourier transform (FFT), 4, 188, 240, 244
Fermi–Dirac distribution, 126
Fermi–Dirac integral, 204, 207
fiber-to-the-home (FTTH), 321, 323
finite difference (FD) method, 4, 172, 200, 220, 221, 223, 231
 FD approach, 173, 238
 step-by-step FD procedure, 214
finite difference time domain (FDTD), 7
finite impulse response (FIR) filter, 354
forward-time backward-space (FT–BS), 174, 176, 178, 231
forward-time explicit center-space (FT–ECS), 179, 233
forward-time forward-space (FT–FS), 178

forward-time implicit center-space (FT–ICS), 180, 234
Fourier component, 60, 61
Fourier transform, 36, 38, 84, 86
  1D, 88
  1D inverse, 89
Fröhlich electron–LO phonon, coupling matrix, 108
free-carrier assumption, 161
free-carrier gain model, 165
frequency domain, 3
full width half maximum (FWHM), 10
fully confined waveguide structure, 44

gain, 1
gain-clamped (GC) SOAs, 31
gain-coupled DFB laser, 277, 278
Gauss–Seidel (G–S) method, 351
Gaussian distributed random process, 49–51
Gaussian distribution, 49
Gaussian–Legendre method, 208
generalized Dirac function, 117
genetic algorithm (GA), 246
governing equation, 2, 9, 12, 31, 34, 118, 121, 122
graded index separate confinement heterojunction (GRINSCH), 259, 260, 262, 263, 267–269, 276, 313, 314
grating shape function, 28
Green's function, 26, 30
GRINSCH, *see* graded index separate confinement heterojunction
group velocity, 38

Hamiltonian, 55–57, 67–69, 76–78, 80, 81, 85, 87, 93, 99
Hamiltonian block, 69
Hamiltonian operator, 54, 55, 57–59, 63
harmonic distortion response, 243
harmonic wave frequency, 9
Hartree self-consistent model, 55
Hartree–Fock form, 56
heat generation, 167
heavy-hole band, 68, 70, 91
Heisenberg equation, 1, 103, 109, 131, 140, 152
Heisenberg's uncertainty relation, 102
heterojunction band edge, 160
hole density, 164
hole SRH decay rate, 163
hot carrier, 156–159
  conduction and valence band, 156, 157
  density, 158
  quasi-Fermi levels, 157, 158
hot–cold carrier coupling, 160
Householder reduction, 200
high-reflective (HR) coat, 276, 278, 281, 282, 305, 307, 313, 315, 323
hyperbolic PDE, 9

in-phase partially gain-coupled DFB laser, 278, 280
index-coupled DFB laser, 278
index-coupled uniform grating DFB laser, 281, 315
indirect discretization method, 182
infinite impulse response (IIR) filter, 354
inhomogeneous spontaneous emission contribution, 40, 42, 183, 240
initial and boundary values, 4
initial quasi-Fermi levels, 207
initial value, 4
integral transformation, 182
integrated semiconductor DFB laser, 313
intensity modulation (IM), 280
  response, 244
  response peak, 280
inverse Fourier transform, 83, 85–87
iterative method, 4
ITU standard, 321

Jacobi method, 351
Jacobian matrix, 229
Jacobian transformation, 200
Joule heating, 166

k–p theory, 54, 65, 80, 84
Kane's model, 65, 71
kinetic energy, 99, 161
Kramers–Kronig transformation, 36

Laguerre polynomials, 65
Langevin noise source, 49
large-signal dynamic performance, 3
laser diode (LD), 5
lattice deformation energy, 99
lattice temperature, 159
Lax–Friedrichs method, 178
Lax–Wendroff scheme, 181
Legendre function, 65
Lie–Trotter–Suzuki product formula, 188
light-hole band, 68–70, 91
Lindhard formula, 132
linear algebraic equation, 230
linear differential equation, 102
longitudinal mode, 45, 46
longitudinal optical (LO) phonon, 108, 109, 121
longitudinal optical mode, 45, 46
longitudinal spatial hole burning (LSHB), 46, 47, 159, 164, 165, 179, 192, 242, 245, 272, 277, 278, 281
  effect of, 255, 272, 274, 275, 277, 280
Lorentzian line-shape function, 127, 135, 136, 144, 238
loss-coupled DFB laser, 277, 278, 280
low pass filter (LPF), 323
Lowdin's renormalization, 71, 81, 332

Lowdin's theory, 72, 74
Luttinger–Kohn's approach, 99, 100
Luttinger–Kohn base, 99
Luttinger–Kohn Hamiltonian, 77, 86, 96, 97
Luttinger–Kohn Hamiltonian matrix, 86, 89, 94, 95, 200, 236
Luttinger–Kohn matrix, 89
Luttinger–Kohn matrix element, 86, 89
Luttinger–Kohn's model, 71, 80
Lyusternik method, 353

macroscopic domain, 157
macroscopic hot carrier density, 157
many-body correlation model, 141
many-body Coulomb effect, 154
many-body gain model, 165
many-body model, 157
material background refractive index, 38
material excitation, 54
material gain, 35, 49
material gain model, 165
material susceptibility, 40, 41
Maxwell equations, 1, 6, 9, 129, 151, 152
Maxwell–Boltzmann distribution, 125
mesh grid, 4
microscopic cold carrier equation, 158
microscopic cold carrier number expectations, 157
microscopic domain, 157
microscopic governing equations, 157, 159, 160
microscopic model, 159
microscopic to macroscopic domain conversion, 158
mixed boundary, 4
modulation bandwidth, 39, 40
monitoring photodetector, 282
monolithic integration, 5
Monte-Carlo integration, 206
Monte-Carlo method, 206, 208
Muller's algorithm, 275

N side material, 157, 159, 164
Neumann boundary condition, 225, 227
Newton's iteration, 4, 5
non-active region, 158
non-periodic QW potential energy, 99
non-physical input parameters, 3
non-radiative plus spontaneous emission recombination rate, 161, 163
non-radiative recombination, 163, 167
non-return-to-zero (NRZ) binary bit stream, 301

octant-wavelength phase-shifted DFB laser, 281
one-dimensional equation, 165
one-dimensional model, 164
one-dimensional space, 87
operator–Hamiltonian commutator, 129
  expansion, 128, 140
  series, 128

optical absorption heating, 166, 167
optical eigenvalue problem, 172
optical field, 162
  2D, 165
optical field distribution, 157, 159
  non-uniform, 159
optical governing equation, 165
optical longitudinal mode distribution, 44
optical loss, 38, 44
optical mode
  1D, 18
  3D, 44, 45
optical rate equation, 1D, 168
optical network unit (ONU), 321, 323
optical wave and carrier governing equation, 165
optical wave equation, 1D, 239
optical wave propagation, 8
optoelectronic modulator, 44
ordinary differential equation (ODE), 4, 43, 46, 103, 109, 116, 157, 168, 191, 200, 206, 209, 210, 222, 232, 347

P side material, 157, 159, 164
Pade approximation, 165
parabolic PDE, 9
parasitic frequency modulation (FM) responses, 244
partial differential equation (PDE), 4, 9, 46, 99, 103, 172, 214, 218, 220, 231, 233
partially corrugated DFB laser, 281, 282, 284
passive optical network (PON), 321, 323
Pauli spin matrix, 67
perfectly matched layer (PML), 200
  boundary condition, 172
periodic Coulomb potential, 80
Petermann's factor, 51
phase, 162
phase noise, 47
photo-luminescent (PL) assessment, 137
photodetector (PD), 137, 200, 238, 321–324, 326, 329
  PD electrode, 322
Planck's constant, 49
plane wave function, 102
Poisson's equation, 129, 152, 153, 157, 158, 162, 219, 220, 226, 232, 238
polariton number expectation, 157
polarization envelope function, 17
Poynting vector, 48
pseudo-random number generation, 4

QL iteration, 200
quantum dot (QD), 36
quantum well (QW), 80–82, 84, 99, 137, 154, 162, 200, 202, 205, 207, 238, 251, 253–257, 259, 268–270, 276, 288, 293, 299, 300, 302, 305, 306, 308, 314, 323, 324

quantum well (QW) (*Cont.*)
  active region, 255
  bandgap energy, 298
  effect, 298
  laser, 270
  material, 288
  potential energy, 99
  scaling rule, 270
  structure, 35, 36, 82, 87–89, 91, 93, 95, 97–100, 137, 162, 200–202, 204–206, 208, 236, 238, 253–256, 258, 269, 317
    1D, 87
  thickness, 246, 257, 305
quarter-wavelength phase-shifted DFB laser, 281–284
quasi-charge neutrality condition, 214
quasi-electrostatic theory, 1
quasi-Fermi distribution, 119–122, 156–158, 204, 238
quasi-Fermi levels, 156–158, 166, 167, 203, 205, 207

radiative and non-radiative processes, 162
radiative recombination, 167
random number generator (RNG), 49
rational (Pade) factorization, 182
recombination heating, 166
recombination rate, 163, 164
refractive index change, 1
ridge waveguide (RW), 12, 259, 260, 262, 268–270
root searching, 4
rotating-wave approximation, 162
Runge–Kutta method, 209, 233
Rydberg energy, 136

s–o energy, 99
scaled Einstein relation, 222
Scharfetter and Gummel approach, 221
Schottky barrier height, 217
Schrödinger equation, 1, 54, 58, 59, 65, 80, 134
  static, 55, 65
  steady-state, 54, 80
self-consistent manner, 169, 181
self-consistent problem, 1D, 165
semi-analytical expressions, 165
semi-vectorial mode, 172
semiconductor band structure, 64
semiconductor laser, 47
semiconductor lattice structure, 64
semiconductor optical amplifier (SOA), 5, 31, 37, 44, 188, 253, 256–258, 299–302, 305
  gain-clamped (GC), 31
semiconductor optoelectronic devices, 172
semiconductor QW structure, 162
Shockley–Read–Hall (SRH) recombination, 163, 164
side mode suppression ratio (SMSR), 29, 276–278, 281, 282, 315, 319, 321, 324, 325
single mode fiber (SMF), 321, 322
single mode waveguide, 45
single-electron band structure, 54
slab waveguide, 1D, 172
Slater determinant, 55
slow-varying envelope, 16, 162
  of the polarization, 162
slow-varying envelope function, 11, 28, 124, 144, 145, 164, 239, 240, 242, 243
  1D, 16
  2D, 20
slow-varying time dependences, 162
small-signal dynamic performance, 3
small-signal IM response, 248, 255, 280
small-signal linear modulation response equations, 243
small-signal linearization, 3
small-signal modulation bandwidth, 323
space Fourier transform, 168
spectral hole burning (SHB), 212
spherical harmonic function, 65, 66
spin effect, 67
spin–orbit coupling, 67
spin–orbit interaction, 69, 75, 76, 81, 93
spin–orbit split, 80
spin–orbit split band, 68–71, 77, 89
split-step method, 182
split-step, alternating-direction implicit (ADI), 182
spontaneous and stimulated emissions, 167
spontaneous emission, 47–51, 163, 164, 166, 167
  and Auger recombination coefficients, 163
  contribution, 38, 39, 48
  gain, 49
  noise, 49–51
  noise coupling coefficient, 51
  noise power, 48, 50, 51
  recombination, 163, 164
  recombination rate, 161, 163
spontaneously emitted photons, 49
SRH time constant,163
staggered leapfrog method, 180
static hot carrier distribution, 158
static quasi-Fermi level, 207
steady state performance, 3
stimulated and spontaneous emission, 162
stimulated emission, 159, 161, 162, 164, 167
  gain, 49
  process, 159
  rate, 161, 168
  recombination, 162, 164
strongly implicit (SI) method, 352
successive over-relaxation (SOR) method, 230, 351
  symmetric (SSOR) method, 352

superluminescent light emitting diode (SLED), 5, 31, 44, 253, 256–258, 305

Taylor expansion, 174
thermal diffusion, 1, 168
  equation, 1, 168, 242
thermal diffusivities, 160
thermal equilibrium, 164
thermal equilibrium state, 163
Thomson heating, 166
3 dB small-signal IM bandwidth, 280
three-dimensional (3D) space, 41, 56, 201
three-dimensional (3D) mode index, 45
three-dimensional (3D) momentum space, 35
three-dimensional (3D) device, 239
three-dimensional (3D) domain, 202
three-dimensional (3D) potential disturbance, 87
three-dimensional (3D) wave vector, 107
top conduction band, 157
transparent boundary condition (TBC), 172
transverse electric (TE) mode, 251, 253, 254, 299, 315, 317
  TE absorption, 299
  TE absorption peak, 299
  TE and TM modes, 294, 297–299, 305, 307
  TE and TM polarization modes, 299
  TE mode absorption peak, 298
transverse magnetic (TM) mode, 253, 254, 299, 315
  TM absorption, 299
  TM mode absorption peak, 298, 299
two-dimensional (2D) cross-section, 46
two-dimensional (2D) domain, 202, 216
two-dimensional (2D) equation, 214
two-dimensional (2D) momentum space, 35
two-dimensional (2D) sheet, 233, 239
two-dimensional (2D) thermal equation, 168
two-dimensional (2D) thermal model, 168
two-dimensional (2D) wave vector, 87, 108

valence band, 66, 71, 75, 78, 80, 81, 85, 86, 88, 89, 91, 156, 157
valence band electron, 81, 86, 89, 95, 96
valence band hydrostatic energy, 94, 97
valence band shear energy, 94
von Neumann analysis, 175, 176, 178, 180–182

Wannier equation, 134
Wannier exciton, 134
wave equation, 7–12
  model, 8, 9
wave envelope function, 11, 20, 33
wave function, 104, 105, 108, 135
wave propagation, 1
wave propagation direction, 8, 159
wave vector, 161
  domain, 156, 157
wave–media interaction, 9
waveguide, 1, 10, 12, 19, 21
  structure, 23, 32
wavelength division multiplexing (WDM), 31, 37

Printed in the United States
by Baker & Taylor Publisher Services